Transforming Global Food Security and Agricultural Sustainability With Geospatial AI (GeoAI)

Walid Al-Shaar
https://orcid.org/0000-0001-6822-8245
Novum Research and Innovation, Lebanon

Satyabrata Mandal
https://orcid.org/0000-0002-7588-9014
Banwarilal Bhalotia College, India

Ghaleb Faour
https://orcid.org/0000-0002-0741-0518
National Remote Sensing Center, CNRS, Lebanon

IGI Global
Scientific Publishing
Publishing Tomorrow's Research Today

Vice President of Editorial	Melissa Wagner
Director of Acquisitions	Mikaela Felty
Director of Book Development	Jocelynn Hessler
Production Manager	Mike Brehm
Cover Design	Jose Rosado

Published in the United States of America by
IGI Global Scientific Publishing
701 East Chocolate Avenue
Hershey, PA, 17033, USA
Tel: 717-533-8845 | Fax: 717-533-7115
Website: https://www.igi-global.com E-mail: cust@igi-global.com

Library of Congress Cataloging-in-Publication Data

Names: Al-Shaar, Walid, DATE- editor | Mandal, Satyabrata, DATE- editor | Faour, Ghaleb editor
Title: Transforming global food security and agricultural sustainability with geospatial AI (GeoAI) / edited by Walid Al-Shaar, Satyabrata Mandal, Ghaleb Faour.
Description: Hershey, PA : IGI Global Scientific Publishing, 2027. | Includes bibliographical references and index. | Summary: "This book explores how the integration of artificial intelligence and geospatial technologies can reshape global food systems toward sustainability and resilience. The book seeks to bridge research and practice by linking data-driven innovation with agricultural policy, environmental management, and socio-economic development"-- Provided by publisher.
Identifiers: LCCN 2026003594 (print) | LCCN 2026003595 (ebook) | ISBN 9798337380377 hardcover | ISBN 9798337380384 paperback | ISBN 9798337380391 ebook
Subjects: LCSH: Precision farming--Geographic information systems | Food security--Data processing | Artificial satellites in agriculture | Land use--Classification--Remote sensing | Vegetation mapping | Water-supply--Data processing
Classification: LCC S494.5.P73 T73 2026 (print) | LCC S494.5.P73 (ebook)
LC record available at https://lccn.loc.gov/2026003594
LC ebook record available at https://lccn.loc.gov/2026003595

British Cataloguing in Publication Data
A Cataloguing in Publication record for this book is available from the British Library.

Table of Contents

Detailed Table of Contents

Despite the rapid adoption of artificial intelligence (AI) in agriculture, a clear understanding of how these technologies enable the transition from precision farming to adaptive, data-driven systems remains limited. This study addresses this gap by examining AI tools such as unmanned aerial vehicles (UAVs), Internet of Things (IoT) sensors, and remote sensing to improve productivity while reducing environmental impacts. The analysis shows that AI enables real-time monitoring, predictive analytics, and input optimization, enhancing resource-use efficiency. It further presents AI as a dynamic system capable of continuous learning and improved decision-making. However, key challenges remain, including fragmented data systems, limited digital infrastructure, high implementation costs, and inadequate technical capacity. The study emphasizes the need for integrated data platforms, supportive policies, and capacity-building to ensure inclusive and scalable adoption. Overall, AI has strong potential to transform agriculture into a more sustainable and resilient system.

Food security is a long-standing concern, prompting international organisations like the Food and Agriculture Organization to categorise it into four key aspects: accessibility, availability, utilisation, and stability. Food security is a key component of the global SDGs. Agricultural methods have a significant impact on the interplay of food security and sustainability. Agricultural sustainability could be achieved through the aggressive use of advanced technologies and innovative practices. One such aggressive and advanced technique is the Geospatial Artificial Intelligence (GeoAI). It helps to forecast for agriculture, soil suitability, soil health, crop health, predict pest spread, monitor environmental impacts, and predict natural calamities to reduce the risk of unforeseen productivity loss. Obviously, the growth of agriculture

Global food systems are now under pressure from a growing population, climate change, and limited resources, necessitating improved sustainability in food production. Precision agriculture (PA) presents a promising pathway to this goal. Geospatial AI can significantly enhance PA by enabling automated, spatially-aware monitoring and decision-making. In practice, the approach used for apple health monitoring mainly relies on manual inspection, which is time-consuming, subjective, and not scalable. This chapter aims to evaluate the effectiveness of an integrated drone and AI system for automating apple health monitoring. The proposed methodology involves constructing a dataset of apple leaf images, training a computer vision model for disease detection and classification, and generating a diagnostic report of detected diseases. The system demonstrates promising accuracy in detecting common apple diseases, significantly streamlining the monitoring process. This approach shows potential for improving orchard management, thereby contributing to agricultural sustainability and food security.

In recent years, GeoAI, an integration of earth observation, GIS systems, and AI, has presented incomparable opportunities for evaluating agricultural monitoring and control. This chapter outlines the review of five projects that are clearly defined and have the primary goal of utilising remote sensing data to derive effective information on crop conditions, land use, and soil moisture. The projects include Google Earth Engine mainly based on Crop Classification Mapping to distinguish and classify various crop types in Madurai, India; Crop Type Change Detection using the AAFC Annual Crop Inventory to identify changes in crop distribution in Canada between 2013 and 2023; Land Use Change Analysis of the Dynamic World dataset to show changes in land cover between 2015 and 2024. These projects combine spatial and temporal data to visualise change detection for sustainable agriculture. This work confirms the necessity of GeoAI, which allows for efficient and scalable data analysis in agriculture and demonstrates possible uses of satellite data in the development of precision agriculture.

Plant diseases significantly impact negatively on agricultural productivity, resulting in huge financial losses and triggering a threat to the security of the world food supply. This study suggests a GeoAI-enabled hybrid deep learning framework that combines CoAtNet for reliable disease classification with Diffusion Models for image restoration in order to overcome these drawbacks. The suggested model was assessed using the PlantVillage dataset and artificially generated occluded images that simulate the real-world agricultural conditions. Based on the experiment findings, the Diffusion CoAtNet framework was able to perform better on degraded datasets compared to baseline CNN and Transformer-based models, with a 96.4% accuracy. Also, the system demonstrated good edge deployment performance, with a rate of approximately 2.4 frames per second (FPS) on Raspberry Pi hardware, and is suitable to real-time field tasks such as localized crop monitoring and disease diagnosis using a smartphone.

This chapter examines how policy frameworks and governance structures shape the developmental outcomes of data-driven agriculture. It argues that digital technologies alone cannot ensure inclusive growth, as the benefits of agricultural data depend on institutional arrangements governing access, ownership, sharing, and commercialization. Using a data value chain perspective, the chapter analyzes how governance decisions from data collection to digital services influence value creation, market power, and farmer autonomy. It highlights two key tensions: openness versus enclosure and innovation versus justice. Applying the safeguards–enablers framework, the chapter shows how regulatory protections, interoperability standards, and participatory institutions promote trust, collaboration, and equitable benefit sharing. The chapter emphasizes the limits of privacy-focused regulation and the need for collective data rights and proposes an integrated governance roadmap to support interoperable, trustworthy, and socially just agricultural data ecosystems that advance sustainable development.

Precision agriculture is fundamentally transforming the management, monitoring, and optimization of agricultural systems through data-driven technologies that enhance productivity, resource efficiency, and environmental sustainability. By quantifying spatial variability in soil properties, crop development, and climatic conditions, it enables site-specific application of inputs, reducing waste and environmental impacts while maximizing yields. Central to this transformation is Geospatial Artificial Intelligence (GeoAI), which integrates remote sensing, GIS, GNSS, UAVs, and AI-driven analytics into a cohesive decision-support framework. Machine learning and deep learning methods extract meaningful patterns from complex geospatial datasets, supporting crop health monitoring, pest and disease detection, yield forecasting, and optimized resource management. Despite challenges related to data quality, infrastructure, and equitable access, GeoAI represents a vital pathway toward resilient, climate-smart, and sustainable agricultural systems capable of addressing growing global food demands.

Sustainable land management (SLM) seeks to balance economic development, environmental protection, and social equity under growing pressure on limited land resources. It promotes efficient land use that maintains productivity, conserves ecosystems, and supports livelihoods for present and future generations. Advances in geospatial data and artificial intelligence have strengthened SLM through Geospatial Artificial Intelligence (GeoAI), which integrates remote sensing, GIS, and machine learning to analyze complex spatial information. GeoAI enables accurate monitoring of land-use change, supports precision agriculture, and helps predict land degradation risks, improving decision-making and resource efficiency. Despite its potential, challenges such as data quality, algorithm transparency, limited institutional capacity, and ethical concerns persist. Addressing these through inclusive governance, capacity building, and transparent frameworks can ensure GeoAI complements human judgment and contributes to resilient, equitable, and sustainable land governance.

To provide high-resolution, useful representations of soil characteristics such as pH, moisture content, and nutrient content, geoAI-powered soil health mapping combines cutting-edge machine learning algorithms with geospatial data from satellites, drones, and Internet of Things sensors. This chapter employs a systematic review of papers published from 2022 to 2026 in Scopus, WOS, and Google Scholar. The finding shows that farmers can switch from uniform field treatments to variable-rate fertilization—applying precise nutrient quantities suited to particular soil conditions—by examining these digital maps. By eliminating excessive nutrient runoff and minimizing chemical inputs, this focused approach reduces the environmental impact of agriculture while optimizing nutrient management, boosting crop yields, and promoting long-term soil sustainability.

Low-cost, edge-computing IoT nodes placed in agricultural areas are integrated into the GAIS architecture. High-resolution cameras and environmental sensors (which track temperature, humidity, and soil moisture) are used to strengthen these nodes. Data is continuously gathered and sent to a central cloud platform, mostly leaf photos. A specially trained KSK (Kutubuddin S. Kazi) Approach, which is ideal for identifying and categorizing dominant illnesses (such as wilt, Pomegranate bacterial blight, and Jawar leaf blight) is the fundamental intelligence. Following classification, the system makes use of a proprietary, rules-based expert system that quickly produces calibrated treatment protocols that take into account the disease stage, environmental factors, and viable fungicide/pesticide substitutes. According to preliminary testing, the GAIS greatly outperforms manual detection techniques in terms of speed and consistency, achieving a disease classification accuracy of over 96%.

Preface

Welcome to *Transforming Global Food Security and Agricultural Sustainability With Geospatial AI (GeoAI)*. This edited reference book is a collaborative effort by Dr. Walid Al-Shaar, Dr. Satyabrata Mandal, and Dr. Ghaleb Faour, designed to examine how Geospatial Artificial Intelligence (GeoAI) is transforming global food security and agricultural sustainability across diverse environmental, economic, and institutional contexts.

Geospatial Artificial Intelligence, commonly abbreviated as GeoAI, refers to the integration of geographic information science with artificial intelligence techniques such as machine learning, deep learning, and computer vision. By combining satellite imagery, unmanned aerial vehicle observations, Internet of Things sensor networks, ground-based surveys, and open geospatial data, GeoAI delivers unprecedented analytical capacity for understanding, monitoring, and managing complex agricultural and environmental systems. It enables researchers and practitioners to characterize landscapes, detect anomalies, predict outcomes, and support decisions at multiple spatial and temporal scales, from individual fields to entire continents.

Agriculture stands as one of the most consequential sectors for the wellbeing of humanity. It is also among the most data-intensive, given the spatial heterogeneity of soils, the temporal dynamics of climate and weather, the biological complexity of crops and pests, and the socioeconomic diversity of farming communities. Traditional agricultural research and extension have long relied on field surveys, statistical sampling, and expert judgement. While these approaches remain indispensable, they are increasingly complemented and, in some applications, transformed by the power of GeoAI. Advances in cloud computing, free and open Earth observation programs such as Landsat, Sentinel, and MODIS, and the rapid maturation of deep learning architectures have collectively created an environment in which sophisticated geospatial intelligence is more accessible than ever before.

In the contemporary world, food systems face unprecedented pressure. The global population continues to grow, projected to approach ten billion by mid-century, while arable land per capita declines. Climate change is intensifying droughts, floods, heatwaves, and shifting growing seasons, with disproportionate impacts on regions already struggling with food insecurity. Soil degradation, water scarcity, biodiversity loss, and the increasing volatility of agricultural input and output markets further complicate the picture. The United Nations Sustainable Development Goal 2, which aspires to end hunger and achieve food security, calls for transformative responses that go beyond incremental improvements in conventional agricultural practice.

Against this backdrop, GeoAI has moved from being a specialized research interest to an essential component of agricultural innovation strategies worldwide. Governments, international organizations such as the Food and Agriculture Organization, the World Food Programme, and the World Bank, as well as private agricultural technology firms, are increasingly deploying GeoAI methods for crop monitoring, yield forecasting, drought early warning, pest and disease surveillance, irrigation scheduling, supply chain logistics, and land use planning.

Smallholder farmers in low and middle income countries are beginning to benefit from advisory services powered by satellite-based assessments and machine learning models. At the same time, precision agriculture in developed nations is being reshaped by computer vision, edge computing, and autonomous machinery guided by spatial intelligence. The convergence of geospatial science and artificial intelligence is therefore positioned at the heart of efforts to build food systems that are simultaneously productive, sustainable, and equitable.

Despite the proliferation of research and development in this area, the literature remains fragmented across disciplines. Geospatial scientists, agronomists, computer scientists, environmental engineers, and policy scholars often publish in separate venues, with limited cross-pollination. Thus, there is a clear need for an integrated reference that brings together foundational concepts, methodological advances, and applied case studies under a coherent thematic umbrella. This book aims to address that need by offering original research and review chapters that span the full breadth of GeoAI applications in agriculture and food security.

We have solicited contributions that advance current methodologies, introduce novel frameworks, and report on real world deployments addressing pressing challenges in food production, environmental monitoring, plant health, soil management, and agricultural governance. The chapters represent a balance of theoretical foundations and applied case studies, with geographic case material drawn from the wider Middle East, Southeast Asia, and beyond. Together, they offer readers a broad yet rigorous perspective on how GeoAI is reshaping the future of agriculture.

Our target audience is intentionally diverse. Primary readers include researchers and graduate students in geospatial sciences, remote sensing, machine learning, agricultural sciences, environmental engineering, and rural development. Data scientists, software engineers, and analysts working in agriculture and sustainability will find both conceptual frameworks and concrete methodological pathways to inform their applied work. Secondary readers include policy makers, professionals in United Nations agencies, the World Bank, non-governmental organizations, agricultural technology firms, and extension services that operate at the interface between science and implementation. The book is designed to function as an academic reference for courses on agricultural technology, spatial data science, environmental informatics, and sustainable development, while also serving as a practical resource for those building, evaluating, and deploying GeoAI solutions in real operational contexts.

The themes addressed in this volume, including remote sensing, machine learning, computer vision, Internet of Things, precision agriculture, plant disease detection, soil health, land use and land cover change, water resource management, climate impact assessment, food security policy, and governance, are designed to cater to a wide array of interests and professional needs. By integrating these themes within a single edited volume, the book fills a gap in the global literature and offers a unifying reference that few comparable resources currently provide.

We hope this book serves as a comprehensive resource for understanding and exploring the vast possibilities of GeoAI in agricultural sustainability and food security. It is our pleasure to present this collection of knowledge and insights to you.

ORGANIZATION OF THE BOOK

Chapter 1: Artificial Intelligence in Agriculture: Pathways to Sustainable and Adaptive Farming Systems

This opening chapter examines how AI technologies facilitate the transition from conventional precision farming to adaptive, data-driven learning farms. It demonstrates how AI supports real-time monitoring, predictive analytics, and input optimization, while identifying persistent challenges such as fragmented data systems, limited digital infrastructure, high implementation costs, and gaps in technical capacity. The chapter argues that integrated data platforms, policy support, and capacity-building initiatives are essential for inclusive and scalable adoption.

Chapter 2: Assessing Agricultural Sustainability through Geospatial AI and Food Security Policy

This chapter analyses the impact of GeoAI alongside government schemes for enhancing agriculture and food security, framed by the four FAO dimensions of food security (accessibility, availability, utilisation, stability) and their linkage to the Sustainable Development Goals. It examines how GeoAI supports forecasting for agriculture, soil suitability and health assessment, crop health monitoring, pest spread prediction, environmental impact monitoring, and natural calamity prediction to reduce productivity loss.

Chapter 3: Smart Agriculture and GeoAI: Integrating Geospatial Technologies for Sustainable Farming and Secure Food Systems

This chapter discusses how smart agriculture, supported by IoT, AI, GeoAI, remote sensing, blockchain, and data analytics, can transform cultivation, harvest, processing, and food supply chain activities. It addresses precision farming, autonomous machines, drone-based monitoring, and smart greenhouses for real-time, data-driven decisions, and outlines a roadmap for integrating smart agriculture with climate-resilient policies and the SDGs to advance equitable and secure food systems.

Chapter 4: From Backyard Herbs to Smart Green Enterprises: Using GeoAI to Optimise Household-Based Medicinal Plant Production in Peri-Urban Indonesia

This chapter analyses the shift from subsistence backyard medicinal plant cultivation to smart green micro-enterprises (SGME) in peri-urban Indonesia, drawing on participatory mixed-methods data from 50 households in Cibinong, Bogor (January to March 2025). Integrating GeoAI, circular economy principles, and community-based innovation, the study finds yield increases of 20 to 30 percent, additional monthly income of approximately IDR 1,000,000, and reductions in health expenditures of around IDR 200,000. The chapter proposes a Backyard-to-SGME model for inclusive and sustainable peri-urban food systems, with particular attention to women-led entrepreneurship.

Chapter 5: GeoAI-Driven Sustainable Precision Agriculture: UAV and Computer Vision for Apple Health Monitoring with a Case Study from Lebanon

This chapter evaluates a two-stage YOLOv8 pipeline integrating UAV imagery and computer vision for automated apple disease detection, with leaf detection followed by classification into Healthy, Scab, Black Rot, and Other classes. On a dataset of approximately 25,000 images with an 80/10/10 split, the framework achieved leaf detection mAP50 of 99.34 percent and Top-1 classification accuracy of 99.93 percent. The authors note that these results were obtained under controlled conditions using publicly available imagery, since field validation in Lebanese orchards and live GPS synchronization were prevented by the conflict that arose in Lebanon in 2024.

Chapter 6: GeoAI-Driven Agricultural Monitoring: Earth Observation Applications for Precision Crop Mapping, Water Resource Management, and Land-Use Change

This chapter reviews five Google Earth Engine based projects that combine remote sensing with GeoAI for agricultural monitoring. Featured case studies include crop classification mapping in Madurai, India; crop type change detection across Canada from 2013 to 2023 using the AAFC Annual Crop Inventory; and land use change analysis using the Dynamic World dataset from 2016 to 2024. The chapter demonstrates how spatial and temporal data can be combined for scalable change detection in support of precision agriculture.

Chapter 7: GeoAI-Enabled Diffusion CoAtNet Framework for Robust Plant Disease Detection under Real-World Field Conditions

This chapter proposes a hybrid deep learning framework combining CoAtNet for disease classification with Diffusion Models for restoring degraded or occluded leaf images, addressing field challenges such as occlusion, noise, uneven illumination, and background clutter. The framework incorporates GIS-based hotspot detection through GPS coordinates and was evaluated on the PlantVillage dataset with synthetically occluded images, achieving 96.4 percent accuracy and approximately 2.4 frames per second on Raspberry Pi hardware, making it suitable for smartphone-based field deployment.

Chapter 8: Policy Frameworks and Governance for Data-Driven Agriculture

This chapter argues that the benefits of agricultural data depend on institutional arrangements governing access, ownership, sharing, and commercialization, not on digital technologies alone. Using a data value chain perspective, it analyses governance decisions from data collection to digital services and highlights two key tensions: openness versus enclosure, and innovation versus justice. Applying a safeguards and enablers framework, it shows how regulatory protections, interoperability standards, and participatory institutions support trust and equitable benefit sharing, and proposes an integrated governance roadmap built on collective data rights rather than privacy-focused regulation alone.

Chapter 9: GeoAI-Enabled Multi-Model Agricultural Intelligence Framework for Plant Disease Detection and Climate-Aware Crop Recommendation

This chapter introduces a GeoAI based multi-model system that combines Vision Transformers for fine-grained plant disease classification using the New Plant Diseases dataset, Temporal Fusion Transformers for climate prediction, and Graph Neural Networks for spatially aware crop recommendation. The framework draws on IoT-sensed soil nutrient and weather data along with relational cultivation patterns across adjacent farms, and includes a chatbot interface to support low-skill access by farmers. Experimental results show improved predictive accuracy and feasibility of deployment in resource-restricted rural settings.

Chapter 10: Remote Sensing and GIS Applications in Assessing Climate Change Impacts on Land Use and Land Cover Changes

This chapter examines how remote sensing and GIS technologies, supported by improved spatial and temporal resolution, time series analysis, multi-sensor integration (optical, radar, and LiDAR), and AI-based change detection, are transforming real-time assessment of climate change impacts on land use and land cover. Cloud-based platforms for big data analytics have democratized access to Earth observation data and enabled global-scale analyses across urban heat islands and city growth, agricultural crop conditions and irrigation, forest deforestation, carbon storage, and species range shifts, and coastal erosion under sea level rise. The chapter also discusses persistent challenges of dataset integration, computational demands, and uncertainty quantification, and surveys emerging tools such as deep learning, IoT sensor networks, hyperspectral remote sensing, advanced LiDAR, and small satellite constellations.

Chapter 11: Precision Agriculture and Crop Monitoring with Geospatial AI

This chapter examines how precision agriculture, anchored in Geospatial AI, integrates remote sensing, GIS, GNSS, UAVs, and AI-driven analytics into a cohesive decision-support framework. By quantifying spatial variability in soil properties, crop development, and climatic conditions, it enables site-specific application of inputs to reduce waste and environmental impact while maximizing yields. The chapter reviews machine learning and deep learning methods for crop health monitoring, pest and disease detection, yield forecasting, and resource optimization, while acknowledging challenges of data quality, infrastructure, and equitable access.

Chapter 12: Sustainable Land Management with Geospatial AI

This chapter situates Sustainable Land Management within the tension between economic development, environmental protection, and social equity, and examines how GeoAI strengthens SLM through the integration of remote sensing, GIS, and machine learning. It shows how GeoAI enables monitoring of land-use change, supports precision agriculture, and helps predict land degradation risks, while identifying persistent challenges of data quality, algorithm transparency, limited institutional capacity, and ethical concerns. The chapter argues that inclusive governance, capacity building, and transparent frameworks are essential so that GeoAI complements human judgment rather than displacing it.

Chapter 13: GeoAI-Powered Soil Health Mapping for Enhancing Nutrient Management and Reducing Environmental Impact

This chapter draws on a systematic review of papers published between 2022 and 2026 in Scopus, Web of Science, and Google Scholar to examine how GeoAI combines satellite, drone, and IoT data with machine learning to produce high-resolution maps of soil pH, moisture, and nutrient content. It shows how such maps enable a shift from uniform field treatment to variable-rate fertilization, reducing nutrient runoff and chemical inputs while optimizing yields and supporting long-term soil sustainability.

Chapter 14: The Guardian Agro-Intelligence System (GAIS): An AI-Driven IoT Framework for Real-Time Jawar and Pomegranate Disease Detection and Precision Treatment Recommendation

This chapter presents the Guardian Agro-Intelligence System (GAIS), an architecture built on low-cost, edge-computing IoT nodes equipped with high-resolution cameras and environmental sensors (temperature, humidity, soil moisture) that stream leaf imagery and field data to a central cloud platform. At the core of the system is the KSK (Kutubuddin S. Kazi) Approach for identifying and categorizing dominant illnesses such as wilt, pomegranate bacterial blight, and jawar leaf blight, coupled with a proprietary rules-based expert system that generates calibrated treatment protocols accounting for disease stage, environmental factors, and fungicide or pesticide substitutes. Preliminary testing reports disease classification accuracy of over 96 percent and significant gains in speed and consistency over manual detection.

IN CONCLUSION

In this comprehensive exploration of GeoAI applications in agriculture and food security, we have traversed a broad spectrum of topics ranging from foundational artificial intelligence concepts and remote sensing methodologies to advanced deep learning frameworks, IoT-enabled real-time monitoring systems, and policy and governance considerations. The chapters collectively illuminate the transformative potential of geospatial artificial intelligence in reshaping how we understand, manage, and safeguard the agricultural systems on which humanity depends.

From precision agriculture and crop health monitoring to climate impact assessment and sustainable land management, this volume underscores the profound impact that GeoAI is having on contemporary and future agricultural practice. Each chapter contributes a distinctive perspective, whether through the introduction of novel computational frameworks, the documentation of region-specific case studies, the integration of multiple data streams into operational decision support tools, or the articulation of policy considerations that shape how these technologies are deployed and governed. Taken together, the chapters demonstrate that GeoAI is not a single technology but rather an evolving ecosystem of methods, data sources, infrastructures, and institutional practices.

The discussions throughout this book also highlight the ethical, social, and governance dimensions of GeoAI in agriculture. As powerful as these technologies are, their deployment raises important questions about data ownership, equity, transparency, environmental responsibility, and the distribution of benefits and risks. The contributions in this volume emphasize the need for responsible innovation, where technical excellence is paired with attention to context, inclusion, and long-term sustainability. Smallholder farmers, indigenous communities, women in agriculture, and vulnerable regions must remain central to the conversation if GeoAI is to fulfill its promise as a tool for global food security.

The future of agricultural transformation, as envisioned through the scholarly contributions in this volume, must be characterized by greater efficiency, enhanced resilience, and new opportunities for interdisciplinary collaboration. Advances in foundation models, edge computing, autonomous platforms, citizen science, and participatory mapping will further expand the GeoAI toolkit. At the same time, the integration of GeoAI with traditional agronomic knowledge, local expertise, and policy instruments will be essential for ensuring that these advances translate into real impact on the ground.

As editors, we hope this reference book serves as a valuable resource for researchers, practitioners, policy makers, and students alike. We believe that by embracing GeoAI thoughtfully, rigorously, and responsibly, we can harness its full potential to support sustainable agriculture, secure food systems, and resilient rural communities around the world. The journey through these chapters has been one of discovery, integration, and forward-looking reflection, and we invite our readers to engage deeply with each contribution, to question, to innovate, and to contribute to the ongoing dialogue on the future of GeoAI in agriculture and beyond.

We extend our sincere thanks to the chapter authors who entrusted us with their original research, to the reviewers whose constructive feedback strengthened every contribution, and to the editorial and production teams whose dedication brought this volume to publication. We also acknowledge the broader communities of geospatial scientists, agronomists, computer scientists, environmental engineers, and policy scholars whose collective work makes a book of this scope possible. It is to these communities, and to the farmers and food system stakeholders they serve, that we dedicate this volume.

Walid Al-Shaar
Novum Research and Innovation Group, Lebanon

Satyabrata Mandal
Banwarilal Bhalotia College, India

Ghaleb Faour
National Center for Remote Sensing, CNRS, Lebanon

Chapter 1
Artificial Intelligence in Agriculture:
Pathways to Sustainable and Adaptive Farming Systems

Kalyani N. G.
http://orcid.org/0000-0003-1651-6696
G. Pulla Reddy Engineering College, India

Satyabrata Mandal
http://orcid.org/0000-0002-7588-9014
BB College, Kazi Nazrul University, Asansol, India

Pallobi Halder
University of Calcutta, India

ABSTRACT

Despite the rapid adoption of artificial intelligence (AI) in agriculture, a clear understanding of how these technologies enable the transition from precision farming to adaptive, data-driven systems remains limited. This study addresses this gap by examining AI tools such as unmanned aerial vehicles (UAVs), Internet of Things (IoT) sensors, and remote sensing to improve productivity while reducing environmental impacts. The analysis shows that AI enables real-time monitoring, predictive analytics, and input optimization, enhancing resource-use efficiency. It further presents AI as a dynamic system capable of continuous learning and improved decision-making. However, key challenges remain, including fragmented data systems, limited digital infrastructure, high implementation costs, and inadequate technical capacity. The study emphasizes the need for integrated data platforms, supportive policies, and capacity-building to ensure inclusive and scalable adoption. Overall, AI has strong potential to transform agriculture into a more sustainable and resilient system.

DOI: 10.4018/979-8-3373-8037-7.ch001

1. INTRODUCTION

Agriculture, one of the fundamental economic sectors sustaining human civilization, is currently undergoing a profound transformation driven by multiple global challenges. Global population growth is projected to reach 9.7 billion by 2050 and therefore it will lead to intensify the demand for food production. However, climate change, land degradation, loss of biodiversity, salinization, water scarcity, and desertification are simultaneously limiting agricultural productivity (IPCC, 2022; FAO, 2023). The increasing frequency of extreme climatic events (agricultural droughts, floods, and heatwaves, etc.) is posing serious uncertainty in the agrarian economy, particularly in climate-sensitive developing countries. Therefore, achieving sustainable agriculture by balancing productivity, environmental protection, and livelihood demands has become an urgent priority. As a result, more attention is being given to technological innovations that improve efficiency while reducing environmental impact.

In recent years, artificial intelligence (AI) has become a powerful tool that can transform agriculture by enabling data-driven, precise, and adaptive farm management. AI integrates advanced techniques such as machine learning, deep learning, and big data analytics. It processes large and complex datasets collected from multiple sources, including satellite remote sensing, Internet of Things (IoT) sensors, unmanned aerial vehicles (UAVs), and ground-based observations. These advancements facilitate real-time monitoring of crop health, soil conditions, and environmental parameters, allowing farmers to make informed decisions that optimize input use and improve productivity (Eli-Chukwu, 2019; Waqas et al., 2025; Li et al., 2026). For example, AI-based precision agriculture allows site-specific management of irrigation, fertilization, and pest control. This reduces resource waste and environmental deterioration while maximizing crop yields (Nitin and Bal Gupta, 2023; Sudha and Loret, 2026). Furthermore, AI-powered predictive models play a crucial role in developing agricultural resilience by forecasting weather variability, pest outbreaks, and crop yields. It can provide early warnings and decision support for critical farm operations such as sowing, harvesting, and agricultural inputs (Deputy et al., 2023). In addition, AI-based farm machinery, such as autonomous machines, robots, and drones, has improved efficiency and reduced labor costs. It also enables precise actions like targeted spraying and selective harvesting (Delgado et al., 2019; Soleimani et al., 2026).

Beyond farm-level applications, AI helps optimize agricultural supply chains by improving logistics, reducing post-harvest losses, and enhancing market insights. It aligns production with demand, reducing food waste and improving food distribution (Olsen, 2026). New advances such as federated learning, digital twins, and AI-based crop breeding are further expanding smart agriculture by improving data sharing, system integration, and climate-resilient cultivar development (Subeesh and Chauhan, 2026; Nitin and Bal Gupta, 2023).

Despite its promising developments, AI adoption in agriculture remains uneven due to several challenges. Key gaps include difficulties in integrating standardized heterogeneous data sources. This often varies in scale, resolution, and quality, which limits its application in smallholder farming, especially in developing countries. High costs, poor digital infrastructure, lack of technical skills, and data privacy issues further restrict its use (Tzachor, 2021; Reitano et al., 2025). Additionally, comprehensive studies covering the entire agricultural value chain across different agro-ecological conditions remain lacking. In light of these challenges and research gaps, the present study aims at contributing to the growing body of knowledge on AI-driven sustainable agriculture by addressing three key objectives: (i) to critically examine the role of artificial intelligence in optimizing agricultural practices for developing sustainability and productivity; (ii) to evaluate the effectiveness of AI-based approaches in improving resource use efficiency and climate resilience; and (iii) to identify the potential pathways for the inclusive, scalable, and ethical adoption of AI technologies in agriculture. By providing a comprehensive and integrative assessment, this study aims to inform both academic research and policy interventions promoting sustainable and climate-resilient agricultural systems in the era of digital transformation.

2. METHODOLOGY

The present study employed an integrated framework to examine AI-driven optimization for sustainable agriculture by combining data acquisition, AI model development, and technology deployment, as shown in Figure 1. It combines artificial intelligence, the Internet of Things (IoT), and remote sensing to improve resource-use efficiency, productivity, and sustainability across agricultural systems (Nitin and Bal Gupta, 2023; Demissie et al., 2026). The methodology begins with the collection of multi-source datasets, including satellite imagery, UAV data, in-situ sensor observations (soil moisture, nutrients, and weather), and historical crop records. These datasets enable high-resolution spatial and temporal monitoring of crop health, soil conditions, and environmental variability (Sharma et al., 2026). Real-time data streams from IoT-enabled sensors are integrated with meteorological and pest/disease data to support predictive modelling (Sinclair et al., 2024; Aijaz et al., 2025).

Figure 1. Methodological flow of the study

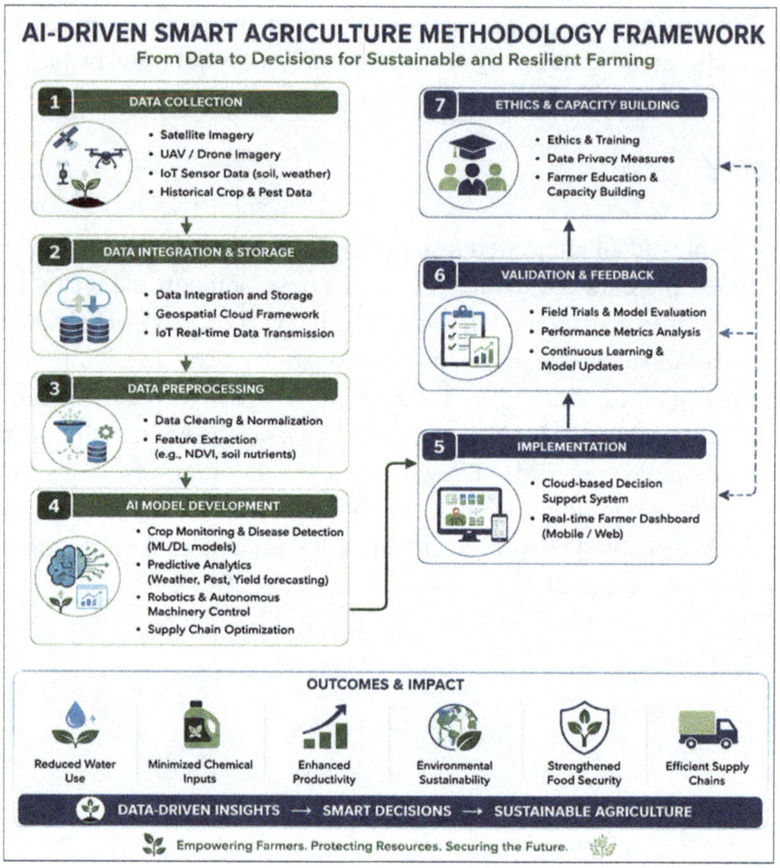

Data are managed within a cloud-based geospatial framework to ensure scalability and interoperability. Preprocessing techniques such as cleaning, normalization, and feature extraction (e.g., NDVI and soil nutrient indices) are applied to improve data quality and model performance (Elbasi et al., 2024; Shahab et al., 2025). AI models are then developed for key applications. Machine learning and deep learning techniques are used for precision agriculture, including crop health monitoring, disease detection, and optimization of irrigation and fertilization, following Oğuztürk (2025) and Kumar et al. (2020). Predictive models, such as time-series and ensemble approaches, are applied to forecast weather variability, pest outbreaks, and crop yields, enabling proactive decision-making (Shawon et al., 2025; Roy et al., 2025). The developed models are deployed on a cloud-based decision-support platform that integrates real-time data streams and provides actionable insights through user-friendly interfaces.

Model performance is evaluated using standard metrics such as accuracy, precision, recall, and mean absolute error, along with sustainability indicators including water-use efficiency, input reduction, and yield improvement (Sánchez-Arnau et al., 2026; Aijaz et al., 2025). Finally, the study incorporates ethical and practical considerations by ensuring data security, transparency, and stakeholder engagement. Farmer training and capacity-building initiatives are emphasized to promote inclusive and scalable adoption of AI technologies in agriculture (Nitin and Bal Gupta, 2023; Zhang et al., 2024).

3. RESULTS

The results shows that the artificial intelligence (AI) is gradually advancing sustainable agriculture by enabling data-driven, precise, and resource-efficient farming systems. It increases agricultural productivity while also minimizing environmental impacts. The integration of AI with precision agriculture technologies such as unmanned aerial vehicles (UAVs), ground-based sensors, and satellite remote sensing facilitates real-time monitoring of crop conditions, soil properties, and pest dynamics. This comprehensive data ecosystem supports site-specific management practices, allowing optimal application of irrigation water, fertilizers, and pesticides, thereby reducing input wastage and ecological degradation (de Smith et al., 2018; Zhang et al., 2021). A major improvement was observed in crop monitoring and early stress detection. AI-enabled sensor networks and UAV-based imaging systems can identify early signs of crop stress impacted by pests, diseases, or nutrient deficiencies. This enables timely and localized interventions, reducing yield losses and limiting the excessive use of agrochemicals. As summarized in Table 1, such AI-driven monitoring systems contribute substantially to both productivity enhancement and environmental sustainability (Jha et al., 2019).

Another key result of AI implementation was found in resource optimization. Conventional irrigation systems, which rely on static scheduling, often result in over- and/or under-irrigation water use. In contrast, AI-based irrigation models integrate soil moisture data, crop water requirements, and weather forecasts to generate adaptive irrigation schedules. This ensures efficient water utilization and prevents crop stress. Similarly, AI-driven nutrient management systems optimize fertilizer application by analysing soil health and plant requirements, thereby minimizing nutrient runoff and preserving soil quality (Lou et al., 2022). The optimization benefits across water and nutrient management are depicted in Table 1.

The application of predictive AI further enhances agricultural decision-making. Machine learning algorithms improve the accuracy of localized weather forecasting. It enables farmers to anticipate and prepare for extreme climatic events, particularly during agricultural droughts and excessive rainfall (floods). Additionally, predictive models facilitate early detection of pest and disease outbreaks, allowing for targeted interventions and reducing reliance on blanket chemical

Table 1. Parameters used in AI technology

Parameter	Drones (UAVs)	Sensors (Soil & Environmental)	Satellite Remote Sensing	Autonomous Machinery	AI/ML Models (Deep Learning, Predictive Analytics)
Code Type Used	Python, C++, ROS (Robot Operating System), OpenCV for vision	Embedded firmware, Python, MATLAB	Python, TensorFlow, CNNs for image analysis	C++, Python, ROS, proprietary AI frameworks	Python, TensorFlow, PyTorch, Scikit-learn, R
Accuracy	85-95% in pest/disease detection	90-98% in soil moisture/nutrient measurement	80-95% in land cover/ use classification	90-99% in precision operations	85-99% depending on model and data quality
Sensors	RGB cameras, multispectral, thermal, LiDAR	Soil moisture sensors, pH, EC sensors, temperature	Multispectral, hyperspectral, radar sensors	Ultrasonic, GPS, inertial sensors, cameras	N/A (data input from multiple sensor types)

continued on following page

Table 1. Continued

Parameter	Drones (UAVs)	Sensors (Soil & Environmental)	Satellite Remote Sensing	Autonomous Machinery	AI/ML Models (Deep Learning, Predictive Analytics)
Weight Lifting Capacity	5-25 kg depending on drone type	N/A (stationary or portable sensors)	N/A (satellites are space-based)	500 kg to several tons (large autonomous tractors)	N/A
Resolution	2-5 cm/pixel (high-res UAV imagery)	High accuracy at point locations	10-30 m spatial resolution typically	Machine precision to 1 cm or better	Depends on input data resolution and model complexity
Covering Area	10-500 hectares per flight	Specific plot or field area	Hundreds to thousands of square kilometers	Entire farm (up to several thousand hectares)	Entire farm or region, depending on data source
Time to Complete Task	15 minutes to 2 hours per flight	Real-time or periodic (minutes to hours)	Days to weeks (satellite revisit time)	Hours to days, depending on task size	Seconds to minutes for prediction once trained
Cost	\$1,000 - \$30,000+ depending on drone sophistication	\$100 - \$5,000 depending on sensor type	Satellite data varies: free to thousands (high-res commercial)	\$50,000 - \$500,000+ for autonomous tractors	Software cost varies; cloud computing fees may apply
Speed	10-40 km/h typical flight speed	Instantaneous data readouts	Slow (dependent on satellite pass)	5-15 km/h for field operations	Real-time or near real-time predictions

applications. These improvements promote environmentally sustainable pest management and reduce overall crop vulnerability (Khan et al., 2021). AI also demonstrates strong potential in yield prediction and harvest planning. By integrating historical datasets with real-time crop health information acquired from remote sensing, AI models generate accurate yield forecasts. These predictions enable farmers to adopt proper planting and harvesting schedules, reducing post-harvest losses and improving labour efficiency. Furthermore, an accurate yield forecast supports supply chain optimization by aligning production levels with market demand, thereby minimizing food wastage (FAO, 2022).

Automation in agriculture has been remarkably enhanced through AI technologies. Autonomous machinery, including tractors, harvesters, and precision sprayers, enables high-accuracy field operations with minimal human intervention. AI-guided systems ensure that inputs such as fertilizers and pesticides are applied only where needed, reducing costs and environmental deterioration. In addition, intelligent drones improve efficiency through targeted aerial surveillance and spraying operations (Tsouros et al., 2019). These technological advancements are categorized in Table 2. Apart from this, AI improved efficiency across agricultural supply chains. By analysing data across production, storage, and distribution stages, AI systems optimize logistics and reduce spoilage and post-harvest losses. Market intelligence tools powered by AI also enable better demand-supply matching, enhancing farmer profitability and strengthening food system resilience (Fuglie et al., 2020).

The cumulative effects of AI-driven optimization are substantial. Increasing input efficiency, reducing environmental degradation, and improving crop productivity collectively contribute to sustainable agricultural development. Smart irrigation reduces water consumption, precision application minimizes chemical runoff, and improved soil management supports long-term agricultural sustainability (Table 1). Its predictive capabilities enable farmers to adapt to climate variability and extreme weather events. Furthermore, AI facilitates the development of climate-resilient crop varieties through the analysis of genetic and phenotypic data, enabling the selection of traits such as drought or flood tolerance and disease resistance (Talaviya et al., 2020).

Overall, the results highlight that AI significantly contributes to global food security by improving resource use efficiency, enhancing crop management, and reducing losses throughout the agricultural value chain. Importantly, AI-based advisory systems provide location-specific recommendations, thereby empowering smallholder farmers and promoting inclusive agricultural development. Finally, the integration of AI into agricultural systems represents a paradigm shift toward more efficient, resilient, and environmentally sustainable farming. As AI technologies continue to evolve and expand (as outlined in Table 2), their role in shaping a sustainable and climate-resilient food future is expected to become increasingly critical.

4. DISCUSSIONS

The results shows that artificial intelligence (AI) is promoting sustainable agriculture through precision, efficiency, and data-driven decision making. This is strongly consonant with the discussion that conceptualizes AI not merely as a tool, but as an adaptive management

Table 2. Application of AI technology in different phases of agriculture

AI Technology/Tool	Application in Sustainable Agriculture	Benefits	Examples/Details
Drones (UAVs)	Crop monitoring, pest/disease detection, aerial spraying.	Early problem detection, targeted interventions, and reduced chemical use.	Real-time imaging, pest identification, and precise pesticide application.
Sensors (Soil & Environmental)	Soil quality assessment, moisture monitoring, nutrient detection.	Optimized irrigation and fertilization, water conservation.	Data-driven irrigation schedules, nutrient management.
Satellite Imagery & Remote Sensing	Crop health monitoring, land use classification.	Large-area coverage, timely data for decision making.	Crop stress identification, yield prediction.
Machine Learning & Deep Learning	Predictive analytics: weather, pest/disease outbreaks, yield forecasting.	Risk mitigation, better planning, and reduced crop loss.	AI breeder genomic predictions, pest outbreak forecasting.
Autonomous Machinery	Automated tractors, harvesters, and sprayers	Labor reduction, precise input application, and cost savings.	AI-driven autonomous tractors, robotic harvesting.
IoT (Internet of Things)	Real-time monitoring, data integration from various farm sensors.	Continuous data flow, smart resource management.	Smart irrigation systems, soil quality sensors.
Predictive Analytics	Weather forecasting, yield, and harvest time prediction.	Adaptation to climate variability, improved resource use.	Climate risk assessment, optimal harvest scheduling.
Supply Chain Optimization	Demand forecasting, logistics planning.	Reduced post-harvest losses and minimized food waste.	AI-enabled market analysis, optimized transportation.
Computer Vision	Weed identification, pest detection.	Targeted herbicide use reduced chemical inputs.	Deep learning for weed management.
Genomic Data Analysis	Crop breeding for drought tolerance and disease resistance.	Climate-resilient crop varieties, improved food security.	AI breeder tools for selecting crop traits.

framework. It transforms agriculture into a feedback-driven system. Both sections agree that precision agriculture using UAVs, sensors, and remote sensing is the main way that AI improves productivity while reducing environmental impacts (Wolfert et al., 2017; Zhang et al., 2021). However, the results focus on measurable improvements in monitoring and input optimization. It places these developments within a broader shift toward 'learning farms.' These farms depend solely on continuous data feedback and algorithmic adaptation.

The agreement lies mainly in early stress detection and crop monitoring. The results highlight that AI-enabled imaging and sensor networks considerably improve crop stress identification, reduction of yield losses, and agrochemical use (Shafi et al., 2019). The discussion strengthens this finding but contributes a critical dimension by highlighting the advantage of predictive deep learning models in detecting stress beyond human visual capacity. Therefore, it established both environmental and economic sustainability. Thus, the results show observed outcomes, while the discussion explains analytical capabilities and system intelligence. In terms of resource optimization, both sections are closely aligned. The results highlight improvements in irrigation and nutrient management through adaptive AI models (Lou et al., 2022), whereas the discussion demonstrates these advancements within integrated IoT-cloud ecosystems and algorithm-based learning frameworks that dynamically optimize inputs. Thus, the discussion widens the scope from individual applications to the interconnected nature of AI applications. It evaluates the progress from static optimization to autonomous decision-making capabilities.

The results demonstrate improved weather forecasting, pest prediction, and yield estimation (Khan et al., 2021; FAO, 2022), directly linking AI to climate resilience and supply chain efficiency. The discussion sufficiently supports these findings, though it introduces a critical contrast by addressing limitations. The 'black box' nature of deep learning models restricts generalization across agro-ecological regions. This shows high predictive accuracy, but model understanding and limited data representation may limit its real-world use. Automation and supply chain optimization are similarly reinforced both in the results and discussion sections. The results highlight the efficiency received from autonomous machinery and AI-driven logistics (Rose & Chilvers, 2018; World Bank, 2021). Although the discussion aligns with the findings, it further evaluates their environmental benefits, including reduced fuel use and lower emissions. It also adds a socio-economic concern. High costs and technological barriers may increase inequality, especially for small and marginal farmers. This marks a clear contrast that the results highlight performance improvements, whereas the discussion explores concerns related to accessibility and fairness.

From a sustainability perspective, both sections strongly correspond in identifying environmental, economic, and food security benefits. The results emphasize reduced water use, minimized chemical inputs, and enhanced productivity, contributing to sustainability and food security (Talaviya et al., 2020). The discussion agrees with the results but incorporates social factors like digital literacy and institutional support. It gives a broader view by including human and institutional aspects. Additionally, the discussion covers data governance, privacy, connectivity challenges, and ethical concerns, including algorithmic bias. While the results highlight the empowering role of AI for smallholders through advisory systems, the discussion contrasts this optimism by cautioning against potential power imbalances between technology providers and farmers, and the risks of data dependency. This adds depth by situating technological advancements within governance and ethical frameworks.

Finally, both sections combine in recognizing AI as a transformative tool for climate-smart and resilient agriculture. The results focus on practical outcomes such as climate adaptation, yield forecasting, and genetic improvement, whereas the discussion extends this toward future pathways, including hybrid modelling, real-time data integration, and Integration with related technologies like blockchain and edge computing. The results show evidence that AI improves productivity, efficiency, and sustainability, while the discussion situates these findings within a broader technological, social, and ethical context. The results show that the AI improves productivity, efficiency, and sustainability, while the discussion outlines these findings in a broader technological, social, and ethical context. The key difference lies in scope: the results emphasize outcomes, whereas the discussion provides a broader, critical perspective that includes limitations and future directions.

5. CHALLENGES AND PATHWAYS

Despite its transformative potential, application of AI in agriculture is limited by several interrelated challenges: (a) fragmented and heterogeneous data from sensors, UAVs, and remote sensing systems, which complicates integration and reduces analytical reliability; (b) inadequate digital infrastructure across locations, high implementation costs, and limited access to advanced technologies, particularly for small and marginal farmers in developing countries; (c) wide gaps in technical capacity, along with data privacy, ownership, and algorithmic transparency. Addressing these challenges requires targeted pathways, including (a) the development of standardized, interoperable data systems supported by investments in rural digital infrastructure; (b) the promotion of affordable, scalable AI solutions through public-private partnerships and policy support; and (c) strengthening farmer training, extension services, and regulatory frameworks; (d) context-specific research and emerging approaches such as explainable AI and federated learning, to ensure inclusive, ethical, and sustainable agricultural transformation.

6. CONCLUSION

This study addresses the gap in the existing literature by providing an integrated perspective on how AI technologies help in transforming agriculture from precision-based interventions to adaptive, feedback-driven learning farms. The primary objective was to analyse both the tangible outcomes of AI applications and the underlying mechanism transformation enabled by continuous data availability and algorithmic intelligence. The findings confirm that AI considerably improves productivity, resource-use efficiency, and environmental sustainability through improved monitoring, predictive decision-making, and optimized agricultural input management. Additionally, the study highlights that the true transformative potential of AI remains in its ability to facilitate self-improving agricultural systems that learn and adapt over time. However, several limitations, including infrastructural shortfall, data integration challenges, cost allocation, and skill gaps, continue to limit widespread adoption, particularly in developing regions. Addressing these challenges requires coordinated efforts that include policy interventions, investment in digital infrastructure, the establishment of a compatible data framework, and the reinforcement of extension services and farmer training. In conclusion, while AI shows a powerful pathway toward sustainable and climate-resilient agriculture, its successful application depends on inclusive, context-specific, and ethically grounded approaches that balance technological innovation with the socio-economic realities of regions.

Competing Interest Statement

The author declares that there are no known competing financial interests or personal relationships that could have appeared to influence the work reported in this paper.

Author Credit Statement

Kalyani NG was primarily responsible for conceptualization, initial design of the research framework, and data collection. Satyabrata Mandal contributed to methodology refinement, data analysis, interpretation of results, and visualization. Kalyani NG worked on writing the original draft, Satyabrata Mandal reviewed and edited the manuscript, and both authors have approved the final version for publication.

REFERENCES

Aijaz, N., Lan, H., Raza, T., Yaqub, M., Iqbal, R., & Pathan, M. S. (2025). Artificial intelligence in agriculture: Advancing crop productivity and sustainability. *Journal of Agriculture and Food Research, 20*, 101762. DOI: 10.1016/j.jafr.2025.101762

de Smith, M. J., Goodchild, M. F., & Longley, P. A. (2018). *Geospatial analysis: A comprehensive guide to principles, techniques, and software tools* (6th ed.). Drumlin Security. https://www.spatialanalysisonline.com

Delgado, J. A., Short, N. M.Jr, Roberts, D. P., & Vandenberg, B. (2019). Big data analysis for sustainable agriculture on a geospatial cloud framework. *Frontiers in Sustainable Food Systems, 3*, 54. DOI: 10.3389/fsufs.2019.00054

Demissie, W. A., Sebastiani, L., & Rossetto, R. (2026). Integration of artificial intelligence and remote sensing for crop yield prediction and crop growth parameter estimation in Mediterranean agroecosystems: Methodologies, emerging technologies, research gaps, and future directions. *European Journal of Agronomy, 127894*(173), 127894. DOI: 10.1016/j.eja.2025.127894

Deputy, K. V., Passi, K., & Jain, C. K. (2023). Crop Disease Detection Using Deep Learning Techniques on Images. *Journal of Computational Science, 19*(12), 1438–1449. DOI: 10.3844/jcssp.2023.1438.1449

Elbasi, E., Mostafa, N., Zaki, C., AlArnaout, Z., Topcu, A. E., & Saker, L. (2024). Optimizing Analysis Techniques through Agricultural Data AI-Powered Decision-Making Processes. *Applied Sciences (Basel, Switzerland), 14*(17), 8018. DOI: 10.3390/app14178018

Eli-Chukwu, N. C. (2019). Applications of Artificial Intelligence in Agriculture: A Review. *Eng. Technol. Appl. Sci. Res., 9*(4), 4377–4383. DOI: 10.48084/etasr.2756

FAO. (2022). Leveraging automation in agriculture for transforming agrifood systems. *Food and Agriculture Organization of the United Nations.*

Food and Agriculture Organization (FAO). (2022). The State of Food and Agriculture 2022: Leveraging automation in agriculture for transforming agrifood systems. *FAO.* DOI: 10.4060/cb9479en

Fuglie, K., Gautam, M., Goyal, A., & Maloney, W. F. (2020). *Harvesting prosperity: Technology and productivity growth in agriculture.* World Bank. DOI: 10.1596/978-1-4648-1393-1

IPCC. (2022). *Climate Change 2022: Impacts, Adaptation and Vulnerability.* Cambridge University Press. DOI: 10.1017/9781009325844

Jha, K., Doshi, A., Patel, P., & Shah, M. (2019). A comprehensive review on automation in agriculture using artificial intelligence. *Artificial Intelligence in Agriculture, 2*, 1–12. DOI: 10.1016/j.aiia.2019.05.004

Khan, A., Sohail, A., Zahoora, U., & Qureshi, A. S. (2021). A survey of the recent architectures of deep convolutional neural networks. *Artificial Intelligence Review, 53*(8), 5455–5516. DOI: 10.1007/s10462-020-09825-6

Kumar, S., Yadav, A., & Sharma, D. K. (2020). Deep Learning and Computer Vision in Smart Agriculture. In Pradeep, N., Kautish, S., Nirmala, C., Goyal, V., & Abdellatif, S. (Eds.), *Modern Techniques for Agricultural Disease Management and Crop Yield Prediction* (pp. 66–88). IGI Global Scientific Publishing. DOI: 10.4018/978-1-5225-9632-5.ch004

Li, Y., Deng, G., Zhao, H., Liu, B., Liu, C., Qian, W., & Qiao, X. (2026, January). A review of UAV remote sensing technology applications in common gramineous crops. *Information Processing in Agriculture*. Advance online publication. DOI: 10.1016/j.inpa.2026.01.009

Lou, S., Hu, R. Q., Liu, Y., Zhang, W., & Yang, S. Q. (2022). The formulation of irrigation and nitrogen application strategies under multi-dimensional soil fertility targets based on preference neural network. *Scientific Reports, 12*(1), 20918. DOI: 10.1038/s41598-022-25133-1 PMID: 36463318

Nitin, & Bal Gupta, S. (2023). Artificial Intelligence in Smart Agriculture: Applications and Challenges. *Current Applied Science and Technology, 24*(2), e0254427. DOI: 10.55003/cast.2023.254427

Oğuztürk, G. E. (2025). AI-driven irrigation systems for sustainable water management: A systematic review and meta-analytical insights. *Smart Agricultural Technology, 11*, 100982. DOI: 10.1016/j.atech.2025.100982

Olsen, T. L. (2026). Artificial Intelligence in Agricultural Supply Chains: Practice and Prospects. In Cohen, M. C., & Dai, T. (Eds.), *AI in Supply Chains. Springer Series in Supply Chain Management* (Vol. 27). Springer. 8_18.DOI: 10.1007/978-3-032-07054-8_18

Reitano, M., Segovia, M. S., & Nayga, R. S.Jr. (2025). A systematic review on the impact of Artificial Intelligence in the agri-food supply chain. *Food Policy, 102983*(137), 102983. DOI: 10.1016/j.foodpol.2025.102983

Roy, S., Hoque, A., Saikia, P., & Padhiary, M. (2025). Climate-Smart Agriculture: AI-Based Solutions for Enhancing Crop Resilience and Reducing Environmental Impact. *Asian Research Journal of Agriculture, 18*(1), 291–310. DOI: 10.9734/arja/2025/v18i1665

Sánchez-Arnau, E., Ferrer-Sapena, A., & Cárcel-Mas, M. C. (2026). Sustainability indicators and data management for small-scale agricultural systems: The case of water resources. *Regional Environmental Change, 26*(2), 90. DOI: 10.1007/s10113-026-02577-z

Shafi, U., Mumtaz, R., García-Nieto, J., Hassan, S. A., Raza Zaidi, S. A., & Iqbal, N. (2019). Precision agriculture techniques and practices: From considerations to applications. *Sensors (Basel), 19*(17), 3796. DOI: 10.3390/s19173796 PMID: 31480709

Shahab, H., Naeem, M., Iqbal, M., Aqeel, M., & Ullah, S. S. (2025). IoT-driven smart agricultural technology for real-time soil and crop optimization. *Smart Agricultural Technology, 10*, 100847. DOI: 10.1016/j.atech.2025.100847

Sharma, M., Kumar, A., Muthuraman, S., & Chaturvedi, P. (2026). Deep learning techniques for crop classification in complex agricultural landscapes. *Scientific Reports, 16*(1), 8831. DOI: 10.1038/s41598-026-37806-2 PMID: 41820421

Shawon, S. M., Ema, F. B., Mahi, A. K., Fahima Lokman Niha, F. L., & Zubair, H. T. (2025). Crop yield prediction using machine learning: An extensive and systematic literature review. *Smart Agricultural Technology*, *10*, 100718. DOI: 10.1016/j.atech.2024.100718

Sinclair, R., Nodi, S., & Kabir, M. A. (2024). Evaluating mobile applications for estimating soil properties: Quality of current apps, limitations and future directions. *Computers and Electronics in Agriculture*, *216*, 108527. DOI: 10.1016/j.compag.2023.108527

Soleimani, A., Abbaspour-Fard, M. H., Sapkota, R., & Sanaeifar, A. (2026). Artificial intelligence in agricultural machinery & robotics: A systematic review and future outlook. *Applied Soft Computing*, *194*, 114889. DOI: 10.1016/j.asoc.2026.114889

Subeesh, A., & Naveen Chauhan, N. (2026). Agricultural digital twin for smart farming: A review. *Green Technologies and Sustainability*, *4*(2), 100299. Advance online publication. DOI: 10.1016/j.grets.2025.100299

Sudha, S. P., & Loret, J. B. S. (2026). A review on machine learning-based precision agriculture techniques for crop farming monitoring with IOT. *Discover Environment*, *4*(1), 10. DOI: 10.1007/s44274-025-00305-8

Talaviya, T., Shah, D., Patel, N., Yagnik, H., & Shah, M. (2020). Implementation of artificial intelligence in agriculture for optimisation of irrigation and application of pesticides and herbicides. *Artificial Intelligence in Agriculture*, *4*, 58–73. DOI: 10.1016/j.aiia.2020.04.002

Tsouros, D. C., Bibi, S., & Sarigiannidis, P. G. (2019). A review on UAV-based applications forprecision agriculture. *Information (Basel)*, *10*(11), 349. DOI: 10.3390/info10110349

Tzachor, A. (2021). Barriers to AI Adoption in Indian Agriculture: An Initial Inquiry. *International Journal of Innovation in the Digital Economy*, *12*(3), 30–44. DOI: 10.4018/IJIDE.2021070103

Waqas, M. A., Humphries, U. W., Hlaing, P. T., Dechpichai, P., & Wangwongchai, A. (2025). Applications of machine learning and deep learning in agriculture: A comprehensive review. *Green Technologies and Sustainability*, *3*(3), 100199. Advance online publication. DOI: 10.1016/j.grets.2025.100199

Wolfert, S., Ge, L., Verdouw, C., & Bogaardt, M. J. (2017). Big data in smart farming: A review. *Agricultural Systems*, *153*, 69–80. DOI: 10.1016/j.agsy.2017.01.023

World Bank. (2021). *Harvesting prosperity: Technology and productivity growth in agriculture*. World Bank Publications.

Zhang, C., & Kovacs, J. M. (2012). The application of small unmanned aerial systems for precision agriculture: A review. *Precision Agriculture*, *13*(6), 693–712. DOI: 10.1007/s11119-012-9274-5

Zhang, J., Guan, K., Peng, B., Pan, M., Zhou, W., Jiang, C., Kimm, H., Franz, T. E., Grant, R. F., Yang, Y., Rudnick, D. R., Heeren, D. M., Suyker, A. E., Bauerle, W. L., & Miner, G. L. (2021). Sustainable irrigation based on co-regulation of soil water supply and atmospheric evaporative demand. *Nature Communications*, *12*(1), 5549. DOI: 10.1038/s41467-021-25254-7 PMID: 34545076

Zhang, R., Feng, Y., Li, Y., & Zheng, K. (2024). Can different information channels promote farmers' adoption of agricultural green production technologies? Empirical insights from Sichuan Province. *PLoS One*, *19*(8), e0308398. DOI: 10.1371/journal.pone.0308398 PMID: 39116048

Chapter 2

Assessing Agricultural Sustainability Through Geospatial AI and Food Security Policy Practices

K. Sivasubramanian
http://orcid.org/0000-0001-6137-0847
Kristu Jayanti College, Bengaluru, India

Siji O. K.
Vel Tech Rangarajan Dr. Sagunthaala Institute of Science and Technology, Chennai, India

R. Kasthuri
A.D.M. College for Women, Bharathidasan University, India

E. Seenivasan
Vel Tech Rangarajan Dr. Sagunthala R&D Institute of Science and Technology, Chennai, India

S. Vighnesh
Kristu Jayanti University, India

V. Vinothina
Kristu Jayanti University, India

S. Viji
Dhanalakshmi Srinivasan University, India

ABSTRACT

Food security is a long-standing concern, prompting international organisations like the Food and Agriculture Organization to categorise it into four key aspects: accessibility, availability, utilisation, and stability. Food security is a key component of the global SDGs. Agricultural methods have a significant impact on the interplay of food security and sustainability. Agricultural sustainability could be achieved through the aggressive use of advanced technologies and innovative practices. One such aggressive and advanced technique is the Geospatial Artificial Intelligence (GeoAI). It helps to forecast for agriculture, soil suitability, soil health, crop health, predict pest spread, monitor environmental impacts, and predict natural calamities to reduce the risk of unforeseen productivity loss. Obviously, the growth of agriculture and productivity promotes food security and sustainability. In this chapter, the authors analyse the impact of GeoAI and various government schemes for enhancing agriculture and food security.

DOI: 10.4018/979-8-3373-8037-7.ch002

INTRODUCTION AND THEORETICAL BACKGROUND

Food security is a long-standing concern, prompting international organisations such as the Food and Agriculture Organisation to categorise it into four key dimensions: accessibility, availability, utilisation, and stability. The agricultural sector is vital to the global economy and food security. Still, it also contributes significantly to environmental degradation and global warming through activities such as pesticide and fertilizer use, water management, and the destruction of forests and wetlands. Natural resources are crucial to agricultural production, and unsustainable farming methods can seriously deplete them. Soil erosion may worsen when forests are cleared for agricultural purposes. Approximately 10-12% of anthropogenic carbon dioxide emissions worldwide originate from agriculture, with farming accounting for the majority of these emissions (Kross et al., 2022). Climate change's global disruption of traditional agriculture and the effects on food security underscore the crucial role of geospatial technologies in addressing these challenges. By facilitating informed decisions through accurate mapping and monitoring, these technologies significantly enhance decision-making and resource optimization, thereby promoting sustainable farming practices. Conversely, geospatial tools enable the collection, analysis, and presentation of spatial data, which are crucial for understanding agricultural systems under shifting climate conditions. These technologies are crucial to the CSA's three pillars: increasing productivity through crop monitoring and precision farming; enhancing adaptation through early-detection systems, land-use mapping, and water-resource management; and supporting mitigation through carbon accounting and sustainable land-use planning. Food security is a component of the goals for sustainable development. Nonetheless, it is agreed that sustainability should be addressed more effectively within the foundations of food security, whether as a new dimension or by incorporating other elements, such as hunger and climate change. Food security is inextricably linked with sustainability. However, the relevant situations are complex and ambiguous. Agricultural methods have a significant impact on the interplay of food security and sustainability. Population growth is particularly alarming in countries such as India. Alternative food sources, such as edible insects, may offer promising alternatives for achieving greater food security. Farming fish is another way to attain more sustainable food security. India scores 102nd out of 123 nations according to the Global Hunger Index (2025), which considers child abuse, stunting, and malnutrition. With food sufficiency nearly attained, India has the potential to eradicate significant hunger and reduce the prevalence of stunting, although this remains a long-term goal (Mahendra & Alakh, 2010).

LITERATURE REVIEW

The availability, access, utilisation, stability, independence, and sustainability are the six primary characteristics that constitute the concept of food security. The first three planes exhibit hierarchical relationships. Ensuring that adequate amounts of food of suitable quality are available through imports or domestic production, including food aid, is the focus of the concept of food availability (McKay et al. 2023). With the world's population predicted to exceed 10 billion by 2050, agriculture is under unprecedented pressure to boost output sustainably. Innovative solutions are urgently needed to increase farm productivity and efficiency, given the additional challenges posed by diminishing water resources, changing weather patterns, and the decline in agricultural land (Babar & Akan, 2024).

With the development of integrated farm management, AI lends modern agriculture, in particular precision farming, new significance. The innovation in agriculture includes robots, automated surveillance systems, and drones. One such application is the use of drones, where they can be used in regions to conduct continuous monitoring of the crops to determine their health and detect any disease, hence managing the field more efficiently. Thus, the technologies promote precision farming by offering information to make informed decisions on pest management, fertilization, and irrigation. Robots also eliminate the use of physical labour by automating work-intensive activities such as weeding, harvesting and seeding (Escobar-Naranjo et al. 2023).

AI is also useful for analyzing and tracking essential environmental parameters, such as rainfall, ambient temperature, and soil conditions. The predictive algorithms, supported by AI, detect patterns in the weather that enable farmers to make the relevant decisions and reduce the negative impact of extreme weather. Constant observation of the environment and soil health has created an optimal environment for crop growth and enhances productivity towards sustainability.

AI and Sustainable Productivity

Large datasets are then further evaluated by using machine learning models. Such models also enable making decisions in real-time, which facilitates a shift towards more sustainable and efficient farming practices. In spite of the existing difficulties, AI can increase productivity, sustainability, and profitability, becoming an essential part of agricultural practices in the future. Sustainable, precise watering is another very crucial measure in attaining food security, besides adopting water-saving measures to counter the uncertainty of rainfall and effects of water scarcity caused by drought in most parts of the world. To balance the loss of water through evapotranspiration, erosion, or deep percolation without over- and under-irrigating, precision irrigation scheduling aims at optimizing water usage to each plant, where and when they need it, and in the right quantities (Aijaz et al. 2025).

Food access is a prerequisite for access, but it is not enough. Access focuses on ensuring sufficient resources, or entitlements, available to purchase the right foods for a healthy diet. For utilisation, access is required but not sufficient. The ability of a person to obtain a sufficient diet, clean water, sanitary conditions, and medical attention to achieve a state of good nutrition is referred to as utilisation (Kendall et al. 1995). Given that the first three criteria can be disrupted by dangers including violence, job loss, epidemic disease, and climate swings, the significance of the other three notions has grown. The first three dimensions' consistency is referred to as stability. The ability of individuals or organisations to make decisions about their food is known as agency. It involves the ability to choose what they consume, what and when they produce, and how that food is allocated within the framework of food systems and government (HLPE, 2020). The capacity of agriculture to supply adequate nutrition and food security over the long term without jeopardising the social, economic, or environmental foundations necessary to ensure food security and food for future generations is known as sustainability (Rautela et al. 2020). It is estimated that 200 million people in India suffer from malnutrition. In India, the main causes of malnutrition are poverty, an absence of safe water to drink, and inadequate sanitation (Olson, 1999). The relationship between malnutrition and food insecurity in India has not received much attention to date, despite substantial levels of malnutrition indicating widespread food insecurity. The urgency of food scarcity as a public health issue is highlighted by the links between household food hardship and infant stunting, wasting, and underweight observed in the limited data available (Dhamija et al., 2021; Pathak et al., 2020).

Geospatial AI integration for Agricultural Sustainability and Food Security

The use of cutting-edge technologies, such as artificial intelligence, to improve food security has been spurred by the quest for a sustainable food system. The goal of artificial intelligence, which spans many areas of computer science, is to build intelligent machines that can perform tasks that typically require human intelligence. Given the uncertainties surrounding the complex and dynamic interplay among the economic, environmental, and social variables that underpin food security, artificial intelligence has become essential (Mendoza-Bernal et al., 2024). More than 80 per cent of smallholder farmers in underdeveloped nations produce food using basic technologies. Concerns over global food security have been heightened by increases in agricultural input costs over the past two years, particularly for fertilizer, which may lead to higher food prices. Optimizing operations to reduce losses and expenses while increasing yields is a challenge for operators in the farming and agriculture sectors (Arora, 2019).

Low crop yields, weather-related losses, insect and disease outbreaks, post-harvest losses during storage and shipment, high production costs, low revenue generation, and market-dynamics uncertainty are among the factors contributing to this dilemma. AI appears to be a tool that can aid in the development of integrated food security management plans, improve crop yields, minimize expenses, and lower operating costs, given the many obstacles that affect stakeholders' efforts to ensure food security (Rani et al., 2023). In terms of efficiency, accuracy, reliability, automation, pattern recognition, accessibility, and scalability, AI models are very beneficial for food security. However, there are drawbacks as well, including issues with data, cost, ethics, and the requirement for continuous management. The particular issue, the resources at hand, and the negotiations involved dictate whether to use AI models or traditional approaches (Muruganandhan et al. 2025). To make well-informed decisions, artificial intelligence (AI) includes analytical tools like machine learning, data analytics, and optimization models that may be used to analyse both historical and current data. The employment of AI models is the main topic of this work (Abioye et al. 2022). Building automated, adaptive, and smart systems that guarantee better, quicker, and more accurate operations requires AI modelling. Stakeholders can approach the complex issue and adjust to shifting circumstances for long-term success by combining numerical optimization, risk analysis, parameterization, and scenario planning (Pandey & Misra, 2024). The scientific community acknowledges the critical role AI models play in guiding decision-making amid growing concerns about food insecurity (Kumari et al. 2025). Even though there is a wealth of models and information available, it is frequently difficult to translate these models into useful policy. Large-scale model output interpretation is a common feature of applying AI in policy-making (Fiedler, 2017).

Impacts of Geospatial AI on Smallholder Farmers, Women, and Marginalized Groups

GeoAI helps in transforming agriculture, especially for small-holding farmers, women workers in agriculture, and marginalized groups. It provides various potential advantages such as enhanced productivity and financial sustainability. AI-driven image optimization helps farmers to take photographs of the crops, diagnose diseases, and help to reduce the crop productivity. There are several prospects for improved resource usage, waste reduction, and ecological responsibility when AI is integrated into agriculture

(Sivasubramanian, 2025). It uses a variety of technologies to make farmland intelligent, precise, and productive, going beyond simple automation. Because these technologies go beyond the optimization of current methods, they will assist in fully managing the production conditions and outcomes. For instance, AI-driven precision agriculture uses data from satellite photography and Internet of Things sensors to predict the ideal planting, irrigation, and harvesting timings. Machine-learning systems predict crop yields with remarkable accuracy and detect plant diseases earlier than they manifest in human vision. However, there are numerous financial, technological, and societal obstacles to the widespread use of AI in agriculture, which affect women and marginalized sections (Savithiri et al. 2026). A wide range of stakeholders and a variety of people are involved in the integration of AI for food security, including small farmers, agricultural researchers and academics, food processing businesses, food retailers and distributors, agencies of government, non-governmental organizations, agriculture and food technology organizations, and consumers. Investors and financial companies, meteorological and climate entities, supply chain and logistics firms, farm laborers and labour forces, and food safety agencies are some of the other stakeholders who use AI to improve food security. Effectively resolving the issues of global food security requires the cooperation and coordination of various stakeholders (Chukwuma et al. 2024).

Although this method is useful, it is crucial to incorporate stakeholders' local knowledge to guarantee the accuracy of model results. It has been suggested that including stakeholders, their demands, and their knowledge into models will improve stakeholder outcomes. Models by themselves can be poor tools; thus, decision-making procedures must involve stakeholders and incorporate their local knowledge (Rebecca et al. 2023); (Bhagat et al. 2022) & (Vinuesa et al. 2020). When AI and robotics are combined with precision agriculture, smart farming ecosystems are created in which autonomous robots collaborate with other technologies, such as drones and sensors, to assess soil and track crop health (Liu et al. 2023; Mkhitaryan et al. 2025).

Public Distribution System

The Public Distribution System (PDS) was developed to distribute food grains at reasonable prices to control scarcity. PDS has grown in importance throughout time as a component of the government's policy for managing the nation's food economy. PDS is supplementary and isn't meant to provide a home or a particular segment of society with all of the goods given under it. The Central Government and the state and territory governments share joint responsibility for PDS operations. India's Public Distribution System (PDS), which provides essential goods at discounted rates and serves as a safety net, is crucial to reducing food insecurity. India still has a sizable population of undernourished and hungry individuals, despite the PDS being a pillar of the country's food and nutrition policies (George & McKay, 2019). Wheat, rice, salt, and paraffin are currently distributed to the States and Union Territories (UTs) under the PDS. Pulses, edible oils, iodine salt, spices, and other mass-consumption goods are also distributed by certain States and Union territories (UTs) through PDS shops (Narain, 2016). Throughout the nation's interwar years, public distribution of necessities was in place. Nonetheless, PDS was born out of the severe food shortages of the 1960s and focuses on the distribution of grain in urban regions of scarcity. PDS has made a significant contribution to ensuring that urban consumers have access to food and to containing the increase in the price of food grains. To improve the PDS's reach in the far-flung, hilly, rural, and inaccessible areas where a sizable portion of the impoverished reside, the Revamped Public Distribution System (RPDS) was launched in June 1992. It included 1775 blocks where area-specific programmes such as the Desert Development Programme (DDP), Integrated Tribal Development Projects (ITDP), and Drought Prone Area Programme (DPAP) were being implemented, as well as certain Designated Hill Areas (DHA) selected for special attention after consultation with state governments.

The States were provided food grains at a 50 paisa lower price than the Central Deal Price for re-distribution in RPDS zones. Up to 20 kg was the extent of the problem. Additional ration cards for the families left out, an area approach to ensuring the PDS commodities' effective reach, the State Governments' delivery of these commodities at the doorstep of FPSs in the designated areas, infrastructure requirements like more Fair Price Shops and storage capacity, and additional commodities like tea, salt, pulses, soap, and other items for distribution via PDS outlets were all included in the RPDS. Families in APL were still allocated at the same level as before TPDS was introduced, but as of that date, the Central Issue Prices (CIPs) for APL had been set at 100% of economic cost, allowing the full consumer subsidy to be directed towards the population of BPL (Bhalla et al. 2001). Hunger and malnutrition still exist in many areas of India, despite vast supplies of food grains and a long-standing public distribution system. To harm the impoverished, state-by-state variations have further muddled matters. To be more exact, food security is defined as the state in which all individuals have physical and financial access to enough safe, nourishing food at all times to suit their dietary requirements and preferences for a healthy and active existence (Ray & Ray, 2011).

A great deal of our population's well-being is directly impacted by the topic of food security. The food-related issues that our nation is currently dealing with are very serious. There is ample data to address the issues at hand (Saxena, 2018). Overcoming poverty and hunger is the biggest problem facing many emerging nations, including India. These are a major threat to the advancement of the country. (Avinash, 2018). In 1995, the Meals on Wheels programme was launched as a nationwide initiative to provide children undergoing basic education with nutritional support. Its two main goals are to raise enrolment, regular attendance, and retention rates in schools while also enhancing the dietary habits of elementary school students (Saxena et al, 2020).

This scheme has been introduced to ensure food security for the vulnerable people. The functions and performance of the scheme are monitored by several innovative technologies, including the application of AI in India. The application of GeoAI technology in agriculture helps to assess and indicate the production capacity and provide predictions to ensure an optimum availability of agricultural produce.

Integrated Child Development Scheme

The Government of India launched the Integrated Child Development Scheme (ICDS), a flagship programme designed to meet the diverse needs of children under six. Since its founding in 1975, ICDS has grown to become one of the biggest and most extensive early childhood development initiatives worldwide. This review aims to explore all facets of ICDS, including goals, tactics for implementation, effects, difficulties, and potential future developments. Fundamentally, the goal of ICDS is to give kids the opportunity for holistic development, emphasising their general well-being, early education, health, and nutrition. (Rajeswari et al. 2025). This scheme is to ensure food security and the nutritional aspects of the children in India. The adoption of GeoAI technology at the root level helps to ensure proper supply and procurement of basic cereals to child development centres (Anganvadi Centre) in both rural and urban areas.

The National Food Security Act,2013

The government of India passed the NFSA in 2013 to address the food security problem in the country. Subsidised food grains are available to qualified households and priority households. The Act provides food grains that are subsidised under TPDS to up to 75% of the rural population and 50% of urban people. Eligible homes receive a maximum of 3 kg of cereal grains per child per month, whereas those with priority have a right to 5 kg at subsidised rates per person per month. Children from six months to fourteen years old are eligible to receive rations at home or hot meals, and expectant mothers and nursing moms are entitled to maternal benefits of at least 6,000. The objectives of this scheme are achieved through the implementation of GeoAI in the agricultural sector for predictions, performance, and monitoring.

Midday Meal Policy

One of India's most important initiatives in the areas of food security and education is the Midday Meal Scheme (MDMS). It was started in the 1960s to feed schoolchildren nationwide with wholesome meals to combat malnutrition and advance education at the same time. The major objectives of the scheme are to improve nutritional levels among children, encourage regular attendance and retention of children in schools, and promote social equity by ensuring that children from all backgrounds receive adequate nutrition. The Indian Ministry of Education, in partnership with the states and union territories, carries out the programme. It includes students in government, state-aided, and local body schools' elementary and upper primary classes, as well as those enrolled in Alternative & Innovative Education centres and the Education Guarantee Scheme. The meals served as part of the programme are planned to supply youngsters with half of their daily protein requirements and at least one-third of their daily energy requirements. Rice, wheat, legumes, vegetables, and fruits are usually on the menu. To make sure it is adequate and balanced, the nutritional information is updated regularly. The Midday Meal Scheme has significantly impacted several areas, and it has made a major contribution to the decline in school-age malnutrition. One of India's main initiatives for encouraging education and guaranteeing food security is the Midday Meal Scheme (Jha, 2007).

The mid-day meal scheme is one of the best and most powerful schemes in promoting food security in India. The adoption of GeoAI predictive technology has highly impacted on ensuring appropriate production and distribution of rice, wheat, ragi, and vegetables for the success of the scheme.

Prime Minister Kisan Sampada Yojana

The Government of India's flagship programme, the Pradhan Mantri Kisan Sampada Yojana aims to modernise and strengthen the food processing industry. The initiative, which was introduced in 2017, aims to increase the competitiveness, decrease waste, and guarantee higher returns for farmers in order to fully utilise the potential of India's agricultural industry and related sectors. To lower post-harvest losses and increase value addition, the scheme focuses on building state-of-the-art food processing infrastructure, such as chilly chains, mega food green spaces, and agro-processing clusters. The programme promotes the conversion of perishable agricultural products into value-added goods, extending their shelf life, making them more marketable, and boosting farmers' profits. Through promoting farmers' integration with food processing companies and retail markets, programmes seek to increase the forward and backward links between farmers and markets. It places a strong emphasis on increasing the food processing industry's capacity and skill set in order to generate jobs, especially in rural areas. (Lanjouw & Rinku, 2009). The food processing industries require a consistent supply of agricultural produce to use as an input. The GeoAI technology helps to monitor, assess, and helps to connect supply chains of agricultural commodities to food processing units. This will ensure the promotion of productivity of the food processing industries and help to support farmers to enhance their income.

Antyodaya Anna Yojana (AAY)

Under AAY, food grains are given out at a heavily subsidised rate of 2 for wheat and 3 for rice per kg. In both rural and urban areas, the programme gives priority to households led by widows, elderly people, people with disabilities, and people who are below the poverty line. State and union territory administrations carry out AAY with financial assistance from the federal government. Food grains must be purchased, stored, and distributed by the states. AAY ensures that the most vulnerable members of society always have access to food, even in the event of natural calamities or economic distress. Notwithstanding its achievements, AAY still has problems with leaks, incorrect identification, and insufficient coverage (Dyson & Hanchate, 2000). The objective of this scheme is achieved through the application of GeoAI in evaluating and improving the productivity and procurement of agricultural commodities from various parts of the country. It supports efficient production and effective distribution.

National Nutrition Strategy (NNS)

The Government of India launched the National Nutrition Strategy (NNS), a comprehensive framework, to address the complex issues of malnutrition and malnutrition in the nation. The 2017 strategy was introduced with the goal of lowering the rate of malnutrition and enhancing the dietary habits of all demographic groups, with an emphasis on women and children in particular. The NNS's multi-sectoral approach, which acknowledges that factors other than food intake, such as healthcare, hygiene, schooling, and livelihoods, can affect nutrition outcomes, is one of its main advantages (Radhakrishna & Ray, 2005). The aims of this scheme were achieved with the support of GeoAI technology adoption in monitoring and assessing the agricultural productivity, quality, and supply.

Pradhan Mantri Krishi Sinchai Yojana (PMKSY)

To ensure water security, improve agricultural output, and provide irrigation infrastructure to all farms, the Indian government introduced the Pradhan Mantri Krishi Sinchai Yojana (PMKSY) in 2015. The government's larger plan to modernise Indian agriculture and accomplish sustainable rural development includes the programme. By emphasizing on developing water-efficient infrastructure, increasing water use efficiency, and supporting sustainable farming practices, this takes a holistic approach to water management. To meet the varied water needs of farmers, it incorporates a number of different elements, including irrigation, building watersheds, rainwater collection, and water conservation. Improving agricultural water use efficiency by implementing micro-irrigation methods like sprinkler and drip irrigation is one of the main goals. The programme seeks to maximise crop yields, minimise water waste, and promote effective water management techniques. The objective of this scheme is to extend the reach of irrigation systems to regions that do not currently have sufficient access to water. To guarantee the fair allocation of water resources, this involves creating new irrigation projects, updating current water infrastructure, and supporting community-based water management programmes. The plan places a strong emphasis on implementing cutting-edge techniques and technology to enhance water management and irrigation effectiveness. This involves encouraging planting patterns and water-saving strategies that are climate resilient, as well as using remote sensing technologies to better design and manage irrigation projects.

GeoAI has been effectively used in this scheme to predict the irrational requirements and water availability. This technology has been widely used by all farmers for the effective utilization of scarce resources and to achieve maximum output. GeoAI helps in environmental predictions, groundwater identification, and automated irrigation systems in agriculture.

CHALLENGES IN APPLICATIONS OF GEOSPATIAL AI IN AGRICULTURE

The lack of standardized datasets and concerns about user data privacy are the main drivers of this restriction (Kpodo & Nejadhashemi, 2025). The effectiveness and usefulness of models based on machine learning and artificial intelligence have been shown in a variety of applications, as was indicated in the preceding section. However, the outcomes produced by these models might not always agree with the evaluations of domain experts, or they might not agree at all. Naturally, we speculate that such behavior creates two problems: problems with AI systems' transparency and end users' reluctance to adopt a technology because they don't trust it or because it's new. These issues have actually been brought up in AI literature as well, but they are still unresolved (Nourani et al. 2020). Furthermore, there is a great deal of potential to improve farming efficiency and managerial skills due to the rapid advancement and widespread adoption of AI technologies in farming on a global scale. However, an extensive awareness of the risks and consequences associated with the application of AI in the agriculture sector is still lacking. In order to customize AI systems' conclusions toward more efficient decision-making mechanisms, a conceptual framework is required to support the cooperation between EEs and machine learning developers. In this sense, some authors support the use of agricultural AI in controlled settings, data cooperatives, accountable and human-focused imaginative thinking, and the participation of rural anthropologists and professional ecologists in the technology design process. Reduced cognitive function, worse work performance, and significant productivity losses might result from food insecurity. These can all impede the expansion and advancement of the national economy. Over the past few decades, India has made significant progress in terms of health determinants. Over time, there has been a steady decrease in vital health indicators, such as infant mortality, maternal mortality ratio, prevalence of diseases, morbidity, and mortality rates. Over the past fifty years, India's life expectancy has increased, and infant mortality has almost halved (Upadhyay & Palanivel, 2011).

There has been remarkable economic growth, with notable contributions from the agricultural sector. India is the world's second-largest farmer. Farmed products and related industries, such as logging, forestry, and fishing, employed sixty per cent of the labour force and contributed 18.6 per cent of the GDP in 2005. Due to issues related to gender inequality, poverty, overcrowding, and lack of education, the children face food insecurity. One of the main causes is poverty, which restricts the quantity of food that kids may eat. Children who are malnourished may result of overcrowding, which is associated with competition for food, particularly in remote regions where food access is scarce. Another area of worry is that moms may not have enough understanding about nutrition, breastfeeding, and parenting. The issue of pay disparities is another. There are disparities in salary rates based on gender in India (Bajpai et al. 2004).

Data Challenges, Limitations, and potential risks of AI in agriculture

Innovative AI technologies such as drones, automated machines, and sensor instruments require significant and high investment. Small and petty farmers, particularly in developing and underdeveloped nations, would find it difficult to afford these technologies. Moreover, AI purely relies on a high volume of accurate data. In many rural areas, data is either unavailable or unreliable. Poor-quality and quantity of data may lead to incorrect predictions. Many small farmers and marginalized women lack digital literacy. Nutritional security can foster healthy, youthful populations that support the nation by increasing literacy, education, and employment. India must address post-harvest expenses, inadequate storage facilities, and poor supply chain issues with governance (Sivasubramanian, 2025). Demand-side concerns, such as changing consumption patterns and per capita consumption, must be addressed alongside supply-side factors. The future goal is zero hunger achieved through not only food availability but also nutritional security, which can be accomplished by sustainable agriculture, resulting in an optimal environment, economic profitability, and healthy farming methods that reduce pollution, make more sustainable and productive lands, and provide food to all, assisted by an effective channel of distribution that reduces hoarding procedures, black marketing, food rotting, and making it accessible, inexpensive, and accessible. Various government welfare schemes in India have significantly promoted food security among rural and urban people. This chapter aims to assess the positive impacts of AI applications and the government's food security schemes on agricultural sustainability in India. Farm waste, including crop residues, animal manure, and chemicals such as pesticides and fertilizers, is discharged into nearby water bodies, contaminating the environment and altering water quality. Nearly 43 percent of the world's species have declined significantly due to practices such as agricultural conversion and the use of fertilizers and pesticides (Thakuri et al., 2025); (Kumar et al., 2012).

STRATEGIC MEASURES FOR FOOD SECURITY

AI-based technological solutions have made it possible for farmers to produce more with fewer ingredients and even improve the quality of their output. They have also made it possible for the yielded crops to reach the market more quickly (Talaviya et al. 2020). The best crop selection and even the selection of seeds that are hybrids that are most appropriate for farmers' needs have been made possible by emerging technologies. It has been put into practice by comprehending how seeds respond to different soil types and weather conditions. Plant diseases are less likely to occur as a result of gathering this data.

Farmers can now effectively maximize crop returns by meeting market trends, annual results, and consumer demands. An integrated policy structure must be adopted by the government to support the rising use of irrigated and modern farming methods. Along with giving farmers better cultivation technology and inputs like irrigation systems, higher-quality seeds, fertilisers, and loans at reduced interest rates, the main goals of the measures should be to rationally distribute cultivable land, increase farm sizes, and offer security to tenant cultivators. A major contributing factor to the unequal distribution of food is the large amount of food waste. Ensuring the affordability of food grains for the general public is imperative. This can be achieved by providing the BPL population with food at a significantly reduced cost through more precise targeting. The BPL listing is connected to an issue. The precise number of people who fall within this category is up for debate. To improve the purchasing power of the lower socioeconomic segment of the population, programmes to alleviate poverty like the Integrated Rural Development Programme (IRDP) and job creation plans like Jawahar Rozgar Yojana, Nehru Rozgar Yojana, etc., need to be reoriented and scaled up. Improving both the amount and quality of work that pays a salary is equally vital. "Crop diversification" is another subject that requires investigation.

Increased profitability and production stability emphasise the value of crop diversity, such as substituting rice and wheat for legumes. It is necessary to promote the growth of non-cereal products such as oilseeds, fruits, and vegetables. A good solution would be to establish decentralised food banks in every village or district block, where residents may get subsidised food grains in exchange for food vouchers. This idea will do away with corruption and enhance the distribution of food grains. The idea of home gardening could be one approach to provide direct access to high-quality food that is simple to grow and prepare. An outstanding illustration of home gardening may be found in Bangladesh, where the non-governmental organisation Helen Keller International carried out a home garden and nutrition awareness project as part of its global initiative to eradicate a lack of vitamin A and nutritional blindness (Talukder et al., 2000).

MAJOR FINDINGS

The review analysis highlights that food security remains a challenge, structured with accessibility, utilisation of resources, availability, and stability, as recognized by the Food and Agriculture Organization. A key finding is that climate change has intensified issues in agricultural systems, making food security more difficult. In response, the integration of Geospatial technologies enables precise monitoring, assessing, predicting, and mapping the resources, thereby improving decision-making and enhancing sustainable farming practices.

(Khattri et al. 2025). It is identified from the present review analysis that the application of Geospatial tools and technologies helped to enhance productivity and food security through government schemes. India's food security programmes are crucial parts of the nation's development agenda since they have, overall, had significant effects on reducing poverty, promoting human development, promoting social fairness, and promoting economic well-being (Uma Maheswari et al. 2025). This study also found various challenges in the application of AI technologies in agriculture, such as data-related challenges and the risk of AI in agriculture. The Geospatial AI helps to for agricultural sustainability and rural development through development schemes of the government.

CONCLUSION

Climate change has intensified vulnerabilities within the agricultural systems across the globe, making food security an excessively complex and urgent challenge. In this scenario, the integration of Geospatial AI emerges as a transformative way toward a sustainable and resilient food security system. Geospatial instruments enhance precision in agriculture by enabling monitoring, efficient resource allocation, and management of input materials, and informed decision-making. Simultaneously, AI contributes through predictive analysis and optimisation models, and adaptive methods that improve productivity, potentially reduce risks, and support data-driven decisions. The synergy among these AI technologies strengthens the three key pillars of food security: agricultural productivity, adaptation, and moderation. While geospatial technologies support various agricultural processes, namely land-use planning, water management, forecasting, crop quality assessment, and supply chain efficiency. India's agricultural and food security policy frameworks are reflecting a strong commitment to addressing hunger, malnutrition, and agricultural rural sustainability. These programmes, when integrated with advanced technologies, could significantly impact by ensuring efficiency. However, the successful adoption of these innovative technologies depends on overcoming challenges such as accessibility, infrastructure, high costs, and ethical concerns. On the whole, to conclude, achieving long-term food security requires a holistic and tech-driven approach that is coupled with innovation and institutional support. By aligning geospatial technologies and AI, nations could build resilient agricultural systems capable of sustaining future generations.

SUGGESTIONS FOR EFFECTIVE FOOD SECURITY

The focus in this area has turned to trends and outcomes in food access and absorption/nutrition. There has been a discernible decrease in household hunger levels, as well as a considerable reduction in the number of homes living below the poverty threshold. Both of these patterns point to improved food access. Though urban employment has improved, the economy in rural areas remains unsatisfactory. Furthermore, there has been a noteworthy increase in the number of people without land in the rural sector, which has been followed by a fall in real earnings, affecting food access in rural areas as opposed to metropolitan areas. Access to the PDS, which contributes to total food access, has increased in almost all regions, which is a positive trend; nonetheless, access to the PDS remains inadequate in numerous states and regions. While overall food access has improved, nutrition trends have not been entirely satisfactory, particularly among the rural poor. The nutritional status of children and women has not improved, while the dietary status of socially vulnerable women has deteriorated. Inadequate policy, poor design, insufficient monitoring and assessment, inadequate governance, and an absence of political appetite all contribute to the challenges of guaranteeing food security and eliminating hunger. Economic expansion alone is insufficient to enhance food security, reduce malnutrition, and increase food intake among the poor. Without extensive reforms on multiple fronts and the necessary policy adjustments, as well as an improvement in implementation effectiveness, food security will remain a formidable task (Saxena, N.C., 2011). Sustainable development is the product of efficient institutions; therefore, the emphasis on development funds must now shift to development results and the efficacy of the delivery of public services. Hunger indicators for marginalised groups and the 300 million poor need to be improved. This necessitates not only money but also additional strategies and a robust delivery system (Venkatesan et al. 2025)

POLICY IMPLICATIONS

The government should provide open and standard geospatial systems at either free or affordable cost. Policies must target bridging the gap of the digital divide by enhancing the availability of AI tools to poor and marginalized people. Public and private partnerships would help to scale up equity. Training programmes must be institutionalized to improve farmers' capacity to use geospatial technologies. Finally, the geospatial AI technology should be embedded with a food security policy.

FUTURE DIRECTIONS FOR AI INTEGRATION

The farmer could use various AI models for sustainable agriculture and food security practices in the future, such as hyper-precision, geospatial intelligence, climate-smart solutions, and AI-driven food security policies. Moreover, international collaborations will enhance the application of innovative and advanced approaches. AI could transform agriculture and the food system, and it helps to address global hunger and climate change.

REFERENCES

Abioye, E. A., Hensel, O., Esau, T. J., Elijah, O., Abidin, M. S. Z., Ayobami, A. S., Yerima, O., & Nasirahmadi, A. (2022). Precision irrigation management using machine learning and digital farming solutions. *AgriEngineering*, *4*(1), 70–103. DOI: 10.3390/agriengineering4010006

Aijaz, N., Lan, H., Raza, T., Yaqub, M., Iqbal, R., & Pathan, M. S. (2025). Artificial intelligence in agriculture: Advancing crop productivity and sustainability. *Journal of Agriculture and Food Research*, *20*, 101762. DOI: 10.1016/j.jafr.2025.101762

Arora, N. K. (2019). Impact of climate change on agricultural production and its sustainable solutions. *Environmental Sustainability*, *2*(2), 95–96. DOI: 10.1007/s42398-019-00078-w

Avinash, T. (2018). Mid-day meal scheme under food security: A study with special reference to upper primary schools of rural Bhadravthi taluk. *International Journal of Research in Social Sciences*, *8*(1). https://indianjournals.com/article/ijrss-8-1-065

Babar, M. I., & Akan, B. (2024). Smart irrigation systems in IoT for sustainable water management. *Journal of Environmental Management*, *316*, 115347.

Bajpai, N., Sachs, J. D., & Volavka, N. H. (2004). Reaching the Millennium Development Goals in South Asia. *Academic Commons*. Columbia University Libraries. https://academiccommons.columbia.edu/doi/10.7916/D8BV7PFH

Bhagat, P. R., Naz, F., & Magda, R. (2022). Artificial intelligence solutions enabling sustainable agriculture: A bibliometric analysis. *PLoS One*, *17*(6), e0268989. DOI: 10.1371/journal.pone.0268989 PMID: 35679287

Bhalla, G. S., Hazell, P., & Kerr, J. (2001). *Prospects for India's Cereal Supply and Demand to 2020, Food, Agriculture and the Environment, Discussion Paper 29*. International Food Security Institute. https://ideas.repec.org/p/fpr/2020dp/29.html

Chukwuma, E. C., Afolabi, O. O. D., Okonkwo, C. C., & Okonkwo, C. E. (2024). Application of geospatial technology and decision model in the development of an improved food security index. *Scientific Reports*, *14*(1), 30204. DOI: 10.1038/s41598-024-77845-1 PMID: 39632976

Dhamija, G., Ojha, M., & Roychowdhury, P. (2021). Hunger and health: The impact of household food insecurity on child malnutrition in India. *The Journal of Development Studies*, *58*(6), 1181–1210. DOI: 10.1080/00220388.2022.2029419

Dyson, T., & Hanchate, A. (2000). India's Demographic and Food Prospects: State Level Analysis. *Economic and Political Weekly, 35*. https://www.jstor.org/stable/i399209

Escobar-Naranjo, J., Caiza, G., Ayala, P., Jordan, E., Garcia, C. A., & Garcia, M. V. (2023). Autonomous navigation of robots: Optimization with DQN. *Applied Sciences (Basel, Switzerland)*, *13*(12), 7202. DOI: 10.3390/app13127202

Fiedler, J. L., & Yadav, S. (2017). How can we better capture food away from home? Lessons from India's linking person-level meal and household-level food data. *Food Policy*, *72*, 81–93. DOI: 10.1016/j.foodpol.2017.08.015 PMID: 29093609

George, N. A., & McKay, F. H. (2019). The public distribution system and food security in India. *International Journal of Environmental Research and Public Health*, *16*(17), 3221. DOI: 10.3390/ijerph16173221 PMID: 31484374

HLPE. (2020). Food security and nutrition: building a global narrative towards 2030. High-Level Panel of Experts on Food Security and Nutrition of the Committee. https://openknowledge.fao.org/server/api/core/bitstreams/8357b6eb-8010-4254-814a-1493faaf4a93/content

Jha, P. (2007). Some Aspects of the Well Being of India's Agricultural Labour in the Context of Contemporary Agrarian Crisis". *The Indian Journal of Labour Economics : The Quarterly Journal of the Indian Society of Labour Economics, 2007*. https://www.macroscan.org/anl/feb07/Agrarian_Crisis.pdf

Kendall, A., Olson, C. M., & Frongillo, E. A.Jr. (1995). Validation of the Radimer/Cornell measures of hunger and food insecurity. *The Journal of Nutrition*, *125*(11), 2793–2801. DOI: 10.1093/jn/125.11.2793 PMID: 7472659

Khattri, P., Sayers, R., Singh, K. K., Slapikas, R., Tamang, C. B., Tamang, D., Sagara, B., & BenYishay, A. (2025). Geospatial impact evaluation of a low-cost agricultural intervention for enhancing environmental resilience. *International Journal of Applied Earth Observation and Geoinformation*, *141*, 104657. DOI: 10.1016/j.jag.2025.104657 PMID: 40599674

Kpodo, J., & Nejadhashemi, A. P. (2025). Navigating challenges and opportunities in developing smart agricultural extension platforms: Multi-media data mining techniques. *Artificial Intelligence in Agriculture*, *15*(3), 426–448. DOI: 10.1016/j.aiia.2025.04.001

Kross, A., Kaur, G., & Jaeger, J. A. G. (2022). A geospatial framework for the assessment and monitoring of environmental impacts of agriculture. *Environmental Impact Assessment Review*, *97*, 106851. DOI: 10.1016/j.eiar.2022.106851

Kumar, A., Bantilan, M. C. S., Kumar, P., Kumar, S., & Jee, S. (2012). Food Security in India: Trends, Patterns, and Determinants. *Indian Journal of Agricultural Economics*, *67*(3). https://oar.icrisat.org/6590/1/IJAgriEco_67_3_445-463_2012.pdf

Kumari, K., Nafchi, A. M., Mirzaee, S., & Abdalla, A. (2025). Ai-driven future farming: Achieving climate-smart and sustainable agriculture. *AgriEngineering.*, *7*(3), 89. DOI: 10.3390/agriengineering7030089

Lanjouw, P., & Rinku, M. (2009). Poverty decline, agricultural wages and non-farm employment in rural India: 1983-2004, Policy Research Working Paper Series 4858. The World Bank. https://openknowledge.worldbank.org/entities/publication/3ee5b391-f80d-5568-95a2-0c8b21c4582e

Liu, Z., Wang, S., Zhang, Y., Feng, Y., Liu, J., & Zhu, H. (2023). Artificial intelligence in food safety: A decade review and bibliometric analysis. *Foods*, *12*(6), 1242. DOI: 10.3390/foods12061242 PMID: 36981168

Mahendra, D., & Alakh, N. S. (2010). Food Security in India: Performance, Challenges and Policies Oxfam India Working Paper Series, September 2010. https://policy-practice.oxfam.org/resources/food-security-in-india-performance-challenges-and-policies-346637/

McKay, F. H., Sims, A., & Pligt, P. (2023). Measuring food insecurity in India: A systematic review of the current evidence. *Current Nutrition Reports, 12*(2), 358–367. DOI: 10.1007/s13668-023-00470-3 PMID: 37022635

Mendoza-Bernal, J., González-Vidal, A., & Skarmeta, A. F. (2024). A convolutional neural network approach for image-based anomaly detection in smart agriculture. *Expert Systems with Applications, 247*, 123210. Advance online publication. DOI: 10.1016/j.eswa.2024.123210

Mkhitaryan, K., Sanamyan, A., Mnatsakanyan, M., Kirakosyan, E., & Ratner, S. (2025). Integrating AI and Geospatial technologies for sustainable smart city development: A case study of Yerevan. *Urban Science (Basel, Switzerland), 9*(10), 389. DOI: 10.3390/urbansci9100389

Muruganandhan, S., Hanish, H., Vinish, H., Thamilmani, R., Kumarasamy, P., Vani, H., & Sivasubramanian, K. (2025). Measuring the application of innovative digital technology in the retail sector for sustainable business. In *Transforming business through digital sustainability models* (pp. 1–22). IGI Global. DOI: 10.4018/979-8-3373-0608-7.ch014

Narain, J. P. (2016). Public health challenges in India: Seizing the opportunities. *Indian Journal of Community Medicine : Official Publication of Indian Association of Preventive & Social Medicine, 41*(2), 85. DOI: 10.4103/0970-0218.177507 PMID: 27051080

Nourani, M., King, J. T., & Ragan, E. D. (2020). The role of domain expertise in user trust and the impact of first impressions with intelligent systems. Proceedings of the AAAI Conference on Human Computation and Crowdsourcing, 8, 112–121. https://ojs.aaai.org/index.php/HCOMP/article/view/7469

Olson, C. M. (1999). Nutrition and health outcomes associated with food insecurity and hunger. *The Journal of Nutrition, 129*(2), 521S–524S. DOI: 10.1093/jn/129.2.521S PMID: 10064322

Pandey, D., & Misra, R. (2024). Towards sustainable agriculture: Harnessing AI for global food security. *Artificial Intelligence in Agriculture., 12*, 72–84. DOI: 10.1016/j.aiia.2024.04.003

Pathak, J., Mahanta, T. G., Arora, P., Kalita, D., & Kaur, G. (2020). Malnutrition and household food insecurity in children attending Anganwadi centres in a district of North East India. *Indian Journal of Community Medicine : Official Publication of Indian Association of Preventive & Social Medicine, 45*(4), 405. DOI: 10.4103/ijcm.IJCM_428_19 PMID: 33623190

Radhakrishna, R., & Ray, S. (2005). *Handbook of Poverty in India: Perspective, Policies and Programmes*. Oxford University Press. https://www.semanticscholar.org/paper/Handbook-of-poverty-in-India-%3A-perspectives%2C-and-Radhakrishna-Ray/17854e567f800a453a9be2b317a947bd1a821acf

Rajeswari, S., Savithiri, D., Usharani, S., Sujatha, P., Kasthuri, R., & Sivasubramanian, K. (2025). Economic Analysis of Trend in Agricultural Price: A Statistical Picture with Reference to India. In Hamdan, A., & Braendle, U. (Eds.), *Harnessing AI, Machine Learning, and IoT for Intelligent Business. Studies in Systems, Decision and Control* (Vol. 555). Springer. DOI: 10.1007/978-3-031-67890-5_50

Rani, S., Mishra, A. K., Kataria, A., Mallik, S., & Qin, H. (2023). Machine learning-based optimal crop selection system in smart agriculture. *Scientific Reports, 13*(1), 15997. DOI: 10.1038/s41598-023-42356-y PMID: 37749111

Rautela, G., Ali, M. K., Prabhakaran, D., Narayan, K. V., Tandon, N., Mohan, V., & Jaacks, L. M. (2020). Prevalence and correlates of household food insecurity in Delhi and Chennai. *India's. Food Security, 12*(2), 391–404. DOI: 10.1007/s12571-020-01015-0 PMID: 33456633

Ray, S., & Ray, I. A. (2011). Role of the effectiveness of the public distribution system in assuring food security in India: An appraisal. *Journal of Economics and Sustainable Development, 2*(4). https://iiste.org/Journals/index.php/JEDS/article/view/359

Rebecca, S., Ulfia, A. C., & Thomas, C. (2023). The application of artificial intelligence models for food security. *Revista de Agricultura (Piracicaba), 13*(10). https://www.mdpi.com/2077-0472/13/10/2037

Savithiri, D., Kasthuri, R., Rajeswari, P., Vinothina, V., & Sivasubramanian, K. (2026). Self-help group model for financial sustainability of rural women in India: An explanatory study. In B. S. Awwad (Ed.), Artificial intelligence and technology: Systems management, decisions and control for sustainability in the digital age (Studies in Systems, Decision and Control, Vol. 645, pp. 969–976). Springer, Cham. DOI: 10.1007/978-3-032-10016-0_77

Saxena, A., Amin, A., Mohan, S. B., & Mohan, P. (2020). Food insecurity in tribal high-migration communities in Rajasthan. *India Food and Nutrition Bulletin, 41*(4), 513–518. https://pubmed.ncbi.nlm.nih.gov/33143470/

Saxena, N. C. (2011). Hunger, Under-Nutrition and Food Security in India. Working Paper 44, Chronic Poverty Research Centre, Indian Institute of Public Administration, New Delhi. https://assets.publishing.service.gov.uk/media/57a08ac540f0b652dd0008cc/CPRC-IIPA44.pdf

Saxena, N. C. (2018). Poverty, chronic poverty, and poverty dynamics. *Springer.* https://link.springer.com/book/10.1007/978-981-13-0677-8

Sivasubramanian, K. (2025). Assessing the impact of public-private partnership on the progress of renewable energy start-up industries for sustainable development. In *Navigating strategic partnerships for sustainable startup growth* (pp. 1–20). IGI Global. DOI: 10.4018/979-8-3693-4066-0.ch005

Sivasubramanian, K., Acharjee, S., Adarsh, R., Kumarasamy, P., Thamilmani, R., & Vani, H. (2025). Application of AI technology in the electronic commerce industry for sustainable business growth in India. In *AI and digital technologies transforming global industries* (pp. 1–28). IGI Global. DOI: 10.4018/979-8-3373-6097-3.ch001

Talaviya, T., Shah, D., Patel, N., Yagnik, H., & Shah, M. (2020). Implementation of artificial intelligence in agriculture for optimisation of irrigation and application of pesticides and herbicides. *Artificial Intelligence in Agriculture, 4*, 58–73. DOI: 10.1016/j.aiia.2020.04.002

Talukder, A., Kiess, L., Huq, N., de Pee, S., Hill, I. D., & Bloem, M. W. (2000). Increasing the Production and Consumption of Vitamin-A Rich Fruits and Vegetables: Lessons Learned in taking the Bangladesh Homestead Gardening Programme to a National Scale. *Food and Nutrition Bulletin, 21*(2), 165–172. DOI: 10.1177/156482650002100210

Thakuri, S., Adhikari, M., Tashi, S., Acharya, P. R., Chauhan, R., Maharjan, B., & Shrestha, M. L. (2025). Integrating geospatial technologies in climate-smart agriculture planning and management: A review focused on South Asia. APN Science Bulletin, https://www.apn-gcr.org/bulletin/wp-content/uploads/2025/10/8-CBA2020-13MY-Thakuri.pdf

Uma Maheswari, V., Balasubramanian, G., Sugatha Kumari, B., Ramya, C., Thamilmani, R., & Sivasubramanian, K. (2025). Business Strategies and Employment Growth of Retail Industry: An Economic Perspective in India. In Hamdan, A. (Ed.), *Achieving Sustainable Business through AI, Technology Education and Computer Science. Studies in Big Data* (Vol. 158). Springer. DOI: 10.1007/978-3-031-70855-8_41

Upadhyay, R. P., & Palanivel, C. (2011). Challenges in achieving food security in India. *Iranian Journal of Public Health*, *40*(4), 31–36. https://pmc.ncbi.nlm.nih.gov/articles/PMC3481742/ PMID: 23113100

Venkatesan, S., Rajkamal, S., Ramesh, K., & Sivasubramanian, K. (2025). An Analytical Review of the Performance of Selected SDGs in South Indian States Using NITI AAYOG Data. In Hamdan, A., & Braendle, U. (Eds.), *Harnessing AI, Machine Learning, and IoT for Intelligent Business. Studies in Systems, Decision and Control* (Vol. 555). Springer. DOI: 10.1007/978-3-031-67890-5_22

Vinuesa, R., Azizpour, H., Leite, I., Balaam, M., Dignum, V., Domisch, S., Felländer, A., Langhans, S. D., Tegmark, M., & Nerini, F. F. (2020). The role of artificial intelligence in achieving the Sustainable Development Goals. *Nature Communications*, *11*(1), 233. DOI: 10.1038/s41467-019-14108-y PMID: 31932590

World Food Summit. (1996). Food and Agriculture Organization of the United Nations; Rome, Italy: 1996. Rome's declaration on the World Food Security and World Food Summit Plan of Action. https://www.fao.org/4/w3613e/w3613e00.htm

Chapter 3

Smart Agriculture and GeoAI: Integrating Geospatial Technologies for Sustainable Farming and Secure Food Systems

Kannan M.
http://orcid.org/0009-0003-9810-3496
Amrita Vishwa Vidyapeetham, India

Parthasarathy Seethapathy
http://orcid.org/0000-0003-0660-9311
Amrita Vishwa Vidyapeetham, India

ABSTRACT

The pressure on global food demand is increasing with a growing population, urbanization, changing climatic conditions, and scarcity of agricultural resources. Conventional agriculture is also becoming inadequate when it comes to satisfying a resilient and sustainable food system. To counteract it, the so-called Smart Agriculture, the use of innovative technologies including the Embedded and Internet of Things (IoT), Artificial Intelligence (AI), blockchain, robotization, and data analytics, has been proposed as a game-changer to streamline all processes in the agricultural value chain, production to consumption. In this chapter, the authors discuss how Smart Agriculture is transforming the way we cultivate, harvest, process, and even distribute food. It is concerned with the combination of digital technologies and traditional activities to increase the crop productivity, the condition of soil, the expected use of water, control of pests, and traceability of supply chains.

1. INTRODUCTION

Agriculture is a vital source of sustainable livelihood for human beings. Due to population growth and instability in climate conditions, faces massive strain all over the world. Also, due to population growth, natural resources like land and groundwater are depleted, which leads to limiting the farm production (Pawlak & Kołodziejczak, 2020). Although traditions have provided generations, they fail to satisfy the increasing demands of the present society, where the concerns lead around the high production as well as

DOI: 10.4018/979-8-3373-8037-7.ch003

the safety and healthy food that is received and continuously produced to suit the new needs of the new society. It is against this background that the application of digital technologies, automation and use of data are creating new possibilities to ensure farming is a more resilient and efficient as well as equitable sector (Costa et al., 2023). In this chapter, the urgency of agricultural transformation towards smart and sustainable practices is presented, and the path to smarter agriculture has been discussed and suggested. The new concept of smart food systems has to define the future of the world's agri-food systems. GeoAI makes it possible to combine artificial intelligence and geospatial data to learn location-aware insights to make agricultural decisions. This chapter is based on a GeoAI approach to combine spatial analytics and smart models to be more precise and sustainable.

1.1 Exigency of the Transformation of Agriculture With GeoAI

Modernization of agriculture has ceased to be an option but rather an urgency spurred by various trends in the world. By 2050, the global population is projected to reach a level of over 9 billion people; meanwhile, agricultural production has to grow drastically without a corresponding escalation in the utilization of land and water resources (Wang, 2022). In addition, due to unstable weather conditions that create unpredictable weather, prolonged dry periods and pest attacks affect the overall productivity. Small-scale farmers are primarily affected by these changes because of inadequate investment provision. Change in the agriculture system is necessary not only to meet food demand but also to save the resources and livelihoods of millions of rural households (de Bruin et al., 2021). Thereby, the deployment of digital technology to agriculture at the field level can improve the productivity and ensure the food supply to the next generation. GeoAI is becoming more important in the transformation of agriculture to interpret spatial variability of soil, climate and crop conditions. With AI and geospatial intelligence, it is now possible to make interventions based on the data to enhance productivity and resilience.

1.2 Evolution of Traditional to Smart Agriculture With GeoAI

The change in the traditional method to modern, often called "smart agriculture," shows the slow deployment in farming methods. The traditional farming methods mainly depend on manual observation and knowledge. It works well in normal situations, but it consumes more resources and is prone to outside shock (Dhanaraju et al., 2022). In the 20th century, the Green Revolution created a turning point in traditional agriculture, which created high productivity with different varieties, but it significantly impacted soil erosion and water pollution. In recent days agriculture has entered into a new era called "smart agriculture." This smart agriculture system mainly looks at the crop health monitoring using different methods to improve the crop yield as well as quality. This smart agriculture uses sensors, drones, IoT and AI techniques to acquire real-time field data and help the farmers to make decisions at the right time (Mohamed et al., 2021). The mobile-based application helps farmers to control and monitor inputs like water and fertilizers to their fields at optimal levels. GeoAI is a combination of remote sensing, GIS, and machine learning that helps in accelerating the transition of traditional to smart agriculture. The convergence makes possible real-time spatial tracking and predictive analytics to optimize farm management.

1.3 Notion of Smart Food and its Applicability in Contemporary Food Systems

Smart agriculture products are called "smart food." It ensures the food safety and quality, including nutrition and consumer protection. Smart agriculture interconnects using digital technology both farmers and consumers to know the exact requirement and quality in the rapidly evolving environment. GeoAI is a technology that helps in the smart food paradigm since it allows traceability and spatial evaluation along the food production and distribution chain. Modern food systems are made transparent, safe and efficient by the use of geospatial technologies.

2. KEY TECHNOLOGIES IN PRECISION FARMING WITH GeoAI

Smart agriculture mainly focuses on the deployment of digital technology solutions to be precise and efficient farming at the field level. Precision farming is a closed-loop system that consists of real-time field monitoring and sensor data evaluation and field-level interventions (Monteiro et al., 2021). In comparison with the traditional methods of farming, where whole fields are usually treated the same, precision agriculture recognizes that some variability is inherent to every field and aims to control this variability with scientific precision. The technologies that support this solution are IoT sensors, AI, drones, robotics, blockchain, big data analytics and smart irrigation systems. They will enable farmers to gather, distribute and utilize data expeditiously, enabling them to increase their output without losing resources and damaging the environment. These technologies and their real-life uses are briefly described in the sub-sections below (Shaikh et al., 2022). The fundamental architecture or framework of smart agriculture technologies and data transfer is shown in Figure 1. The sensors and drones are used to acquire real-time field data, and then it is transferred to a cloud server using the internet. Data analysis can be performed with blockchain, and mobile applications are developed and used by the farmers to view results. GeoAI is becoming the next wave of precision farming technologies, which interfere with spatial datasets and advanced analytics to manage sites. A combination of GIS, satellite imagery, and AI models will help to improve the quality of decisions and resource utilization.

2.1 Smart Agriculture With IoT Sensors and Wireless Communication

The key to precision farming is the IoT, or connecting a real-life device like a soil sensor, a weather station, cameras or livestock trackers with a digital network. These detectors produce a continuous data feed that records the health of crops and farm soil and environment. Indicatively, soil moisture sensors enable farmers to detect the amount of water in the soil at varying depths, and they only irrigate when they are sure of need (Pramanik et al., 2022). They monitor the nutrient deficiencies or excess, and the targeted use of fertilizers is performed. Multispectral and hyperspectral imaging enables plant health sensors to be designed to detect the symptoms of stress before the human eye can, allowing farmers to respond immediately. Wireless communication will be an essential factor in the success of the IoT systems. It is now possible to afford long-range data transmission with low power consumption using such technologies as Wi-Fi, GSM, 4G/5G, and new Low Power Wide Area Networks (LPWANs) when compared to other technologies like LoRaWAN and Sigfox. In developed countries, 5G networks operate at high speed, allowing the waves to establish large-scale deployments of IoT devices, thus providing high-resolution monitoring on large fields (Shafique et al., 2020). Conversely in resource-limited or rural systems, LPWAN technologies provide cost-efficient solutions since they enable thousands of sensors to attain tens of kilometres of range without frequent battery changes. IoT sensors and wireless networks of communication points make farms a connected ecosystem where the information circulates easily between the fields, warehouses and decision platforms. GeoAI complements IoT agriculture with the addition of spatial maps to monitor fields in real-time with sensor data. This geo-centric system of communication allows accurate and place-specific agricultural responses.

2.2 Decision Support Systems Based on Artificial Intelligence and Machine Learning

Data collected by IoT sensors can only be processed into actionable insights when analyzed properly, and this is where AI and machine learning (ML) come in. Focusing on ML, it is possible to run years of weather, soil and crop performance data and forecast yield results; suggest when to plant; or propose pest management strategies (Domingues et al., 2022). Deep learning (DL) is one of the subsets of AI that works on the concept of artificial neural networks. Different DL models are used in image classification applications like crop disease classification and crop leaf identification. This AI-based system helps farmers to make crucial decisions, predicting risks and securing crops. This AI-based system works like a real-time farm management system, which helps farmers to make timely decisions.

Figure 1. Data flow framework of smart agriculture technologies

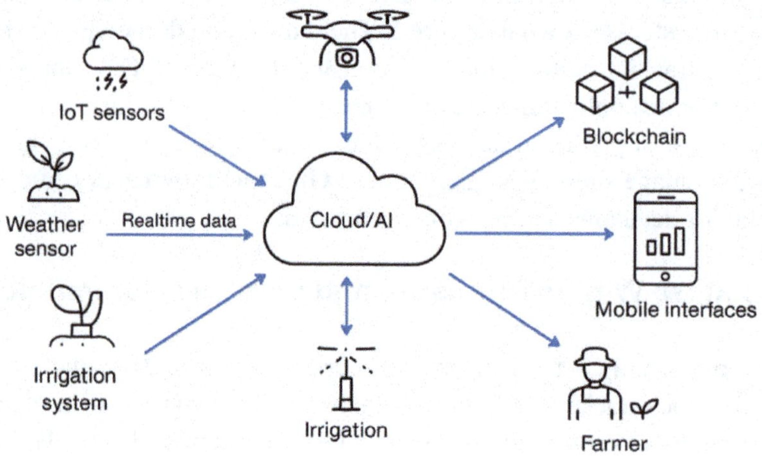

(Authors)

2.3 Automated Monitoring and Operations by Drones, Robotics and GeoAI

Automation is another one of the most evident forms of smart farming. Multispectral, thermal and high-definition camera drones allow farmers to have a bird's-eye perspective of their fields and create detailed maps of the crop's health, water and pest infestations. Such an aerial view assists in establishing the variability in the field, and it also provides the opportunity of conducting specific interventions, e.g., by using fertilizers only in those areas that lack nutrients. In addition to surveillance, drones may have spraying mechanisms, which would spray pesticides or nutrients directly where it is necessary to avoid excessive use of chemicals and pollution of the environment (Hafeez et al., 2023). Robotics is an extension of automation to physical labor in agriculture. Autonomous tractors, controlled using GPS and machine vision, are capable of ploughing, planting and cultivating fields with supervision by few human beings. Robotic weeders differentiate between crops and weeds and either cut them down mechanically or by micro-dosing them with herbicides. The robots that harvest crops especially apples and vineyards, can detect the ripe fruits and pick them with the delicacy needed so that they do not damage the fruits. Robotic milking systems are used in livestock farming to observe cow health and productivity and regulate milking schedules to meet the needs of cows. Not only does this integration of drones and robotics help solve the labor shortage, but scalability, consistency and efficiency in agricultural tasks are also enhanced. Combining drones and robotics with GeoAI, high-resolution spatial data to track crop health and field conditions are offered. Geospatial analysis with the participation of AI will allow conducting autonomous activities and selective interventions.

2.4 Blockchain and Big Data based Food Traceability and Logistics

Transparency and trust in food supply chains are major issues as the globalization of agriculture takes place. The blockchain technology provides a decentralized, immutable system of ledgers in which all purchases made since the manufacturing of a seed to its end user can be traced and verified. Food tracing helps the consumer to know the details of the food product, including authenticity and safety (Walaszczyk & Galińska, 2020). In recent days most of the food products are labeled with a QR code that contains information about the farmhouse, cultivation process, and storage history, including the packed month and year. Big data technology is an enhancement to blockchain; it works with the large amount of real-time agriculture data. Big data stands to improve the supply chain by inspecting consumption outlines, weather patterns and logistic outlines. Real-time tracking systems in logistics also help to inform the consumer that the products are supplied under the best temperature and humidity levels (Ding et al., 2023). Big data can be used for the policymakers for their research in agriculture design and specific interventions. Integrating big data and blockchain creates smarter, more effective food systems to meet consumer expectations.

2.5 Intelligent Irrigation, Fertigation, and Weather Prediction

Water shortage has been a major problem all over the world. To solve the water resource shortage, intelligent irrigation systems are essential in ensuring sustainable food production. A smart irrigation system consists of soil moisture sensors to measure real-time soil moisture value, and microcontrollers are used to transfer the sensor data to a cloud server or locally to take actions (Nikolaou et al., 2020). ESP32 microcontrollers are widely used in smart irrigation systems because of their low power consumption and low cost for field-level deployment. These systems also combined with weather monitoring sensors to predict microclimate information about their fields. Smart irrigation and smart weather prediction create a strong connection between technology and sustainability and the challenges of climate variability. A summary of major technologies driving smart agriculture is given in table 1. GeoAI supports smart irrigation and fertigation by assessing spatial differences in moisture and nutrient distribution of soils. It helps in managing resources effectively and adaptively, coupled with geospatial weather forecasting.

Table 1. Key digital technologies driving precision and smart agriculture

Technology	Description & Application	References
IoT Sensors & Wireless Networks	Real-time monitoring of soil, crops and environment; enables data-driven decisions and automation.	(Alahmad et al., 2023)
AI & ML	Predictive analytics for yield, disease and resource optimization; supports decision support systems.	(Mansoor et al., 2025)
UAVs/Drones & Robotics	High-resolution field mapping, targeted spraying, automated monitoring and labour-saving operations.	(Gagliardi et al., 2021)
Big Data Analytics	Aggregates and analyzes large-scale sensor and operational data for actionable insights and forecasting.	(Mansoor et al., 2025)
Blockchain	Ensures food traceability, transparency and trust in supply chains through secure, immutable records.	(Alahmad et al., 2023)

2.6 GeoAI-Based Conceptual Framework for Smart and Sustainable Food Systems

The offered conceptual framework based on GeoAI is the integrated perspective on the ways in which geospatial technologies and artificial intelligence can be used together to create smart and sustainable food systems. The framework is designed to have three layers that are related to each other, namely data acquisition, spatial analytics, and intelligent decision-making. At the data acquisition level, there are heterogeneous geospatial data that are gathered using a variety of data collection instruments, such as remote sensing by satellites, unmanned aerial systems, and IoT-based field sensors. These data sources are able to record spatial and temporal changes in environmental parameters like soil moisture, temperature, vegetation health and climatic conditions. When these technologies are integrated, they will ensure that the landscape of farms is continuously and high-resolution monitored. The integration allows monitoring in real-time, location-specific interventions, and data-based decision-making throughout the whole agricultural and food supply chain, enhancing efficiency, sustainability, and resilience. Figure 2 illustrate geoAI-based conceptual framework for smart and sustainable food systems.

Figure 2. GeoAI-based conceptual framework for smart and sustainable food systems

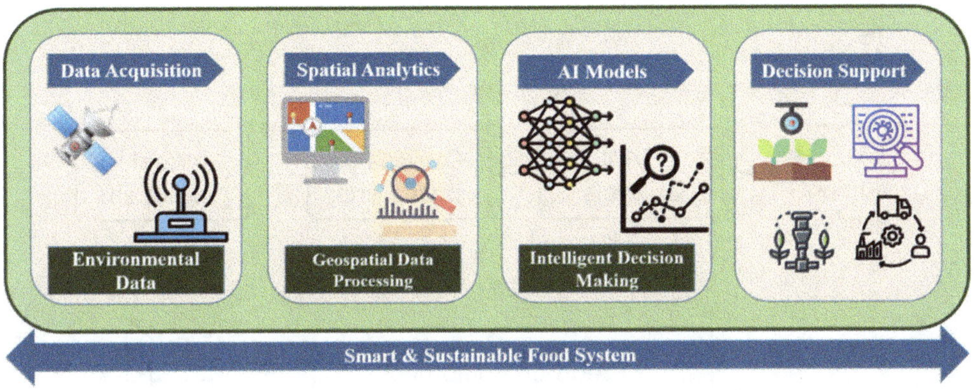

(Authors)

3. CROP PRODUCTION BY PRECISION FARMING TECHNIQUES WITH GeoAI

In recent years AI-based smart systems like mobile applications and edge device deployments have been used to monitor crop diseases and nutrient levels (Qazi et al., 2022). In addition, more technologies like vertical farming and variable rate applications (VRT) are involved to maximize crop production. The recent innovative technologies are used to monitor crop health and improve yield and also make the farming system more environmentally friendly. The dataflow diagram of smart or modern crop production is shown in Figure 3. The soil data can be measured with the help of moisture and nutrient sensors and then forwarded to an AI system for further analysis. Predicted or classified results are sent to the end user for making the right decision at field conditions. GeoAI-driven precision farming leverages spatial datasets to optimize crop production at micro-level scales. The integration of AI with geospatial technologies enables accurate yield prediction and resource allocation.

Figure 3. Workflow of smart crop production using IoT and AI systems

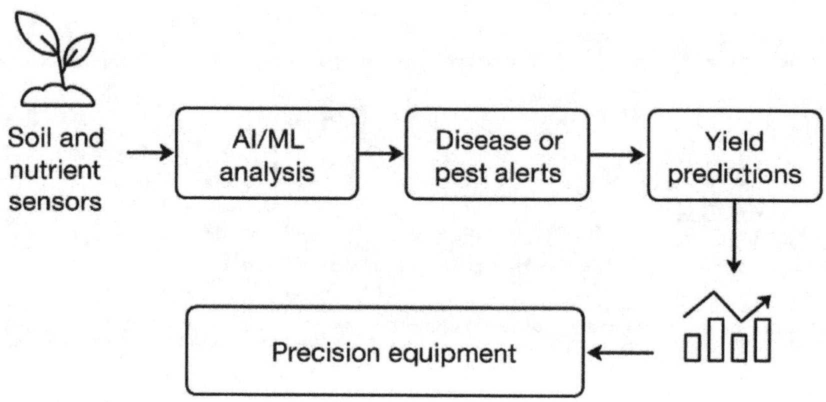

(Authors)

3.1 Variable Rate Technologies in Precision Farming

Variable Rate Technologies (VRT) allow farmers to apply irrigation, fertilizers and pesticides at varying rates in their fields. By using this VRT, inputs are supplied to the whole field, ensuring each crop gets its required amount of water or fertilizers (Fabiani et al., 2020). In VRT, equipment will be directed to provide the exact number of inputs to each site by the use of sensors, GPS mapping and geospatial data. Geospatial Artificial Intelligence (GeoAI) uses AI and data-driven insights to improve the sustainability and intelligence of resource and environmental management (e.g., Krishi Decision Support System). These maps regulate automated equipment to adapt seeding depth, level of fertilizer or concentration of pesticides on the fly (Fabiani et al., 2020). The advantages are dual: farmers save money on their inputs, and the environment is also benefited by reduced chemicals leaking. Additionally, VRT also helps increase yield stability since every section of the field gets precisely what they should grow optimally.

3.2 Vertical Farming and Smart Greenhouses

Food shortage, scarcity of arable land, and the problem of climate change have prompted the consideration of controlled-environment agriculture, in particular, vertical farming and smart greenhouses. Vertical farming is a method of cultivating agricultural products in layers, each layer resting atop the previous one, in a controlled in-door environment, commonly in an urban environment (Vatistas et al., 2022). Through this system, the land requirements are significantly lowered, as well as the water requirements, and the production can occur throughout the year irrespective of the outside weather conditions. Vertical farming methods such as hydroponic, aeroponic and aquaponic produce nutrients to plant roots, passing on all nutrients to the plant with no reliance on soil and without wastage of resources (Vatistas et al., 2022). Smart greenhouses, however, are a blend of conventional greenhouse technologies with high-quality sensor networks, automated climate control and AI. These greenhouses do keep the temperature, humidity, light intensity and carbon dioxide levels under constant control and ventilation, irrigation and nutrient delivery are automatically adjusted to keep the growing conditions exactly where they are needed (Vatistas et al., 2022). An example of this is when sensors record that the humidity is high and may cause fungi to thrive, the automated systems regulate air moving through the crops and dehumidification. Vertical farms and smart greenhouses will be especially useful in areas with severe weather conditions, urbanization or scarce water resources. Due to the decoupling between crop production and natural variability, they are a futuristic approach to secure a high-quality and sustainable food supply.

3.3 AI in Prediction of Disease and Pests

Crop diseases and pests are still one of the greatest threats to agricultural productivity and, in some areas, can cause losses of up to 40% of yield. In the past, farmers used their eyes and experience to detect issues, and this always delayed taking corrective measures before things were too late (Tonnang et al., 2022). This process is being transformed by AI and ML, with computer vision (CV), which allows the prediction of diseases and pests early and accurately. AI models consider big data with historical weather conditions, soil health, records related to crop health, and real-time sensor data to predict the risk of disease outbreak or pest infestation. As an example, CNNs could be trained on thousands of plant images to detect individual symptoms of a disease, e.g. leaf spots, discoloration, or curling and it is often possible to detect the symptoms before the naked eye. This capability is continued by drones and mobile phone applications synched with image recognition, which allow farmers on the ground to monitor disease, making it scalable and accessible. Through early warning and personalized recommendations, AI decreases cover-blanket use of pesticides and supports integrated methods of pest management (Tugrul et al., 2022). This focused method reduces the cost of production and lowers the effects on the environment and makes crops more resistant to environmental conditions. As the AI is incorporated into the decision support systems, pest and disease management is not reactive but proactive, taking control to protect the yields and increase ecological balance. GeoAI improves prediction of diseases and pests by integrating spatial patterns and the environment into AI models. This allows early identification and geospatially based intervention strategies.

3.4 Nutrient and Soil Health Monitoring

Productive practices in farming are based in healthy soil, but the problems of soil degradation and nutrient depletion are significant global issues. The production of modern crops is based on the use of high-level technologies in the field of nutrient and soil health monitoring. A conventional laboratory-based soil testing system provides a generalized report and is a time-consuming process. In recent times 7-in-1 NPK sensors are used to measure nutrient level (N, P, K), pH, soil moisture, soil temperature and conductivity (Dattatreya et al., 2023). The sensor-based system is portable field-testing equipment, and results can be obtained once sensors are inserted in the soil. Remote sensing is another area for monitoring the soil health with the help of satellite images and sensor data. ML algorithms like random forest and decision trees are used to train the soil data and predict the soil health and crop recommendations. With the help of digital technology, conducting soil tests before each cycle of farming helps to improve productivity. Summary of innovations in crop production systems is given in Table 2.

Table 2. Smart crop production techniques and their key advantages agriculture

Techniques	Key Advantages	References
Variable Rate Technology (VRT) for Resource Optimization	Reduces fertilizer and pesticide use by up to 80%, increases yield and improves water efficiency	(Neupane & Guo, 2019)
Vertical Farming for Urban Agriculture	Enables year-round production, saves water and land, reduces transport costs and minimizes pesticide use	(Petrovics & Giezen, 2022)
Smart Greenhouses for Controlled Environments	Enhances yield and quality through precise climate and resource control, reduces labor and energy costs	(Farooq et al., 2022)
AI-Based Pest and Disease Prediction	Enables early detection, reduces chemical use and improves crop health and yield	(Domingues et al., 2022)
IoT-Enabled Soil and Nutrient Monitoring	Provides real-time data for targeted fertilization and irrigation, boosting productivity and sustainability	(Senapaty et al., 2023)

3.5 Geospatial AI and Spatial Data-Driven Frameworks for Precision Crop Production

Geospatial AI is the future of the most potent technology in precision farming because it gathers and compiles satellite data, drone data, and spatial data (GIS-based) to optimize crop production. One of them is high-resolution imagery of the Sentinel-2 platforms that will enable the farmers to monitor the health of crops at 10-20-meter resolution that is making it possible to detect any stress condition. Geospatial analysis models entail combining machine learning and spatial interpolation analysis models to predict differences in yields across different regions of a field. Spatially explicit regression approaches such as Geographically Weighted Regression (GWR) have demonstrated the capacity to capture local variability in soil-water-energy interactions, offering a methodological foundation for site-specific resource management in precision agriculture (Al-Shaar et al., 2022). The zonal management approach also helps farmers apply water, fertilizer and pesticides where it is needed and this increases the efficiency of the resources. IoT sensors can also be integrated to provide real-time weather data, which can be applied to make better decisions and offer hyperlocal climate data. These models are also conducive to sustainable agriculture through low environmental effects and fewer greenhouse gases. The combination of geospatial artificial intelligence with mobile-based advisory systems can help smallholder farmers to obtain actionable insights in real time. Precision agriculture is a powerful ecosystem comprising remote sensing, AI models and spatial databases (Pokhariyal et al., 2023). This will not only enhance the yield of crops but also enhance food security systems since there will be proper utilization of resources. Spatial prediction models are used in areas that experience climatic variations so that farmers can adjust their cropping plans according to predictions. The use of these kinds of technologies is slowly closing the divide between conventional farming and intelligent agriculture. On the whole, geospatial AI-based precision farming is a revolutionary move toward a sustainable crop production system and the resilience of food systems.

4. INTELLIGENT SYSTEMS IN LIVESTOCK AND FISHERY

Livestock and aquaculture industries are also imperative in attaining worldwide food security. Population increase and urbanization are creating demand for animal protein (Michalk et al., 2019). In recent times, animal farming is affected by significant problems such as the outbreak of diseases and environmental sustainability. Smart technologies are being used in livestock and fish farming to achieve precise farm management. IoT and AI-based systems are combined to help farmers to identify the problems in the early stage. In India, sensor and IoT-based automated systems are partially implemented in poultry farms. Many startup companies are offering solutions in farm management. These technologies ensure the economy growth animal and fish resource conservation, and productivity.

4.1 Livestock Health and Tracking

Wearable sensors and biometric equipment have been used for tracking the health of animals a data-driven and real-time procedure (Neethirajan, 2020). An example is cows that can be monitored with smart collars or ear tags that constantly monitor the vital signs of body temperatures, heart rate and activity levels. Such devices identify when the person is about to become sick, in oestrus to breed, or have management and alterations in the feeding behavior. Radio Frequency Identification (RFID) tags have wide usage in identification and traceability as they are used by farmers to monitor individual animals throughout their lives. GPS technology is built in on more sophisticated systems, which give the location information to the free-range or grazing herds, minimizing the chance of loss or theft. With the integration of health and location data and cloud application management systems, farmers have the chance to be alerted and receive recommendations on their mobile gadgets (Neethirajan, 2020). This is a proactive method, which limits the diseases, lowers death rates and optimizes the number of veterinary treatments. Besides the advantages gained in productivity, livestock tracking technologies enhance traceability in the food supply chain and give the consumer more confidence in the quality and safety of animal products. Livestock system location data in the form of satellite and GPS are combined with spatial mapping platforms to track the distribution of animals in grazing fields. Spatial analysis can be used to detect areas of disease hotspots, grazing activities and areas of environmental stress on livestock health. Location data can be used to make decisions that are specific to the region, including the optimal use of pasture and monitoring seasonal movement trends. These geospatial understandings complements real time tracking by correlating information about the health of the animals with environmental and geographical elements.

4.2 Animal Behaviour Analysis Using AI and GeoAI

Animal behavior is also important to enhance efficiency in livestock farming. Behavioral changes, which you can hardly notice, are the initial signs of health issues or stress in time before symptoms start showing up (Chen et al., 2024). AI led to the creation of new possibilities in this field, as it provides the opportunity to analyze behavior with the help of video surveillance, sound recognition and sensor-based data automatically. DL-equipped CV systems can be used to monitor animals in barns or feedlots and identify animal behaviors, including feeding rate, lying down time, or social behavior. As an example, the poor appetite or increased fidgeting may be symptoms of illness, and the atypical gait might be an index of lameness (Chen et al., 2024). Nonstop operation of these systems eliminates human observation constraints and can handle data on hundreds or thousands of animals at once. Another new tool is acoustic monitoring, especially in poultry farming. AI can be used on chicken sounds in order to detect indicators of respiratory distress or other health issues (Chen et al., 2024). Likewise, identical patterns in piggery may be used to detect levels of stress or aggressive behavior, and thus the farmer is able to step in before the problem gets out of control. GeoAI assists in the analysis of the behavior of animals and combines the data of the location tracking with intelligent models. Patterns of spatial movements and environmental situations enhance monitoring and welfare management.

4.3 Smart Monitoring in Aquaculture Systems

Fish and related aquatic livelihoods are the rapidly developing areas of food products in the world. It also contributes significantly to the supply of protein energy sources globally. These types of farming experiences face challenges like water quality, disease outbreak and environmental conditions (Golden et al., 2021). To overcome these problems, smart monitoring systems are helping the farmer to know the real-time data of aquatic environments. The health and growth of fish can be improved by measuring dissolved oxygen, temperature, salinity and pH, and this can be achieved by implementing the smart monitoring systems. These systems are automated feeders that are driven by sensor data and AI algorithms to optimize both timing and quantities fed and reduce waste and optimize feed-to-weight ratios. In illustration, fish cameras can record when the feed pellets have not been consumed, and this can be translated to mean feeding should cease. It lowers the cost of feed, the largest proportion of the cost of aquaculture, and limits pollution through surplus feed in water bodies. In aquaculture, AI-surveillance of diseases is also becoming a trend. CV is used to analyze pictures and videos of fish in order to determine signs of infection at an early stage or abnormal fish behavior or lesions on the skin. The fishermen are able to avoid the use of antibiotics, and the mortality rate can be minimized by corrective actions being taken; thus, the problem is detected before it goes wrong. Moreover, there is the application of satellite and drone technologies to survey aquaculture farms along the coastal area, which gives a wider perspective at the environs and performance of farms (Ubina & Cheng, 2022). Smart aquaculture systems are also primarily concerned with sustainability. Live monitoring allows growers to keep water quality within a safe range about which there are no adverse ecological effects on the surrounding ecosystems. The spatial data collected by satellite and drones are combined with mapping platforms to track changes in water quality and farm conditions in various aquaculture areas. Spatial Analysis allows the determination of areas of high risk in disease outbreaks. spreading of pollutants, and environmental stress in aquatic systems. Information on location is used in site-specific management, such as the best feeding areas and distribution of resources in farms. Such geospatial knowledge enhances real-time monitoring through the correlation of aquatic health parameters and environmental and spatial variability.

5. INNOVATIONS IN POST-HARVEST TECHNOLOGIES AND FOOD SUPPLY CHAIN

In recent times, the Food and Agriculture Organization (FAO) reported that one third of food products are being wasted or lost before reaching the consumer end. These problems arise due to inefficient processing of post-harvest products and supply chains (Fernandez et al., 2021). To overcome this challenge, smart agriculture provides solutions, including introducing intelligent packaging and traceability through blockchain. AI-based ML and DL models and cold chain logistics also help the farmers to overcome these losses. These smart agriculture tools, including AI and big data, create consumer trust through the provision of food safety and transparency (Fernandez et al., 2021). GeoAI enhances post-harvest control, making it possible to track and optimize the supply chain spatially. Geospatial analytics enhance logistics, minimize losses, and create more efficient distribution.

5.1 Intelligent Identification of Package Spoilage

Packaging is a part of preserving the food physically and it has become more of a smart system that has the ability to track freshness, prolong shelf life and offer real-time data to consumers and retailers. Smart packaging is embedded with sensors, indicators and trackers that respond to the products being held and the outside environment (Drago et al., 2020). Indicators such as time-temperature indicators (TTIs) are an example, and they transform their color to indicate whether a product has experienced improper storing conditions to avoid selling spoiled items. Changes in the levels of oxygen or carbon dioxide can be detected by gas sensors inside the packaging, which is evidence of a living organism and a predictor of spoilage (Drago et al., 2020). One of the other potential new technologies is using biosensors to detect individual pathogens like *E. coli* or *Salmonella.* These sensors are able to give alerts right away concerning contamination and minimizing cases of foodborne diseases. In packaging materials, nanotechnology is being used to enhance barrier properties, oxidation-slowing properties and delay ripening in fruits and vegetables. To consumers, smart packaging will give them transparency, whereas to retailers, smart packaging will eliminate the possibility of economic loss through the sale of unsafe quality products, and only safe ones will be kept in stores (Drago et al., 2020). When these technologies are incorporated as ingredients in our daily food, this will be a great milestone towards sustainable, safe food distribution. The architecture of a smart food supply chain enabled by blockchain and AI is shown in Figure 4. The field is monitored by IoT sensors, and real-time data is sent to a cloud server for data analysis. Blockchain technology helps in secure food transmission from the supplier's end to the consumer's end. Location-based tracking information of packaged food systems is combined with spatial mapping systems to track the movement of products throughout the supply chain. Spatial analysis is used in determining areas that are at a greater risk of spoilage as a result of temperature changes, delays, or handling conditions. Location information helps to make route optimization and storage decisions to ensure that products are fresh. throughout transportation and distribution. This geospatial knowledge improves the spoilage detection process by connecting the data on the package condition with the environmental and location-related factors.

Figure 4. Architecture of smart food supply chain enabled by blockchain and AI

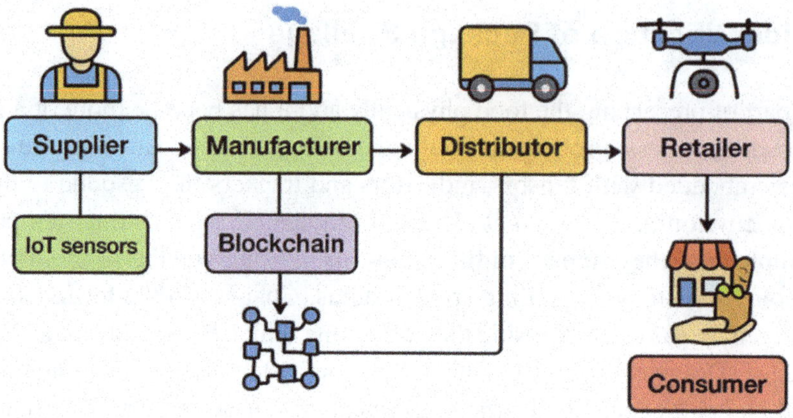

(Authors)

5.2 Food Traceability and Transparency With Blockchain

Food supply chains have become globalized, which has necessitated traceability and transparency on the global front to maintain safety, authenticity and processing of regulations. Blockchain technology, a decentralized and tamper-proof registry mechanism, can be used to track the complete food journey of agricultural productions to food processing, transportation and retail businesses (Baralla et al., 2021). A transparent and unchangeable record is created with every transaction or activity being stored in a digital block that is linked to the last one being created. As an example, by purchasing packaged fish, blockchain systems can give more information regarding the fishing place, date of harvest, storage and routes, which can be viewed with a QR code on the package. Such transparency does not only assure consumers of quality and sustainability. It also gives regulators and businesses the ability to swiftly respond to contamination or recalls (Baralla et al., 2021). In many countries smart agriculture tools like blockchain are used to improve market access, including for small-scale farmers. It creates products of smallholder farmers with guarantees to market prices and increases seller confidence. GPS and logistics systems are used to track the location of food products in various regions, which are integrated with blockchain records. The spatial mapping allows the visualization of the supply chain routes to identify delays, areas with a high risk of delays and differences in storage conditions. Information that is local Sensitive can be used to verify the origin and guarantee authenticity of the products and mislabelling of products in international markets. Such geospatial understanding enhances traceability through the connection of the records of transactions to the physical movement and the environment.

5.3 Sorting, Grading, and Quality Assurance Using AI

AI systems may be used to communicate whether a tomato is ripe or not, or whether a grain is of high or low quality. The accuracy of ML models is also enhanced with time as they get trained on bigger datasets. Already the FAO offers large datasets and resources https://www.fao.org/agroinformatics/training

-and-resources/data-sets/en. Such technologies are known to provide a standard level of quality and minimize wastage of consumable produce, and the farmers receive better prices on their harvests (Farhadi et al., 2020). In addition to sorting, AI is also using it in quality assurance of processed foods. Through processing in terms of image, spectral data and even sound, AI systems identify any abnormalities of a texture, packaging or ingredient composition. This will assure uniformity of food products and minimization of the chances of defects being passed to the consumers (Farhadi et al., 2020). Notably, automated grading and quality assurance also lead to a decrease in reliance on the human factor as workers are fewer in a lot of agricultural territories. The data on farms and processing units are location-based data, which are combined with spatial mapping platforms to track the quality differences in various regions. Spatial analysis is used in determining patterns in the geography of quality of crops, soil conditions, and environmental conditions that influence the results of grading. Location awareness information aids in location-specific quality measurement and enhances the decision-making in the sorting and distribution procedure. These geospatial understandings can be used to improve quality control through attachment of product features to environmental and spatial variability.

5.4 Optimization of Cold Chain and Logistics

In most of the countries, the cold chain infrastructure is inadequate and facing substantial post-harvest losses, which compromises food security and economic profits. Cold chain management and logistics are now being optimized using smart technologies (Chen et al., 2021), with the help of IoT and sensor technology to monitor temperature, humidity and airflow in the system. If there is any fault, an immediate notification is sent to the operators, which ensures fast corrective action. An embedded system integrated with GPS helps in real-time monitoring the position of shipments and their conditions over large distances. AI-based algorithms are used to analyze traffic information and weather data; based on that, it produces the most efficient delivery schedules (Chen et al., 2021). This AI-based system combines both reduction in air pollution (carbon emissions) and food waste. The blockchain is providing a combined solution to meet the enhanced safety standards in cold chain logistics for both customs and retailers. Table 3 represents how smart technologies improve post-harvest management and supply chains. The data of location-based shipments are combined with spatial mapping systems to track cold chain operations and storage environments in different regions. Spatial analysis is useful in determining the paths that are more prone to temperature changes, delays, and other infrastructural constraints. Location-based information assists in planning routes and distribution to ensure that the storage conditions are optimum during transportation. These geospatial understandings enhance the efficiency of cold chains by connecting the environment and real-time location monitoring.

Table 3. Smart technologies in food supply chain and post-harvest management agriculture

Techniques	Key Advantages	References
Intelligent packaging for freshness detection	Real-time monitoring of food quality and spoilage, reduces waste and enhances consumer trust	(Jafarzadeh et al., 2024)
AI-driven grading and sorting systems	Automated, accurate quality assessment; reduces labor, increases speed and minimizes human error	(Jafarzadeh et al., 2024)
Blockchain-based transparency and traceability	Immutable records, end-to-end product tracking, fraud reduction and increased supply chain trust	(Salah et al., 2019)
Cold chain monitoring for perishable goods	Real-time temperature/humidity tracking, reduces spoilage, ensures food safety and compliance	(Zhang et al., 2022)

6. SMART FOOD: SAFETY, NUTRITION AND CONSUMER AWARENESS WITH GeoAI

Smart agriculture mainly focuses on the technologies in the production and supply chain phases of the process. In real-time requirements, the technologies also ensure the food delivered to consumers is safe, healthy, and specific to diets of consumers (Corallo et al., 2019). Food quality is always dependent on quality, healthy and the form of its originality. Smart food systems provide online quality checking systems and AI-enhanced dietary advice and authentication used to evaluate the food quality. This kind of system guarantees both food security and food quality (Corallo et al., 2019). GeoAI has a role to play in food safety through mapping out dangers of contamination and tracing food safety across locations. Spatial intelligence also boosts consumer awareness by using transparent and data-driven food systems.

6.1 Online Food Quality Control

Traditional laboratory-based food quality testing is a precise but time-consuming process in real-time decision-making. To fill this gap, online food quality control systems are emerging that provide rapid and non-invasive monitoring (Wang et al., 2025). The sensors that are placed in storage units, processing lines and even in packaging give real-time information on the parameters like temperature, humidity, microbial load, and chemical residues, among others. As an example, near-infrared (NIR) spectroscopy and hyperspectral imaging may be incorporated into processing lines and be used to identify the chemical composition of a food product instantly, whether it is adulterated, spoiled, or contains certain nutrients without destroying the sample (Wang et al., 2025). This data is then compiled in cloud-based platforms and alerted to producers, retailers and regulators in case deviations are experienced. This prevents those batches from being compromised by making sure that they are identified and extracted before getting to the consumer's end. The data on distribution, processing, and storage units are stored, processed and distributed and then integrated with spatial mapping platforms to monitor the quality of food in various regions. Spatial analysis assists in determining regions where contamination, temperature changes or storage problems are more likely to occur, and this affects the quality of the product. Location-based information assists in quality control measures in the region and enhances decision-making in supply chain management. These geospatial perceptions boost the monitoring online by connecting the quality parameters to the condition of the environment and location.

6.2 AI in Dietary Suggestions and Customized Nutrition

In recent times, with the development of AI and cloud-based systems, dietary regimens are provided based on individual preferences. For example, sensor- and AI-based wearable fitness trackers and health applications are used to collect real-time data like exercise, heart rate and calories burned (Tsolakidis et al., 2024). These real-time data are processed by AI systems to provide individual health-related issues like diabetes. Combined dietary information can allow food concerns and farmers to accumulate production in orientation with consumer demand. This creates a closed-loop system between agriculture and nutrition related to healthier humanities (Tsolakidis et al., 2024). Notably, AI-based nutrition platforms can also contribute to population health by helping direct its population towards healthier eating, decreasing the number of non-communicable diseases, and decreasing healthcare costs. GeoAI allows individualized nutrition whereby regional food habits and spatial health information are used to drive AI models. The method helps to recommend dietary guidelines based on location and enhance the overall health of the population.

6.3 Food Authentication and Fraud Detection

Adulteration, mislabelling and substitution are some of the food frauds that have a high level of danger to consumer safety and confidence. The mislabelling of meat, the thinning of milk, or fake organic certifications are examples of how this issue is worldwide (Magdas et al., 2025). To fight such challenges, smart technologies have already been implemented based on superior authentication and fraud-detection technologies. As explained above, traceability based on blockchains will give a clear and unaltered history of food provenance, making it harder to alter (Magdas et al., 2025). DNA barcoding is used for accurate determination of food sources and species. Also, this DNA barcoding is used to confirm a given fish that is sold as salmon to be genuine. CV and image processing with the help of AI models could also be used to identify fake packaging or labelling. This system not only preserves consumers against the economy but also saves the producers and the honest businesses. Location-based data of production, processing and distribution locations are merged with spatial mapping platforms to confirm the source of food products. Spatial analysis is useful in detecting geographic anomalies between reported and actual sources, which can be used to detect fraud. The location-aware information can be used to cross-check the supply chain routes to identify unauthorized replacements or mislabelling.

7. CHALLENGES AND ETHICAL ISSUES

Food insecurity and unsustainable farming practices are inherently spatial phenomena, forced by geographic variability in climate, soil fertility, irrigation availability, land use patterns and socio-economic conditions. These spatial heterogeneities influence crop productivity, vulnerability to pests and diseases, and access to food resources, thereby necessitating geographically explicit analytical approaches in a smart and sustainable way by AI. Here, the GeoAI offers a powerful platform by incorporating geospatial technologies, including remote sensing and GIS, into advanced AI models to understand spatial patterns and make location-specific decisions. This combination increases the predictability, real-time monitoring, and efficient management of resources of sustainable agricultural systems.

7.1 Barriers to Entry and High-Cost Arenas

AI-friendly decision support systems, IoT-enabled sensors, drones and robotized machinery are known to be highly capital-intensive technologies. To most smallholder farmers, especially those in the developing world, these prices are usually prohibitive. The initial acquisition of high-resolution drones for imaging, accurate irrigation, and AI analytics programs can yield returns that exceed the annual earnings of a small farmer. Besides the cost of hardware and software, it has recurrent expenditures on the maintenance of systems, cloud data storage, software licenses, updates and technical support. These economic inequalities are due to capital and operating expenses, which favor big commercial farms and agribusinesses and discriminate against smallholders (Deichmann et al., 2016). This economic hindrance does not only reduce the speed of the diffusion of innovation and can also worsen the occurrence of inequalities within rural agricultural societies. These barriers are also enhanced by infrastructure constraints. In most of the rural places, there is still no dependable electricity, internet connection, or phone connections, which are indispensable in the effective operation of the digital agricultural systems (Deichmann et al., 2016). In some cases, advanced technologies can be ineffective or impractical due to lack of proper infrastructure even in cases where the finances are available. Besides, the fast rate of technological change can lead to obsolescence, thus necessitating constant reinvestment in order to remain competitive, and such costs may discourage adoption further. The need to access location-based data and spatial mapping tools also involves investing in satellite services, GPS devices, and data processing platforms, which add to the overall expenses. Spatial data collection and maintenance may be costly, particularly to the smallholder farmers who have low financial capabilities. In rural areas, there is also limited infrastructure, which further limits the effective use of location-aware systems in making decisions and monitoring. The challenges further pose more obstacles because they connect the expensive technologies with the necessity to have stable geospatial information and connectivity.

7.2 Rural Inclusion and Digital Literacy

The other important challenge is the problem of rural inclusion and digital literacy. Even in the case of highly sophisticated equipment, a degree of technical expertise is needed to make proper use of it, and some rural farmers might lack such expertise (Nogueira et al., 2022). The concept of AI-powered insights, operating within a mobile or web-based interface, processing sensor data and the combination of digital guidance into conventional behavior may be overwhelming unless one is educated accordingly. The digital exclusion between the city and the country is pronounced. In urbanized areas or commercial practices, farmers may be used to technology and can quickly learn and use new technology, but smallholders in remote areas may find it difficult to trust or understand automated advice (Nogueira et al., 2022). Lack of digital literacy can lead to low use, abuse, or total avoidance of new technology. The inclusion of rural populations cannot be achieved on its way using technical training alone. Location-based tools like GPS devices and spatial mapping platforms also demand special skills that most rural users might not possess. Without appropriate training and awareness, it may be difficult to interpret maps, spatial data and location-based outputs. The lack of easy access to easy-to-use mapping interfaces also lowers the uptake of location-based systems in the rural regions. Such spatial literacy gaps contribute to digital barriers since it becomes challenging to leverage location-based information in agriculture.

7.3 Ownership, Privacy, of Data and Algorithmic Bias

In the absence of clear deals and excellent governance structures, farmers could not be in control of crucial information that directly impacts their livelihoods (Nyamekye et al., 2018). Another current ethical problem is algorithmic bias. AI models can be as trustworthy as the training data they are trained on. In case a training dataset represents biased behavior that reflects certain regions, crop types, or agricultural farming practices, then the resulting models would make incorrect or unfair suggestions in various agricultural settings. To illustrate this, a pest detection algorithm, which is mostly trained on commercial farms in the temperate climates, might not detect an outbreak in the smallholder farms in the tropical climates (Nyamekye et al., 2018). These prejudices may not be favored among the farming communities, which are already at risk, which could instead increase inequality, not reduce it. The privacy of the farmer also needs to be safeguarded by ethical data management. Unless sufficient protective measures have been taken, the set of comprehensive geolocation data, data on crop health, and data on finances may be used in ways that are detrimental to the commercial interests, insurance evaluation, or surveillance. To ensure that there is trust among the farming communities and technology providers, it is important to create strong privacy policies, transparent consent systems and secure data storage systems. Location-based data, including GPS position and spatial farm data, ownership, and control are also issues of concern to farmers. Unless well secured, spatial datasets may provide sensitive information on the location of farms and patterns of production and use of resources. Spatial data that is inconsistent or region-specific can also result in biased results when used in other geographic conditions. To guarantee trust and prevent abuse in agricultural systems, it is necessary to ensure safe usage and processing of location-aware data.

7.4 Environmental and Ethical Consequences

The use of advanced techno-agriculture can raise all the environmental and ethical problems, though efficiency gets ensured. A great number of computerized and automated systems are energy-consuming. Computations made by AI, analytics in the cloud, and battery-powered drones are also the causes of carbon emissions and environmental degradation. Electronic waste and depletion of resources are other environmental burdens that the production and disposal of electronic equipment generate (Ghulam & Abushammala, 2023). Ethically, automation and the involvement of AI in decision-making can lead to the inferiority of other systems of knowledge and diminish the autonomy of farmers. Precision farming can in no instance encourage monocultures or intensive production at the expense of biodiversity, soil health or water overexploitation in other cases. Moreover, unequal access to technologies may increase the social differences and marginalize those farmers who are not able to afford modern technological devices and utilize them properly. On top of ecological and social aspects, there are more general issues about the long-term effect of technological dependency (Ghulam & Abushammala, 2023). Complete dependency on AI can cause less human control, and mistakes of automated technologies can cause the loss of crops or natural resource misuse. Location-based monitoring systems and spatial data collection are also a source of energy consumption through the use of satellites, data processing and constant tracking devices. Spatial data storage and processing on a large scale have the potential to raise carbon footprint and resource demand in the long term. Improper or imbalanced spatial data coverage in some areas can result in biased decisions impacting some agricultural communities. The location-aware data should be ethically utilized with a close control to avoid the environmental load and provide fair results in various geographic regions.

7.5 Limitations of Geospatial Data and Location-Based Systems

One of the greatest limitations is spatial data quality that can influence the results of decisions due to the imprecision of satellite images or sensor measurements. The reliability of spatial datasets can be diminished by variations in resolution, time gaps and cloud interference. The lack of uniformity in data collection procedures in different regions can cause inaccuracies in mapping and monitoring of agricultural conditions. Another issue is spatial bias in datasets whereby the data is usually concentrated in certain geographical areas. The models created based on the data of one region might not work well in other agro-climatic regions. Location-based models are not as transferable due to differences in soil type, climate, and the type of crops. This inability to be flexible may limit the usefulness of the decision support systems in novel settings. Homogeneity in the form of standardization of spatial datasets across regions has been an issue with regard to homogenous analysis and interpretation. Limited ground-truth validation further affects the accuracy of spatial predictions and recommendations. The management of satellite-based information presents issues of ownership, access, and control of information. Accessibility for smallholders may be restricted by restrictions on data sharing and the high cost of high-resolution imagery. The problem of data privacy also occurs when the exact location data is associated with the farm-level operations. Absence of proper regulatory frameworks may result in abuse or unequal sharing of spatial data resources. These limitations should be addressed to make the use of geospatial technologies in agriculture reliable, fair and effective.

8. FUTURE PERSPECTIVES

Advancement in digital technology changes the traditional agriculture practice to smart agriculture. Smart agriculture implementation is necessary because of increasing food demand, global population and resource limitations. This section discusses the key points of the future of agriculture, of smart agriculture compatibility with the United Nations Sustainable Development Goals (SDGs). GeoAI will play a more and more critical role in the future developments of the agricultural sector, as it will provide a means to combine multi-source geospatial data and intelligent systems. This will facilitate scalable, autonomous and climate-resilient agricultural solutions.

8.1 Climate-Smart Integrated Agriculture

Climate Smart Agriculture (CSA) is a new innovation in agricultural farm practice and helps to manage food security and climate change. The primary advantage of CAS is being productive with a lesser environmental impact (Yu et al., 2025). It is believed that precision agriculture uses analytics based on AI and IoT technology, which helps in optimizing water usage, nutrient management and crop protection. Recent technology innovation along with CSA makes the farmers take the right decision at the right time (Yu et al., 2025). Also, CSA makes the integration of local knowledge with scientific innovation. Addition of traditional agricultural practices with sophisticated monitoring techniques will enable farmers to come up with adaptive supervision towards certain agro-ecological frameworks. The CSA increases resilience to the ever-increasing weather variability, pest attack and other problems triggered by the climate condition. Satellite and field sensors provide location data, which are combined with spatial mapping platforms to observe climate variability in agricultural areas. The use of spatial analysis can be used to determine patterns of rainfall, temperature changes, and soil conditions that influence the performance of crops. The location-based information helps in region-specific adaptation strategies, such as the choice of crops and practices of resource management. These geospatial understandings improve climate-smart agriculture as they connect environmental conditions to the variability of space to make improved decisions. The list of ethical challenges and future perspectives in smart agriculture is given in table 4.

Table 4. Ethical challenges and future perspectives in smart agriculture

Challenges	Key Issues and Future Directions	References
Data ownership and privacy issues	Concerns over who controls farm data, privacy risks and need for clear data governance and trust frameworks	(Dembani et al., 2025)
Cost and accessibility barriers for smallholders	High technology costs, digital literacy gaps and limited infrastructure hinder adoption among small-scale farmers	(Cesco et al., 2023)
Environmental impact and e-waste	Potential for increased energy use, e-waste and unintended ecological effects from digital technologies	(Cesco et al., 2023)
Alignment with UN Sustainable Development Goals	Smart agriculture must support food security, equity and environmental sustainability in line with SDGs	(Saikanth et al., 2024)

8.2 Contribution to UN Sustainable Development Goals

The agricultural revolution directly affects a number of United Nations Sustainable Development Goals (SDGs) (Husaini & Sohail, 2024). Climate sensitivity and technology will lead to SDG 2 (Zero Hunger), as they will aid in improving crop yields and decreasing post-harvest losses. AI-based mobile applications help farmers to predict plant diseases at the early stage to save the crop (Husaini and Sohail, 2024). Supply chains ensure that food gets into the markets in an effective way and improve nutritional and food security. Sustainable agriculture works in conjunction with SDG 12 (Responsible Consumption and Production) and SDG 13 (Climate Action) to enable efficient use of natural resources and lower greenhouse gas emissions. Digital technologies that monitor soil health, carbon sequestration and water use aid in the conservation of biodiversity to assist SDG 15 (Life on Land) (Sharma et al., 2024). Mobile advisory communities give empowerment to smallholder farmers, including in distant areas, thereby benefiting SDG 8 (Decent Work and Economic Growth) and SDG 10 (Reduced Inequalities). Integrating digital innovation within the outline of the SDGs makes agriculture a sustainable development that aligns the economic, social and environmental goals. Spatial mapping and other location-based data sources can be used to monitor agricultural activities in various regions to determine how various development objectives are being achieved. The spatial analysis assists in determining regions where there is food insecurity, shortage of resources and environmental stress to implement specific interventions. Information on locations can enhance policy decisions and planning through the connection of agricultural performance and geographic and environmental factors. These geospatial understandings enhance sustainable development initiatives in that the regional Information is linked to the global development aspirations.

9. CONCLUSION

Smart agriculture is currently in progress due to IoT, sensor technology, AI and climate-smart practices, which are the beginning of the smart agriculture revolution. This will support farmers in taking decisions that are supported by real-time data and enhance productivity. Also, it can minimize wastage of resources and impact on the environment. IoT sensors, drones, and sensor-based irrigation systems are transforming conventional methods of farming to be more efficient, precise and sustainable. This research paper identifies GeoAI as a unifying model, which brings together geospatial technologies and artificial intelligence in smart farming. With the implementation of GeoAI, there will be better decision-making, sustainability, and efficiency at the system level. GeoAI can map spatial patterns of crop stress, identify hotspots of food insecurity, optimize resource allocation and support precision farming related to location-specific needs. By integrating technology with traditional farming experiences, one can manage crop health and exploit resources, which results in improved quality and quantity of products. By using recent technology innovation, it is possible to develop more responsive and resilient agri-food systems. With the help of AI-based technologies like DL, it helps to detect pests and diseases in a timely manner. Online resources can also help smallholder farmers with practical insights or close knowledge gaps. The technology-based solutions are also used to streamline supply chains and make a direct contribution to food security and the economy. This is the efficiency and accessibility as well as the sustainability, which is at the core of the smart agriculture vision. The sustainable agri-food systems are focused on collaboration, policy support and capacity building. Governments, research units, private businesses and agricultural societies need to work together to ensure equal access to technology, improve digital literacy and create climate resiliency. The policy outlines should contain incentives towards the implementation of climate-smart practices and promote the creation of low-cost technological solutions. Knowledge sharing and scaling of innovations can be speeded up in terms of the domestic scale of collaboration between public and private sectors. International coalitions are enhancing the collaboration globally on some of the issues, such as climate change and insecurity in food. Smart agriculture makes a transition towards sustainable, inclusive and resilient agri-food systems. Incorporation of innovation and sustainability will make sure that agriculture keeps growing. This technology innovation not only provides profit to the farmers but also to the consumers and the environment, a base compatible with long-term coherence and success in food worldwide.

ACKNOWLEDGMENT

The authors are grateful to the ANRF, New Delhi, India, for funding the Early Detection and Spatiotemporal Monitoring of Plant Diseases research (Grant No. SUR/2022/004268).

REFERENCES

Al-Shaar, W., Bonin, O., de Gouvello, B., Chatellier, P., & Hendel, M. (2022). Geographically weighted regression-based predictions of water–soil–energy nexus solutions in Île-de-France. *Urban Science (Basel, Switzerland)*, *6*(4), 81. DOI: 10.3390/urbansci6040081

Alahmad, T., Neményi, M., & Nyéki, A. (2023). Applying IoT sensors and big data to improve precision crop production: A review. *Agronomy (Basel)*, *13*(10), 2603. DOI: 10.3390/agronomy13102603

Baralla, G., Pinna, A., Tonelli, R., Marchesi, M., & Ibba, S. (2021). Ensuring transparency and traceability of food local products: A blockchain application to a Smart Tourism Region. *Concurrency and Computation*, *33*(1), e5857. DOI: 10.1002/cpe.5857

Cesco, S., Sambo, P., Borin, M., Basso, B., Orzes, G., & Mazzetto, F. (2023). Smart agriculture and digital twins: Applications and challenges in a vision of sustainability. *European Journal of Agronomy*, *146*, 126809. DOI: 10.1016/j.eja.2023.126809

Chen, J., Liao, W., & Yu, C. (2021). Route optimization for cold chain logistics of front warehouses based on traffic congestion and carbon emission. *Computers & Industrial Engineering*, *161*, 107663. DOI: 10.1016/j.cie.2021.107663

Chen, Y., Jiao, T., Song, J., He, G., & Jin, Z. (2024). AI-enabled animal behavior analysis with high usability: A case study on open-field experiments. *Applied Sciences (Basel, Switzerland)*, *14*(11), 4583. DOI: 10.3390/app14114583

Corallo, A., Latino, M. E., Pizzi, R., Spennato, A., Fortunato, L., & Menegoli, M. (2019, September). Human factor in smart labelling: How enhance food awareness in consumers. In *2019 6th International Conference on Frontiers of Industrial Engineering (ICFIE)* (pp. 89-93). IEEE. https://ieeexplore.ieee.org/abstract/document/8907769/

Costa, F., Frecassetti, S., Rossini, M., & Portioli-Staudacher, A. (2023). Industry 4.0 digital technologies enhancing sustainability: Applications and barriers from the agricultural industry in an emerging economy. *Journal of Cleaner Production*, *408*, 137208. DOI: 10.1016/j.jclepro.2023.137208

Dattatreya, S., Khan, A. N., Jena, K., & Chatterjee, G. (2023). Conventional to modern methods of soil npk sensing: A review. *IEEE Sensors Journal*, *24*(3), 2367–2380. DOI: 10.1109/JSEN.2023.3334243

de Bruin, S., Dengerink, J., & van Vliet, J. (2021). Urbanisation as driver of food system transformation and opportunities for rural livelihoods. *Food Security*, *13*(4), 781–798. DOI: 10.1007/s12571-021-01182-8 PMID: 34221191

Deichmann, U., Goyal, A., & Mishra, D. (2016). Will digital technologies transform agriculture in developing countries? *Agricultural Economics*, *47*(S1), 21–33. DOI: 10.1111/agec.12300

Dembani, R., Karvelas, I., Akbar, N. A., Rizou, S., Tegolo, D., & Fountas, S. (2025). Agricultural data privacy and federated learning: A review of challenges and opportunities. *Computers and Electronics in Agriculture*, *232*, 110048. DOI: 10.1016/j.compag.2025.110048

Dhanaraju, M., Chenniappan, P., Ramalingam, K., Pazhanivelan, S., & Kaliaperumal, R. (2022). Smart farming: Internet of Things (IoT)-based sustainable agriculture. *Agriculture*, *12*(10), 1745. DOI: 10.3390/agriculture12101745

Ding, H., Tian, J., Yu, W., Wilson, D. I., Young, B. R., Cui, X., & Li, W. (2023). The application of artificial intelligence and big data in the food industry. *Foods*, *12*(24), 4511. DOI: 10.3390/foods12244511 PMID: 38137314

Domingues, T., Brandão, T., & Ferreira, J. C. (2022). Machine learning for detection and prediction of crop diseases and pests: A comprehensive survey. *Agriculture*, *12*(9), 1350. DOI: 10.3390/agriculture12091350

Drago, E., Campardelli, R., Pettinato, M., & Perego, P. (2020). Innovations in smart packaging concepts for food: An extensive review. foods, 9(11), 1628. https://www.mdpi.com/2304-8158/9/11/1628

Fabiani, S., Vanino, S., Napoli, R., Zajíček, A., Duffková, R., Evangelou, E., & Nino, P. (2020). Assessment of the economic and environmental sustainability of Variable Rate Technology (VRT) application in different wheat intensive European agricultural areas. A Water energy food nexus approach. *Environmental Science & Policy*, *114*, 366–376. DOI: 10.1016/j.envsci.2020.08.019

Farhadi, M., Abbaspour-Gilandeh, Y., Mahmoudi, A., & Mari Maja, J. (2020). An integrated system of artificial intelligence and signal processing techniques for the sorting and grading of nuts. *Applied Sciences (Basel, Switzerland)*, *10*(9), 3315. DOI: 10.3390/app10093315

Farooq, M. S., Javid, R., Riaz, S., & Atal, Z. (2022). IoT based smart greenhouse framework and control strategies for sustainable agriculture. *IEEE Access : Practical Innovations, Open Solutions*, *10*, 99394–99420. DOI: 10.1109/ACCESS.2022.3204066

Fernandez, C. M., Alves, J., Gaspar, P. D., & Lima, T. M. (2021). Fostering awareness on environmentally sustainable technological solutions for the post-harvest food supply chain. *Processes (Basel, Switzerland)*, *9*(9), 1611. DOI: 10.3390/pr9091611

Gagliardi, G., Lupia, M., Cario, G., Cicchello Gaccio, F., D'Angelo, V., Cosma, A. I. M., & Casavola, A. (2021). An internet of things solution for smartagriculture. *Agronomy (Basel)*, *11*(11), 2140. DOI: 10.3390/agronomy11112140

Ghulam, S. T., & Abushammala, H. (2023). Challenges and opportunities in the management of electronic waste and its impact on human health and environment. *Sustainability (Basel)*, *15*(3), 1837. DOI: 10.3390/su15031837

Golden, C. D., Koehn, J. Z., Shepon, A., Passarelli, S., Free, C. M., Viana, D. F., & Thilsted, S. H. (2021). Aquatic foods to nourish nations. *Nature*, *598*(7880), 315–320. DOI: 10.1038/s41586-021-03917-1 PMID: 34526720

Hafeez, A., Husain, M. A., Singh, S. P., Chauhan, A., Khan, M. T., Kumar, N., & Soni, S. K. (2023). Implementation of drone technology for farm monitoring & pesticide spraying: A review. *Information Processing in Agriculture*, *10*(2), 192–203. DOI: 10.1016/j.inpa.2022.02.002

Husaini, A. M., & Sohail, M. (2024). Agrochemical-free genetically modified and genome-edited crops: Towards achieving the United Nations sustainable development goals and a'greener'green revolution. *Journal of Biotechnology*, *389*, 68–77. DOI: 10.1016/j.jbiotec.2024.04.015 PMID: 38663518

Jafarzadeh, S., Yildiz, Z., Yildiz, P., Strachowski, P., Forough, M., Esmaeili, Y., & Abdollahi, M. (2024). Advanced technologies in biodegradable packaging using intelligent sensing to fight food waste. *International Journal of Biological Macromolecules*, *261*, 129647. DOI: 10.1016/j.ijbiomac.2024.129647 PMID: 38281527

Magdas, D. A., Hategan, A. R., David, M., & Berghian-Grosan, C. (2025). The Journey of Artificial Intelligence in Food Authentication: From Label Attribute to Fraud Detection. *Foods*, *14*(10), 1808. DOI: 10.3390/foods14101808 PMID: 40428587

Mansoor, S., Iqbal, S., Popescu, S. M., Kim, S. L., Chung, Y. S., & Baek, J. H. (2025). Integration of smart sensors and IOT in precision agriculture: Trends, challenges and future prospectives. *Frontiers in Plant Science*, *16*, 1587869. DOI: 10.3389/fpls.2025.1587869 PMID: 40438737

Michalk, D. L., Kemp, D. R., Badgery, W. B., Wu, J., Zhang, Y., & Thomassin, P. J. (2019). Sustainability and future food security—A global perspective for livestock production. *Land Degradation & Development*, *30*(5), 561–573. DOI: 10.1002/ldr.3217

Mohamed, E. S., Belal, A. A., Abd-Elmabod, S. K., El-Shirbeny, M. A., Gad, A., & Zahran, M. B. (2021). Smart farming for improving agricultural management. *The Egyptian Journal of Remote Sensing and Space Sciences*, *24*(3), 971–981. DOI: 10.1016/j.ejrs.2021.08.007

Monteiro, A., Santos, S., & Gonçalves, P. (2021). Precision agriculture for crop and livestock farming - Brief review. *Animals (Basel)*, *11*(8), 2345. DOI: 10.3390/ani11082345 PMID: 34438802

Neethirajan, S. (2020). Transforming the adaptation physiology of farm animals through sensors. *Animals (Basel)*, *10*(9), 1512. DOI: 10.3390/ani10091512 PMID: 32859060

Neupane, J., & Guo, W. (2019). Agronomic basis and strategies for precision water management: A review. *Agronomy (Basel)*, *9*(2), 87. DOI: 10.3390/agronomy9020087

Nikolaou, G., Neocleous, D., Christou, A., Kitta, E., & Katsoulas, N. (2020). Implementing sustainable irrigation in water-scarce regions under the impact of climate change. *Agronomy (Basel)*, *10*(8), 1120. DOI: 10.3390/agronomy10081120

Nogueira, V. B., Teixeira, D. G., de Lima, I. A. C. N., Moreira, M. V. C., de Oliveira, B. S. C., Pedrosa, I. M. B., de Queiroz, J. W., & Jeronimo, S. M. B. (2022). Towards an inclusive digital literacy: An experimental intervention study in a rural area of Brazil. *Education and Information Technologies*, *27*(2), 2807–2834. DOI: 10.1007/s10639-021-10711-z PMID: 34493924

Nyamekye, A. B., Dewulf, A., Van Slobbe, E., Termeer, K., & Pinto, C. (2018). Governance arrangements and adaptive decision-making in rice farming systems in Northern Ghana. *NJAS Wageningen Journal of Life Sciences*, *86*(1), 39–50. DOI: 10.1016/j.njas.2018.07.004

Pawlak, K., & Kołodziejczak, M. (2020). The role of agriculture in ensuring food security in developing countries: Considerations in the context of the problem of sustainable food production. *Sustainability (Basel)*, *12*(13), 5488. DOI: 10.3390/su12135488

Petrovics, D., & Giezen, M. (2022). Planning for sustainable urban food systems: An analysis of the up-scaling potential of vertical farming. *Journal of Environmental Planning and Management*, *65*(5), 785–808. DOI: 10.1080/09640568.2021.1903404

Pokhariyal, S., Patel, N. R., & Govind, A. (2023). Machine learning-driven remote sensing applications for agriculture in India—A systematic review. *Agronomy (Basel)*, *13*(9), 2302. DOI: 10.3390/agronomy13092302

Pramanik, M., Khanna, M., Singh, M., Singh, D. K., Sudhishri, S., Bhatia, A., & Ranjan, R. (2022). Automation of soil moisture sensor-based basin irrigation system. *Smart Agricultural Technology*, *2*, 100032. DOI: 10.1016/j.atech.2021.100032

Qazi, S., Khawaja, B. A., & Farooq, Q. U. (2022). IoT-equipped and AI-enabled next generation smart agriculture: A critical review, current challenges and future trends. *IEEE Access : Practical Innovations, Open Solutions*, *10*, 21219–21235. DOI: 10.1109/ACCESS.2022.3152544

Saikanth, D., Ragini, M., Tripathi, G., Rajesh, K., Giri, A., Pandey, S. K., & Verma, L. K. (2024). The impact of emerging technologies on sustainable agriculture and rural development. International Journal of Environment and Climate Change, 14(1), 253-263. https://hal.science/hal-05135258/

Salah, K., Nizamuddin, N., Jayaraman, R., & Omar, M. (2019). Blockchain-based soybean traceability in agricultural supply chain. *IEEE Access : Practical Innovations, Open Solutions*, *7*, 73295–73305. DOI: 10.1109/ACCESS.2019.2918000

Senapaty, M. K., Ray, A., & Padhy, N. (2023). IoT-enabled soil nutrient analysis and crop recommendation model for precision agriculture. *Computers*, *12*(3), 61. DOI: 10.3390/computers12030061

Shafique, K., Khawaja, B. A., Sabir, F., Qazi, S., & Mustaqim, M. (2020). Internet of things (IoT) for next-generation smart systems: A review of current challenges, future trends and prospects for emerging 5G-IoT scenarios. *IEEE Access : Practical Innovations, Open Solutions*, *8*, 23022–23040. DOI: 10.1109/ACCESS.2020.2970118

Shaikh, F. K., Karim, S., Zeadally, S., & Nebhen, J. (2022). Recent trends in internet-of-things-enabled sensor technologies for smart agriculture. *IEEE Internet of Things Journal*, *9*(23), 23583–23598. DOI: 10.1109/JIOT.2022.3210154

Sharma, P., Sharma, P., & Thakur, N. (2024). Sustainable farming practices and soil health: A pathway to achieving SDGs and future prospects. *Discover Sustainability*, *5*(1), 250. DOI: 10.1007/s43621-024-00447-4

Tonnang, H. E., Sokame, B. M., Abdel-Rahman, E. M., & Dubois, T. (2022). Measuring and modelling crop yield losses due to invasive insect pests under climate change. *Current Opinion in Insect Science, 50*, 100873. DOI: 10.1016/j.cois.2022.100873 PMID: 35051620

Tsolakidis, D., Gymnopoulos, L. P., & Dimitropoulos, K. (2024, August). Artificial intelligence and machine learning technologies for personalized nutrition: a review. In Informatics (Vol. 11, No. 3, p. 62). MDPI. https://www.mdpi.com/2227-9709/11/3/62

Tugrul, B., Elfatimi, E., & Eryigit, R. (2022). Convolutional neural networks in detection of plant leaf diseases. *Revista de Agricultura (Piracicaba), 12*(8), 1192. https://www.mdpi.com/2077-0472/12/8/1192

Ubina, N. A., & Cheng, S. C. (2022). A review of unmanned system technologies with its application to aquaculture farm monitoring and management. *Drones (Basel), 6*(1), 12. DOI: 10.3390/drones6010012

Vatistas, C., Avgoustaki, D. D., & Bartzanas, T. (2022). A systematic literature review on controlled-environment agriculture: How vertical farms and greenhouses can influence the sustainability and footprint of urban microclimate with local food production. *Atmosphere (Basel), 13*(8), 1258. DOI: 10.3390/atmos13081258

Walaszczyk, A., & Galińska, B. (2020). Food origin traceability from a consumer's perspective. *Sustainability (Basel), 12*(5), 1872. DOI: 10.3390/su12051872

Wang, X. (2022). Managing land carrying capacity: Key to achieving sustainable production systems for food security. *Land (Basel), 11*(4), 484. DOI: 10.3390/land11040484

Wang, X., Feng, Y., Wang, Y., Zhu, H., Song, D., Shen, C., & Luo, Y. (2025). Enhancing optical non-destructive methods for food quality and safety assessments with machine learning techniques: A survey. *Journal of Agriculture and Food Research, 19*, 101734. DOI: 10.1016/j.jafr.2025.101734

Yu, L., Du, Z., Li, X., Zheng, J., Zhao, Q., Wu, H., & Huang, X. (2025). Enhancing global agricultural monitoring system for climate-smart agriculture. Climate Smart Agriculture, 2(1), 100037. https://www.sciencedirect.com/science/article/pii/S2950409024000376

Zhang, X., Sun, Y., & Sun, Y. (2022). Research on cold chain logistics traceability system of fresh agricultural products based on blockchain. *Computational Intelligence and Neuroscience, 2022*(1), 1957957. DOI: 10.1155/2022/1957957 PMID: 35154298

Chapter 4

From Backyard Herbs to Smart Green Enterprises:

Using GeoAI to Optimise Household-Based Medicinal Plant Production in Peri-Urban Indonesia

Triyani Sumiati
http://orcid.org/0000-0002-7153-5374
School for Industrial Technology and Pharmacy (STTIF), Bogor, Indonesia

Rimarsa Haninnakhonsa Margono
http://orcid.org/0009-0000-4133-8057
Kitami Institute of Technology, Hokkaido, Japan

Tri Margono
National Research and Innovation Agency (BRIN), Jakarta, Indonesia

Rinieta Sausan Margono
http://orcid.org/0009-0003-6531-4051
University of National Development "Veteran" (UPN), Jakarta, Indonesia

Koko Kusnanto
http://orcid.org/0009-0000-2133-3777
University of Prof. Dr. Mustopo (Religious), Indonesia

Rimaira Sanaurairba Margono
University of National Development "Veteran" (UPN), Jakarta, Indonesia

ABSTRACT

This chapter analyses the shift in household-based medicinal plant (MP) cultivation in peri-urban Indonesia (PUI) from subsistence backyard practices to smart green micro-enterprises (SGME). This approach is underpinned by the integrated use of GeoAI (geospatial artificial intelligence), circular economy (CE) principles, and community-based innovation (CMI). The analysis draws on participatory mixed-methods data from 50 households in Cibinong, Bogor (January–March 2025). The study examines the socio-economic and ecological dynamics shaping women-led medicinal plant entrepreneurship in urban fringe areas. The analysis suggests that applying GeoAI as spatial decision support is associated with improved awareness of microscale land use and microclimatic conditions. This improvement leads to yield increases of 20–30% across 50 households, along with additional monthly income of approximately IDR 1,000,000 and reductions in health expenditures of around IDR 200,000. The chapter proposes the Backyard-to-SGME model to support inclusive and sustainable peri-urban food systems.

DOI: 10.4018/979-8-3373-8037-7.ch004

1. INTRODUCTION

1.1 Background

Household-based MP cultivation has long played a strategic role in supporting health security and supplementary income in peri-urban communities in Indonesia, traditionally managed as backyard or small-scale herbal gardens but increasingly recognised for their potential to evolve into SGME, particularly where urban expansion intersects with residual agricultural livelihoods (Mulya et al., 2023). However, the transition from subsistence production to more efficient, market-based systems remains hampered by fragmented knowledge, management inconsistencies, and limited access to spatially based agronomic information. Variations in household practices impact profitability, which is strongly influenced by integration into the value chain. GeoAI and digital decision support systems offer opportunities to increase MP production by integrating spatial and environmental data and adaptive recommendations. Information extracted from GeoAI satellite imagery patterns, including soil data and climate variables, is crucial for supporting cultivation decisions (Gupta & Kumar Pal, 2025). In Indonesia, the increasing adoption of drones for agriculture is influenced by behavioural and institutional factors within a green economy framework. This is intended to strengthen data collection capacity to support agricultural productivity, such as crop monitoring, spraying, and more targeted input distribution (Khofiyah et al., 2025; Singh et al., 2025).

In principle, this technology has the potential to improve the quality of planting decisions, the efficiency of input use, and the stability of crop yields in peri-urban areas characterised by heterogeneous land conditions and microclimate variations. However, evidence in the precision agriculture literature suggests that adoption still faces various socio-economic, technical, and institutional barriers, particularly among smallholder farmers and households (Manzoor et al., 2025). Although households generally perceive digital tools as useful, particularly when information is presented clearly, easily accessible, and through interactive features, their ability to operate the technology independently and internalise recommendations remains uneven. Decision-making autonomy is often limited by low digital literacy, difficulties in interpreting results, and reliance on interpersonal guidance (Chen, 2025). These constraints are even more pronounced in medicinal plant systems, where decision-making processes are closely tied to experiential knowledge, cultural practices, and livelihood strategies characterised by risk aversion. Several studies have shown that integrating local knowledge with participatory approaches can improve system design and strengthen the relevance of implementation in the field (Astutik et al., 2025). Therefore, this chapter emphasises that optimising household-based MP production through GeoAI cannot be understood solely as a technical issue, but rather as a socio-technical transformation process. Based on evidence from the household level, this discussion positions GeoAI-based systems within a community-based empowerment framework and a social work perspective, with an emphasis on facilitation, capacity building, and participatory learning processes. In line with the IGGI theme, this chapter highlights that digital innovation can solely contribute to sustainable development when integrated into social processes rooted in local contexts. GeoAI is positioned as a context-dependent supporting tool, rather than a stand-alone solution.

1.2 Problem Statement and Knowledge Gap

Although household-based medicinal plant (MP) cultivation is increasingly recognised as a potential strategy to strengthen health independence and increase income in peri-urban areas, its development practices are still dominated by informal and suboptimal patterns. This is due to limited and fragmented land, which leads to the cultivation of a variety of crops, including flowers, medicinal plants, and plantation crops. This mixed planting system is largely based on local experience and knowledge, without much systematic planning. As a result, the production of certain crops tends to be unstable, land use is less efficient, and scaling up beyond subsistence remains limited (Astutik et al., 2025). Efforts to preserve medicinal plants as part of traditional heritage also face increasing challenges in the context of rapid modernisation. At the same time, tools such as the Indo-Herb database and transfer learning models have improved plant identification by combining local knowledge with digital technology (Musyaffa et al., 2024). However, the use of GeoAI in digital agriculture still tends to focus on large-scale monoculture systems, making it less relevant for household-based medicinal plant systems (Gupta & Kumar Pal, 2025). This creates a structural mismatch with household farming systems in peri-urban areas, which are fragmented, multispecies, and integrated within residential spaces. Furthermore, findings in the literature on digital agriculture adoption indicate that the presence of technology does not automatically guarantee effective utilisation. Effective use depends heavily on digital literacy levels, trust in the technology, facilitation support, and the existence of local social networks (Manzoor et al., 2025). Most home-grown medicinal plant producers (*Tanaman Obat Rumah Tangga*/TORT) in Indonesia are women's communities. They do not manage TORT in particular, but they need to meet household health needs with medicines while also fulfilling their domestic responsibilities as housewives.

Conceptually, existing literature often treats GeoAI-based optimisation and the process of social adoption as two separate domains. Technical studies focus more on algorithm development and data integration, while social research focuses more on livelihoods, empowerment, and traditional knowledge systems. As a result, there is limited integrative research examining how GeoAI can be operationalised within community-based, household-level green enterprise models, particularly in peri-urban contexts of the Global South (Mulya et al., 2023). Three key gaps emerge: limited empirical evidence on adapting GeoAI for small-scale, multi-species MP systems; the absence of frameworks linking spatial intelligence with household decision-making and socio-cultural practices; and insufficient policy attention to how GeoAI can support the transition from backyard cultivation to resilient, market-oriented green micro-enterprises without undermining inclusivity or local knowledge. Addressing these gaps is essential for aligning GeoAI innovation with inclusive and sustainable health–food systems. In response, this chapter develops and empirically grounds a household-scale GeoAI-enabled optimisation framework that integrates spatial analysis, MP production, and community-based enterprise development within a PUI context.

1.3 Objectives and Chapter Contributions

This chapter aims to examine how GeoAI can be strategically applied to optimise household-based MP production in PUI and facilitate its transformation into smart, sustainable green micro-enterprises. Building on empirical household-level evidence, the chapter addresses the gap between spatially advanced digital technologies and the socio-cultural realities of small-scale MP producers. Specifically, the objectives of this chapter are threefold. *First*, it aims to analyse existing TORT practices in peri-urban settings to identify spatial, ecological, and managerial constraints that limit productivity and enterprise development. This includes examining how households currently utilise limited yard space, respond to microclimatic variability, and make production decisions under conditions of uncertainty. *Second*, the chapter examines the potential role of GeoAI-enabled SDS tools in enhancing household-level optimisation. By integrating geospatial data, microclimate indicators, and production patterns, this study evaluates how GeoAI can be applied to support more accurate decision-making on crop selection, spatial planning, input use, and harvest cycles in multi-species household systems. This chapter examines socio-technical conditions for GeoAI adoption at the household level, emphasising digital literacy, community facilitation, participatory learning, and institutional support to enable meaningful engagement and benefits, especially for women.

Beyond these objectives, this chapter also makes several key contributions. Empirically, this chapter presents one of the first in-depth studies of the application of GeoAI in a household-based MP system in peri-urban Indonesia. Conceptually, it develops an integrated socio-technical framework that links spatial optimisation with community-based empowerment and green economy development. Methodologically, it demonstrates how household ethnographic data can be combined with spatial intelligence to generate digital innovations that are more sensitive to local contexts. Thus, this chapter contributes to the broader discourse on sustainable agriculture, digital transformation, and inclusive innovation by demonstrating how advanced technologies such as GeoAI can be adapted to support small-scale production systems rooted in cultural practices. This contribution aligns with the thematic orientation of the IGGI volume, which emphasises the role of science-based innovation in transforming food systems and agricultural sustainability through responsible, equitable, and spatially oriented approaches.

2. CONCEPTUAL AND THEORETICAL FOUNDATIONS

2.1 GeoAI and Spatial Decision Support for Household-Based Agriculture

GeoAI is widely used in machine learning and pattern recognition through processing geospatial data derived from remote sensing, spatial databases, and environmental monitoring systems to analyse location-specific variations in land characteristics, microclimate, and crop performance. However, most applications are still concentrated in large-scale commercial agriculture with relatively homogeneous land conditions and standardised production cycles, such as optimising fertilisation, irrigation, and harvest timing for major food crops or monoculture systems. In contrast, applying GeoAI to enhance the cultivation of food crops and medicinal plants at the household level in peri-urban areas entails considerably greater complexity. The reasons for this are limited land area, irregular landforms, small and fragmented production units, and the cultivation of diverse species within the same space. These conditions indicate that conventional household farming does not adequately capture spatial heterogeneity. In this context, GeoAI is not intended for large-scale optimisation, but rather for providing micro-scale spatial intelligence support. GeoAI has the potential to help households identify suitable planting zones, improve spatial efficiency, and better assess crop suitability. To make this work, GeoAI needs to be used as a tool that supports decision-making rather than replacing local knowledge, so that it remains relevant to everyday practices, needs, and social contexts.

In this sense, GeoAI in household farming is better understood as a decision-support tool, not as a fully automated decision-making system. In practice, MP producers' decisions are influenced by various factors, including traditional knowledge, health needs, risk perceptions, and household labour availability.

Therefore, GeoAI will be more effective when combined with local knowledge and the lived experiences of households, enabling spatial information to be effectively interpreted and adapted within their social and cultural contexts. Theoretically, this study adopts a Decision Support System (DSS) perspective for households. This approach aligns with critiques of technology-centric agricultural innovation, which emphasise that digital technologies are only effective when they strengthen capacity, are easily understood, and are trusted by users in decision-making processes. In the peri-urban MP context, GeoAI serves as a supporting layer that connects spatial intelligence with household autonomy for the purposes of strengthening health and medicinal plant availability, business development, and sustainable resource use. By framing GeoAI in this way, the chapter places household producers, rather than algorithms, at the centre of decision-making, treating spatial technology as a means to strengthen adaptive capacity, resilience, and the transition from backyard cultivation to smart green enterprise systems. This perspective can be further strengthened by engaging with spatially explicit modelling approaches, such as geographically weighted regression and spatial-temporal predictive modelling, which enable fine-scale analysis of environmental variability and decision-support scenarios in complex urban and peri-urban systems (Al-Shaar et al., 2022, 2025).

Figure 1 illustrates the conceptual logic of this chapter, highlighting how GeoAI functions as an SDS tool within household-based MP systems in PUI.

Figure 1. Conceptual framework of GeoAI-enabled spatial decision support for household medicinal plant systems in peri-urban contexts. Source: Author's own conceptualisation based on relevant GeoAI and spatial decision support literature

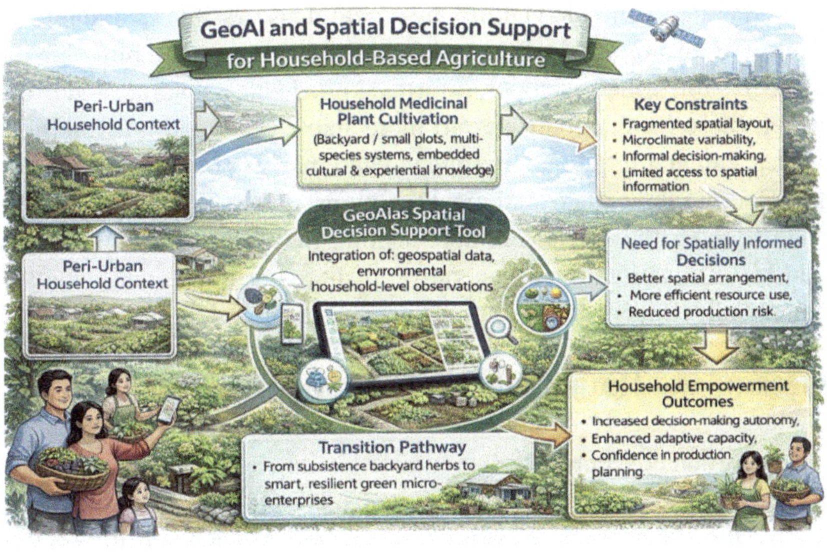

The figure highlights that GeoAI does not substitute household decision-making but instead strengthens spatial awareness and adaptive capacity, supporting the shift from subsistence-based backyard cultivation to smart, resilient green micro-enterprises.

2.2 Circular Economy Theory and Household Green Enterprises

CE theory provides a critical lens for understanding how household-based MP cultivation can transition from subsistence to sustainable green micro-enterprises. Unlike linear production systems centred on extraction, use, and disposal, CE emphasises resource efficiency, waste minimisation, and value retention through reuse, recycling, and regeneration. The MP management system in a rural community on the outskirts of Cibinong shows emerging circular economy (CE) patterns, particularly through the reuse of organic waste and crop residues as compost and natural fertilizer. This practice reflects CE principles and helps improve soil quality (Ghisellini et al., 2016). However, these practices are still informal, have not been formally adopted by the village government, and are not yet integrated into broader planning processes. As a result, the system has limited economic value and remains unsustainable. From a CE perspective, these activities can be reinterpreted as a closed-loop household micro-enterprise, where biological resources are regenerated locally and added value is created through processing, diversification, and upgrading of MP products (Cahyadi et al., 2024). This shift opens opportunities for households to move from subsistence-based activities towards more structured and productive green businesses, in line with circular bioeconomy strategies in small-scale agriculture and short supply chains (Korhonen et al., 2018; Neyra et al., 2025).

Applying CE principles to household-based MP production can be understood both theoretically and practically as an effort to frame home garden cultivation as a regenerative, knowledge-based micro-ecosystem. In this system, input use is minimised, waste streams are reused, and product diversification—such as dried herbs, extracts, and simple formulations—can improve both ecological and economic performance. When combined with GeoAI, the CE approach can further support productivity, environmental efficiency, and economic viability by providing accessible spatial information tailored to household conditions. CE also highlights the social dimension of sustainability, as these practices are closely linked to women's roles, family labour, informal knowledge networks, and everyday community interactions. Together, these factors contribute to more stable production, diversified yields, stronger resilience, and reduced vulnerability to environmental and market pressures (Wirjodirdjo et al., 2025). In this framework, CE serves three main analytical functions: understanding resource recycling dynamics, identifying value-added pathways, and complementing GeoAI-based spatial decision support. This positions household MP cultivation as a regenerative, resource-efficient, and locally grounded system that supports sustainable livelihoods in Indonesia's peri-urban areas.

2.3 CBI, Facilitation, and Household Empowerment

CBI provides an important conceptual lens for understanding how GeoAI-based optimisation can be meaningfully applied in household-managed MP systems. Evidence from digital agriculture literature consistently shows that the effectiveness of advanced technologies depends not only on their technical design but also on the social processes that enable knowledge to be shared, interpreted, and applied at the local level (Manzoor et al., 2025). In peri-urban household contexts, MP cultivation is strongly shaped by local knowledge and inherited practices, both as cultural traditions and livelihood strategies related to health and safety. As a result, households often find it difficult to interpret spatial analysis and algorithm-based recommendations on their own, even when they recognise the value of the technology. Without facilitation, these digital decision-support systems tend to be underused or only partially applied, especially among small-scale producers and household entrepreneurs (Klerkx et al., 2019). Based on this situation, community-based facilitation plays a key role in bridging the gap between digital outputs and practical decision-making in the field. Smallholder farmers' engagement with digital technologies is strongly influenced by local social dynamics, inter-farmer interactions, and community support, which enable the interpretation and adaptation of complex information to be more easily understood and implemented (Gumbi et al., 2023).

In agrarian contexts, community networks often function as informal extension systems. The goal is to help improve access to information, resolve technical issues, and contextualise local knowledge according to household conditions and capacities (Abdulai et al., 2023). Furthermore, social institutions such as seed banks, farmer groups, and farmer forums are crucial because they can accelerate technology diffusion and increase adoption and innovation through learning mechanisms and social networks (Okori et al., 2022). Therefore, the implementation of GeoAI must be integrated with these community programs. The goal is to strengthen the decision-making capacity of social institution managers. Furthermore, these social institution managers will facilitate the community. Through this integration, it is hoped that TORT production can be well-managed, productivity increased, community adaptive resilience also enhanced, and the transition from small-scale subsistence farming to smarter and more sustainable green micro-enterprises can continue to accelerate gradually. This framework aligns with the IGGI volume's emphasis on science-based innovation that is socially grounded, spatially oriented, and contributes to sustainable development outcomes.

2.4 Integrated Conceptual Framework for GeoAI-Enabled TORT Production

In this chapter, GeoAI is positioned as a decision-support mechanism embedded within household socio-ecological systems, rather than as a stand-alone technology. The framework brings together three interconnected elements: (i) GeoAI-based spatial intelligence, (ii) household food and medicinal crop cultivation practices, and (iii) community-based facilitation grounded in circular economy (CE) principles. At its core, the system is designed to support household decision-making in food crops and medicinal plants under conditions of limited land, diverse microclimates, risk-averse livelihood strategies, and mixed cropping practices that are common in peri-urban settings (Abdulai et al., 2023). These households require clear and accessible spatial information to support decisions related to land suitability, microclimate conditions, and planting arrangements. GeoAI integrates high-resolution environmental data with household land characteristics and production records to produce location-specific recommendations. This becomes especially important in areas with high spatial variation and micro-scale environmental differences that affect plant health, yield potential, and resource allocation (Gumbi et al., 2023). However, GeoAI outputs cannot be directly implemented without interpretation. They need to be translated into locally meaningful knowledge. Here, community-based facilitation through local institutions plays an important role as a bridging layer, helping households understand, adapt, and apply spatial recommendations in practice. In this process, peer-to-peer learning and local institutional engagement are important for strengthening user understanding and improving the practical use of digital tools (Abdulai et al., 2023).

In addition, CE principles are reinforced through socio-technical integration, where household medicinal plant production is seen as a small-scale, efficient resource management system. Practices such as composting organic waste, reusing biomass, and closing nutrient loops contribute to improving the sustainability of small food systems (Neyra et al., 2025). In the TORT context, these practices not only reduce reliance on external inputs but also create added value through resource recycling. GeoAI supports this process by helping identify suitable spatial arrangements for circular practices, while community facilitation ensures that these insights are consistently integrated into household production decisions. Overall, this framework seeks to foster the transformation of TORT and SGME production as a learning-based adaptive system, where spatial intelligence, social facilitation, and circular resource cycles are integrated in a sustainable manner. By placing the household at the centre of analysis, this framework aligns with IGGI's focus on science-based innovation rooted in local social processes, thereby offering a pathway to transform backyard herb cultivation into more resilient and sustainable green microenterprises.

2.5 GeoAI-Specific Implications for Household Decision-Making and Green Enterprise Development

This integrated framework demonstrates that GeoAI serves not only as a general-purpose digital support tool but also facilitates the transformation of how households understand and manage space within MP production systems. GeoAI-based spatial analysis redirects decision-making processes from intuition-driven decision-making to more site-specific strategies informed by microspatial variability. By translating diverse environmental conditions, such as differences in soil characteristics, moisture distribution, and microclimate variations, into interpretable spatial recommendations, GeoAI helps reduce uncertainty while enhancing households' ability to optimise the use of limited peri-urban land (Gumbi et al., 2023). It is important to note that GeoAI does not directly determine household decisions but rather influences decision-making through socially mediated interpretation mechanisms. In this process, households collectively interpret the results through community-based facilitation, allowing spatial recommendations to be tested, adjusted, and integrated into production practices that are already embedded in the local cultural context, while the ongoing learning process strengthens household capacity. These changes in social dynamics, along with the strengthening of local knowledge through GeoAI adoption, contribute to the development of a more inclusive and sustainable digital agricultural system(FAO, 2023).

Furthermore, CE principles supported by GeoAI at the household level further enhance the efficiency of practices such as composting, biomass reuse, and nutrient cycling within spatially limited production spaces. This integration contributes to increased resource efficiency, reduced dependence on external inputs, and strengthened household resilience (Neyra et al., 2025). From an entrepreneurial perspective, GeoAI also enables the gradual transformation of garden herbal cultivation into SGME through more optimal spatial planning, increased production efficiency, and a risk-aware scale-up process. Rather than encouraging rapid and uncontrolled expansion, GeoAI supports gradual, adaptive, and context-specific growth that aligns with household capacity and environmental constraints, thus strengthening the system's long-term sustainability and viability.

3. METHODOLOGY

This chapter examines how household-based MP production in PUI areas can transition to SGME through a qualitative-participatory mixed-methods approach integrated with GeoAI-based spatial analysis. This approach is designed to explore the perspectives of local actors and understand their adaptive practices in the face of spatial limitations, access to health services, and the pressures of urbanisation. Methodologically, this framework is designed to systematically integrate qualitative data with geospatial technology in accordance with applicable academic standards.

3.1 Research Design and Context

The research focus in this chapter is directed at the household environment in the densely populated area of Cibinong, Bogor Regency, Indonesia. This peri-urban area is characterised by rapid spatial growth, but still has limited infrastructure to support sustainable economic development. The location was selected purposively based on the existence of local initiatives in household farming and the established practices of MP within the local community. The research was conducted over three months, from January to March 2025.

3.2 Data Collection Techniques

The triangulation approach used in the data collection process in this chapter was carried out to ensure depth and validity in the process, including: (1) In-depth interviews. Semi-structured interviews were conducted with 50 households (primarily women-led, as primary actors), 10 community leaders, 5 local herbal entrepreneurs, and 5 green economy facilitators. These interviews (45–60 minutes each), explored motivations, strategies, challenges, and the cultural and health values underpinning their micro-enterprises; (2) Participatory observation: This involved observing MP cultivation practices in backyards, production, and processing of herbal products (e.g., herbal teas, infused oils, spice powders), and sales in local markets or informal networks. Observations documented the creative use of limited space and the adoption of circular economy practices, including composting, vertical farming, and eco-friendly packaging. Initial spatial data for GeoAI analysis was gathered concurrently, using high-resolution satellite imagery (e.g., Sentinel-2) and drone data to map microclimate variations and land use patterns at a fine spatial scale (10–50 m^2 plots); (3) Focus Group Discussions (FGDs): FGDs were held in 3 sessions (20–30 participants each) with local green communities, comprising household members, community mobilisers, and herbal artisans. These discussions aimed to identify shared values, social networks, and collective knowledge developed among green micro-enterprise actors.

All data collection procedures adhered to ethical guidelines for Community-Based Participatory Research (CBPR), ensuring informed consent, participant anonymity, and member-checking to validate findings.

3.3 GeoAI Integration and Data Analysis

GeoAI was employed as a spatial decision-support layer rather than a prescriptive tool. Machine learning models, such as random forest classifiers applied to Sentinel-2 satellite imagery, were used to analyse spatial relationships relevant to TORT production, including land suitability and microclimate variability such as shade and moisture distribution. Given the relatively coarse spatial resolution of Sentinel-2 imagery, satellite-derived insights were complemented with drone-based observations to enhance sub-plot resolution and enable validation at the household scale (10–50 m^2). These integrated datasets improved the identification of micro-zones and spatial heterogeneity within constrained peri-urban environments. GeoAI analysis was further used to uncover patterns and optimisation opportunities in household decision-making under resource-limited conditions, including identifying suitable planting areas, improving space utilisation, and aligning crop selection with local environmental characteristics. Simple spatial maps were developed to facilitate GeoAI interpretation at the household level and to support community discussions and reflection processes. The effectiveness of adoption was assessed through participatory cross-checking during focus group discussions (FGDs), while success was evaluated by comparing estimated harvest productivity before and after implementation.

The research in this chapter uses a thematic analysis approach to identify key patterns, meanings, and narratives related to household-based SGME. Field findings are then mapped onto a sustainable business model framework based on the Triple Bottom Line (TBL: People, Planet, Profit) concept. This framework is used to assess the balance between social benefits, such as improved management of MP and health, environmental benefits such as resource efficiency and circular practices, and economic benefits, such as household income and local market opportunities. The integration of GeoAI indicators, such as estimated yield increases and optimised input use, provides quantitative support to the qualitative findings, thereby strengthening the empirical validity of the study. To improve spatial accuracy in the context of households, small-scale and heterogeneous SGME, verification was conducted using drones to confirm classification results from satellite imagery at the sub-plot level. Validation was conducted through a verification process based on field findings with participating households.

3.4 Ethical and Analytical Considerations

GeoAI outputs generated with the active involvement of community household participants are used as informational input rather than as directives. This approach is intended to ensure that households maintain autonomy in making decisions related to agricultural production based on their local knowledge. This ethical position aligns with the principles of participatory and emancipatory research, where participants are involved as co-interpreters of spatial data. Digital innovation in agriculture must strengthen local decision-making and social structures (FAO, 2025).

4. FINDINGS

The empirical findings presented within the TBL framework are derived from in-depth interviews, observations, and FGDs with 50 households (January-March 2025). These findings relate to the integration of GeoAI interventions with innovations in CE practices within women's communities. They also relate to their contribution to the transformation of resilient SGMEs from small-scale food crop cultivation in their yards.

4.1 People: Empowering Women and Enhancing Health Autonomy through GeoAI-Supported CBI

Female informants, the primary managers of their backyard gardens, continually strive to integrate ethnobotanical knowledge into their daily routines. Their consistent cultivation of medicinal plants is a strategy to reduce food expenses, ensure continued access to reliable medicinal resources, and maintain daily health. One of them stated, "The herbs we grow are not just for sale, but to ensure our families always have medicine and healthy food, without having to buy them." (Female head of household, suburban community). Eighty per cent of households experienced a significant increase in confidence in decision-making. This occurred following the integration of GeoAI-enabled spatial insights using simplified microclimate and land suitability mapping. For instance, in ginger and bitter melon cultivation, GeoAI identifies more systematically defined optimal planting zones for each crop, with ginger being more heat-tolerant and bitter melon showing stronger adaptability to shaded conditions. Savings in healthcare costs through the empowerment of household communities using GeoAI have increased noticeably. Coordinators of women's groups reported an average reduction in monthly healthcare expenses of around IDR 200,000, mainly through the use of home-produced herbal remedies. In this context, CBI networks, particularly women-led herbal groups, play an important role in mobilising local, culturally grounded knowledge and in helping households interpret and make sense of GeoAI outputs. Informants also noted that working in groups helps reduce individual risk. As one women's group coordinator explained, 'When we work as a group, it's easier to try new things because we share experiences and learn from the mistakes we make.

Table 1 summarises how CBI networks, particularly women-led herbal groups, play a role in mobilising ethnobotanical knowledge and collective action within peri-urban household TORT, as observed in 50 households in Cibinong. Based on FGDs and in-depth interviews, the table demonstrates the significant contribution of CBI in interpreting spatial information from GeoAI and facilitating low-risk experimentation, enabling more operational implementation of circular practices within environmental constraints. These findings also align with IGGI Global's focus on socially integrated GeoAI applications to support sustainable agriculture and food security.

Table 1. Community networks mobilise cultural knowledge and collective action (n = 50 households)

CBI network component	Description and function	GeoAI integration	Observed impact
Women's herbal groups	Informal innovation hubs sharing ethnobotanical knowledge, processing techniques, and market information	Collective interpretation of micro-zone maps (soil moisture, shade suitability)	Coordinated production schedules; reduced individual risk ("When we work as a group, it is easier to try new things" – Women's group coordinator) [From Backyard Herbs to Smart Green Enterprises.docx]
Peer learning networks	Knowledge exchange through shared experiences and troubleshooting GeoAI outputs	Validation of spatial recommendations via group discussions and field testing	80% increased decision confidence; strengthened social cohesion
Community facilitation	Local advisory support bridging digital outputs with cultural practices	Translation of GeoAI composting zones into practical backyard layouts	Enhanced circular economy adoption; 70% resilience to urban shocks

Overall, Table 1 shows that the CBI network played a role in transforming GeoAI from a mere technical output into a practice that can be implemented at the household level, thus confirming its role in the Backyard-to-SGME model and extending the findings (Abdulai et al., 2023) regarding the adoption of digital agricultural technologies in smallholder contexts in PUI areas. Through the mobilisation of culturally based knowledge facilitated by the role of women, this network contributed to the achievement of TBL People dimensions, such as empowerment and health autonomy, and positioned community structures as key intermediaries in supporting the scalability of GeoAI in food systems in the Global South.

4.2 Planet: GeoAI-Enabled Circular Economy Practices for Enhanced Resource Efficiency

CE practices at the household level are quite prominent, particularly through women-led processing activities, which include the reuse of organic waste and crop residues for compost, mulch, and natural fertiliser. Informants consistently stated that these circular practices initially emerged out of necessity but were further strengthened through collective learning and the support of community networks: "We learned together how to reuse waste from the kitchen and garden. "This is a significant cost-saving measure because we maintain and care for the plants to keep them healthy" (Women's herbal group member). These practices contribute to significant cost reductions in the long-term, resulting from reliance on external inputs.

In this context, GeoAI plays a crucial role in optimising CE practices through more structured spatial planning. The integration of diverse GeoAI spatial data, which provides concrete information on efficient space allocation and resource cycling support, significantly assists household communities in decision-making. Decisions regarding optimal plant placement, composting locations, and microclimate management have significantly impacted production cost savings. This spatial integration has been shown to reduce external input costs by up to 20%, with some households achieving a significant return on investment (ROI) within approximately six months of investing in improved composting facilities. Furthermore, GeoAI-based spatial planning support also encourages more efficient use of limited land through the implementation of vertical farming and strategic crop arrangement, resulting in a 25% increase in space efficiency.

As summarised in Table 2, circular economy practices at the household level, supported by the GeoAI-based decision-making tool, contribute to strengthening women's capacity, increasing economic resilience, and improving environmental sustainability by simplifying spatial management on small and fragmented plots of land. The table also shows how GeoAI transformed previously informal circular economy practices into a more spatially optimised system, with increased resource efficiency through a 20% reduction in external inputs and a 25% increase in space utilisation in 50 households in Cibinong. These findings extend the short-chain circular economy concept proposed by Neyra et al. (2025) to the peri-urban context in Indonesia, while positioning GeoAI as a critical driver in achieving Triple Bottom Line (TBL) outcomes on the Planet and sustainable land management that can be replicated in the Global South.

Table 2. GeoAI-supported CE practices in household-based MP systems

CE practice	Description & GeoAI contribution	Observed impact
Organic waste composting	GeoAI identified optimal locations for composting within limited backyard spaces, reducing spatial uncertainty.	20% reduction in external fertiliser costs; improved soil health.
Biomass reuse	Spatial planning guided by GeoAI for efficient reuse of plant residues for mulching and soil enrichment.	Enhanced nutrient cycling; reduced dependency on new inputs.
Low-input cultivation	GeoAI-informed plant placement and microclimate matching, optimising resource use from the start.	Reduced water and nutrient loss; 25% space efficiency.

4.3 Profit: Transforming Backyard Cultivation into Economically Viable SGMEs

Table 3. Guantifiable TBL outcomes from GeoAI-supported TORT in Cibinong (n = 50 households)

TBL aspect	Key outcome	GeoAI contribution	Quantifiable impact
People	Enhanced decision confidence & health autonomy	Microclimate maps, optimal planting recommendations	80% increased confidence; IDR 200,000/month health savings
Planet	Improved resource efficiency & circularity	Spatial zoning for composting, efficient space utilisation	20% reduction in external inputs; 25% space efficiency
Profit	Increased income & economic resilience	Yield prediction, production optimisation	IDR 500k–1m/month additional income; 20–30% yield increase

In the women-led composting practices of the CBI network, the integration of GeoAI-based spatial insights has been shown to significantly transform local household cultivation systems from subsistence-oriented to more economically viable SGME. At the household level, several socioeconomic improvements were observed, including additional monthly incomes of around IDR 500,000–1,000,000 from value-added herbal products. This diversification of income sources has also strengthened household economic resilience, particularly during periods of financial difficulty, with 70% of informants reporting greater ability to adapt to environmental stressors such as urban flooding. In addition, a 20–30% increase in medicinal plant yields was reported, resulting from the combination of microclimate analysis, GeoAI-based production planning, and more efficient processing practices, as well as collective market access through CBI networks, particularly for ginger and turmeric. These improvements contributed directly to higher profitability and greater sustainability of these small-scale herbal enterprises. Table 3 summarises the key empirical findings from 50 households in Cibinong, organised using the Triple Bottom Line (People–Planet–Profit) framework. The table illustrates how the interaction between GeoAI, circular economy (CE) practices, and CBI networks has generated measurable outcomes in terms of women's empowerment, resource efficiency, and household income growth. Data were obtained through a triangulation of methods, including in-depth interviews, participant observation, and focus group discussions (FGDs), and are supported by GeoAI indicators showing a 20–30% increase in production yields.

Overall, these findings demonstrate that the Backyard-to-SGME model operates as a dynamic socio-technical system, where GeoAI, circular practices, and CBI networks interact to strengthen household resilience and support the development of women-led green micro-enterprises. The table also confirms the model's ability to generate integrated social, economic, and environmental impacts, while positioning GeoAI as a spatial driver facilitating the transition from household subsistence farming to more resilient micro-enterprises. Furthermore, Table 3 presents a summary of the integrated socio-economic and environmental outcomes across the entire study sample. These findings not only quantify the impact of TBL but also extend GeoAI's application to the context of small-scale agroforestry, as reported in Bali (Liu et al., 2023), providing additional empirical evidence on production outcomes and livelihoods that have not been widely measured in previous studies. Table 3 provides an important foundation for further discussions in supporting food security in the Global South.

4.4 GeoAI Mechanisms for TORT Optimisation and Supported Spatial Decision-Making at Household Scale

Consistent with the chapter's aim to highlight the strategic application of GeoAI, Figure 2 shows the mechanisms through which GeoAI-enabled tools generate actionable insights for optimising TORT production and facilitating socio-economic benefits in peri-urban settings (Yao et al., 2025). Ultimately, these GeoAI mechanisms, when coupled with community-based interpretation and adaptation, enable households to transition towards more efficient, resilient, and economically viable medicinal plant micro-enterprises, aligning with broader sustainable agriculture and digital innovation pathways.

Figure 2. Validated causal chain of GeoAI-enabled spatial decision-making for TORT optimisation at household scale in peri-urban contexts. Source: Author's own conceptualisation based on field observations and relevant literature

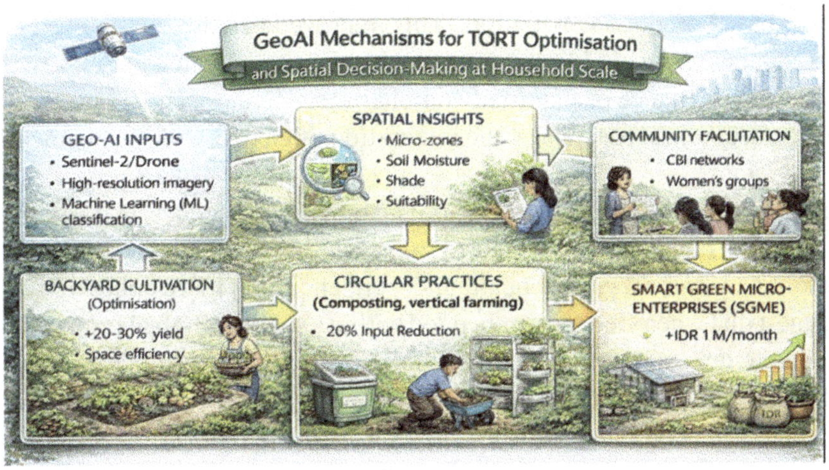

Figure 2 illustrates a validated causal chain in which GeoAI inputs generate spatial insights that are interpreted through community facilitation, transforming intuitive backyard management into data-driven precision cultivation in PUI. High-resolution satellite or drone imagery combined with machine-learning models classifies micro-zones (10–50 m² plots) based on soil moisture, shade, and suitability, enabling households to optimise planting decisions. As a result, 80 per cent of households reported increased decision confidence, while medicinal plant yields for species such as ginger and turmeric improved by 20–30 per cent, based on field observations in this study; related GeoAI applications in agroforestry systems in Bali similarly utilise spatial data for land-use classification and decision support, although without directly reporting comparable yield outcomes (Liu et al., 2023).

Using women-led household production and circular economy principles, GeoAI functions as a household decision-support tool by translating environmental conditions into clear spatial guidance. It helps identify land potential, assess microclimate suitability, optimise planting layouts, and define zones for composting and processing. Informants stressed that GeoAI strengthens rather than replaces local knowledge. As one female herbal producer noted, spatial maps explain plant performance across garden areas, building household confidence in experimentation and improving peri-urban agricultural productivity and medicinal plant systems.

Thus, simplifying GeoAI's spatial complexity through adjustments to small, fragmented, and diverse land management will become easier, making the resulting information more accessible and interpretable. This technology significantly assists household communities in making key spatial decisions, such as determining planting locations, analysing microclimate suitability, planning production, and categorising environmental zones, without replacing local knowledge and household experience. In this context, GeoAI functions as an interpretive decision support tool that complements local expertise and serves as an experiential learning process, rather than as a system that automates production decisions.

4.5 Economic Resilience, Health Autonomy, and Social Cohesion

The study results show that GeoAI-enabled herbal SGME contributed to increased household economic resilience through additional income, strengthened health autonomy, and enhanced social cohesion through cooperative-based production networks. This improved health was achieved through the use of home-produced medicines. Household communities also reported greater stability when facing income shocks, as well as reduced reliance on externally purchased medicines. These outcomes were not driven by technological advancement alone, but by the integration of GeoAI within a women-led, community-based, and circular system.

Following the integration of GeoAI-based spatial insights, CE practices, and community facilitation, households experienced a range of interconnected impacts. These included not only higher incomes, but also stronger economic resilience, greater health independence, and more cohesive social relationships. These findings suggest that when GeoAI is embedded within a socio-technical framework, the technology can support sustainable development at the household level by transforming traditional practices into more resilient and economically valuable SGME.

These improvements in economic and health aspects also strengthen the relevance of GeoAI applications that have real benefits for communities in peri-urban areas, where the CBI network acts as an intermediary in transforming collective interpretations of GeoAI results into production decisions that are appropriate to the local context.

4.6 Core Contribution: The Backyard-to-SGME

The development of the Backyard-to-SGME model is a key contribution of this study, particularly through the integration of GeoAI and household-based SGME in peri-urban Indonesia. This model combines local ethnobotanical knowledge, household resource mapping, the application of CE principles, and spatially based decision-making support through GeoAI to strengthen decision-making processes and increase household resilience.

Table 4 summarizes the main components of the Backyard-to-SGME model, as an integration of GeoAI and household-based SGME in PUI. The table shows the results of interactions among 50 households on technological (GeoAI), ecological (CE), and social (CBI) aspects. Collectively, these outcomes contribute to improved livelihoods, health autonomy, and household economic resilience. This model also directly supports the IGGI Global agenda, which focuses on GeoAI-based sustainable land management and precision agriculture to support food security transformation.

Table 4. Core components of the backyard-to-SGME model and associated household impacts (n = 50 households)

Model component	Function	GeoAI contribution	Perceived household impacts
Ethnobotanical knowledge	Cultural foundation for plant selection and health applications	Validation through microclimate suitability mapping	Health autonomy (IDR 200,000/month savings); preserved cultural heritage
Household resource mapping	Spatial inventory of backyard constraints/opportunities (10–50 m² plots)	High-resolution classification (Sentinel-2/ML) of soil moisture, shade, suitability	25% space efficiency gains; optimal vertical planting
Circular economy practices	Resource regeneration (composting, biomass reuse)	Spatial zoning for nutrient cycling and composting locations	20% external input reduction; ROI <6 months
GeoAI spatial decision support	Micro-scale precision agriculture recommendations	ML classification → actionable planting/harvesting insights	20–30% yield increase (ginger, turmeric); 80% decision confidence
Community-Based Innovation (CBI)	Social learning and risk-sharing networks	Collective interpretation of spatial outputs via women's herbal groups	Economic resilience (70% shock resistance); +IDR 500,000–1M/month income

Overall, Table 4 confirms that the Backyard-to-SGME model is an integrated socio-technical system, where GeoAI, CE practices, and CBI networks simultaneously contribute to strengthening household resilience and fostering women-led SGME development. This approach, in addition to enhancing the economic value of conventional precision agriculture, is also relevant to peri-urban contexts characterised by multi-species systems. This model also fills the literature gap previously identified by Mulya et al. (2023) on peri-urban agriculture and Astutik et al. (2025) on MP systems in Indonesia, while offering a replicable framework for food and health systems in the Global South, with strong relevance to SDGs 2, 3, and 5, particularly in terms of women's empowerment and inclusive agricultural development (Quisumbing et al., 2023).

4.7 Quantifiable Socio-Economic Impacts

The GeoAI integration resulted in a positive increase in household income of IDR 500,000–1,000,000 per month from processed herbal products. Furthermore, there was an average reduction in healthcare expenditures of IDR 200,000 per month through self-care practices and increased resilience to shocks such as urban flooding, experienced by approximately 70% of households. Combined, these outcomes contribute to strengthening food security, improving health outcomes, and empowering women economically, all of which align with SDGs 2, 3, and 5 (Quisumbing et al., 2023). This 20–25% increase in medicinal plant yields, coupled with an increase in household income of approximately IDR 1,000,000 per month, significantly strengthens household economic stability. These results demonstrate that SGME through MP businesses significantly supports adaptive livelihood resilience and increases household income value-added in peri-urban Indonesia through more optimal management practices (Astutik et al., 2023).

5. DISCUSSION AND ETHICAL IMPLICATIONS

The interpretation of the key findings in this chapter includes a discussion comparing them with previous studies, both in Indonesia and globally. This discussion highlights the specific contribution of the Backyard-to-SGME model supported by GeoAI.

5.1 Repositioning Household Backyards as Strategic GeoAI-Enabled Food–Health–Enterprise Systems

The findings of this study suggest that backyards in PUI (Indonesian Infrastructure Development Centre) are not merely marginal spaces. With the support of GeoAI, these spaces can be used more effectively to enhance food production, health outcomes, and household economic activities. This view is in line with global research that emphasises the multifunctional role of home gardens in supporting food security and community well-being (Santos et al., 2022). Building on this literature, this study argues that GeoAI can also support the transformation of household spaces into semi-commercial micro-enterprises, particularly through women-led cultivation and processing activities. Unlike urban agriculture models that typically require high capital investment, in the Indonesian context, this suggests the possibility of a less expensive yet knowledge-rich pathway for peri-urban development, relying on local ethnobotanical expertise. This indicates that resilience does not necessarily depend on formal commercialisation, while also reinforcing the relevance of household-based approaches in addressing the pressures of rapid urbanisation. Similarly, evidence from home garden studies in Europe also suggests that small-scale backyards can significantly contribute to meeting household food needs while reducing environmental impacts (Sanyé-Mengual et al., 2018).

5.2 GeoAI as a Socio-Spatial Enabler: Bridging Local Practices and Global Precision Agriculture Debates

This chapter emphasises GeoAI as a socio-spatial driver reshaping how households understand and manage limited space. Microscale information on climate and land suitability reduces uncertainty while strengthening women's experiential knowledge on small plots, in contrast to GeoAI's dominance in large-scale precision agriculture (Gupta & Kumar Pal, 2025) and demonstrates its relevance in the household context. The 20–30% increase in crop yields found in Cibinong through GeoAI-based decision-making demonstrates the technology's significant contribution to improving productivity and resilience. Similar experiences in agroforestry systems across Indonesia, including Bali (Liu et al., 2023), also demonstrate the complementary role of spatial analysis in small-scale contexts. However, GeoAI's effectiveness is largely determined by community-based facilitation, which reinforces the argument that a human-centred AI approach is key for small-scale agribusiness systems (Gumbi et al., 2023), especially in the face of limited digital literacy (Manzoor et al., 2025). Thus, GeoAI does not replace experiential knowledge, but rather enhances it, thus aligning with the notion of human-centred AI supporting improved practical decision-making (Holzinger et al., 2024). This approach is also consistent with an ethical framework that emphasises the principles of fairness, transparency, and accountability in the development of AI systems in the agricultural sector (Dara et al., 2022).

5.3 CE Theory and Women-Led Regenerative Practices

This chapter strengthens the argument that resource circularity will have the most optimal impact when integrated into daily household activities, such as composting and biomass reuse, which aligns with circular economy (CE) principles that emphasise efficient resource use and waste reduction (Geissdoerfer et al., 2017). This chapter expands on CE studies by demonstrating that GeoAI operationalises circularity spatially on limited peri-urban land by identifying optimal locations for composting, regeneration, and low-input cultivation. Spatial maps of CE are scarce in the literature. Unlike studies in rural Indonesia (Astutik et al., 2023), GeoAI empowers peri-urban spatial decisions. These findings also enrich the global study of circular economy strategies in small-scale agricultural systems (Neyra et al., 2025). Most importantly, when combined with spatial-based decision-making tools, the circular economy not only acts as an environmental approach, but also develops into a livelihood mechanism that strengthens the sustainability, resilience, and adaptive capacity of micro-enterprises among small-scale producers (Hilmi et al., 2024).

5.4 CBI and the Mobilisation of Cultural Knowledge

The CBI framework emphasises the role of women's herbal groups in transforming ethnobotanical knowledge into business opportunities through collective learning. In PUI, they serve as knowledge centres that standardise traditional medicines without weakening social ties. GeoAI strengthens collective planning, reduces risk, and expands the theory of community-based innovation. In comparison to Astutik et al. (2025) which discusses MP production system management in rural Java through a participatory rural appraisal approach, distinctively shows how GeoAI acts as a catalyst within these community structures to convert traditional knowledge into measurable economic and health outcomes in a peri-urban context. This is in line with global references that sustainable development can be achieved effectively when local knowledge systems with scientific innovation are jointly applied.

5.5 The Backyard-to-SGME Model: A Multidimensional Analytical Lens for Global Food Security

The Backyard-to-SGME model is a key contribution integrating GeoAI with household micro-enterprises in PUI through ethnobotany, resource mapping, CE, and AI-based spatial DSS. GeoAI, CE, and CBI mutually reinforce ecological, technological, and social decision-making, supporting a sustainable food-health system that can be replicated in the Global South. An income of IDR 500,000–1,000,000/month and a 20–30% increase in yield strengthen economic and food security, as well as women's empowerment (Quisumbing et al., 2023).

Access to and understanding of GeoAI spatial information is heavily influenced by women-led community facilitation. CBPR principles such as culturally appropriate informed consent, confidentiality protection, and benefit sharing foster trust and align interventions with local priorities. Experience in rural Africa demonstrates that GeoAI's sustainability depends on negotiating cultural norms, equitable researcher-participant relationships, and collaborative interpretation of results with communities, thus strengthening the relevance and sustainability of local implementation (Appiah, 2020; Bastida et al., 2010). Furthermore, inclusive decision-making will be more effective if women's access, participation, and leadership in technology implementation are strengthened, in line with a gender-responsive digital agriculture framework (Banks et al., 2013; Cornwall & Jewkes, 1995; Hidrobo et al., 2024; Quisumbing et al., 2023).

The Backyard-to-SGME model developed in peri-urban Indonesia has cross-regional relevance. For example, in Alas Mertajati, Bali, the use of GeoAI has been shown to support small-scale multispecies agroforestry. Meanwhile, in Peru, the Backyard-SGME approach can be applied through community-based agroecology supported by circular economy (CE) strategies and short supply chains. The difference lies in the explicit integration of GeoAI as a spatial driver that complements traditional knowledge and collective action. This integration can increase scalability and relevance in the Global South, which faces rapid urbanization and the need for sustainable food (Mulya et al., 2024).

5.6 GeoAI-Driven Sustainable Livelihood Transformation in PUI

Women in households can interpret GeoAI through community facilitation. Eighty percent of informants reported that their understanding of GeoAI maps would be better if they were simplified to reflect their local ethnobotanical knowledge (e.g., how to predict microclimates based on personal experience), and then communicated to each other to ensure the information was more relevant. This human-centred approach bridges digital literacy gaps, enabling incremental adoption (Gupta & Kumar Pal, 2025).

Causal Chain to Economic Empowerment: 1) Production: Agricultural output increased 20–30%, consistent with field observations across the study sample. Further processing, such as drying or powdering crops (e.g. chilli), generated additional income of more than IDR 500,000 per month; 2) Inputs: The GeoAI-informed CE composting zone reduced input costs by approximately 20%, with an ROI of less than six months; 3) Resilience: Health autonomy reduced household medical expenditures by an estimated 15–25%. (Mulya et al., 2023); and 4) Theoretical integration: GeoAI operationalises CE (spatial nutrient cycling) and Community Innovation (facilitated adoption), fostering resilient micro-enterprises amid peri-urban pressures.

6. CONCLUSION

This chapter argues that contextualised GeoAI can transform peri-urban household medicinal plant cultivation from subsistence to SGME through socio-technical processes influenced by household and community capacity. The Backyard-to-SGME model integrates GeoAI, ethnobotany, household assets, and the circular economy to improve backyard efficiency, resilience, and multifunctionality.

GeoAI is more effective when combined with community facilitation, particularly the role of women in translating spatial information into field practices, thus emphasising the importance of inclusive digital agriculture. This framework supports scalable green initiatives, but is still limited by short observation periods and self-reported data, necessitating validated quantitative longitudinal studies.

7. POLICY, PRACTICE, AND FUTURE RESEARCH DIRECTIONS

7.1 Policy Recommendations

7.1.1 Integrating GeoAI Into Household-Scale Agri-Food and Health Policies

National and local governments should recognise peri-urban household backyards as strategic units in food, health, and livelihood policies. Household-scale GeoAI enhances productivity and sustainability, aligning with calls for human-centred digital innovation that prioritises smallholders (FAO, 2023), and supporting the integration of GeoAI into national digital agriculture strategies through accessible tools for urban agriculture and traditional medicine programmes (FAO, 2025).

7.1.2 Recognising Women's Herbal Groups as Innovation and Enterprise Actors

Policies should move beyond treating women as programme beneficiaries and formally recognise women-led herbal groups as actors in innovation and small-scale enterprise. This study demonstrates the role of women in mobilising ethnobotanical knowledge and community networks, in line with agroecological evidence that collective action strengthens knowledge exchange, innovation, and local resilience (FAO, 2018), and calls for legal support, microfinance, and digital inclusion.

7.1.3 Embedding CE Principles in Peri-Urban Development Policy

Household CE practices through GeoAI emphasise composting, biomass recycling, and low-input MP cultivation. This aligns with the FAO policy that CE principles should be emphasized in waste management processes in peri urban areas (FAO, 2021), while GeoAI supports the identification of spatial regenerative opportunities.

7.2 Practice-Oriented Recommendations

7.2.1 Designing GeoAI Tools for Household Usability

The study in this chapter confirms that GeoAI adoption is highly dependent on the simplicity of the information generated (uncomplicated), ease of delivery (clarity), and relevance to local knowledge and culture. Since the informants are peri urban female households who produce MP and SGME, GeoAI must be able to reduce spatial uncertainty. This is in line with Gumbi et al. (2023), who argue that GeoAI design should emphasise clear visuals, be relevant to local knowledge, and facilitate learning.

7.2.2 Strengthening Community-Based Facilitation and Learning

To support GeoAI, actors in local social institutions should strengthen household community facilitation, such as women's herbal groups and local forums. FAO (2023) recommends collective interpretation of spatial adaptation to strengthen the sustainability of small-scale agricultural technology adoption.

ACKNOWLEDGMENT

The author would like to thank Mr Tri Margono from BRIN as the author's supervisor for completing this manuscript. The author gratefully acknowledges the Head of STTIF for the trust placed in enabling the independent completion of this manuscript. The author would also like to thank my beloved children, Nita, Hana, and Ira.

REFERENCES

Abdulai, A. R., Tetteh Quarshie, P., Duncan, E., & Fraser, E. (2023). Is agricultural digitization a reality among smallholder farmers in Africa? Unpacking farmers' lived realities of engagement with digital tools and services in rural Northern Ghana. *Agriculture & Food Security, 12*(11), 1–14. DOI: 10.1186/s40066-023-00416-6

Al-Shaar, W., Bonin, O., de Gouvello, B., Chatellier, P., & Hendel, M. (2022). Geographically weighted regression-based predictions of water–soil–energy nexus solutions in Île-de-France. *Urban Science (Basel, Switzerland), 6*(81), 1–30. DOI: 10.3390/urbansci6040081

Al-Shaar, W., Lehmann, X., Saad, N., Elmazi, G., Al-Shaar, M., & Tohme, C. (2025). Urban morphology and climate vulnerability assessment in Kuwait: A spatio-temporal predictive analysis utilizing deep neural network-enhanced markov chain models for 2050 and 2100. *PLoS One, 20*(8), 1–27. DOI: 10.1371/journal.pone.0318604 PMID: 40824951

Appiah, R. (2020). Community-based participatory research in rural African contexts: Ethico-cultural considerations and lessons from Ghana. *Public Health Review, 41*(27), 1–13. DOI: 10.1186/s40985-020-00145-2 PMID: 33292760

Astutik, S., Ahimbisibwe, V., Hintz, K. S., Purwanto, P., & Humaedi, M. A. (2025). Medicinal Plant Production System Management in Rural Java, Indonesia: Views of Local Actors from a Participatory Rural Appraisal Approach. *Forest and Society, 9*(1), 20–47. DOI: 10.24259/fs.v9i1.31352

Astutik, S., Pretzsch, J., Kimengsi, J. N., & Kapp, G. (2023). Medicinal plants production systems in rural Indonesia: Management practices and performance insights. *Forest Policy and Economics, 153*(March), 1–17. DOI: 10.1016/j.forpol.2023.102972

Banks, S., Armstrong, A., Carter, K., Graham, H., Hayward, P., Henry, A., Holland, T., Holmes, C., Lee, A., McNulty, A., Moore, N., Nayling, N., Stokoe, A., & Strachan, A. (2013). Everyday ethics in community-based participatory research. *Contemporary Social Science, 8*(3), 263–277. DOI: 10.1080/21582041.2013.769618

Bastida, E. M., Tseng, T.-S., McKeever, C., & Jr, L. J. (2010). Ethics and Community-Based Participatory Research: Perspectives From the Field. *Health Promotion Practice, 11*(1), 16–20. DOI: 10.1177/1524839909352841 PMID: 20038649

Cahyadi, E. R., Hidayati, N., Zahra, N., & Arif, C. (2024). Integrating Circular Economy Principles into Agri-Food Supply Chain Management: A Systematic Literature Review. *Sustainability (Basel)*, *16*(16), 1–17. DOI: 10.3390/su16167165

Chen, X. (2025). The role of modern agricultural technologies in improving agricultural productivity and land use efficiency. *Frontiers in Plant Science*, *16*(September), 1–19. DOI: 10.3389/fpls.2025.1675657 PMID: 41036393

Cornwall, A., & Jewkes, R. (1995). What is participatory research? *Social Science & Medicine*, *41*(12), 1667–1676. DOI: 10.1016/0277-9536(95)00127-S PMID: 8746866

Dara, R., Hazrati Fard, S. M., & Kaur, J. (2022). Recommendations for ethical and responsible use of artificial intelligence in digital agriculture. *Frontiers in Artificial Intelligence*, *5*, 1–11. DOI: 10.3389/frai.2022.884192 PMID: 35968036

FAO. (2018). *Scaling up agroecology initiative: Transforming food and agricultural systems in support of the SDGs*. https://www.fao.org/3/I9049EN/i9049en.pdf

FAO. (2021). *The State of Food and Agriculture 2021: Making agrifood systems more resilient to shocks and stresses*. DOI: 10.4060/cb7351en

FAO. (2023). *Promoting the digitalization of small and medium-sized agrifood enterprises in Asia and the Pacific*. DOI: 10.4060/cc8826en

FAO. (2025). *Agro-informatics*. https://www.fao.org/agroinformatics/en

Geissdoerfer, M., Savaget, P., Bocken, N. M. P., & Hultink, E. J. (2017). The Circular Economy - A New Sustainability Paradigm? *Journal of Cleaner Production*, *143*, 757–768. DOI: 10.1016/j.jclepro.2016.12.048

Ghisellini, P., Cialani, C., & Ulgiati, S. (2016). A review on circular economy: The expected transition to a balanced interplay of environmental and economic systems. *Journal of Cleaner Production*, *114*, 11–32. DOI: 10.1016/j.jclepro.2015.09.007

Gumbi, N., Gumbi, L., & Twinomurinzi, H. (2023). Towards Sustainable Digital Agriculture for Smallholder Farmers: A Systematic Literature Review. *Sustainability (Basel)*, *15*(16), 1–20. DOI: 10.3390/su151612530

Gupta, G., & Kumar Pal, S. (2025). Applications of AI in precision agriculture. *Discover Agriculture*, *3*(61), 1–19. DOI: 10.1007/s44279-025-00220-9

Hidrobo, M., Kosec, K., Gartaula, H. N., Van Campenhout, B., & Carrillo, L. (2024). Making complementary agricultural resources, technologies, and services more gender-responsive. *Global Food Security*, *42*(June), 1–14. DOI: 10.1016/j.gfs.2024.100778

Hilmi, Y. S., Tóth, J., Gabnai, Z., Király, G., & Temesi, Á. (2024). Farmers' resilience to climate change through the circular economy and sustainable agriculture: A review from developed and developing countries. *Renewable Agriculture and Food Systems*, *39*, 1–11. DOI: 10.1017/S1742170524000097

Holzinger, A., Fister, I., Fister, I., Kaul, H.-P., & Asseng, S. (2024). Human-Centered AI in Smart Farming : Toward Agriculture 5.0. *IEEE Access : Practical Innovations, Open Solutions*, *12*(April), 62199–62214. DOI: 10.1109/ACCESS.2024.3395532

Khofiyah, N. A., Nugraha, I., & Windyatri, H. (2025). An extended TAM approach to understanding behavioral and institutional drivers of drone technology adoption in Indonesia's green economy. *Journal of Advances in Information and Industrial Technology*, *7*(2), 227–240. DOI: 10.52435/jaiit.v7i2.732

Klerkx, L., Jakku, E., & Labarthe, P. (2019). A review of social science on digital agriculture, smart farming and agriculture 4.0: New contributions and a future research agenda. *NJAS Wageningen Journal of Life Sciences*, *90–91*(October), 1–16. DOI: 10.1016/j.njas.2019.100315

Korhonen, J., Honkasalo, A., & Seppälä, J. (2018). Circular Economy: The Concept and its Limitations. *Ecological Economics*, *143*, 37–46. DOI: 10.1016/j.ecolecon.2017.06.041

Liu, J., Iryadi, R., & Böhlen, M. (2023). Agroforestry in the Alas Mertajati of Bali, Indonesia. A Case Study in Sustainable Small-Scale Farming Practices and GeoAI. *International Archives of the Photogrammetry, Remote Sensing and Spatial Information Sciences - ISPRS Archives, 48*(4/W7), 101–108. DOI: 10.5194/isprs-archives-XLVIII-4-W7-2023-101-2023

Manzoor, F., Wei, L., Siraj, M., Lu, X., & Qiyang, G. (2025). Digital agriculture technology adoption in low and middle-income countries—A review of contemporary literature. *Frontiers in Sustainable Food Systems*, *9*(September), 1–11. DOI: 10.3389/fsufs.2025.1621851

Mulya, S. P., Hudalah, D., & Prilandita, N. (2024). Developing a new methodology for determining typologies of peri-urban agriculture: A case study of the Jakarta Bandung Mega Urban Region Indonesia. *Land Use Policy, 146*(May), 1–19. DOI: 10.1016/j.landusepol.2024.107310

Mulya, S. P., Putro, H. P. H., & Hudalah, D. (2023). Review of peri-urban agriculture as a regional ecosystem service. *Geography and Sustainability*, *4*(3), 244–254. DOI: 10.1016/j.geosus.2023.06.001

Musyaffa, M. S. I., Yudistira, N., Rahman, M. A., Basori, A. H., Mansur, A. B., & Batoro, J. (2024). IndoHerb: Indonesia medicinal plants recognition using transfer learning and deep learning. *Heliyon*, *10*(23), 1–16. DOI: 10.1016/j.heliyon.2024.e40606 PMID: 39660181

Neyra, J. M. V., Schmitt, V. G. H., & Cequea, M. M. (2025). Sustainable circular strategies and short supply chains among smallholder producers: The case of Peru's Agroferias Campesinas. *Agricultural and Food Economics*, *13*(77), 1–24. DOI: 10.1186/s40100-025-00423-0

Okori, P., Munthali, W., Msere, H., Charlie, H., Chitaya, S., Sichali, F., Chilumpha, E., Chirwa, T., Seetha, A., Chinyamuyamu, B., Monyo, E., Siambi, M., & Chirwa, R. (2022). Improving efficiency of knowledge and technology diffusion using community seed banks and farmer-to-farmer extension: Experiences from Malawi. *Agriculture & Food Security*, *11*(38), 1–14. DOI: 10.1186/s40066-022-00375-4

Quisumbing, A., Cole, S., Elias, M., Faas, S., Galiè, A., Malapit, H., Meinzen-Dick, R., Myers, E., Seymour, G., & Twyman, J. (2023). Measuring Women's Empowerment in Agriculture: Innovations and evidence. *Global Food Security*, *38*, 1–15. DOI: 10.1016/j.gfs.2023.100707 PMID: 37752898

Santos, M., Moreira, H., Cabral, J. A., Gabriel, R., Teixeira, A., Bastos, R., & Aires, A. (2022). Contribution of Home Gardens to Sustainable Development: Perspectives from A Supported Opinion Essay. *International Journal of Environmental Research and Public Health, 19*(20), 1–26. DOI: 10.3390/ijerph192013715 PMID: 36294295

Sanyé-Mengual, E., Gasperi, D., Michelon, N., Orsini, F., Ponchia, G., & Gianquinto, G. (2018). Eco-efficiency assessment and food security potential of home gardening: A case study in Padua, Italy. *Sustainability (Basel), 10*(7), 1–25. DOI: 10.3390/su10072124

Singh, T., Mukhopadhyay, R., Ravita, , Ks, S., Deep, N., Tarannum, , Chaudhary, S., Paramjita, D., Goyal, A., & Singh, G. (2025). Drone-assisted agronomy: Monitoring, spraying and precision application in the field. *International Journal of Research in Agronomy, 8*(6), 717–727. DOI: 10.33545/2618060X.2025.v8.i6i.3094

Wirjodirdjo, B., Fudhla, A. F., Singgih, M. L., & Ridho, H. (2025). Driving forces for sustainable development of smallholder sugarcane agribusiness: A system dynamics approach to circular economy strategies. *Agricultural and Resource Economics: International Scientific E-Journal, 11*(2), 32–64. DOI: 10.51599/are.2025.11.02.02

Yao, W., Liu, C., Liu, Y., Zheng, Q., Wang, J., Yu, H., Chen, C., & Guo, S. (2025). Unmanned aerial vehicle payload technology applications in agriculture and other low-altitude scenarios: A review. *Frontiers in Plant Science, 16*, 1–27. DOI: 10.3389/fpls.2025.1721484 PMID: 41459473

Chapter 5
GeoAI-Driven Sustainable Precision Agriculture:
UAV and Computer Vision for Apple Health Monitoring With a Case Study From Lebanon

Makram Zouheir Chehayeb
http://orcid.org/0000-0002-1738-6291
Islamic University of Lebanon, Lebanon

Walid Ali Fahs
Islamic University of Lebanon, Lebanon

Mohamed Samhat
Islamic University of Lebanon, Lebanon

Ali Abed AlHussein Sleiman
http://orcid.org/0009-0008-0122-3798
Islamic University of Lebanon, Lebanon

Ghaleb Faour
National Center for Remote Sensing, CNRS-L, Lebanon

ABSTRACT

Global food systems are now under pressure from a growing population, climate change, and limited resources, necessitating improved sustainability in food production. Precision agriculture (PA) presents a promising pathway to this goal. Geospatial AI can significantly enhance PA by enabling automated, spatially-aware monitoring and decision-making. In practice, the approach used for apple health monitoring mainly relies on manual inspection, which is time-consuming, subjective, and not scalable. This chapter aims to evaluate the effectiveness of an integrated drone and AI system for automating apple health monitoring. The proposed methodology involves constructing a dataset of apple leaf images, training a computer vision model for disease detection and classification, and generating a diagnostic

DOI: 10.4018/979-8-3373-8037-7.ch005

report of detected diseases. The system demonstrates promising accuracy in detecting common apple diseases, significantly streamlining the monitoring process. This approach shows potential for improving orchard management, thereby contributing to agricultural sustainability and food security.

INTRODUCTION AND PROBLEM STATEMENT

On the international scale, the global population has increased significantly, and it is expected to reach the limit of 10 billion by 2050 (the United Nations Department of Economic and Social Affairs “UNDESA”, 2024). The total global food demand is expected to increase by 35% to 56% between 2010 and 2050, while the population that is at risk of hunger is expected to change by −91% to +8% over the same period. This situation comes up with what is commonly referred to as the ‘yield gap’ which is the difference between the maximum possible yield (what could be grown under ideal conditions) and the actual yield farmers achieve.

Taking into consideration that, in many regions globally, we have reached ‘peak farmland’, so the only remaining way to close the gap would be through ‘sustainable intensification’ i.e. using the same amount of land to grow more.

In addition, atmosphere change, shown through increasing temperatures, warmer winters, drought and change in rainfall patterns, is raising biotic stresses, such as pests and diseases. At the same time, increased climate variability exacerbates disease outbreaks and makes them less predictable. When climate change is considered, the ranges mentioned above change slightly (+30% to +62% for total food demand and −91% to +30% for population at risk of hunger) (Van Dijk et al. 2021).

Also, agriculture is currently facing significant problems with the resources. On one hand, essential resources such as land and water are not only limited but also increasingly reduced by phenomena like soil erosion, salinization, aquifer depletion, and pollution from competing uses. On the other hand, the production process is getting more expensive, driven by the unstable prices of key inputs, such as synthetic fertilizers, pesticides, and diesel fuel, due to geopolitical tensions, supply chain disruptions, and energy market instability.

All these factors combined threaten food security worldwide, and calls for a serious move to find new solutions that ensure the sustainability of food production. In this perspective, the United Nations Sustainable Development Goals (SDGs) and especially the second target of SDG 12 aims to “achieve the sustainable management and efficient use of natural resources” (Chan et al. 2018). This target calls for moving agriculture from a model of extractive resource use to one of regenerative resource management. The goal is to produce more food with less resource input and less environmental degradation.

Many initiatives and techniques are being developed worldwide, as a result of these strains. The agricultural field is noticing a digital transformation process characterized by satellite imagery, IoT soil sensors, drone-based scouting, and AI-powered decision support systems. This model, often known as “Agriculture 4.0”, aims to improve every step of production, transforming farming from a standard practice to a data-driven, site-specific management system. This digital layer is essential for achieving the 'efficient use' required by SDG 12.2, allowing data to be applied as needed. Precision Agriculture is a response to these strains as well, it is not experimental anymore, and it is being used in developed and developing regions. Precision agriculture technologies aren’t used consistently because they can be expensive and complex. Big farms often implement entire systems, while smaller farms usually pick and choose specific tools, like GPS guidance or soil mapping, based on their needs. This field is limited not only by large upfront cost, and lack of technical expertise, but also by limited remote internet access, data ownership concerns, and a lack of specific solutions for diverse cropping systems. In this case, using new technologies can help address these challenges.

Some tools, like Unmanned Aerial Vehicles (UAVs), make collecting data easier. AI can look at data and help predict when disease outbreaks might happen. It becomes even more powerful when combined with location-based insights, which is how the new field of Geospatial AI (GeoAI) has developed.

Apples are one of the most widely grown and economically valuable fruit crops in the world. Their long production cycles make early disease detection critical. In addition, orchards are spatially complex because of tree-level variability, so errors in orchard management grow across multiple seasons. In addition, diseases like apple scab and black rot happen all over the world, reducing yield, fruit quality, and the potential for exports. The traditional reliance on chemical control is costly and environmentally damaging. While early detection is critical, it is difficult to achieve at scale because visual inspection requires a lot of work and is subjective. Satellite imagery lacks the detail level for leaf-level analysis in most operational contexts, and ground-based sensors cannot directly capture visual disease symptoms. Therefore, the development of location-aware, UAV-based, and AI-driven monitoring systems has become a necessity. These challenges are strongly felt in regions like Lebanon, where apple planting is a major agricultural sector and an important cultural element, with thousands of hectares dedicated to orchards producing thousands of tons of fruit annually. However, this vital industry is threatened by diseases, such as apple scab, which reduce crop yield and quality. These losses directly impact apple exports and pose a substantial risk to the economic livelihood of local communities.

Within this context, this chapter addresses the following research question: How can a UAV-based system, integrated with computer vision and AI, contribute to sustainable agricultural management and support early disease detection in apples. The primary objective is to design and test a drone-based GeoAI framework capable of automatically detecting apple leaf diseases through computer vision and providing actionable reports via a web-based GIS platform.

This chapter contributes to the field of precision agriculture by presenting a modular GeoAI framework specifically designed for automated apple health monitoring. A central technical contribution is the development of a two-stage computer vision pipeline optimized for complex orchard environments:

1. Stage 1 (Detection): Employs a YOLOv8m detection architecture to accurately isolate apple leaves and distinguish them from other vegetation or environmental elements.
2. Stage 2 (Classification): Utilizes a YOLOv8 classification architecture to analyze the isolated leaves and categorize them into four distinct states: healthy, scab, black rot, or other.

Finally, this work introduces an efficient web-based decision support system. This platform integrates the vision pipeline with geospatial data to generate actionable diagnostic reports, facilitating data-driven orchard management.

CONCEPTUAL FOUNDATIONS

Precision Agriculture: Evolution, Paradigms, and Limitations

In traditional farming, agricultural fields were treated as 'homogenous units', meaning that all parts of a field were treated uniformly. This approach led to inefficiency, such as over-fertilizing healthy areas while under-treating struggling ones. However, agricultural lands are not uniform; spatial variability causes soil nutrients, moisture, and pest pressure to vary meter-by-meter. To draw an analogy with imagery, agricultural fields were once managed like low spatial resolution satellite images, but now, thanks to new technology, they can be managed like high spatial resolution images, where each small pixel is treated according to its specific characteristics. This shift was made possible by Global Navigation Satellite Systems (GNSS), which allows machinery to know its precise location within a field, thereby enabling the application of spatial heterogeneity rather than the traditional averaging approach.

This paradigmatic shift is referred to as 'Precision Agriculture', which enables the customization of treatments according to the needs of each small area. The main principles of this field can be summarized in three pillars: Observation, Evaluation, and Response. These pillars can also be conceived as three stages of the precision agriculture process. The observation pillar concerns in-situ sensing, the process of taking measurements at the exact location of the plant. This can be done using remote sensing via satellites or UAVs, and through proximal sensing, where sensors measure soil pH or leaf chlorophyll on-the-go. The evaluation pillar consists of Decision Support Systems (DSS), which analyze and model the data collected in the first phase to transform raw data into actionable information. This process can be carried out using Geographic Information System (GIS) tools by analyzing and combining several layers of data. Traditionally, this stage required a human expert (agronomist) to interpret the maps. The response phase is the action phase, where the analysis results are used to deliver a precisely targeted solution, known as Variable Rate Technology (VRT). VRT-enabled machinery adjusts the application of inputs in real-time based on the evaluation map. For example a smart sprayer can be set up to turn on its nozzle when it is right over a specific tree that the evaluation map says is diseased. This means it does not spray the row of trees just the one that needs it.

This approach, despite being powerful, has some limitations that affect its overall application. First, the technologies necessary to implement this approach are quite expensive, requiring a considerable budget for proper deployment. Second, analyzing the data that is generated can be tough. This is because a large amount of data does not always give answers. This is often called the "Rich Data, Poor Information" problem. Third, combining types of data like GPS logs, images from satellites and soil test results can be technically difficult. These problems show that we need ways to collect and analyze data.

The Convergence of GIS and AI: GeoAI Theory

In the century there was a big cholera outbreak in the United Kingdom. People then thought cholera was caused by air. In 1854 Dr. John Snow, a doctor looked into the cholera outbreak on Broad Street, in London. He started mapping where each person who died of cholera lived.

This mapping process revealed a distinct spatial cluster centered around a specific water pump. After noticing this pattern, some water tests were done. The results showed that the water had cholera in it. This finding changed what people thought about the disease. It turned out that cholera was spread through the system, not the respiratory system. It also demonstrated that location was the key to identifying the source of a biological threat.

This experience is not an isolated observation made by coincidence; these relationships are formalized by the principle of "Spatial Autocorrelation." This principle was later simplified into what's known as the first law of geography. It was proposed by Waldo Tobler in 1970. Most people quote it like this: "Everything is related to everything else, but near things are more related than distant things." (Tobler, 1970) This law helps us make maps. It also helps us fill in gaps in data like guessing rainfall between weather stations. The law is used in statistics too. It gives us a reason to use maps and spatial statistics. The law is very useful. It also underpins regionalization, the process of delimiting boundaries around areas with similar characteristics.

These ideas are used in modern spatial modeling methods like Geographically Weighted Regression (GWR). What's different here is that it doesn't assume relationships between variables are the same everywhere. Instead, they can change depending on the location. This matters a lot in environmental and agricultural studies, since conditions are rarely identical from one place to another. For example, Al-Shaar et al. (2022) show how this type of method can handle complex relationships between environmental factors and give predictions that better fit each area. In the end it just shows how important location is in modeling. Location is really important when we are trying to guess what will happen next. That is why GeoAI systems focus on capturing these local differences in location. The law is not about understanding data. It is also very important for models like kriging in geostatistics. For example, kriging in geostatistics uses points to predict values at unknown locations. So location plays a role in predictive models like kriging in geostatistics where values at unknown locations are predicted based on nearby points.

Furthermore, it informs models of interaction,such as trade, migration, or communication,where flow between two places is inversely related to the distance between them. This idea is used in many areas, like real estate, studying diseases, and cultural research. It's useful, but some people point out problems, it's not clear what 'near' really means, it depends on scale, and today's technology can connect places that are far apart.

A classical Geographic Information System (GIS) can be described as a system of record,a largely static system focused on storing geographic data and displaying it through thematic layers (such as water, soil, or vegetation).

Although capable of performing analytical geometric operations like buffering and overlay, traditional GIS often lacks the ability to "learn" from data. While it can effectively show where things are, it does not necessarily explain what they mean or predict what will happen next. This is not to say that GIS entirely omits predictive capabilities; some analytical tools within GIS can generate forecasts, such as land cover predictions based on historical data. However, these predictions have typically relied on statistical models that do not fully capture spatial dimension and complexity.

For this reason, there emerged a need for an intelligent system capable of accounting for spatial complexity. In this context, Geospatial Artificial Intelligence or GeoAI is a type of intelligence system. It combines data like coordinates and topology with advanced Artificial Intelligence algorithms. These algorithms include learning and data mining. The goal is to find knowledge. GeoAI is different from systems. It is dynamic. Learn from data. It also considers dimensions. This allows it to move from tasks like mapping a location. It can do complex tasks like predicting the spread of something based on how things move. GeoAI also helps interpreting results. This means it can do more on its own and needs human input. The process is often called reducing the need for a human to be in the loop. GeoAI makes it possible to get insights without needing human involvement. Instead of relying solely on an agronomist to interpret data, GeoAI enables automated analysis.

GeoAI in Agriculture: The "Spatial Intelligence" Bridge

In agriculture, Geospatial Artificial Intelligence can play an important role. First, GeoAI can be applied to yield forecasting by comparing high-resolution drone imagery from the blossom season to the fruit-set season, allowing models to analyze spatial-temporal trends. Such approximations are essential for planning and logistics, such as securing export contracts and organizing cold-storage logistics in advance. Second, GeoAI can automatically detect, count, and measure the canopy diameter of individual trees, as well as recognize "missing trees" or "dead zones" where a tree has died and not been replaced, enabling precise replanting schedules. Third, by integrating data from multiple sources, GeoAI can support soil evaluation and risk identification, for instance, evaluate erosion risk by combining Digital Elevation Models (DEM) with land cover data.

When focusing on crop diseases, it is important to consider that, unlike mechanical issues, biological diseases in an orchard spread through contagion. For example, a single apple tree infected with scab acts as a source of spores. To show how the "Geo" affects the "Bio," we need to look at key environmental factors. Wind direction can carry spores along certain paths, spreading infection in patterns rather than randomly across an orchard. Slope and drainage matter too—low areas may hold more moisture or have less airflow, creating small climates that help fungi grow.

Orchards are also complex because individual trees vary. Some trees may be more vulnerable due to their age or closeness to water. The buffer effect is important as well: a healthy tree surrounded by infected trees is much more at risk than one far from the infection.

One big advantage of Geospatial Artificial Intelligence is its ability to use context. Unlike traditional computer vision, which treats each image on its own, GeoAI considers the surrounding geospatial information to make smarter predictions. Furthermore, integrating the geospatial dimension can increase identification accuracy. For instance, if a YOLO model detects an unclear brown spot on a leaf and assigns it a 50% confidence score, geospatial data indicating that the five surrounding trees are confirmed to have "Black Rot" can increase the probability that the unclear spot is also Black Rot.So Geospatial data improve decision-making by enabling risk evaluation based on spatial relationships between diseased and healthy trees. For example, even if Tree X is not yet infected, its closeness to a disease cluster 10 meters away may justify protective treatment now.

Geospatial artificial intelligence enables actionable mapping that moves beyond mere classification to deliver operational intelligence. While image classification can identify leaves with scab, geospatial intelligence goes further by determining where the problem is located, estimating its extent, and even predicting the direction of its spread.

Unlike conventional image classification, which evaluates each frame independently, GeoAI integrates spatially referenced outputs into a unified analytical layer. Rather than producing an itemized list of individual detections, GeoAI aggregates results into spatial density surfaces, enabling disease intensity to be visualized as a continuous field rather than a discrete catalogue of occurrences. This representation is operationally more actionable: a georeferenced hotspot map delineating, for example, a two-hectare zone of elevated scab incidence provides a directly implementable basis for targeted pesticide application, whereas a list of GPS coordinates without spatial context offers limited decision support. Furthermore, multi-temporal comparison of disease distribution maps enables the monitoring of spread velocity and directionality over successive survey periods, a capability that non-spatial AI systems cannot replicate.

Unmanned Aerial Vehicles (UAVs) in Precision Agriculture

Unmanned aerial vehicles (UAVs) are kind of a middle solution between satellite images and checking plants by hand. Satellites can cover large areas, but even the high-quality ones usually show images at about 30 cm to 1 meter per pixel. That's fine for seeing a whole forest, but not detailed enough to capture something very small, like a small 2 mm spot on a leaf. Also, satellites are often blocked by cloud cover, whereas UAVs can fly between the clouds, making sure data is collected precisely where needed. On the other hand, human inspection,whether by eye or handheld camera,though high-resolution, is spatially disconnected; the observer sees only what is immediately in front of them, missing the top of the canopy or broader patterns across, for example, 50 hectares.

In this context, UAVs flying at low altitudes (10–30 m) can achieve sub-centimeter resolution, which is essential for computer-vision models to discern fine textures and small lesions.

Unlike handheld cameras, UAVs are equipped with GNSS and IMU sensors that tag each photograph with precise coordinates and orientation angles,data that serve as the essential "fuel" for GeoAI.

Recently, unmanned aerial vehicles have gained significant temporal flexibility. Previously, a UAV capable of such work cost tens of thousands of dollars, but today drones equipped with high-quality 4K cameras are accessible even to regional agricultural cooperatives. This affordability also enables daily flights if needed, greatly enhancing the temporal resolution of monitoring. In disease management, timing is critical. For instance, after a rain event, a farmer can deploy a drone within two hours to assess conditions that may trigger apple scab germination.

Several additional features are now making drones an even more compelling option, as they are no longer merely "flying cameras." Increased onboard processing power is driving the shift toward Edge AI, enabling detection tasks (such as YOLO-based models) to be performed directly on the drone or a local field laptop, bypassing the delays of cloud uploads. Furthermore, improvements in flight time and autonomous path-planning software allow for "set-and-forget" mapping of entire orchards, eliminating the need for specialized piloting skills.

RELATED WORK AND LITERATURE REVIEW

UAVs and Remote Sensing for Crop Health

Remote sensing has established itself as an essential, non-invasive technology for modern agriculture, enabling the rapid acquisition and analysis of the spectral properties of the Earth's surface from various distances. According to Gogoi et al. (2018), the science behind this technique relies on three distinguished spectral domains: the visible domain, where pigments like chlorophyll control light absorption; the near-infrared (NIR) domain, which is influenced by internal leaf scattering; and the shortwave-infrared (SWIR) domain, determined primarily by foliar moisture. When a crop is under stress or diseased, these biophysical and biochemical attributes are modified, leading to detectable changes in reflectance, such as the breakdown of internal leaf structures or reductions in chlorophyll activity. To effectively spatialize these diagnostic results and render agriculture more sustainable, the choice of imaging platform is critical for ensuring that disease detection occurs early enough to allow for site-specific management.

Traditional satellite remote sensing, utilizing platforms such as Landsat and Sentinel-2, provides broad and regular temporal coverage that is invaluable for monitoring agricultural production at a regional or governmental scale. However, as noted by Hunt and Daughtry (2018), these platforms are significantly constrained by "coarse pixel resolutions," typically ranging from 10 to 30 meters. Such large pixel sizes result in "mixed pixels," where the spectral signatures of leaves are blended with those of bare soil, residue, and shadows, making individual tree or leaf-level diagnosis impossible.

Furthermore, while commercial high-resolution satellites offer improved clarity, they are often hindered by infrequent coverage and slow delivery of information, meaning they are not enough for the real-time requirements of disease detection and precision management. Consequently, satellite imagery often fails to provide the site-specific detail necessary to identify pathogens before they spread through a field.

Manned aircraft offer a significantly better spatial resolution than satellites and have been utilized for agricultural surveillance for over six decades. Despite this, their utility for modern precision agriculture is constrained by high operational costs and significant scheduling difficulties. Because aircraft operations require specialized pilots and carry heavier payloads, it is often unfeasible to perform the frequent, multiple flights necessary to capture dynamic disease progression. Furthermore, manned aircraft are generally unable to fly at the low altitudes (under 50m) required for the detailed identification of specific plant diseases. As discussed by Garcia-Ruiz et al. (2013), aircraft-based sensors typically yield resolutions around 0.5m per pixel, which may still be too coarse to detect the subtle, pre-symptomatic spectral changes required for early pathological intervention.

Unmanned Aerial Vehicles (UAVs), or drones, have revolutionized remote sensing by effectively bridging the "sensing gap" left by satellite and manned platforms. These systems provide ultra-high spatial resolution, often achieving sub-centimeter to a few-centimeter Ground Sampling Distance (GSD), which allows for the discrimination of individual objects and textures using Object-Based Image Analysis (OBIA). UAVs offer unparalleled high temporal flexibility, as they can be deployed "on-demand" by farmers to capture data during critical growth stages or immediately following weather events. This high-resolution capability significantly improves diagnostic accuracy; for instance, Garcia-Ruiz et al. (2013) found that UAV-based sensors achieved classification accuracies for citrus diseases of up to 85%, compared to only 74% for aircraft-based sensors. By providing quick delivery of information at a lower cost, UAVs enable farmers to implement site-specific disease control and avoid the expensive, over-application of pesticides.

A major limitation of satellite-based remote sensing is that it is "frequently hindered by cloud cover," which can result in long periods where no reliable data is acquired. Furthermore, even in seemingly clear skies, second-to-second variations in atmospheric transmittance due to invisible clouds can degrade image quality and lead to inaccurate reflectance data. Because UAVs operate "under the clouds," they bypass the majority of these atmospheric obstacles, ensuring data continuity during critical disease germination windows when monitoring is most vital. Flying at low altitudes,typically between 60m and 150m,minimizes the atmospheric path length between the sensor and the crop. This proximity allows for more accurate sensor calibration, often utilizing up-looking incident-light sensors to correct for lighting variations in real-time, thereby ensuring that the resulting maps accurately reflect the health of the vegetation rather than atmospheric noise.

Sensor Technologies for Crop Health Monitoring

Non-destructive data acquisition through imaging sensors constitutes the foundation of remote crop health monitoring in precision agriculture. These sensors detect disease-related changes by measuring the spectral reflectance properties of plant tissue. Sensor selection involves a fundamental trade-off between diagnostic accuracy, acquisition cost, and data processing complexity.

RGB Sensors

RGB sensors capture imagery in the red, green, and blue spectral bands, providing a visible-spectrum representation directly analogous to human visual perception. Their low cost, operational simplicity, and compatibility with deep learning architectures such as YOLO make them well suited for automated disease detection workflows. Rouša et al. (2023) demonstrated that YOLOv5 trained on RGB imagery achieved competitive or superior performance compared to multispectral sensors for apple scab detection, an outcome attributable in part to the relative robustness of RGB sensors under variable outdoor illumination.

Multispectral Sensors

Multispectral sensors acquire data across a discrete set of spectral bands, typically between three and ten, commonly including the near-infrared region. This spectral range enables the computation of vegetation indices such as the Normalized Difference Vegetation Index (NDVI) and facilitates the detection of chlorophyll-related stress signatures that are invisible to RGB sensors. However, their relatively coarser spatial resolution can limit the discrimination of early-stage, fine-scale disease symptoms. Additionally, radiometric imbalances across spectral channels under high irradiance conditions can introduce noise that complicates downstream AI model inference.

Hyperspectral Sensors

Hyperspectral sensors acquire continuous spectral data across hundreds of narrow contiguous bands, enabling the detection of subtle biochemical changes in plant tissue that precede visible symptom expression. When combined with three-dimensional deep learning architectures capable of exploiting both spectral and spatial dimensions simultaneously, these sensors offer high sensitivity for early disease identification. Nagasubramanian et al. (2019) demonstrated near-perfect detection of soybean disease using hyperspectral imaging with explainable 3D deep learning models. Their primary constraints are high acquisition cost, large data volumes, and the substantial computational and specialist expertise required for processing and analysis.

Drone Photos and Flight Tips

The quality of UAV-acquired imagery is determined by a combination of flight parameters and sensor configuration. Key variables include flight altitude, image overlap ratio, illumination conditions, and platform stability, all of which directly influence the suitability of the resulting data for AI-based disease analysis.

Flying Height

Flying low, about 25–40 meters, is preferred if a lot of details are needed. Flying too high, over 100 meters, makes it hard to see small details like leaves or twigs. Medium heights, 60–90 meters, are okay for general crops but might miss early disease signs.

Overlap Ratio

When making 3D models, your photos need to overlap a lot. Around 80–90% overlap works best because the software can match the same spot from different angles, making results more accurate. You can sometimes reduce the side overlap to about 70% if the front overlap stays high, and it usually still works fine.

Lighting and Weather Conditions

Environmental factors like sunlight and wind have a big impact on the accuracy of AI analysis and spectral measurements. Most studies say to fly drones around noon. This helps avoid shadows that confuse AI models. Too much direct sunlight causes glare. This is especially true for leaves like those, on apple trees. In such cases, a cloudy day with soft light can actually give more balanced light across the field. Also, Yang et al. (2026) point out that environmental turbulence can affect the drone's stability and can lower image quality because moving leaves and branches create noise and make it harder to align overlapping photos.

Motion Blur

Motion blur is a major challenge when processing UAV data. It happens when the camera moves too much while taking a photo. Sieberth et al. (2016) explains that blur from drone movement, wind, or sudden operator actions can reduce the performance of detection models like YOLOv8 by softening the edges that these models need to detect. To prevent this, drone speed should be balanced with the camera's shutter speed. Typical agricultural flights move at 4–8 m/s with fast shutter speeds (1/500 s to 1/1000 s) so that the drone doesn't move too far during exposure. If the motion is too much, small details, like tiny disease spots on leaves, can become impossible to see.

Orchard-Specific Disease Monitoring

Orchards are really tricky to deal with because they are more complicated than things like wheat or corn. This makes it harder to keep an eye on them. With wheat or corn it is easy to use cameras from above and control the lighting.. Apple trees are different. We need to use methods to monitor apple trees like using drones to get a view from above or putting cameras on the ground to get a view, from the side of the apple trees (Rouša et al. 2023).

The problem is made worse by natural conditions—changing light, leaves blocking each other, and shadows make automatic detection tricky (Rouša et al. 2023). Orchards also have more varied backgrounds, with soil, branches, and shadows that can confuse disease detection (Chao et al. 2020). Drones are especially useful because they can reach places satellites or regular equipment can't, like hills or dense trees (Kouadio et al. 2023).

Researchers have focused on major orchard diseases, especially Apple Scab (Venturia inaequalis) and rot-related problems. Apple Scab is a big concern because it damages leaves and fruit, lowering both yield and quality (Rouša et al. 2023). Studies testing deep learning on regular (RGB) and multispectral images found that RGB often works better, since multispectral sensors struggle in bright orchard light (Rouša et al. 2023). Other diseases, like Black Rot, Cedar Rust, and Alternaria leaf spot, are also being studied (Kouadio et al. 2023). Deep learning models can identify these diseases accurately if trained to recognize the visual signs, like dark spots or circular patterns (Chao et al. 2020).

A lot of research uses small datasets, which makes models less reliable in real orchards (Rouša et al. 2023). Larger datasets help a lot. Some studies have gathered thousands of images from labs and orchards at different times and lighting (Chao et al. 2020). These datasets can grow to around 25,000 images by using tricks like rotating images or flipping images. Big datasets are really important for training models so that the models work well in the real-world conditions of commercial orchards. The datasets are key for training models so that the models work well in orchards as we see in the work of Chao and others, from the year 2020.

AI and Deep Learning Architectures

The move from manual scouting to automated plant health monitoring came from fast progress in deep learning models made for image recognition. In the early "CNN Era," researchers stopped using older machine vision methods, which depended on manually designing features like texture and color, and started using Convolutional Neural Networks (CNNs) that learn features directly from raw images. Models like AlexNet (Krizhevsky et al. 2012) played an important role by introducing GPU use and ReLU activation, while VGGNet (Simonyan and Zisserman, 2014) showed that going deeper with smaller filters could improve how complex leaf patterns are classified. The release of ResNet (He et al. 2016) was a big step, as its residual blocks solved the vanishing gradient problem, allowing much deeper networks to capture small details of plant diseases. At first, these models were often used as feature extractors with classifiers like Support Vector Machines (SVM), but later they became full end-to-end systems used widely today (Lu et al. 2021).

As agricultural work moved from labs to fields research started focusing on real time object detection led by the You Only Look Once models (YOLO).

The YOLO models are really good at this. Unlike methods that look at parts of a picture the YOLO models look at the whole picture at once and predict boxes around objects and what they probably are in one step. The YOLO model has gotten better over time.

Now the YOLO version 8 model is seen as a choice for use because it is fast, it works well and it is good on different devices like computers or phones. According to Ali and Zhang the YOLO version 8 model is liked because it is good at finding things and it is also accurate especially when there are many diseases or pests in one picture. The YOLO version 8 model does not need a lot of computation like versions did. This makes the YOLO version 8 model for use on small devices, like mobile phones or drones which need to give answers quickly even when the internet is not strong.

Research also separates one stage and two stage detection methods, which affects how well an agricultural monitoring system performs. Two stage models like Faster R-CNN first find possible regions using a Region Proposal Network (RPN) and then classify them; they are accurate but usually too slow for real time use. On the other hand, one stage models like YOLO and SSD are faster because they detect and classify objects at the same time. Still, some researchers prefer a pipeline approach, where one model finds the leaf or plant area and another model classifies the disease (Liu and Wang, 2021). This method can work better in complex orchard scenes where background elements like soil or branches may cause mistakes. While one stage models are quicker, the pipeline approach helps the classifier focus only on the important region, improving accuracy but slightly reducing speed.

GeoAI and Spatial Decision Support Platforms

One of the main challenges in modern precision agriculture is turning AI-generated insights into something that can be mapped and used on the ground. This requires a strong connection between machine learning results and Geographic Information Systems (GIS). While AI is very good at handling large amounts of data and finding patterns, research that links these results to practical maps has been limited. Most studies have focused more on classifying individual pixels rather than helping with real land management (Espinel et al. 2024). Recently there have been some developments in Geospatial Artificial Intelligence also known as GeoAI. This is a deal because GeoAI is starting to help us with some important issues.By putting Convolutional Neural Networks into GIS platforms researchers can now make maps that show what they think will happen with crop yield and how we will use resources. This is really helpful because it is more practical and useful. For example some people found this out when they did a study in 2023 like Zhou and his team.

For instance systems like MicroLEIS show us how powerful it can be to combine models and expert systems with GIS. This helps us make decisions about how to use land in a way that is sustainable. It does this by looking at things like the environment and biology of areas like what some researchers did in 2010, such as De la Rosa and Anaya-Romero.

This is important because it helps us turn a lot of data into maps that show us things like what the soil's like, how the land is being used and what crops are growing at different times of the year like what Espinel and his team did in 2024.

To make sure that these complicated spatial models are useful to everyone there is a move to show the results in a way that is easy to use, such as, with interactive WebGIS dashboards and mobile applications. This way Geospatial Artificial Intelligence or GeoAI can really help people.

Traditional methods of data delivery often suffered from a "technical gap" where sophisticated analytics remained inaccessible to small-scale farmers due to high data costs and complexity (Mafenya, 2024). Smartphone accessibility has recently narrowed this divide, facilitating the development of user-centric platforms that offer real-time, location-specific insights (Panda et al. 2025). Researchers have demonstrated the efficacy of these tools through real-time dashboards that allow farmers to visualize IoT sensor data (parameter factors such as temperature, humidity, soil moisture, and light intensity) in order to make informed decisions (Athesan and Ghazi, 2024). Other digital solutions, such as the Macho Sauti platform, further democratize this knowledge by allowing small-scale farmers to connect with researchers to share simple agroecological practices (Petraki et al. 2025).

The transition from raw AI data to "actionable advice" relies on sophisticated decision support logic, often derived from expert systems and iterative consensus methods. Rather than merely presenting raw indices, modern platforms like the Site-Specific Fodder Management Decision Support System (SSFM-DSS) utilize the Delphi Scaled Value Determination process to weigh multi-criteria environmental parameters,such as soil pH, drainage, and climate extremes,to generate suitability scores (Panda et al. 2025). The dashboard can alert farmers when certain aspects of their land need attention to grow crops better (Panda et al. 2025). Tools like RASKAZ look at soil data and give farmers simple, practical advice, like how to fix nutrient shortages or improve the soil (Tikhonova and Buzylev, 2024). By combining the farmer's own observations with outside data and expert tips, these tools suggest different ways to manage the land, while still letting the farmer make the final decisions (Petraki et al. 2025).

METHODOLOGY

Architecture Overview

The system is made up of three parts: the data acquisition part, which has the UAV and GPS, the processing part built using Node.js and Python/Flask and the interaction part made with the Flutter web platform. The drone is like a camera that can move around. It has a high-definition camera to record video and a separate GPS device to keep track of where it's when it records the video. The backend is, like a team: Node.js takes care of the routing and user accounts and a Python/Flask server does the work of computing the AI things. Detected disease types and frequencies are processed to generate structured diagnostic reports, including treatment recommendations cross-referenced against an internal database of pesticides approved for use in Lebanon. The system also includes a geospatial element by mapping disease detections using timestamped GPS data, producing a "Geographical Report."

Data Flow Explanation

Video from the drone enters the system in two ways: a live RTMP (Real-Time Messaging Protocol) stream from the drone's controller, or by manually uploading recorded videos through the web interface. The second method was added to allow testing and to ensure the system works even when connectivity is unstable.

Disease detection is done in two AI stages. In the first stage (detection), each video frame is analyzed by a YOLOv8m model to find and separate apple-leaf areas from the background. This step filters the frames so that only those containing apple leaves are used. The selected leaf images then go through another YOLOv8 model to figure out what disease they have. This YOLOv8 model looks at the leaf images to classify the diseases into four groups: Healthy Scab, Black Rot or Other. To make sure the results are correct we use a 50% confidence threshold for the YOLOv8 model. This means that if the YOLOv8 model is not at least 50% sure we ignore those results to make sure the report is reliable, for the YOLOv8 model.

For live processing, the system gets GPS location data through an API connected to the cloud service of the GPS device.

It aligns each detection's timestamp in the video with the nearest GPS timestamp to determine precise field coordinates. After analysis, the post-processing results, disease types and frequencies, are processed to generate treatment recommendations cross-referenced against an internal database of Lebanese-approved pesticides. Finally, processed video frames, now annotated with bounding boxes and labels, are returned in real time to the graphical user interface, allowing users to observe detections as they occur.

To support edge-computing capabilities, the system leverages NVIDIA RTX 4060 GPU acceleration (CUDA), ensuring the hardware can handle the intensive parallel processing demands of real-time video feeds.

Figure 1. System architecture overview

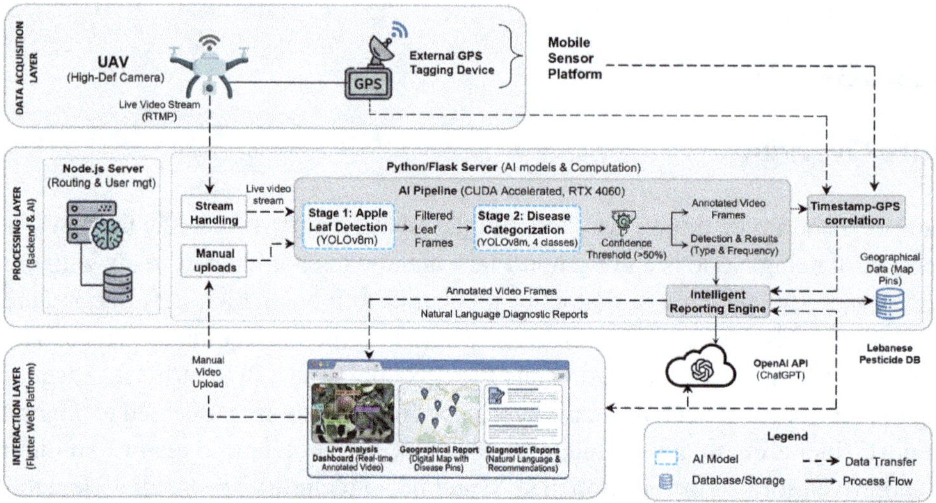

Component Responsibilities

The components of the system and their responsibilities are summarized in the following table (Table 1):

Table 1. System components and their responsibilities

Component	Responsibility
UAV & GPS Unit	The UAV is responsible for high-resolution visual data capture and the GPS unit ensures the generation of a continuous log of geographical coordinates with precise timestamps.
Node.js (Backend Orchestrator)	Manages server-side communication, data routing, user account verification, and the storage of reports in the SQL database.
Python/Flask (AI Engine)	Specifically dedicated to managing the machine learning inference pipeline and the execution of YOLOv8 scripts.
Flutter (Frontend Interface)	Provides a cross-platform Graphical User Interface (GUI) compatible with mobile and desktop, allowing users to initiate analysis, manage fields, and visualize disease hotspots.
SQL Database	Handles structured data storage across interconnected tables for Admins, Clients, Fields, and analysis Reports.
Report Generation Process	Translates detection outputs into structured diagnostic reports and treatment recommendations, cross-referenced against the Lebanese-approved pesticide database.

Data Sources and Preparation

The selection process began with identifying a target crop that offered both regional importance and technical feasibility. Apples were selected as the main focus due to their high economic importance in Lebanon, their sensitivity to different pathogens, and the high availability of established image datasets.

Our initial data acquisition involved collecting approximately 25,000 images from various public repositories, primarily "Kaggle". We then performed a precise data-cleaning phase to eliminate noise, removing blurred images and samples where the leaf features were not clearly distinguishable. Following this, the dataset was organized into a hierarchical structure where images were first sorted into "Apple Leaves" and "Other" for non-apple species. Within the apple group, we established four specific categories: Healthy, Scab, Black Rot, and Other for diseases that did not fit the primary focus.

To ensure model stability, we maintained a balanced distribution of images across the disease classes, though we intentionally maintained a higher number of healthy leaf samples to more accurately reflect real-world orchard conditions. The annotation process utilized a semi-automated pipeline, applying Roboflow's AI-assisted labeling tool to generate initial bounding boxes and class assignments. To ensure data reliability, every label was manually reviewed, and the necessary enhancements were performed.

To ensure the integrity of the experimental results, the dataset was partitioned into an 80/10/10 split. The training set (80%) was utilized for model weight optimization, while the validation set (10%) served as the basis for monitoring training loss and hyperparameter tuning. Crucially, a held-out test set (10%) was strictly reserved to provide an unbiased evaluation of the model's generalization performance on entirely unseen data, thereby mitigating the risk of data leakage.

It should be noted that all training and evaluation data were sourced from publicly available repositories, primarily Kaggle, and do not reflect the specific lighting conditions, canopy densities, soil backgrounds, or disease appearance characteristics of Lebanese orchards. The reported performance figures must therefore be interpreted as indicative of model capability under controlled dataset conditions rather than as validated field accuracy.

Model Selection and Training

In the model selection process, our main objective was to balance resource utilization with high detection accuracy. YOLOv8 was chosen due to its lightweight architecture and its ability to perform high-speed, real-time leaf localization and classification to enhance performance. We implemented a two-stage processing manner. The model first performs object detection to isolate the leaves within the frame, followed by a classification step to identify the specific leaf type. By separating these tasks, we effectively filtered out background noise, allowing the model to focus mainly on the relevant biological features and significantly improving overall predictive accuracy.

Evaluation Metrics

Since the AI workflow was divided into two parts, detection and classification, the evaluation metrics are similarly grouped into Detection metrics and Classification metrics. These metrics were used to assess the performance of the detection model and are further divided into two categories: Success metrics, which quantify how well the model performed (summarized in Table 2), and Loss metrics, which quantify the model's errors (summarized in Table 3).

Detection Evaluation Metrics

Table 2. Detection evaluation metrics (success metrics)

Metric	Measure
Precision	Measures the accuracy of the bounding box detections, indicating the proportion of predicted apple leaves that were actually correct.
Recall	Measures the model's ability to find all the actual apple leaves in the frames, showing the proportion of ground-truth leaves successfully detected.
mAP50	The Mean Average Precision at an Intersection over Union (IoU) threshold of 0.5. This is the primary benchmark for the detection model's overall performance.
mAP50-95	The average mAP calculated across multiple IoU thresholds (from 0.5 to 0.95). This represents the precision and robustness of the bounding box localization.

Table 3. Detection evaluation metrics (loss metrics)

Metric	Measure
Box Loss (val/box_loss)	Evaluates how precisely the model localized the centers of the leaves and how well the predicted bounding boxes covered the actual objects.
Class Loss (val/cls_loss)	Measures the model's effectiveness in identifying the object as an "apple leaf" versus the background or other non-target objects.
DFL Loss (val/dfl_loss)	Distribution Focal Loss, used to refine the box boundaries when the edges are not clearly defined.

Classification Metrics

The classification metrics were also divided into two categories: Success metrics, which quantify how well the model performed (summarized in Table 4), and Loss metrics, which quantify the model's errors (summarized in Table 5).

Table 4. Classification evaluation metrics (success)

Metric	Measure
Top-1 Accuracy	This represents the percentage of cases where the model's highest-probability prediction exactly matched the ground-truth label (Healthy, Scab, Black Rot, or Other).
Top-5 Accuracy	This measures the percentage of cases where the correct label was among the model's top 5 predictions. Since your model only has four classes, this metric serves as a robustness check and achieved a perfect score.

Table 5. Classification evaluation metrics (loss)

Metric	Measure
Train Loss	The aggregate error value calculated on the training dataset, which the AI model minimized during its "intensive training operations".
Validation Loss (val/ loss)	The error value calculated on the 10% reserved validation set, used to ensure the model generalizes well to new, unseen leaf images.

Report Generation and Web Platform

The frontend was developed using Flutter, enabling a single cross-platform codebase compatible with both desktop and mobile devices. Server-side routing, user account management, and report storage are handled by a Node.js backend orchestrator, which interfaces with a SQL database structured around four interconnected tables for administrators, clients, fields, and analysis reports. The database schema is designed to accommodate additional crop types beyond apples with minimal structural modification. AI inference is managed by a dedicated Python/Flask server running the YOLOv8 pipeline, kept architecturally separate from the backend to allow independent scaling.

Fields function as the primary organizational unit within the platform, acting as persistent containers for all assessments, historical records, and reports associated with a given plot of land.

The system generates reports through two workflows depending on data source. In the Live Mapping workflow, disease detections from a live drone feed are aligned with concurrent GPS timestamps, producing a georeferenced map of affected areas in real time. In the Manual Upload workflow, users assign recorded footage to a specific field, enabling comparison with past and future assessments. In both workflows, the system processes detection outputs to generate structured reports summarizing identified disease types, their frequency, and their spatial distribution. Treatment recommendations are included in each report, derived by cross-referencing the detected diseases against an internal database of pesticides approved for use in Lebanon. These reports are intended as decision-support outputs and should be reviewed by a qualified agronomist before any field intervention.

RESULTS

Detection Performance

By the final training epoch (100), the model achieved very robust performance, achieving a precision of 98.7% and a recall of 97.88%. Figure 2 clearly shows these results, highlighting the model's almost perfect balance between false positives and false negatives.

Figure 2. F1-confidence curve and precision-confidence curve for the apple leaf detection phase

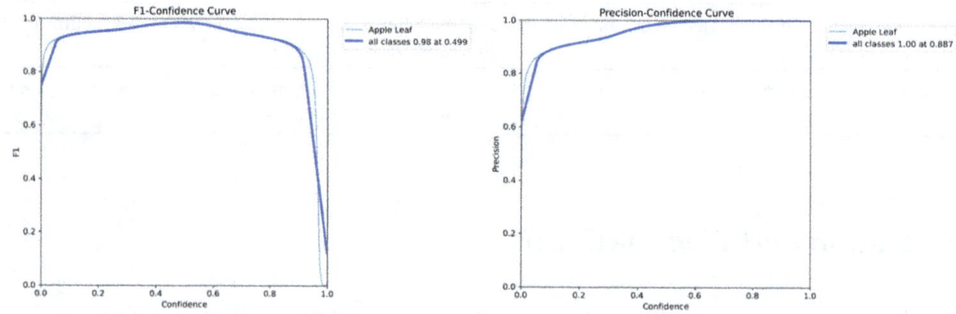

These figures demonstrate that the detection model achieves both high precision in distinguishing apple leaves from background elements and high recall in identifying the full population of leaves present within each frame. At the same time, it effectively distinguishes leaves from the background. This ability to separate leaves from other elements is further confirmed by the confusion matrix in Figure 3, which shows very little overlap between the leaf and background classes.

Figure 3. Normalized confusion matrix for the detection phase

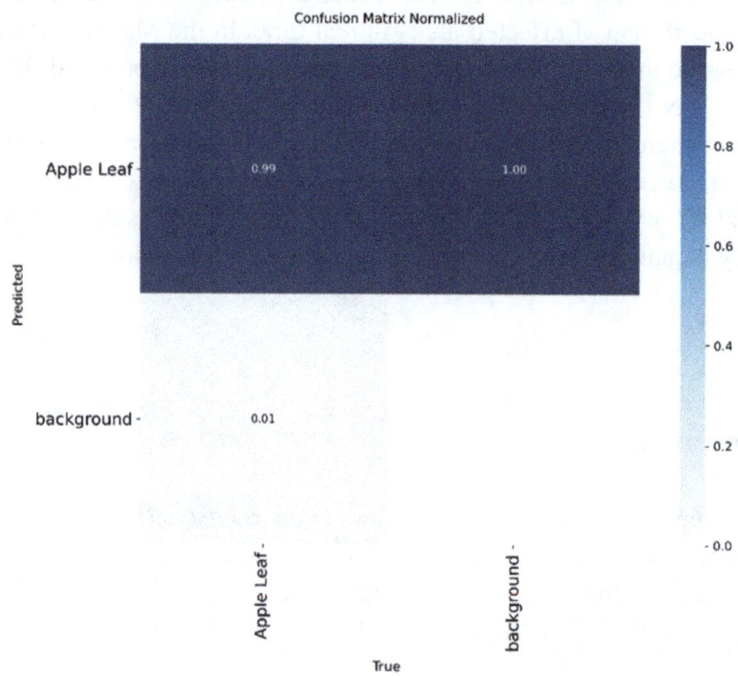

The mAP50-95, which evaluates bounding box localization precision across multiple IoU thresholds, reached 96.35%. This shows that the YOLOv8m model can accurately pick out leaf edges, even in tricky natural backgrounds. The mAP50 value, the standard detection benchmark, finished at 99.34%, indicating the model has nearly mastered the visual signature of apple leaves.

During training, the model exhibited notable "spatial intelligence" from an early stage; by the first epoch it had already attained an mAP50 of 97.40%. However, the subsequent 99 epochs were essential for refining bounding-box accuracy, as evidenced by the rise in mAP50-95 from 80.62% to 96.35%. The training process effectively minimized localization errors, with Box Loss declining substantially from an initial 0.6939 to a final 0.1690,a reduction of approximately 75%.

All the results are summarized in the following figure (Figure 4):

Figure 4. Model performance metrics across training and validation datasets

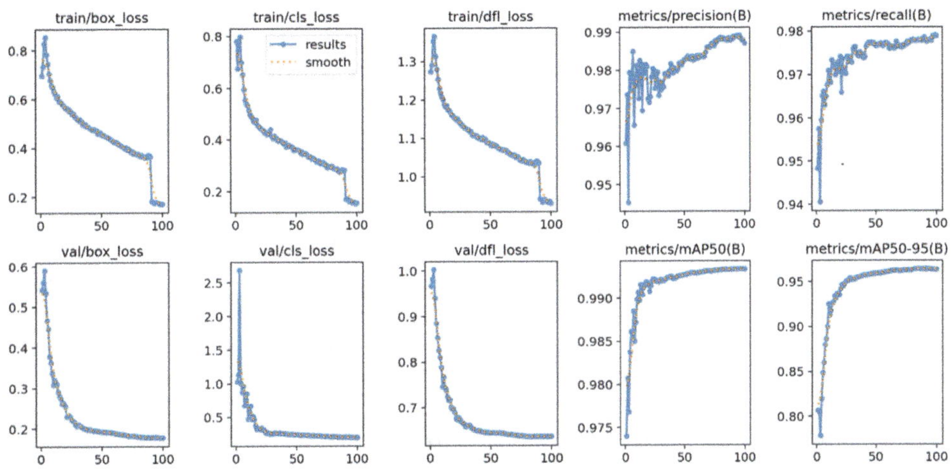

For a qualitative comparison of this accuracy, Figure 5 below represents a comparison between ground truth (val_batch0_labels, on the left) and model prediction (val_batch0_pred, on the right) which show virtually identical bounding box placements, confirming the model's high localization precision.

Figure 5. Qualitative comparison of ground truth annotations (a) and model predictions (b)

(a) val_batch0_labels, ground truth (b) val_batch0_pred, model prediction

Classification Performance

By the final epoch (35), the classification model achieved a Top-1 Accuracy of 99.93%. This indicates the model can reliably distinguish among Healthy leaves, Scab, and Black Rot once leaves are isolated by the detection stage. As shown in the results plot (specifically the metrics/accuracy_top1 graph) (Figure 6), the model-maintained a near-perfect trajectory throughout the training.

Figure 6. Training and validation loss with top-1 and top-5 accuracy metrics over 35 epochs

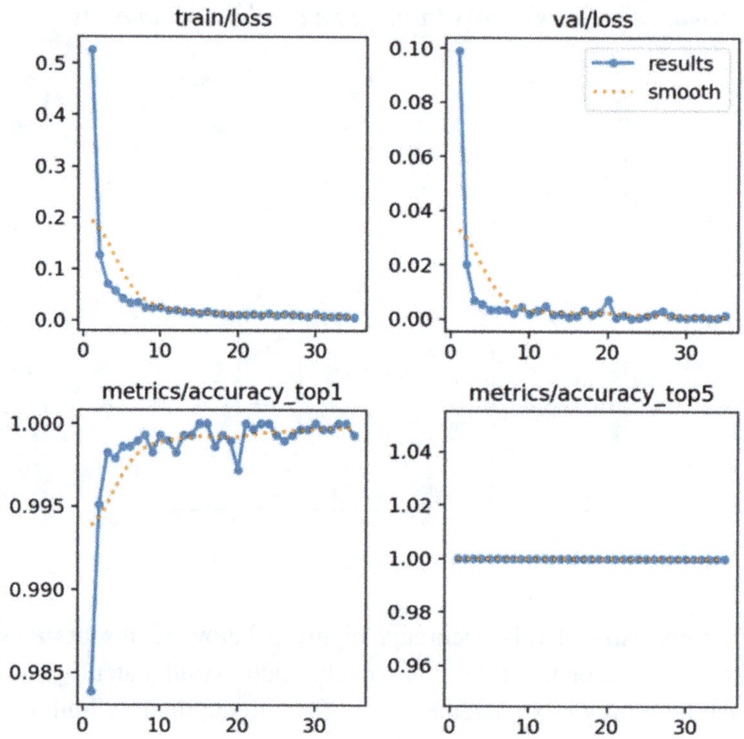

The Top-5 Accuracy reached 100%, demonstrating that even in the most ambiguous cases, the correct disease label consistently appeared among the model's top predictions. This performance is further validated by the confusion matrix (Figure 7), which shows clear diagonal control with nearly zero misclassification between the three disease classes and the background.

Figure 7. Normalized confusion matrix showing near-perfect classification performance with 100% top-5 accuracy

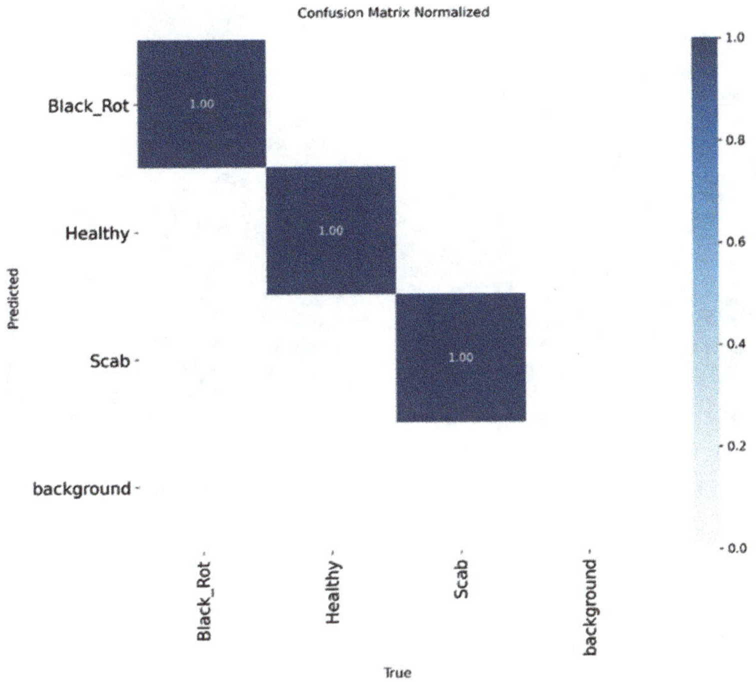

Regarding the learning process, the model achieved a Top-1 Accuracy of 98.39% at Epoch 1. This shows strong transfer-learning efficiency and indicates that the YOLOv8 architecture was suitable to the visual complexity of apple leaves from the outset. As shown in the results dashboard (Figure 6), Training Loss decreased significantly, dropping from 0.5272 to a near-zero value of 0.0060. More significantly, Validation Loss concluded at an even lower 0.0014, a strong statistical indicator that the model did not overfit.

A qualitative review of the prediction images (Figure 8, (a) val_batch0_pred and (b) val_batch1_pred) confirms this high confidence, as the predicted labels on individual leaf crops align perfectly with the actual disease characteristics visible in the images.

Figure 8. Qualitative comparison of model predictions on validation batches

(a) val_batch0_pred (b) val_batch1_pred

Runtime and System Performance

For the detection phase, despite the large dataset size, the entire 100-epoch training process was completed in approximately 15.103 hours (54371.13 seconds) using NVIDIA RTX 4060 hardware, demonstrating the computational efficiency of the YOLOv8m framework for agricultural applications.

In the classification phase, the model achieved a high level of accuracy within 35 epochs, with total training requiring approximately 0.395 hours (23.71 minutes) on the same NVIDIA RTX 4060 hardware.

Results Discussion

The results demonstrate strong model performance under the controlled dataset conditions used for training and evaluation. These figures represent an upper bound on expected accuracy, as field deployment under real Lebanese orchard conditions, including variable illumination, occlusion, and Mediterranean climate disease patterns, was not possible during this study and constitutes a necessary direction for future validation.

For detection, the high mAP50-95 score of 96.35% is very important because it means the model draws precise boxes around the leaves. Accurate boxes are key so that the next stage, classification, works on clean, clear leaf images without distractions.

The steady drop in Validation Loss, from an initial 0.54 to a final 0.17, along with the Training Loss, shows the model did not overfit and should perform well on new orchard images. Furthermore, the mAP50 of 99.34% and the 98.7% precision confirm that the model has nearly mastered the visual signature of apple leaves while maintaining an almost perfect balance between false positives and false negatives. These detection results establish a high-quality input for the subsequent classification stage, ensuring that disease categorization operates on accurately localized and background-filtered leaf regions.

During classification, the strong detection model helped by filtering out background noise, letting the classifier focus entirely on the leaf's texture and color. This contributed to the excellent results, with Top-1 classification accuracy reaching 99.93% and Top-5 accuracy reaching a perfect 100%. The high classification accuracy provides a reliable diagnostic basis to the reporting process. The progressive decrease in Training Loss from 0.5272 to 0.0060 across 35 epochs confirms consistent model convergence and supports the dependability of the generated treatment recommendations as a decision-support output. The extremely low final Validation Loss of 0.0014 also shows that the system is strong and could be used to detect other diseases or work with different crops without losing consistency.

CONCLUSION, LIMITATIONS, AND FUTURE WORK

The main objective of this chapter was to advance precision agriculture by designing and validating a GeoAI-driven UAV system for automated apple disease detection and spatial reporting. Technically, a two-stage YOLOv8 pipeline was implemented for leaf detection and disease classification, enabling the integration of computer vision outputs with geospatial intelligence. Furthermore, the system was designed to translate AI results into actionable, GIS-based decision support. The use of UAV-based RGB imagery proved fitting for leaf-level disease monitoring in orchards, offering a practical compromise between traditional satellite imagery (macro-scale) and manual scouting (micro-scale).

The developed framework contributes to sustainable precision agriculture by enabling early disease detection, which can help limit pathogen spread. Because the system generates precise location data, interventions can be targeted to specific areas rather than applied across entire fields. Consequently, these localized treatments help reduce overall chemical usage.

This system constitutes a modular and adaptable framework that can be modified according to regional needs. It is especially relevant for areas with limited infrastructure. Moreover, GeoAI serves as a bridge between the analytical accuracy of AI systems and the operational relevance of spatial context, thereby transforming simple detection into actionable agricultural intelligence.

The original intent of the work was to design a system to be used on the field to support sustainable agriculture and precision agriculture. This system should have been capable of capturing a video and analyze it directly (live video analysis) by doing so it should be adopting real-time GPS synchronization and be able to generate accurate location data. This location data should then be used to identify locations on the fly and be able to generate hotspot disease maps. However, several external constraints prevented the full achievement of the intended goals. These stemmed from the conflict that arose in Lebanon in September 2024. Prior to the outbreak of hostilities, substantial signal jamming and electronic interference caused unreliable drone telemetry and faulty GPS logs. Consequently, stable video transmission was problematic, and geolocation data were often inaccurate or missing. This situation deprived the project of the ability to validate live GeoAI mapping under field conditions.

To proceed with the work, it was necessary to shift from real-time video processing to offline video analysis. Additionally, the planned geospatial analysis, which would have enabled the evaluation of location accuracy and disease spread, was replaced by timestamp-based correlation where feasible, albeit with reduced spatial certainty. As a result, detailed geospatial analysis could not be conducted, and validation of spatial accuracy at the tree level remained limited.

This work will not stop here; it will be advanced in the future toward achieving 'GeoAI maturity' that addresses all current limitations and fulfills the intended objectives. Technically, robust GNSS solutions, such as Real-Time Kinematic (RTK), Post-Processed Kinematic (PPK), and sensor fusion with IMU, will be adopted, possibly supplemented by additional positioning techniques to verify location accuracy. Moreover, live analysis will be enabled through improved communication resilience, and Edge AI will be deployed on field laptops in the near term, with potential onboard UAV deployment in the longer term.

Spatial intelligence will evolve beyond point-level detection and single-tree identification to include disease-density surfaces and hotspot modeling. Advanced spatial-statistical techniques, such as spatial autocorrelation and neighborhood risk metrics, will also be integrated. At the data level, inputs will not be limited to UAV imagery; they will also incorporate Sentinel-2 satellite data, weather records, and terrain models to enable more sophisticated spatial analysis. Furthermore, analysis will expand beyond static disease identification to incorporate temporal tracking of disease progression and spread velocity.

At the AI-model level, the disease taxonomy will be extended to include additional pathogens, and generalization will be pursued across diverse orchards, seasons, and lighting conditions. At the interpretation level, field-based validation with agronomists will be conducted to test model robustness under real-world conditions. At the intervention level, the system will be linked to variable-rate spraying equipment to evaluate its impact compared to conventional blanket treatments.

Together, these improvements will enable a truly operational GeoAI-driven UAV system to serve as a scalable decision layer that supports resilient agriculture, especially in regions facing instability.

REFERENCES

Adamchuk, V. I., Hummel, J. W., Morgan, M. T., & Upadhyaya, S. K. (2004). On-the-go soil sensors for precision agriculture. *Computers and Electronics in Agriculture*, *44*(1), 71–91. DOI: 10.1016/j.compag.2004.03.002

Al-Shaar, W., Bonin, O., de Gouvello, B., Chatellier, P., & Hendel, M. (2022). Geographically weighted regression-based predictions of water–soil–energy nexus solutions in Île-de-France. *Urban Science (Basel, Switzerland)*, *6*(4), 81. DOI: 10.3390/urbansci6040081

Ali, M. L., & Zhang, Z. (2024). The YOLO framework: A comprehensive review of evolution, applications, and benchmarks in object detection. *Computers*, *13*(12), 336. DOI: 10.3390/computers13120336

Athesan, A. I., & Ghazi, M. F. E. M. (2024). Innovative dashboard designs for real-time smart farming data visualization. *National EngiTech Digest, 1*(1). Retrieved from https://jktss.puo.edu.my/jurnalpuo/index.php/NED/article/view/143

Chan, S., Weitz, N., Persson, Å., & Trimmer, C. (2018). SDG 12: Responsible consumption and production. A review of research needs. Technical annex to the Formas report *Forskning för Agenda 2030.* https://www.sei.org/wp-content/uploads/2018/11/review-sdg12-research-needs-final.pdf

Chao, X., Sun, G., Zhao, H., Li, M., & He, D. (2020). Identification of apple tree leaf diseases based on deep learning models. *Symmetry*, *12*(7), 1065. DOI: 10.3390/sym12071065

De la Rosa, D., & Anaya-Romero, M. (2010). MicroLEIS DSS: For planning agro-ecological soil use and management systems. In *Decision Support Systems in Agriculture, Food and the Environment: Trends, Applications and Advances* (pp. 339–376). IGI Global. https://doi.org/DOI: 10.4018/978-1-61520-881-4.ch016

Espinel, R., Herrera-Franco, G., Rivadeneira García, J. L., & Escandón-Panchana, P. (2024). Artificial Intelligence in Agricultural Mapping. *Agriculture*, *14*(7), 1071. DOI: 10.3390/agriculture14071071

Garcia-Ruiz, F., Sankaran, S., Maja, J. M., Lee, W. S., Rasmussen, J., & Ehsani, R. (2013). Comparison of two aerial imaging platforms for identification of Huanglongbing-infected citrus trees. *Computers and Electronics in Agriculture*, *91*, 106–115. DOI: 10.1016/j.compag.2012.12.002

Gebbers, R., & Adamchuk, V. I. (2010). Precision agriculture and food security. *Science*, *327*(5967), 828–831. DOI: 10.1126/science.1183899 PMID: 20150492

Gogoi, N. K., Deka, B., & Bora, L. C. (2018). Remote sensing and its use in detection and monitoring plant diseases: A review. *Agricultural Reviews (Karnal)*, *39*(4). Advance online publication. DOI: 10.18805/ag.R-1835

He, K., Zhang, X., Ren, S., & Sun, J. (2016). Deep residual learning for image recognition. In *Proceedings of the IEEE conference on computer vision and pattern recognition* (pp. 770–778). https://arxiv.org/abs/1512.03385

Hunt, E. R.Jr, & Daughtry, C. S. (2018). What good are unmanned aircraft systems for agricultural remote sensing and precision agriculture? *International Journal of Remote Sensing*, *39*(15–16), 5345–5376. DOI: 10.1080/01431161.2017.1410300

Janowicz, K., Gao, S., McKenzie, G., Hu, Y., & Bhaduri, B. (2020). GeoAI: Spatially explicit artificial intelligence for geographic knowledge discovery. *ABACUS*, *16*(2), 1–25. DOI: 10.1080/13658816.2019.1684500

Kouadio, L., El Jarroudi, M., Belabess, Z., Laasli, S. E., Roni, M. Z. K., Amine, I. D. I., Mokhtari, N., Mokrini, F., Junk, J., & Lahlali, R. (2023). A review on UAV-based applications for plant disease detection and monitoring. *Remote Sensing (Basel)*, *15*(17), 4273. DOI: 10.3390/rs15174273

Krizhevsky, A., Sutskever, I., & Hinton, G. E. (2012). Imagenet classification with deep convolutional neural networks. *Advances in Neural Information Processing Systems, 25*. https://proceedings.neurips.cc/paper/2012/file/c399862d3b9d6b76c8436e924a68c45b-Paper.pdf

Li, M., Imou, K., Wakabayashi, K., & Yokoyama, S. (2009). Review of research on agricultural vehicle autonomous guidance. *International Journal of Agricultural and Biological Engineering*, *2*(3), 1–16. https://ijabe.org/index.php/ijabe/article/view/160/82

Liu, J., & Wang, X. (2021). Plant diseases and pests detection based on deep learning: A review. *Plant Methods*, *17*(1), 22. DOI: 10.1186/s13007-021-00722-9 PMID: 33627131

Lu, J., Tan, L., & Jiang, H. (2021). Review on convolutional neural network (CNN) applied to plant leaf disease classification. *Agriculture*, *11*(8), 707. DOI: 10.3390/agriculture11080707

Mafenya, K. (2024). Determining the uptake of ICT, GIS, RS, and 4IR technologies in improving food security in the agricultural sector in Gauteng Province, South Africa (Master's thesis, University of South Africa). Retrieved from https://search.proquest.com/openview/9f7b2f5d25bd961acb89d3bbce9dc2c8/1

Mulla, D. J. (2013). Twenty five years of remote sensing in precision agriculture: Key advances and remaining knowledge gaps. *Biosystems Engineering*, *97*(4), 66–85. DOI: 10.1016/j.biosystemseng.2012.08.009

Nagasubramanian, K., Jones, S., Singh, A. K., Sarkar, S., Singh, A., & Ganapathysubramanian, B. (2019). Plant disease identification using explainable 3D deep learning on hyperspectral images. *Plant Methods*, *15*(1), 98. DOI: 10.1186/s13007-019-0479-8 PMID: 31452674

Panda, S. S., Siddique, A., Terrill, T. H., Mahapatra, A. K., Morgan, E., Pech-Cervantes, A. A., & Van Wyk, J. A. (2025). Decision support system for *Lespedeza cuneata* production and quality evaluation: A WebGIS dashboard approach to precision agriculture. *Frontiers in Plant Science*, *16*, 1520163. DOI: 10.3389/fpls.2025.1520163 PMID: 40747527

Patrício, D. I., & Rieder, R. (2018). Computer vision and artificial intelligence in precision agriculture for grain crops: A systematic review. *Computers and Electronics in Agriculture*, *153*, 69–81. DOI: 10.1016/j.compag.2018.08.001

Petraki, D., Gazoulis, I., Kokkini, M., Danaskos, M., Kanatas, P., Rekkas, A., & Travlos, I. (2025). Digital tools and decision support systems in agroecology: Benefits, challenges, and practical implementations. *Agronomy (Basel)*, *15*(1), 236. DOI: 10.3390/agronomy15010236

Rouš, R., Peller, J., Polder, G., Hageraats, S., Ruigrok, T., & Blok, P. M. (2023). Apple scab detection in orchards using deep learning on colour and multispectral images. *arXiv preprint arXiv:2302.08818.*

Seifert, E., Seifert, S., Vogt, H., Drew, D., Van Aardt, J., Kunneke, A., & Seifert, T. (2019). Influence of drone altitude, image overlap, and optical sensor resolution on multi-view reconstruction of forest images. *Remote Sensing (Basel), 11*(10), 1252. DOI: 10.3390/rs11101252

Sieberth, T., Wackrow, R., & Chandler, J. H. (2016). Automatic detection of blurred images in UAV image sets. *ISPRS Journal of Photogrammetry and Remote Sensing, 122*, 1–16. DOI: 10.1016/j.isprsjprs.2016.09.010

Simonyan, K., & Zisserman, A. (2014). Very deep convolutional networks for large-scale image recognition. *arXiv preprint arXiv:1409.1556.* https://arxiv.org/abs/1409.1556

Szeliski, R. (2022). *Computer Vision: Algorithms and Applications* (2nd ed.). Springer., DOI: 10.1007/978-3-030-34372-9

Tikhonova, M. V., & Buzylev, A. V. (2024). Application of DSS in agroecological assessment of the potential of use of chernozem soils in the conditions of the Penza region. In *BIO Web of Conferences* (Vol. 82, p. 02004). EDP Sciences. DOI: 10.1051/bioconf/20248202004

Tobler, W. R. (1970). A computer movie simulating urban growth in the Detroit region. *Economic geography, 46*(sup1), 234-240. .DOI: 10.2307/143141

United Nations Department of Economic and Social Affairs (UNDESA). (2024). *World Population Prospects 2024: Summary of Results.* Stylus Publishing, LLC. https://population.un.org/wpp/assets/Files/WPP2024_Summary-of-Results.pdf

Van Dijk, M., Morley, T., Rau, M. L., & Saghai, Y. (2021). A meta-analysis of projected global food demand and population at risk of hunger for the period 2010–2050. *Nature Food, 2*(7), 494–501. DOI: 10.1038/s43016-021-00322-9 PMID: 37117684

Yang, C., Suh, C. P. C., & Fritz, B. K. (2026). Effects of flight and processing parameters on UAS image-based point clouds for plant height estimation. *Remote Sensing (Basel), 18*(2), 360. DOI: 10.3390/rs18020360

Zhang, N., Wang, M., & Wang, N. (2002). Precision agriculture, a worldwide overview. *Computers and Electronics in Agriculture, 36*(2–3), 113–132. DOI: 10.1016/S0168-1699(02)00096-0

Zhou, Z., Zheng, F., Wen, J., Chen, Y., Li, X., Liu, Y., Wang, Y., & Chen, W. (2023). A user-driven sampling model for large-scale geographical point data visualization via convolutional neural networks. *IEEE Transactions on Human-Machine Systems, 53*(5), 885–894. DOI: 10.1109/THMS.2023.3296692

Chapter 6
Earth Observation Applications for Precision Crop Mapping, Water Resource Management, and Land-Use Change:
GeoAI-Driven Agricultural Monitoring

Shilpa Mahajan
http://orcid.org/0000-0001-8975-8114
The NorthCap University, India

Aryan Dahiya
The NorthCap University, India

Pankaj Rathi
Directorate of Urban Planning, India

ABSTRACT

In recent years, GeoAI, an integration of earth observation, GIS systems, and AI, has presented incomparable opportunities for evaluating agricultural monitoring and control. This chapter outlines the review of five projects that are clearly defined and have the primary goal of utilising remote sensing data to derive effective information on crop conditions, land use, and soil moisture. The projects include Google Earth Engine mainly based on Crop Classification Mapping to distinguish and classify various crop types in Madurai, India; Crop Type Change Detection using the AAFC Annual Crop Inventory to identify changes in crop distribution in Canada between 2013 and 2023; Land Use Change Analysis of the Dynamic World dataset to show changes in land cover between 2015 and 2024. These projects combine spatial and temporal data to visualise change detection for sustainable agriculture. This work confirms the necessity of GeoAI, which allows for efficient and scalable data analysis in agriculture and demonstrates possible uses of satellite data in the development of precision agriculture.

DOI: 10.4018/979-8-3373-8037-7.ch006

1. INTRODUCTION

Agriculture is a vital industry that is significant to economic development, food supply and environmental sustainability. This sector is experiencing numerous problems because of the differences in climatic conditions, growth in population and inadequacy of resources. Such problems require prompt action and an information-intensive surveillance system which transcends the traditional field methods. Geospatial Artificial Intelligence (GeoAI) has been noted in this respect to become a significant paradigm to begin tracking and controlling the agriculture sector.

GeoAI combines geospatial data through earth observation satellites, remote sensing technologies and Geographic Information Systems (GIS) in a combination with artificial intelligence and machine learning. The required infrastructure and computing resources in the form of satellite systems (MODIS, Landsat, Sentinel and SMAP) and cloud-based computing solutions (Google Earth Engine) give the power to carry out geo-intelligent assessments at regional and global levels (Li, 2025).

Real-time agricultural monitoring of the health of crops, estimation of water use through evapotranspiration, and management of the health of soil are enabled by GeoAI because stakeholders are able to monitor it. These insights can be used by farmers, legislators, and environmental managers to make decisions on how best they can utilize the resources, make predictions, mitigate the environmental impacts, and enhance the resilience of agriculture to climate change. Precision and sustainable agriculture is one of the main enablers because of the ability to analyse a substantial amount of spatio-temporal data.

1.1 Background of the Study and Gaps Identified

The rapid evolution of geospatial technologies and Artificial Intelligence has transformed traditional agricultural monitoring methods worldwide. Earlier approaches rely on manual observations and ground knowledge, which are overtaken by data-driven earth observation technologies. Nowadays, satellite-based remote sensing platforms such as Landsat, Sentinel, MODIS, and SMAP are used to monitor large-scale agriculture at high temporal and spatial resolution. Despite these advancements, there are various challenges faced in implementing GeoAI in the real world. The farmers often lack access to high-resolution data, variations in climatic conditions and a lack of a standardised data integration framework lead to ineffective decision making.

The case studies from regions such as India, Sub-Saharan Africa, and Southeast Asia indicate the importance of remote sensing-based monitoring for improving crop assessment and water management. However, this highlights the importance of having a unified analytical framework like GeoAI, where multiple data sources, machine learning techniques and a cloud-based platform support sustainable agriculture practices.

1.2 Major Contributions

This study presents a unified GeoAI-based analytical framework that uses remote sensing and artificial intelligence approaches to address multiple dimensions of agriculture monitoring. The work includes crop classification mapping, crop type change, land use change analysis, estimation of evapotranspiration, and soil moisture monitoring. It also demonstrates how GeoAI generates timely and actionable insights using spatio-temporal data for sustainable Agriculture. This study makes several significant contributions to the field of GeoAI-based agricultural monitoring, as listed below.

- It offers an integrated and holistic analytical framework that combines crop properties, land use dynamics, evapotranspiration and soil moisture in comparison to the previous research, which focuses on individual components at a time. This study examines multiple dimensions together under a unified, coherent GeoAI-driven methodology.
- The work uses diverse datasets, including GFSAD1000, AAFC Annual Crop Inventory, Dynamic World, MODIS, and SMAP in an environment that is well integrated on a cloud platform. The ability to perform consistent processing, temporal consistency, and reproducibility across geographically disparate areas ensures consistent use of Google Earth Engine.
- This study contributes by applying machine learning and deep learning models, such as Random Forest, CNN, LSTM, and CNN-LSTM, for spatio-temporal modelling in agriculture. The combination of change detection, time-series analysis and predictive modelling enhances the depth of the analysis of the framework.
- The research shares empirical insights into agriculture dynamics across multiple agro-climatic conditions, including the dynamics of rainfed, irrigated agriculture, crop-crop transition, land use change, water use, and soil moisture variability. These insights show that the offered framework can be flexible for both data-rich and data-constrained areas

2. LITERATURE REVIEW

Remote sensing and GIS studies have been common in agricultural research, and many publications on crop types, land use changes, evapotranspiration rates, and soil moisture changes have been available. This section integrates prior literature and practices, which form the context for the study conducted in this research.

2.1 GeoAI in Agriculture

Remote sensing and geographic information systems (GIS) have been used extensively in agricultural management as they aid in monitoring the health and location of crops over a wide area, the soil fertility distribution, and land use in general. Remote sensing technology aids in gathering information about regions of the ground surface through satellites and aerial photography (Mavisclara et al., 2025). This is especially true where there is a need to track alterations in farm production, environmental effects and management of natural resources. It can be spatially applicable to support the witnessed interventions to enhance farmer utilisation, policymaking and enhanced crop examination in an effort to support agriculture when combined with GIS. These include land use mapping, crop classification, yield forecasting, soil moisture, and evapotranspiration, which are unavoidable in an era of population growth and climate change. The use of these applications is enhanced in combination with a GeoAI Framework that allows automated data-driven insights beyond conventional remote sensing and GIS approaches.

Google Earth Engine (GEE) has been used as a platform for implementing GeoAI-based agriculture research. It provides access to a vast volume of online data and does not require cumbersome image-processing tools. The GEE pipeline combines AI-based tools with large-scale Earth observation datasets to rapidly analyse data and develop new applications within a test environment (Afuye et al., 2024). This engine performs efficiently when applied to large–scale research across large coverage areas, such as in agriculture, particularly in correlating crop health, land use, water, and soil, which are pertinent to the decision-making of smallholder farmers and commercial farming conglomerates (Mahajan, 2024).

Figure 1. Conceptual framework of GeoAI agricultural monitoring

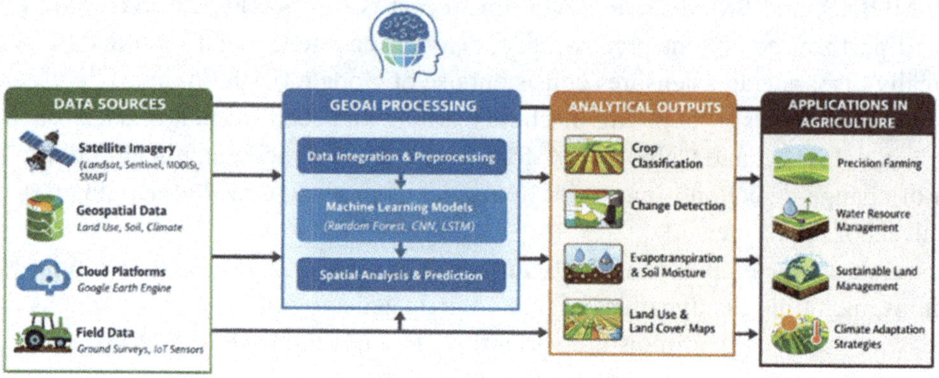

The objective of this study is to develop an integrated, unified GeoAI framework for Agricultural monitoring by combining Earth observation data, geospatial analytics, and artificial intelligence in a cloud-based environment, as shown in Figure 1. The scope of this work spans multiple regions and datasets, providing a comprehensive exploration of how remote sensing can be applied to support sustainable agricultural practices. This research spans various remote sensing applications in crop farming, land-use change analysis, water-use efficiency, and soil conditions, to enhance agricultural productivity through data compiled from field studies.

2.2 Crop Classification

The major techniques of crop classification based on the data collected by remote sensing systems play a key role in monitoring agriculture and evaluating yield. The potential classification algorithms to be utilised in the research are decision trees, support vector machines (SVMs), and neural networks to optimally classify the type of crop based on multispectral and hyperspectral images. Indicatively, Melichar et al. (2023) uses methods and data fusion approaches using Landsat 8 OLI data and phenology metrics to produce an RF classification model capable of classifying 31 different land cover class types within the Sonoran and Mojave Deserts BCR. The final land cover map produced from this study captured the spatial distribution of vegetation at a much finer resolution than that found in existing land cover maps for both the US and MX. The most recent advances in machine learning have gone a step further to improve the efficiency of crop classification by adopting the application of time series analysis of satellite images, as shown by Simón Sánchez et al. (2022), who conducted study in Albacete, Spain and Eisfelder et al. (2024) who conducted research in Ethopia using Sentinel-2 data and a convolutional neural network. Simón Sánchez et al. (2022) is conducted in Albacete, Spain. Eisfelder et al. (2024) is about Ethiopia

2.3 Land Use Change Analysis

It is imperative to monitor changes in land use to evaluate the extent and degree of environmental impacts and support. Some of the most common techniques in land-use change analysis are image differencing, post-classification comparison, machine learning, and change detection. For example, Fu et al. (2024) outlined and explained the transition of land use in Southeast Asia over 20 years using the Landsat data and image differencing. Using machine learning on MODIS data, Sarif et al. (2022) and Ullah et al. (2024) obtained the expansive agricultural sector alterations and the rise of cropland with less natural vegetation. Al-Shaar et al. (2022) proposed Geographically Weighted Regression (GWR), which explains how local variables can lead to accurate predictions in land water-energy research

2.4 Soil Moisture Mapping

Soil moisture analysis is crucial for evaluating irrigation and drought. NASA recently initiated a significant mission known as the Soil Moisture Active Passive (SMAP), which has provided researchers with rich information about soil moisture at the global level. Soil moisture mapping has been performed through the analysis of the vegetation index, radar backscatter, and microwave remote sensing. Zha et al. (2025) yielded improved predictions of soil moisture in sub-Saharan Africa by incorporating SMAP data in machine learning models to serve drought-infested regions and improve the availability of water resources.

2.5 Evapotranspiration Estimation

In the estimation of water used in agriculture, evapotranspiration (ET) must be estimated correctly (Tang et al., 2024). The techniques to determine ET using remote sensing data are usually thermal and multispectral image-based, with the models being SEBAL and Penman-Monteith. The SEBAL model was used by Awada et al. (2021) to assess ET in irrigation-dominated areas using Landsat images to enable effective water management. Liu et al. (2021) used the data of MODIS and machine learning to compute the changes in the agricultural areas of the United States across the year. The following table summarises prominent methodologies used in remote sensing studies similar to the studies conducted in this chapter, as shown in Table 1.

Table 1. Comparative study of related work

Study	Reference	Methodology	Data Source	Key Findings
Crop Classification	Melichar et al. (2023)	Random Forest Classification	Landsat-8	Classify 31 different land cover class types within the Sonoran and Mojave Deserts BCR
	Simon et al. (2022)	Convolutional Neural Network	Sentinel-2	High accuracy in crop classification using time-series data in Albacete, Spain
Land Use Change Analysis	Fu et al. (2024)	Image Differencing	Landsat	Revealed deforestation and urban expansion in Southeast Asia

continued on following page

Table 1. Continued

Study	Reference	Methodology	Data Source	Key Findings
	Sarif et al. (2022)	Machine Learning on MODIS	MODIS	Identified significant land use shifts from vegetation to cropland
Soil Moisture Mapping	Zha et al. (2025)	Machine Learning with SMAP data	SMAP	Improved soil moisture predictions for drought-prone areas
Evapotranspiration	Awada et al. (2021)	SEBAL Model	Landsat	Used for ET estimation in irrigation areas, aiding water resource planning
	Liu et al. (2021)	Machine Learning	MODIS	Assessed seasonal ET variations across agricultural zones

3. METHODOLOGY

The methodological approach adopted in this study follows a unified GeoAI approach in which multiple geospatial datasets are integrates processed using machine learning techniques on a cloud platform. The overall methodology encompasses remote sensing data acquisition, preprocessing, and feature extraction; machine learning and spatiotemporal analysis to examine crop classifications, land-use dynamics, evapotranspiration, and soil-moisture variability.

3.1 Study Area and Data Sources

The framework was applied to diverse geographical regions, including India, Canada, Europe, Thailand, and South Sudan, each with varied agricultural and climatic conditions. The study utilised datasets such as MODIS, Landsat, Sentinel-2, SMAP, GFSAD1000, AAFC Annual Crop Inventory, and Dynamic World. These datasets provide complementary spatial, spectral and temporal information necessary for agriculture analysis across diverse geographic contexts. These datasets are useful for various aspects of agricultural surveillance and monitoring, including crop classification, soil moisture content, land-use status, and evapotranspiration. The datasets used in this study are briefly discussed in Table 2.

Table 2. Key datasets

Dataset	Spatial Resolution	Temporal Coverage	Application
GFSAD1000	1 km	Static	Crop classification
AAFC Annual Crop Inventory	30 m	Annual	Crop change detection
MODIS	500m–1km	8-day/monthly	ET estimation
Dynamic World	10 m	Near real-time	LULC
SMAP	10 km	2–3 days	Soil moisture

3.2 Crop Classification Mapping

This work aimed to classify and map the types of crops in the Madurai district of India with emphasis on categorizing the different types of croplands, forests, and other vegetation. Figure 2 gives the methodology used in the classification of crops.

Figure 2. Crop classification methodology

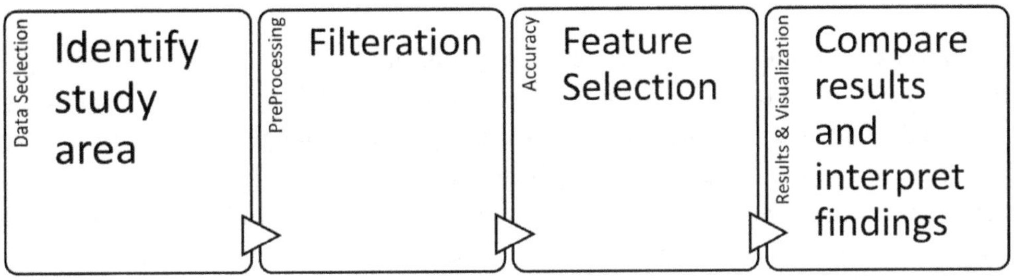

This analysis was based on the South Indian district of Madurai which is a region with a variety of agricultural operations. The classification of crops also used the GFSAD1000 dataset which has the global coverage of croplands at a 1-kilometre resolution, classifying different types of land cover such as rainfed or irrigated croplands, forest and other types of vegetation. In order to clip the dataset with Madurai area of interest, there was a rectangular geometry (longitude: 77.9 to 78.4, latitude: 9.5 to 10.2) used to define the boundary of the district. The data was filtered and preprocessed to obtain the cover layer of croplands and isolate the agricultural and vegetation classes of interest to the area.

The land cover of the GFSAD1000 dataset was pulled out and re-classified to identify significant agricultural and vegetation classes that comprise rainfed cropland, irrigated cropland and forest. To make the visualization scheme more interpretable, a colour-coded visualization scheme was used (Yu et al., 2025). Machine learning models such as the Random Forest and the Convolutional Neural Network were utilized in pixel-level classification. The model performance is compared in terms of classification measures such as accuracy. This has been discussed in the results section.

3.3 Crop Type Change Detection

The Crop Type Change Detection has been done using the AAFC Annual Crop Inventory dataset of the years 2013-2023 for Canada. Crop maps were analyzed for 2013,2016 and 2023 years. This paper gives an insight and representation of these modifications about succession, distribution of crops and land use pattern by environmental and/or economic factors. Figure 3 describes the methodology used in detecting the change in the crop.

Figure 3. Crop type change detection methodology

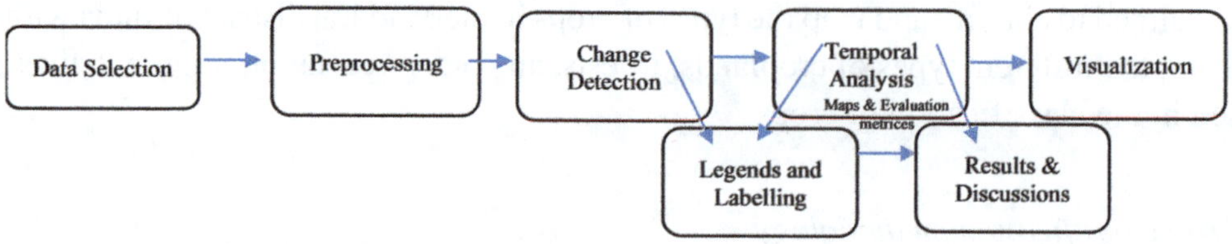

The area of interest in this research was diverse agricultural lands in Canada that were aiming at using well-prepared information on the AAFC Annual Crop Inventory (ACI). The data about the yearly crop that is reflected in this data set is highly accurate in identifying the changes in the same over an extended period of time. The study was aimed at covering three years (2013, 2016 and 2023). The research sought to record changes in a time interval of three years, and this gave the overall view of the changes in crop distribution across Canada during a period of time. The ACI data set was further reduced by year to get crop information of each target year (2013, 2016 and 2023). The data from the two years was then subsampled to the agricultural land of interest in Canada, and this allowed the analysis to be limited to the agricultural land of interest and not the non-cultivated land.

Change detection was performed by subtracting the crop data from one year from the data of another crop. Specifically:

- 2013 data was subtracted from 2016 data to capture three-year changes.
- 2016 data was subtracted from 2023 data to observe seven-year changes.
- 2013 data was subtracted from 2023 data to visualise a ten-year transition.

Colour-coded visualisation: The change maps were visualised using a red-white-green palette to represent various types of changes:

- Red represented a loss in crop coverage.
- White indicated no change.
- Green showed an increase in crop presence or other positive changes in the land cover type.

The temporal aspect involved identifying change patterns at the intervals of 2013-2016, 2016-2023, and 2013-2023, making it possible to gain short-run and long-run information on the crop distribution. The intervals made it possible to determine the regions of stable crops, the areas of crop expansion, and areas of attributions for crop rotation or shift in land cover for agricultural practices a decade apart. A detailed legend was created to help interpret the crop change map accurately. The legend included colour codes for different land cover classes, such as cereals, pulses, oilseeds, and other specific crops, as well as categories for non-agricultural land like forests and urban areas. Each land cover category was associated with a unique colour in the legend, enabling easy differentiation between crops and other land uses.

Similar to the crop classification mapping, the accuracy of the change detection could be evaluated by cross-referencing with ground truth data or using independent crop inventories for each year. Suggested metrics for evaluation include change detection accuracy and confusion matrices for each time interval, which would validate the reliability of observed transitions in crop types. An LSTM-based temporal crop detection model was trained for 10 years on the ACI dataset, which achieved an MSE of 0.14, and it also forecasted an oilseed expansion area by approximately 15% by 2030.

The results were presented in a results section as a sequence of maps, each illustrating the change between consecutive intervals, as well as an overarching map displaying the cumulative change from 2013 to 2023. A title panel and scale bar were added to enhance map readability, along with labels indicating the regions where the most significant crop type changes occurred.

3.4 Land Use Change Analysis

Land Use Change Analysis was aimed at identifying the changes of the land cover during nine years of study, 2015-2024, with the help of the Dynamic World dataset. This discussion illuminates the effects of various forms of land cover that include forests, agricultural land, and built-up areas (Vitale & Lamonaca, 2025; Al-Shaar et al., 2025). These have been changed in the course of time by human activities, climate, and environmental policies. Figure 4 presents the methodology used.

Figure 4. Land use change methodology

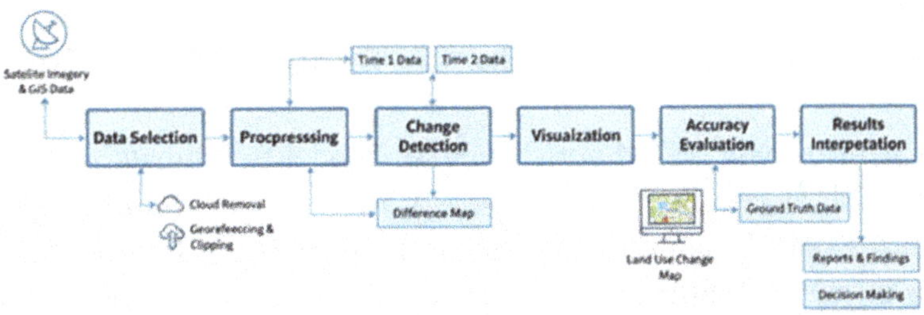

The central location of the study was a particular point in Europe (latitude: 40.4305, longitude: 20.6729), which primarily belongs to the region of mixed urban, agricultural, and natural land classes. In order to generate the data of the years 2015 and 2024, the land classification information of the Dynamic World (DW) dataset classifying the entire planet land surface into finer geographical resolution was utilised. By the fact that this dataset has clear cut classes, it is specifically applicable in land use analysis whereby different classes consist of water, trees, grass, flooded vegetation, crops, shrubland, built-up area and barren land. The DW data was narrowed down to include only the years 2015 and 2024 and then a mosaic operation was done on the data of the years to produce a single and unified image of the land covers. A dataset band that contained the labels was chosen and that has comprehensive land cover classes necessary in determining the alteration in land use patterns.

The 2015 image of the land cover was detected to change and the outcome was subtracted with the 2024 image. This is an easy way of determining regions that have undergone a change in land cover since the changes are presented in form of positivity and negativity.

Colour-Coded Visualisation. The resulting and final change map was visualised using a three-colour palette:

- Red indicated a loss in vegetation or other types of cover, which suggested deforestation or urban expansion.
- White represented areas with no change in land cover.
- Green depicted gains in vegetation or other land covers, which could indicate reforestation or agricultural expansion.

DW dataset contains different land cover classes and each of these classes was given a different colour in the end map in order to enable easy identification of the changes. Each type of land cover was assigned a legend that described the colour palette that was used to categorize the type of land cover, such as water (blue), trees (dark green), grass (light green), crops (yellow), and built-up areas (red). This myth was a key to the interpretation of the visual outcomes more precisely. The nine-year interval analysis between the two years (2015 and 2024) was possible due to the difference, and it indicated the long-term land use changes. Lastly, trend relationships were specified with reference to critical regional changes that encompassed the changes in time in urbanisation, conversion of agricultural land, and the changes in the natural vegetation. This facilitated easier knowledge of the spatial and temporal distribution of the intensity and direction of the change in land use and most locations of the change (Moumane et al., 2025). GeoAI Change Modelling (2016-24) analysis was done using Markov Transition modelling and CNN-LSTM Spatio-Temporal Prediction. Table 3 demonstrates the crop-wise transition of 2016 24.

3.4.1 CNN–LSTM and Markov-Based Spatio-Temporal Modelling

This paper uses a hybrid CNN LSTM with a Markov model to estimate spatio-temporal dynamics of land use transitions. This GeoAI hybrid methodology allows the learning of space and time relationships of land use patterns. The CNN part is used to obtain the spatial characteristics of multi-temporal satellite images. The Dynamic World dataset consists of land cover classes, Vegetation indices and spatial texture data. The CNN consists of 3x3 layers with ReLU activation and max-pooling layers for the extraction of land cover hierarchically. This data is further fed into an LSTM to learn time-varying dependencies at several time steps. The LSTM model consists of a single memory unit (a hidden layer) of 64 memory units that store long-term temporal dependencies in land use change between 2016 and 2024. The sequential input design enables the model to acquire patterns like slow urbanisation and the change of cropland to urban in the course of time. The model was trained in a supervised learning way in which the input sequences were historical land use maps (2016-2020), and the target outputs were the subsequent land use states (2021-2024). A 70:30 training-validation division was embraced. It was trained on 50 epochs with an optimizer of Adam with a learning rate of 0.001 and a batch size of 32. The loss function was the Mean Squared Error (MSE), which is appropriate in continuous spatial prediction tasks. The Markov transition model was employed to estimate the likelihood of transitions of land use classes to supplement the deep learning approach. The transition probability matrix was calculated with regard to the changes in classes between periods of observation, which are the probability of a transition, e.g., cropland to built-up or forest to shrubland. Having this probabilistic structure gives a minimum knowledge of the land dynamics and improves interpretability. The fact that CNN-LSTM and Markov modelling are integrated is that both data-driven learning and probabilistic transition dynamics are used to enhance the robustness of predictions. Standard measures of performance, such as Mean Squared Error (MSE), Root Mean Square Error (RMSE) and classification accuracy of predicted classes of land use, were used to evaluate the model performance. This type of hybrid modelling is especially useful in the representation of nonlinear spatial patterns and long-term temporal changes in the urban development trend, and so can be applied to the prediction of urban sprawl, agricultural land use transformation, and changes in vegetation structure in heterogeneous landscapes The related work was done by Al Shaar (2025b) to study the spatio-temporal dynamics of urban growth in Kuwait and assesses its climate change vulnerability using a Multi-Layer Perceptron Markov Chain Model (MLPMCM) to forecast land use and land cover (LULC) changes for the years 2050 and 2100. .

Table 3. Crop-wise transition is shown from 2016 to 2024

From / To	Built-up	Cropland	Forest	Shrubland	Water	Barren	No Change (%)
Built-up	92.3	1.1	0.0	2.8	0.0	3.8	High stability
Cropland	6.4	81.2	1.7	8.1	0.6	2.0	Moderate stability
Forest	0.9	2.1	84.5	10.7	0.1	1.7	Forest → Shrub
Shrubland	3.2	4.7	5.8	83.6	0.3	2.4	Expansion & fragmentation
Water	0.0	0.4	0.0	0.6	95.8	3.2	Seasonal fluctuations
Barren	4.1	3.2	1.2	6.5	1.0	84.0	Urban-edge conversion

To enhance map readability, several additional UI elements were integrated:

- A scale bar to indicate spatial reference.
- A north arrow for orientation.
- A title, "Dynamic World Land Use Change Detection 2015 to 2024," to contextualise the map.

The accuracy of the detected changes in land use could be assessed by comparing the findings with other independent sources of land cover information, for example, national or regional land surveys or by actual ground truthing wherever feasible. This method can be used to provide urbanization and agricultural impact on land transformations.

3.5 Evapotranspiration

The Evapotranspiration (ET) was based on an analysis of water usage in Thailand. This work focuses on the identification of trends in ET over some years and contributes to long-term water resource management and/or irrigation in arid areas (Gong et al., 2024). The workflow of the methodology adopted can be seen in Figure 5.

Figure 5. Workflow of the evapotranspiration methodology

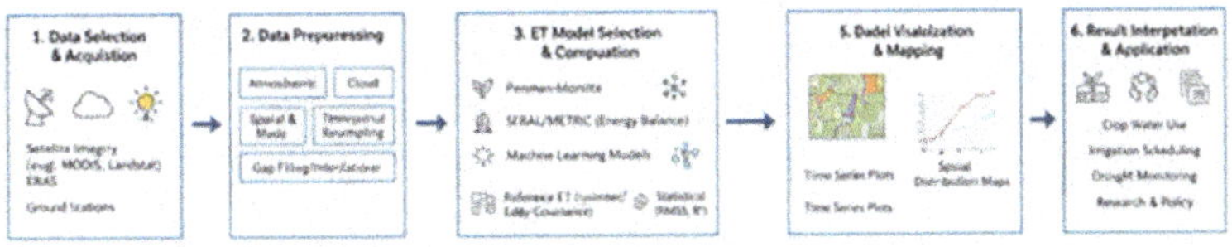

The study area is focused on Thailand, where agriculture plays a major role, and climate variation affects water availability. MODIS (Moderate Resolution Imaging Spectroradiometer) data from the MOD16A2 collection were used for selecting the ET (evapotranspiration) band. MOD16A2 provided the ET data every eight days at a spatial resolution that was suitable for regional water management studies. The data was filtered for the period of interest that covered multiple years (2015 to 2019) to analyse ET trends over time.

The dataset was clipped to the boundary of Thailand, which was defined using a country's boundary layer from the USDOS LSIB SIMPLE dataset. So that we used the filter date on each of the ET images to obtain the needed seasonal and or annual ET variations. The secondary data collected for this study are secondary source data that have been collected and cross-sectional data that took several years (2015–2019) into consideration, giving a broad temporal perspective of the dynamics of ETs.

The ET data was visualised using a colour-coded palette that represented different ET intensity levels, from low ET values (white) to high ET values (dark green). This colour palette enabled an easy interpretation of regions with high water use versus those with minimal water demand. The map was centred on Thailand to provide a regional overview of evapotranspiration distribution across the different land types, like agricultural, forested, and urban areas.

To understand the seasonal and annual trends in the ET, the MOD16A2 data were used to create a time series chart of mean ET values for Thailand from 2015 to 2019. The ui.Chart.image.doySeriesByYear function was employed to generate a time series chart, which visualised the annual variation of ET for Thailand by averaging ET values across the country for each day of the year. This time series analysis enabled pinning of the peak ETs in tune with Thailand's cropping calendar and climate, giving important insights into water usage that are vital to irrigation planning. The results have been discussed in Section 4.

3.6 SMAP Soil Moisture

The SMAP Soil Moisture initiative would help in tracking the soil moisture content with time in the South Sudan areas. Information on the Soil Moisture Active Passive (SMAP) mission was applied. An important variable in the agricultural and water resource management is soil moisture that determines the health of crops, drought risk and irrigation (An et al. 2025). The study gives monthly soil moisture trends and is very useful in the optimisation of water use and in the drought vulnerability evaluation. The methodology adopted is shown in Figure 6.

Figure 6. SMAP soil moisture methodology

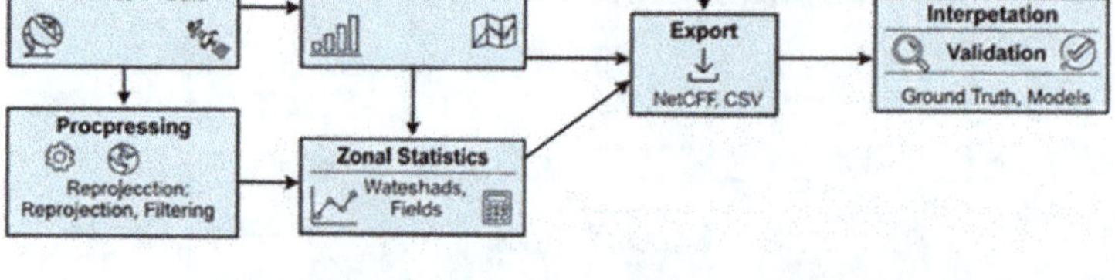

South Sudan was the target of the study, and high seasonal rainfall and agricultural dependence on the availability of seasonal water make soil moisture monitoring relevant in the region. The SMAP 10 km Soil Moisture dataset gave the data on soil moisture at the spatial resolution of 10 km that was adequate to conduct regional analysis within the context of agriculture. The dataset included the surface soil moisture content, as it was essential in determining the short-term water conditions and drought conditions.

SMAP data were narrowed down based on the date to reflect data between 2016 and 2021 that gave a six-year record of trends of soil moisture in South Sudan. The data was trimmed to the South Sudan border and was characterised through the use of the USDOS LSIB SIMPLE countries boundary layer, presenting the opportunity to narrow the analysis to this AOI.

The data on soil moisture was displayed with a colour scheme, which was a colour scale between blue (low moisture) and red (high moisture), which was useful in distinguishing the areas with different moisture levels in the region. The SMAP average of soil moisture was calculated and superimposed on South Sudan to initiate the average layer of soil moisture as one way of determining the overall condition of the moisture in the area.

The data on SMAP were summarised by month and year in order to have a closer look at the trends of soil moisture. All the images in the dataset were classified by the month and year in which they were taken, and this aided in forming a time series of monthly mean soil moisture values across the six years. The aggregation of this was done with the help of the byYearMonth function, which calculated the average of soil moisture over the months of the year. This provided the possibility of isolating the temporal patterns of soil moisture in the sense that the effect of the wet and dry seasons in South Sudan would be observed in relation to moisture.

To summarise the level of soil moisture in the geographic area of South Sudan, zonal statistics were carried out. The reduceRegions function estimated the mean soil moisture per month, which gave a monthly overview of the levels of soil moisture in the country. The zonal analysis made it possible to extract regional soil moisture values that were made available in order to guide water management practices that were country-specific to agricultural needs.

The information about soil moisture on a monthly basis was downloaded in a CSV format and stored on Google Drive so as to enable further analysis. The information exported was in the form of text since the fields which were contained in the exported data were the year, month and the average soil moisture to provide a structured time series to be used in the analysis that was not part of the GEE system. This data, which has been exported, is useful in the incorporation of soil moisture data into policy planning or local models in South Sudan.

Data on soil moisture taken at monthly intervals showed the existence of a direct relationship between moisture regimes and the rainfall season of South Sudan, with the high moisture level in the rainy season and the low moisture level in the dry season. This information proved useful in predicting irrigation requirements, assessing drought risks and the temporal distribution of water resources with the aim of agricultural.

Concerning validation, soil moisture SMAP measurements might have been compared to ground-truth soil moisture measurements or soil moisture SMAP models of the region. Reliability measures, which are likely to contain the mean absolute error (MAE) and the coefficient of determination, substantially determined the accuracy of SMAP data in reflecting moisture variation in the area. The water resource managers, agronomists and policymakers in South Sudan would find such findings useful since they could use such a method to monitor the temporal variation in soil moisture and use such information to make decisions regarding irrigation methods, planting, and drought preparedness (Xu et al.,2024) The results have been discussed in section 4.3

4. RESULTS AND DISCUSSIONS

4.1 Crop Classification

The crop classification map was centred on Madurai to provide an overview of land use types at the district level. A legend was provided to the users to give them details on the designation of the land cover categories and the colours used in the map-making process. This legend was used to differentiate the categories, such as rainfed and irrigated areas, croplands and forested areas, and those that have both croplands and other covers (Goyal et al., 2025).

The GFSAD1000 dataset is used as a reference dataset for identifying cropland extent and class-level interpretation. Since this dataset is a single, static, and global crop land product, it lacks temporally explicit ground-truth data. It was, therefore, used in the classification only to mask the cropland and guide the classification and not to validate it year by year. Two machine learning methods were used in crop classification, which include Random Forest (RF) and Convolutional Neural Network (CNN). The classification process was used separately in the years 2013, 2016, and 2023 with corresponding images of the satellite. The data preprocessing, spectral features extraction of both the Normalised Difference Vegetation Index (NDVI) and Enhanced Vegetation Index (EVI), model training, and pixel-level classification were all part of the workflow implemented in the Google Earth Engine environment. The internal validation method was used to assess model performance. A 70:30 split was chosen to divide the dataset into the training and testing subsets randomly. The performance values obtained are the results of internal validation of this train test framework. External validation was not done using standard measures of accuracy in the absence of temporally consistent and independently collected ground-truth datasets in the target years. Thus, the values reported are to be seen as the measures of model performance in the experimental context and not as the separately verified classification accuracies. The classification shows that the major land cover types of the Madurai region are the rainfed croplands and forested areas, with the irrigated croplands having a relatively lesser percentage as depicted in Figure 7.

Figure 7. Crop classification mapping

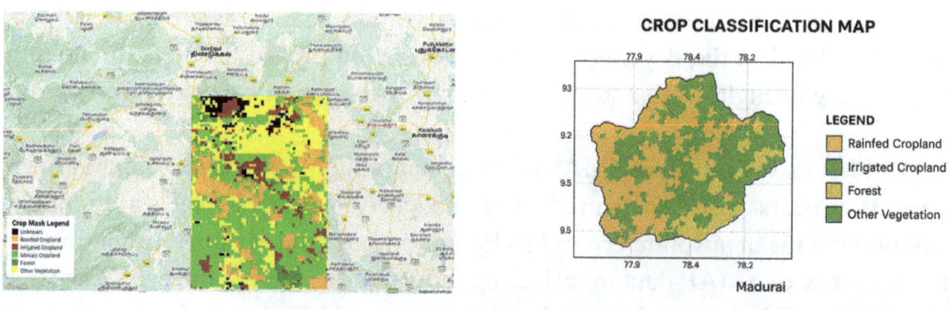

The internal validation performance of the Random Forest and CNN models across the selected years is presented in Table 4.

Table 4. Random and CNN Classifier accuracy for crop detection

Year	RF	CNN-Refined
2013	84.2%	90.2%
2016	86.6%	93%
2023	89.1%	94.2%

4.2 Crop Type Change Detection

The Crop Type Change Detection highlights the annual crop inventory of the years 2013, 2016, and 2023 as shown in Figures 8, 9, and 10 respectively and identifies significant changes in crop types across Canada over three intervals, i.e., 2013-2016, 2016-2023, and 2013-2023 as shown in Figures 11, 12, and 13 respectively. Between 2013 and 2023, notable shifts were observed. This included a decrease in cereal crop areas in some regions and an increase in oilseed crops. This can likely result in changes in market demand and crop rotation practices. Regions exhibiting stable crop cover suggest established agricultural practices with minimal crop type variation. These established findings embrace the fluidity of agricultural land use in the Canadian territory due to the economic prospects, nature of the climate and agronomic innovations. Crop-type analysis that focuses on the prolonged shift can contribute significantly to the policy adjustment process because it demonstrates the requirements of both local agricultural production and international markets.

Figure 8. 2013 Annual crop inventory

Figure 9. 2016 Annual crop inventory

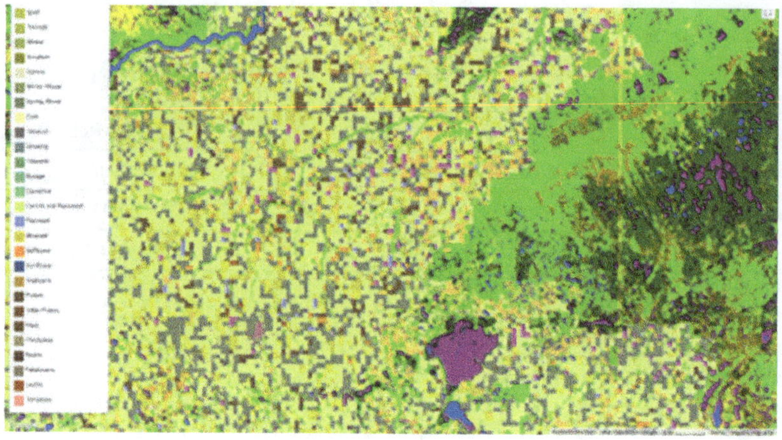

Figure 10. 2023 Annual crop inventory

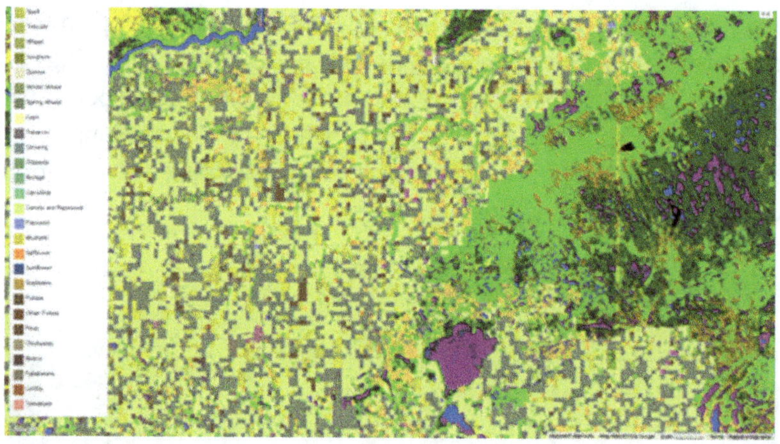

Figure 11. Crop type change between 2013 and 2016

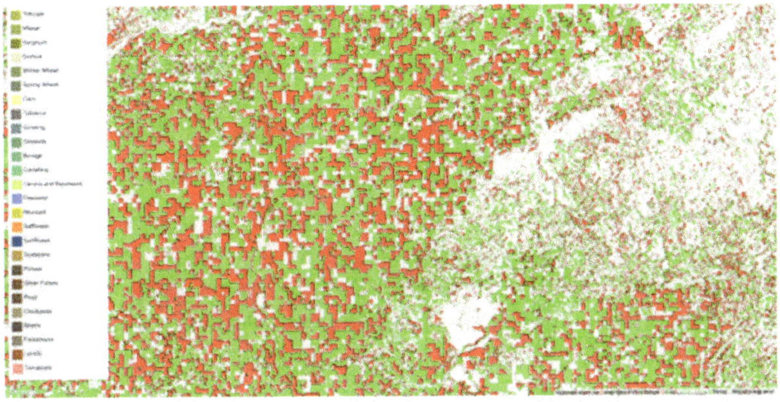

Figure 12. Crop type change between 2016 and 2023

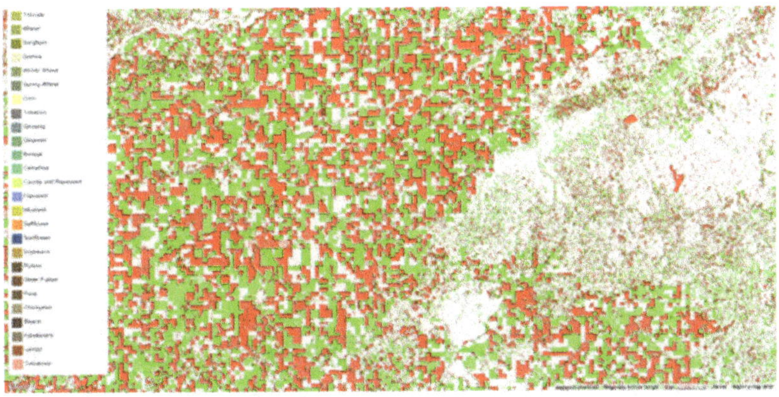

Figure 13. Crop type change between 2013 and 2023

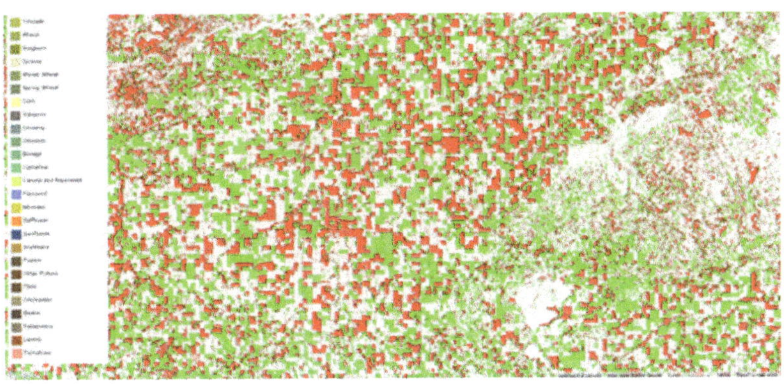

4.3 Evapotranspiration

The ET analysis provided an insight into the changes in the water demand in Thailand that depend on the change of seasons, the nature of crops and climate in the region. Higher ET levels were determined to be in correspondence with the wet season in Thailand as compared to the dry season, when there was lower values. This kind of variation depicts the significance of incremental water management to provide the required quantity of water to be used in ensuring the growth of crops, and also a means of conserving water during the dry season.

The Evapotranspiration (ET) study using MODIS data for Thailand, as shown in Figure 14. This represented the seasonal trends of water consumption over the years. As per the time series plot shown in Figure 15, we could also notice that ET is also a cyclical variable, taking high values in the wet season and low values in the dry season. The high levels imply that it was likely places that are highly agricultural in terms of activity, which require numerous water supplies. These results justify the need for seasonal water management in Thailand. Both time-related and location shifts of ET can guide farmers and water authorities who desire to match the duration of irrigation periods, water scarcity and crop productivity stability. The same kind of analysis can be sought after other water-limited streams at longer timescales, and the analysis technique described here can be applied.

Figure 14. MODIS evapotranspiration (ET) time series data of Thailand

Figure 15. ET annual variation between 2015-2019

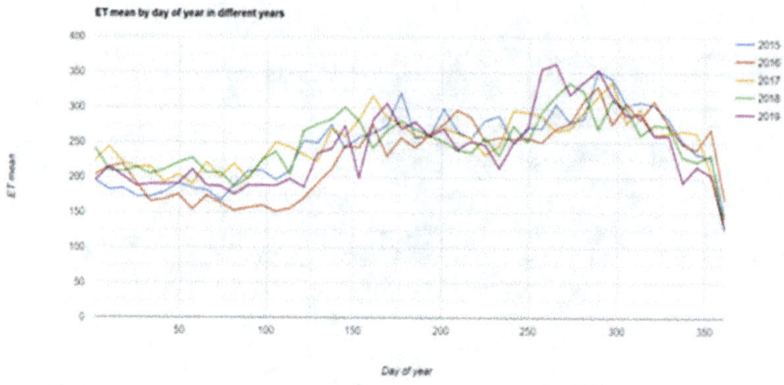

In the present study, the discourse focused on the representation of the temporal dynamics of ET, despite similarity tests being conducted using ground-based ET data or other high-quality ET models, and the MODIS ET map in Thailand. The conclusion drawn from this ET analysis is essential for water and agriculture stakeholders and environmental planning stakeholders. Therefore, in instances of excessive water demand, irrigation mechanisms should be optimised to ensure sustainable agriculture in Thailand. The work has demonstrated the value of remote sensing in the management of agricultural water resources and has provided practical implications to support efficient irrigation and reduce water stress. This study also incorporates a machine-learning-based approach, such as a random forest, for satellite-derived variables. This regression model was trained using MODIS spectral bands, NDVI, EVI, land surface temperature, and meteorological variables such as humidity and wind speed. Using the RF model reduces the ET estimation error by 22% and achieves an RMSE of 0.68 mm/day. All ET values are expressed in mm/8-day after applying the MOD16A2 scale factor (0.1).

4.4 Land Use Change Analysis

The Land Use Change Analysis highlighted the land use in 2015, as shown in Figure 19 and in 2024, as depicted in Figure 17 and examined transformations in land cover between 2015 and 2024, as shown in Figure 18, in a European region that highlights the shifts among categories like trees, grass, built-up areas, and croplands. The change detection analysis showed that there were fewer trees in that area, signifying either expansion of built-up areas, which also signifies the expansion of urban areas. Moderate changes to grassland were observed in agricultural regions and could be related to the abandonment of arable land or to crop rotation processes. The changes observed in land use give insight into the effects of such drivers of change as urbanisation and altered management of the land resource. It is imperative, therefore, to track such changes when undertaking sustainable land use planning, especially for agricultural expansion and conservation. The integration of GeoAI with LULC enhances precision and scalability. It has been seen from the results that Spatio-temporal change modelling using Markov transition probabilities and a CNN-LSTM model revealed a 13.2% expansion of built-up areas and a 6.4% decline in cropland by analysing data from 2016-24.

Figure 16. Land use in 2015

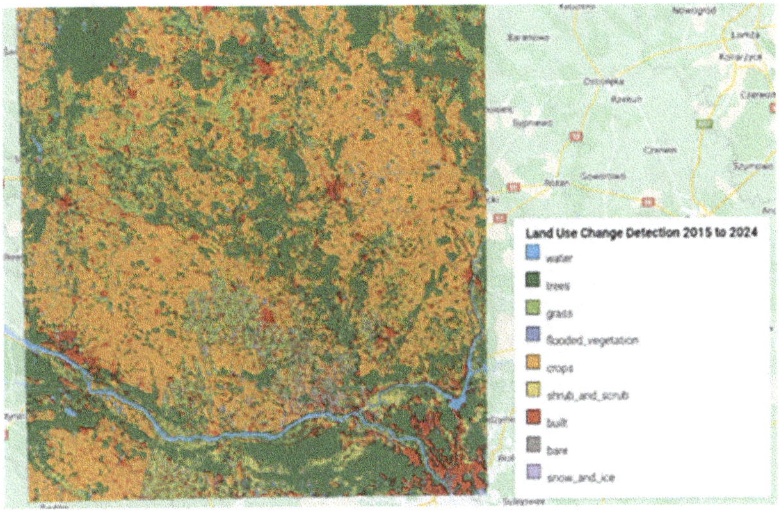

Figure 17. Land use in 2024

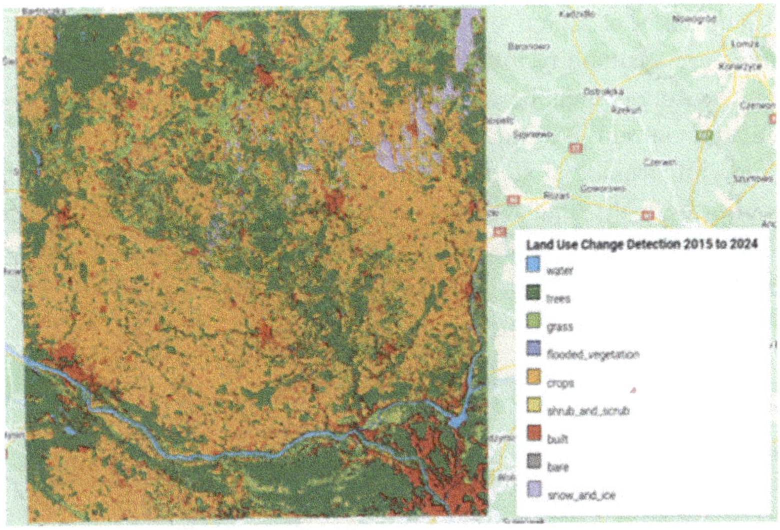

Figure 18. Land use change between 2015 and 2024

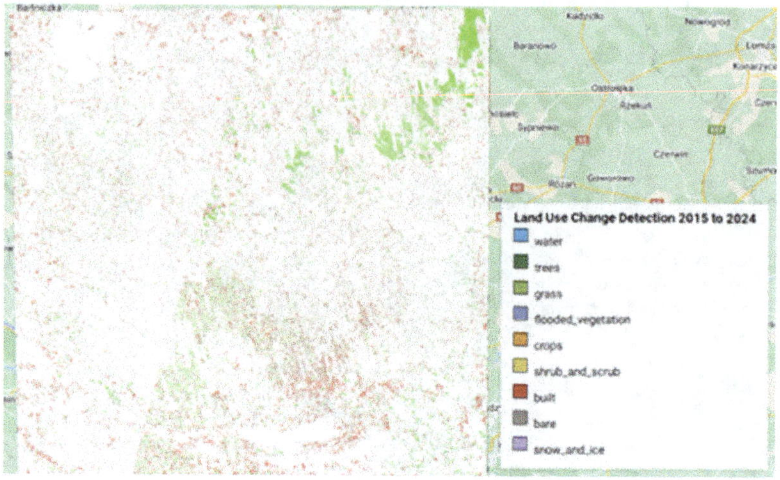

4.5 SMAP Soil Moisture

The SMAP soil moisture data were a monthly time series of soil moisture across South Sudan, as shown in Figure 19. The analysis depicts the temporal changes in six years (20162021) and highlights the differences in moisture availability in the region. These findings suggest that seasonal variations are high with the soil moisture content being high in rainy season and low in dry season. It is observed to be a pattern of reliance on rain in agricultural systems in South Sudan. The fluctuations which are observed are of critical importance in studying the dynamics of drought, as well as in underpinning irrigation planning. Moreover, the time series analysis presents the capacity to determine the possible stressful water intervals and the policy makers and stakeholders to prepare to manage crops and allocate water resources proactively

Figure 19. SMAP soil moisture

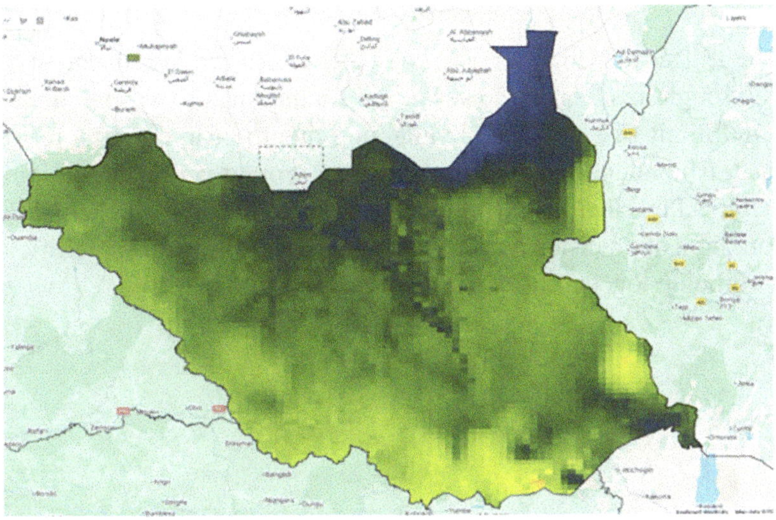

Soil moisture has been used as an important indicator of agricultural productivity, drought detection and hydrological planning (Leizel G. de la Cruz, 2024). Nonetheless, SMAP has a low-quality spatial resolution (10 km), which restricts its use in the field-level analysis in South Sudan where the landscape is heterogeneous and rainfall distribution is highly variable. A GeoAI-based forecasting and downscaling framework was created to overcome this weakness, in which a Convolutional Neural Network (CNN) was used. It uses the inputs provided at a fine resolution, like SAR and NDVI, to create soil moisture maps with a spatial resolution of 1 km. The reduced scale results enhance the depiction of local variability of moisture, allowing for better estimate field-scale conditions when assessing irrigation and drought conditions. The model has an R^2 of 0.79 and a Mean Absolute Error (MAE) of 0.04 m^3/m^3, which has good predictive performance and can be used in operational management of water resources.

5. CONCLUSION

This chapter presents a unified GeoAI framework that incorporates satellite data from multiple sources using machine learning methods for agricultural monitoring in a varied agroclimatic environment and regions.

The chapter discusses crop classification, crop type transition analysis, land use change modelling, estimation of evapotranspiration, and soil moisture monitoring to present a scalable, interoperable Geo-AI architecture. The effectiveness of this study brings forth the importance of using machine learning models for some important empirical findings. CNN refinement was used to classify crops with 94.2% accuracy and LSTM generated an MSE of 0.14 for timely crop production. The use of land use models showed that built-up areas had increased by 13.2 per cent and the cropland reduced by 6.4 per cent within the period of study. The evapotranspiration model was enhanced by machine learning, and the resulting model minimized the error of estimation by 22 per cent with an RMSE of 0.68 mm/day. Additionally, the GeoAI-based downscaling of the SMAP soil moisture yielded an R^2 of 0.79 and an MAE of 0.04

m^3/m^3, which further enables the representation of moisture at the field level. These results prove that GeoAI could contribute to agricultural intelligence substantially, as it can provide multi-scale, spatio-temporal analytics.

This study could be utilized to plan water resources, predict crop yields, manage the urban-agricultural interface, and create drought resilience. However, the limitations of the spatial resolution of datasets like SMAP and MODISlimit fine-scale field validation due to coarse spatial resolution. To enhance transparency and successful adoption of AI in the future, ground-truth datasets, finer resolution SAR images, and explainable AI mechanisms should be included in future work. This chapter makes GeoAI a groundbreaking paradigm of precision agriculture where Earth observation, artificial intelligence, and cloud computing are brought together as one decision-support system.

REFERENCES

Afuye, G. A., Nduku, L., Kalumba, A. M., Santos, C. A. G., Orimoloye, I. R., Ojeh, V. N., & Sibandze, P. (2024). Global trend assessment of land use and land cover changes: A systematic approach to future research development and planning. *Journal of King Saud University. Science, 36*(7), 103262. DOI: 10.1016/j.jksus.2024.103262

Al-Shaar, W., Bonin, O., de Gouvello, B., Chatellier, P., & Hendel, M. (2022). Geographically weighted regression-based predictions of water–soil–energy nexus solutions in Île-de-France. *Urban Science (Basel, Switzerland), 6*(4), 81. DOI: 10.3390/urbansci6040081

Al-Shaar, W., Lehmann, X., Saad, N., Elmazi, G., Al-Shaar, M., & Tohme, C. (2025). Urban morphology and climate vulnerability assessment in Kuwait: A spatio-temporal predictive analysis utilizing deep neural network-enhanced Markov chain models for 2050 and 2100. *PLoS One, 20*(8), e0318604. DOI: 10.1371/journal.pone.0318604 PMID: 40824951

An, H., Ouyang, C., & Chen, X. (2025). Real-time estimation of SMAP soil moisture in mountainous areas and its impact on rainfall-runoff simulation. *Journal of Hydrology (Amsterdam), 133487.* Advance online publication. DOI: 10.1016/j.jhydrol.2025.133487

Awada, H., Di Prima, S., Sirca, C., Giadrossich, F., Marras, S., Spano, D., & Pirastru, M. (2021). Daily actual evapotranspiration estimation in a Mediterranean ecosystem from Landsat observations using SEBAL approach. *Forests, 12*(2), 189. DOI: 10.3390/f12020189

de la Cruz, L. G. (2024). *Optimizing sustainable agriculture with EO-derived data for soil health assessment* (Doctoral dissertation, Central European University). https://www.etd.ceu.edu/2024/de-la-cruz_leizel.pdf

Eisfelder, C., Boemke, B., Gessner, U., Sogno, P., Alemu, G., Hailu, R., & Huth, J. (2024). Cropland and crop type classification with Sentinel-1 and Sentinel-2 time series using Google Earth Engine for agricultural monitoring in Ethiopia. *Remote Sensing (Basel), 16*(5), 866. DOI: 10.3390/rs16050866

Fu, Y., Zhu, Z., Liu, L., Zhan, W., He, T., Shen, H., & Ao, Z. (2024). Remote sensing time series analysis: A review of data and applications. *Journal of Remote Sensing, 4*, 0285. DOI: 10.34133/remotesensing.0285

Gong, Z., Gao, F., Chang, X., Hu, T., & Li, Y. (2024). A review of interactions between irrigation and evapotranspiration. *Ecological Indicators, 169*, 112870. DOI: 10.1016/j.ecolind.2024.112870

Goyal, A., Aggarwal, R., Mahajan, S., & Saran, S. (2025). Automated GUI-based system for in-season crop classification and acreage estimation using multi-temporal SAR data and machine learning. *Archives of Agriculture Research and Technology, 6*, 1082. DOI: 10.54026/AART/1082

Li, W. (2025). Artificial intelligence in Earth science: A GeoAI perspective. *Journal of Geophysical Research: Machine Learning and Computation, 2*(3), e2025JH000691. https://doi.org/DOI: 10.1029/2025JH000691

Liu, Y., Zhang, S., Zhang, J., Tang, L., & Bai, Y. (2021). Assessment and comparison of six machine learning models in estimating evapotranspiration over croplands using remote sensing and meteorological factors. *Remote Sensing (Basel), 13*(19), 3838. DOI: 10.3390/rs13193838

Mahajan, S. (2024). An irrigation system: Design, implementation, and benefits. In *Smart agritech: Robotics, AI, and Internet of Things (IoT) in agriculture* (pp. 309–328). https://doi.org/DOI: 10.1002/9781394302994.ch11

Mavisclara, O. A., Esekie, J. O., Atoyebi, T. O., Atumah, P. E., Akadiri, O. O., Jimoh, R. A., & Oshobugie, I. (2025). GeoAI at the forefront of climate action: Mapping mitigation and adaptation with artificial intelligence. *Global Journal of Engineering and Technology Advances*, *24*(2), 217–234. DOI: 10.30574/gjeta.2025.24.2.0248

Melichar, M., Didan, K., Barreto-Muñoz, A., Duberstein, J. N., Jiménez Hernández, E., Crimmins, T., & Nagler, P. L. (2023). Random forest classification of multitemporal Landsat 8 spectral data and phenology metrics for land cover mapping in the Sonoran and Mojave deserts. *Remote Sensing (Basel)*, *15*(5), 1266. DOI: 10.3390/rs15051266

Moumane, A., Elmotawakkil, A., Hasan, M. M., Kranjčić, N., Batchi, M., Karkouri, J. A., & Youssef, Y. M. (2025). Integrating GIS, remote sensing, and machine learning to optimise sustainable groundwater recharge in arid Mediterranean landscapes: A case study from the Middle Draa Valley, Morocco. *Water (Basel)*, *17*(15), 2336. DOI: 10.3390/w17152336

Sarif, M. O., & Gupta, R. D. (2022). Evaluation of seasonal ecological vulnerability using LULC and thermal state dynamics using Landsat and MODIS data: A case study of Prayagraj City, India (1987–2018). *Environmental Science and Pollution Research International*, *29*(51), 77502–77535. DOI: 10.1007/s11356-022-21225-7 PMID: 35676584

Simón Sánchez, A. M., González-Piqueras, J., de la Ossa, L., & Calera, A. (2022). Convolutional neural networks for agricultural land use classification from Sentinel-2 image time series. *Remote Sensing (Basel)*, *14*(21), 5373. DOI: 10.3390/rs14215373

Tang, R., Peng, Z., Liu, M., Li, Z. L., Jiang, Y., Hu, Y., & Fisher, J. B. (2024). Spatial-temporal patterns of land surface evapotranspiration from global products. *Remote Sensing of Environment*, *304*, 114066. DOI: 10.1016/j.rse.2024.114066

Ullah, S., Qiao, X., & Abbas, M. (2024). Addressing the impact of land use and land cover changes on land surface temperature using machine learning algorithms. *Scientific Reports*, *14*(1), 18746. DOI: 10.1038/s41598-024-68492-7 PMID: 39138301

Vitale, A., & Lamonaca, F. (2025). Advancing built-up area monitoring through multi-temporal satellite data fusion and machine learning-based geospatial analysis. *Remote Sensing (Basel)*, *17*(11), 1830. DOI: 10.3390/rs17111830

Xu, J., Su, Q., Li, X., Ma, J., Song, W., Zhang, L., & Su, X. (2024). A spatial downscaling framework for SMAP soil moisture based on a stacking strategy. *Remote Sensing (Basel)*, *16*(1), 200. DOI: 10.3390/rs16010200

Yu, J., Zhao, L., Liu, Y., Chang, Q., & Wang, N. (2025). Automatic crop type mapping based on crop-wise indicative features. *International Journal of Applied Earth Observation and Geoinformation*, *139*, 104554. DOI: 10.1016/j.jag.2025.104554

Zha, X., Jia, S., Han, Y., Zhu, W., & Lv, A. (2025). Enhancing soil moisture prediction in drought-prone agricultural regions using remote sensing and machine learning approaches. *Remote Sensing (Basel)*, *17*(2), 181. DOI: 10.3390/rs17020181

Chapter 7
GeoAI-Enabled Diffusion:
CoAtNet Framework for Robust Plant Disease Detection Under Real-World Field Conditions

G. Vennila
http://orcid.org/0000-0002-2205-518X
Mohan Babu University, India

Prabhu Shankar B.
http://orcid.org/0000-0003-4394-9171
Vel Tech Rangarajan Dr. Sagunthala R&D Institute of Science and Technology, India

Kanniappan Elumalai
Parul University, India

ABSTRACT

Plant diseases significantly impact negatively on agricultural productivity, resulting in huge financial losses and triggering a threat to the security of the world food supply. This study suggests a GeoAI-enabled hybrid deep learning framework that combines CoAtNet for reliable disease classification with Diffusion Models for image restoration in order to overcome these drawbacks. The suggested model was assessed using the PlantVillage dataset and artificially generated occluded images that simulate the real-world agricultural conditions. Based on the experiment findings, the Diffusion CoAtNet framework was able to perform better on degraded datasets compared to baseline CNN and Transformer-based models, with a 96.4% accuracy. Also, the system demonstrated good edge deployment performance, with a rate of approximately 2.4 frames per second (FPS) on Raspberry Pi hardware, and is suitable to real-time field tasks such as localized crop monitoring and disease diagnosis using a smartphone.

INTRODUCTION

Agriculture is important to global food security, economic stability and rural livelihoods. Crop health is critical to sustainable development since a large percentage of the world population is directly or indirectly reliant on agriculture. The risk of plant diseases, nevertheless, remains highly problematic

DOI: 10.4018/979-8-3373-8037-7.ch007

to the agricultural output and, according to the estimates, the global crop yield loses 2040% every year. These losses not only impact the income of farmers, but also destabilize food chain, are raised in food prices and global food security becomes a challenge.

Traditional methods that have been used to diagnose plant diseases largely rely on the manual inspection of the diseases by agricultural experts. This is a lengthy, subjective and difficult to scale in large or geographically dispersed farmlands, even though it is effective in the controlled or small scale situation. Other challenges that accompany real agricultural environments are obstruction by overlapping leaves, shadows caused by uneven light, clutter in the backgrounds and noises in the environment, which significantly reduce the accuracy of diagnosis.

Recent advances in deep learning (DL) and artificial intelligence (AI) have allowed applying image-based analysis to automated plant disease detection. Convolutional Neural Networks (CNNs) such as AlexNet, VGG16, ResNet, and DenseNet have demonstrated great accuracy on benchmark datasets such as PlantVillage. Nevertheless, due to their susceptibility to image degradation and their inability to infer relationships arising to a global context, such models often cannot effectively generalize in the field. Despite the improved global feature models with Vision Transformers, it is not always possible to deploy them on low-power edge devices commonly used in agriculture, as they have a large dataset and consume a lot of computational resources.

This study suggests a hybrid deep learning framework that combines CoAtNet and Diffusion Models to overcome these drawbacks. Diffusion models utilise noise-reversal probabilistic process to rebuild missing visual details to restore images of degraded or occluded images. CoAtNet is an attention-based hybrid architecture that integrates convolutional architecture with attention mechanisms to obtain stable disease classification by grasping local texture characteristics and the global spatial relationship. This two-step method improves the strength of the models in harsh environmental conditions and is computationally efficient when it comes to the deployment of edges.

This not only improves the performance of classification, but also looks at the problem of plant disease detection in the light of GeoAI (Geospatial Artificial Intelligence). The proposed system connected geographic location on devices with GPS-based capabilities, such as smartphones, drone, and Internet of Things sensors, to every disease prediction. A combination of these spatially referenced outputs with Geographic Information Systems (GIS) can enable disease mapping, hotspots analysis, and spatial decision-making. Such integration facilitates precision agriculture practices by facilitating targeted interventions such as localized application of pesticides, crop surveillance and optimization of resources.

The proposed DiffusionCoAtNet architecture makes the transition between accuracy on the laboratory scale and the usefulness of artificial intelligence in agriculture by combining image restoration, hybrid deep learning, and geospatial intelligence. The system helps in sustainable farming practices and improved management of crop yield through the provision of a scalable, effective, and field-deployable system that detects early plant diseases.

LITERATURE REVIEW

The agricultural sector has experienced a significant change with the rapid development of Artificial Intelligence (AI) and Deep Learning (DL) particularly in the field of automated plant disease detection. There has been a combination of large-scale datasets, advanced neural network structures, and computer vision to produce intelligent systems capable of diagnosing plant diseases accurately. This chapter will provide an in-depth discussion of existing solutions, such as hybrid designs, generative systems, deep learning-based solutions, traditional machine learning algorithms, and recently developed GeoAI-based systems. Additionally, it draws attention to their shortcomings and points out research gaps that spur the development of the Diffusion–CoAtNet framework.

CONVENTIONAL MACHINE LEARNING METHODOLOGIES

The early research on plant disease detection mainly involved conventional image processing and machine learning techniques. These methods involved extraction of manually generated features such as color histograms, shape features, and texture features such as Gray-Level Co-occurrence Matrix (GLCM) and Local Binary Patterns (LBP). These features were then classified using algorithms such as Support Vector Machines (SVM), k-Nearest Neighbors (KNN), Decision Trees and Random Forests.

In one of the first studies, Burks et al. (2006) showed promising results on small datasets when using color-texture features and discriminant analysis for citrus disease detection. These techniques were not without drawbacks despite establishing the way to automated disease detection. The system was also not as scalable and adaptable to other crops and diseases since feature extraction was manually tuned and had to rely on domain knowledge.The system was less scalable and adaptable to other crops and diseases as feature extraction needed domain expertise and manual tuning. Also, hand-crafted features were very sensitive to environment variations including light, occlusion, and background noise, which made them poor in the generalization in the real world environment.

DEEP LEARNING AND CNN-BASED METHODS

The introduction of Convolutional Neural Networks (CNNs) was a significant breakthrough in the field of plant disease detection because it opened up the possibility of automatically deriving features of unprocessed images. CNNs perform image classification quite well due to the fact that they learn hierarchical representations of visual features, starting with low-level edges, and complex disease patterns.

Mohanty et al. (2016) demonstrated the effectiveness of CNNs by training them on the PlantVillage dataset, achieving classification accuracies of over 99% on various plant disease classes. Subsequent studies enhanced performance and stability on classification tasks in controlled settings with more elaborate architectures such as VGG16, ResNet50, and DenseNet121.

However, CNN-based models face significant challenges when used in real agricultural environments even though they are very precise. Their ability to acquire global contextual information, which is essential when the symptoms of the disease are partially observed or are spatially spread, is limited because they depend on the local receptive fields. Moreover, in non-laboratory application, CNNs deteriorate to a large degree because of their vulnerability to degradation aspects of images such as occlusion, noise, and illumination variations. Moreover, deep CNNs are not well suited to running on edge computing devices in precision agriculture since they often require substantial computing power.

RNN AND LSTM-BASED METHODS

To simulate the temporal dynamics of plant disease progression, Recurrent Neural Networks (RNNs) and Long Short-Term Memory (LSTM) networks have been investigated. These models have been applied to follow the changes in diseases with time using time-series image data since they are able to identify consecutive patterns.

Zhang et al. (2019) showed improved early detection by suggesting an RNN-based system to monitor disease progression based on sequential leaf images. Still, the most common use case of assets of agricultural diagnostics, i.e. the classification of static images, does not fit well with RNN-based techniques. They also have other limitations to their practical applicability in real-time systems, such as slow training, high computational complexity, and vanishing gradients.

IMAGE ENHANCEMENT TECHNIQUES AND GENERATIVE MODELS

Recurrent neural networks (RNNs) and long short-term memory (LSTM) networks were investigated to capture dynamic aspects of plant disease. Since these models are able to detect sequential patterns, they have been applied to monitor how diseases have evolved over time with time-series image data.

IMAGE RESTORATION WITH DIFFUSION MODELS

The new type of generative models called Diffusion Models has been presented, which is able to create high-quality images with the help of probabilistic noise reversal method. Ho et al. (2020) proposed diffusion models based on the iterative noise addition approach and training to undo this noise to recover original images.

Unlike GANs, diffusion models can be trained in a stable environment and have higher convergence rates and image reconstruction qualities. Moreover, they are not affected by mode collapse. The recent research has shown that diffusion models have the potential to restore damaged images. In this regard, Song et al. (2022) have indicated that diffusion models have the capacity to recover images with high-level information, and this aspect can be used to enhance plant disease images.

HYBRID ARCHITECTURES AND VISION TRANSFORMERS

In their proposal, Dosovitskiy et al. (2021) introduce Vision Transformer (ViTs) models that use self-attention models to learn about dependency in images on a global scale. Their ability to achieve remarkable results with many image classification problems, however, renders them inapplicable to agricultural problems because of the large amount of data and computations they require.

To alleviate these issues, a number of models including CoAtNet have been suggested. The concept underpinning CoAtNet is that, local regions use convolutional layers to learn features efficiently and global dependencies are learned using attention models. The CoAtNet has proven to be very efficient and performant in many problems as demonstrated by Dai et al. (2021).

SPATIAL INTELLIGENCE AND GEOAI IN AGRICULTURE

The recent advancements of GeoAI (Geospatial Artificial Intelligence) have offered a novel aspect to agricultural analytics by introducing AI models having geospatial information and Geographic Information System (GIS) technologies. GeoAI has offered the use of spatial analysis of agro data, disease distribution, crop health, and environment.

The incorporation of AI models to detect diseased and GPS to gather data has offered features of disease distribution maps in geographical locations. This has assisted in the distribution of the disease, and has given the ability to perform precision agriculture.

The majority of the current models of plant disease detection are limited to image classification, and none of them is combined with geospatial data. Precision agriculture becomes difficult due to the absence of integrating it with geospatial data.

RESEARCH GAPS

An analysis of the literature available indicates that there are a number of gaps in research in the field. Existing algorithms are usually difficult to cope with the reality, especially when the images are subject to occlusion, noise, or lighting changes. Moreover, the amount of work that integrates image restoration methods with classifier models is minimal, which may enhance the overall performance. Most deep learning methods are also characterized by high computational complexity, which makes them less feasible in extensive applications. Besides, the use of geospatial intelligence is highly ignored in the plant disease detection systems, although it is potentially valuable. On the whole, it is evident that a consistent framework is required, one that can successfully incorporate image restoration, classification and geospatial intelligence into one, powerful solution.

The gaps of the current methods indicate that the holistic framework of image restoration, robust classification, and geospatial intelligence should be taken into account, which is the cornerstone of the proposed GeoAI-enabled Diffusion-CoAtNet framework.

PROBLEM STATEMENT AND OBJECTIVES

Problem Description

Food security and economic stability of the world depend on agriculture. Nevertheless, the occurrence of the diseases in plants has continuously been a significant setback in ensuring the productivity of crops thereby resulting in a decline in productivity and economic stability.

Despite the recent breakthroughs of deep learning that contributed to the creation of automated systems of the detection of diseases in plants, the currently in use approaches are largely experimented in a laboratory setting with clean and structured data, e.g. the PlantVillage. Nevertheless, the reality might not be identical, and the pictures can be distorted by various aspects like occlusion, noise, and light.

The existing approaches have a poor performance in the real world, as CNN-based approaches are not capable of classifying the images that are not visible in the dataset. Also, the existing approaches may not be feasible for practical applications, as most of the existing approaches are not capable of working on low-power devices such as smartphones and drones, due to the high computational cost of the existing approaches.

The second major weakness of the existing systems is that it has not been integrated with geospatial intelligence. The existing methods in the majority of cases just depend on the categorization of images without taking into consideration the distribution of diseases in the agricultural fields.

Rationale Behind the Study

This research is motivated by the inefficiency of the existing plant disease detection systems and the increasing demand of the solutions working well in the real-life agricultural environment. Most of the existing models have been shown to perform well under controlled or simulated environments but cannot keep their accuracy when they are exposed to real-world challenges like occlusion, noise and different illumination. Moreover, partially concealed and damaged images also lower the reliability of classification, which underlines the role of considering powerful image-restoring algorithms. A demand of lightweight and efficient models that do not rely heavily on cloud infrastructure to run on low-power edge devices is another major issue. Further, majority of the current methods fail to use spatial or geospatial data, which have the potential to greatly improve the accuracy of detection and decision-making. Lastly, recent advances in hybrid learning methods, especially by incorporating diffusion models and CoAtNet structures, are a chance to create a more effective and robust structure that will overcome these shortcomings.

Problem Definition

The study aimed at addressing the problem of developing an effective, powerful and scalable plant disease detector system, which could be applied in the field.

The system was to:

a. Process pictures of leaf in a real-world setting, such as those that can be influenced by other effects such as occlusion, noise, and lighting.
b. Improve images that can be deteriorated or partially covered by a diffusion-based image enhancer.
c. Effectively identify plant diseases using a hybrid CoAtNet model.
d. Work well on edge devices, where efficiency is a must in real time processing.
e. Add geospatial characteristics to make spatial decisions and precision agriculture.

Study Objectives

The main goals of this study focus on creating an effective and practical system to detect plant diseases. To start with, the research sought to review the available methods of detection and see their limitations especially in the field of practicality. It further aims at creating a diffusion-trained image restoration model to improve the quality of deteriorated leaf images. Moreover, the study suggests the creation of a hybrid system that combines the diffusion models with CoAtNet to classify diseases accurately. The proposed system was tested in terms of benchmark datasets as well as simulations that represent real-life situations. The other target is to streamline the system to allow it to run on resource-constrained platforms like Raspberry Pi and smartphones. Lastly, the study also aimed to integrate GeoAI features through the incorporation of spatial mapping applications, which will enhance the overall intelligence and usefulness of the system.

Key Outcomes

The research has resulted in some important conclusions in the area of detecting the disease in plants. The method proposed finds correct classification of the images even in cases of image occlusion or degrading images and this is a great advantage in the real world. It is also made strong with the addition of diffusion-based image restoration methods. The system is also able to perform effectively in real time and thus is well integrated to run on edge devices with scarce computing capabilities. Moreover, geospatial intelligence allows successful spatial surveillance of plant diseases. On the whole, the suggested approach demonstrates a significant enhancement of the performance in relation to the conventional CNN-based approaches.

IMPORTANCE OF THE RESEARCH

Importance of Plant Disease Detection

In the world, agriculture plays a very crucial role in ensuring food security, economic growth and the livelihoods of people. In such countries as India, a significant portion of the population depends on the agricultural industry whether directly or indirectly. The availability of food, livelihood of the people and the economy of the country are significantly affected by crop health. Nevertheless, the occurrence of plant diseases such as fungi, bacteria, viruses and pests has always been a significant contributor to the productivity of the agricultural sector.

According to reports, around 20-40% of the crop yield is lost every year due to the presence of diseases in plants, causing a major loss to the economy of the country. The small and marginal farmers of the country end up bearing the brunt of these losses since they lack the means to acquire the most recent technology and expertise in the area of agriculture.

Contribution to Precision Agriculture and GeoAI

The proposed Diffusion–CoAtNet model is beneficial in the field of precision agriculture in terms of the accurate and real-time prediction of plant diseases in the field. By using the GeoAI concept, the proposed system is not only limited to image classification but is also extended to provide spatial intelligence.

Each prediction of the disease can be assigned with geographic coordinates using GPS-enabled devices such as smartphones, drones, and IoT sensors. This can be incorporated further with Geographic Information Systems (GIS) to determine the following:

a. Determine the hotspot of the diseases in the field.
b. Compare the spatial distribution of the diseases.
c. Spray the field with the pesticide.
d. Help in decision-making in the field of precision agriculture

Technical Contributions and Innovation

The current study suggests a hybrid deep learning method, which combines Diffusion Models and CoAtNet in detecting plant diseases.

a. Diffusion-Based Image Restoration:

The diffusion model recovers damaged and covered images of leaves and thus it is easier to see the symptoms of plant diseases and enhance the classification process.

b. Hybrid Feature Extraction using CoAtNet:

The CoAtNet model is a hybrid of local feature extraction (convolutional operations) and global feature understanding (attention mechanisms). This enables the model to learn about the local and global spatial association among various sections of an image.

c. Stability to Real-World Conditions:

The approach proposed is contrasting to traditional approaches being trained on a clean dataset. It is robust to different conditions that can be present in a real-world scenario.

d. Computational Efficiency:

The given method is computationally efficient and can be carried out with the help of edge devices.

Socioeconomic Impact

The advantages of the proposed system have a lot of socioeconomic implications especially on the farmers in the rural and resource-starved areas. In practice, a lack of access to professional advice on agriculture compels farmers in most instances to rely on trial and error, which can result in excessive use of pesticides and rise in the production expenses. The automated disease detection system that has been developed in this study helps in solving these challenges since it allows the detection of plant diseases early hence contributing to crop losses reduction. It also minimizes the un-needed use of pesticides, which will encourage more sustainable agricultural systems. Moreover, the system improved farmer profitability by reducing the cost of inputs. Moreover, it enhances confidence of farmers to make informed decisions, which ultimately contributes to the effective and sound agricultural management.

Environmental Sustainability

One of the issues of contemporary agriculture is environmental sustainability. Excessive or improper utilization of pesticides leads to environmental degradation, contamination of soil, biodiversity loss and health risks to both the farmer and the consumer.

the suggested system encouraged environmental sustainability in farming by the use of pesticides wisely. It is also possible to determine with accuracy the area of the disease which leads to less use of pesticides thereby making the system environmental friendly.

The proposed edge computing approach also minimized the usage of cloud services, resulting in low energy consumption.

Academic and Research Relevance

Regarding the research, the current study makes a contribution to the field in the following manner: the establishment of a single framework which is a synthesis of image restoration, classification, and geospatial intelligence in the agriculture field.

The proposed approach has the following potential for opening new research directions:

a) Combination of different types of multimodal data in the sphere of agriculture.
b) Expansion of the suggested approach to the other fields of application in the sphere of agriculture.
c) Lightweight forms of the proposed hybrid approach development.
d) Research on the possibilities of GeoAI in decision support systems.

The Industrial and Practical Applications

The suggested system exhibits great deployment opportunities in a wide range of applications in agriculture. It may be employed to create disease detection applications in smartphones, which will allow farmers to detect diseases easily in their plants in real-time. Crop monitoring on the system can also be optimized to use the drone and ensure that the fields are analyzed in large scales and efficiently. Moreover, it can be combined with the IoT-based smart farming solutions to facilitate the automated and data-driven farming. Moreover, the framework can also aid in the creation of agricultural monitoring systems by the government that could enable effective policy implementation and management of large crops.

The proposed system has promising opportunities of application in various areas since it consumes little power.

Ethical and Social Issues

The reason is that the system is designed in such a way that it will guarantee the ethical application of AI. This can be achieved since processing is local and does not involve transfer of sensitive information to the cloud. This guarantees privacy and security of the data.

Furthermore, the system is transparent by using interpretability tools like Grad-CAM. This is made available since the users are able to understand the decision making process of the system. This builds confidence of the farmers and motivates them to use the AI system.

The proposed system is to help the farmers and not to substitute the traditional knowledge of the farmers..

METHODOLOGY

System Overview

The suggested system is a two-phase hybrid deep learning system, which specifically targets the detection of plant diseases in the real world. The model combines Diffusion Models to restore images, and CoAtNet to classify images, improving their performance in real-world scenarios, such as with occlusion, noise, and illumination.

The main difference between this proposed framework and traditional end-to-end classification models is that this framework is divided into two stages: image enhancement and classification. This makes sure that the images undergo an enhancement prior to classification leading to increased accuracy.

The overall pipeline of this framework includes the following stages:

Input Image, Preprocessing, Diffusion-Based Restoration, CoAtNet Classification, GeoAI Mapping and Output Prediction.

System Architecture and Workflow

The system architecture has the following components:

Input Layer:

The images of the leaves are captured by the smartphone, drones, or IoT-based agricultural sensors. These images can be distorted, obscured or blurred by the environment.

Preprocessing Module:

The pictures are rescaled to a standard size (224 x 224) and are normalized to improve the quality of the pictures through simple filtering processes.

Diffusion Restoration Module:

This aspect reconstructs the images by eliminating the noise in the images, and hence enhances the quality of the images.

CoAtNet Classification Module:

These pictures are then fed into the CoAtNet network where the local and global features are extracted to categorize the diseases appropriately.

GeoAI Mapping Layer:

The classified disease label is mapped to the geographic location (latitude and longitude) by the GPS device.

Output Layer:

The resulting product will contain the estimated disease category and the confidence scores and optional spatial mapping data.

Figure 1. System architecture diagram

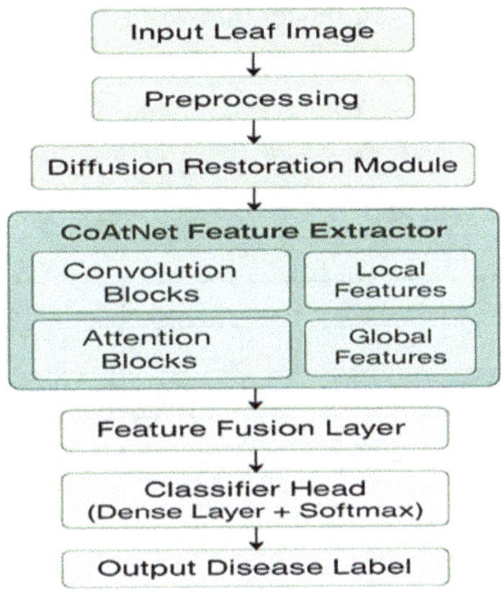

The proposed system has the following workflow depicted in the diagram below: input leaf image - preprocessing - diffusion-based restoration - CoAtNet feature extraction (convolution + attention blocks) - feature fusion - classifier - final illness label. It shows the combination of classification and enhancement to identify plant diseases at the right time.

Dataset Preparation

The model was trained and evaluated using the PlantVillage dataset, which consists of more than 50,000 labeled images spanning 38 different plant disease categories. However, in this study, a subset of 7 representative disease classes was selected to simulate real-world deployment scenarios and reduce computational complexity. The selection was based on commonly occurring diseases and visual distinctiveness among classes.es To better reflect real-world conditions, an additional dataset was created by applying various transformations to simulate practical challenges. These include random occlusion to represent leaf overlap, Gaussian blur to mimic motion and camera noise, shadow overlays to introduce lighting variations, and contrast reduction along with color distortion to emulate changes in visual quality. The images were resized to 224x224 and normalized between [0,1]. Mean variance normalization is used.

Diffusion models are a type of probabilistic generative model which consists of reversing a noise process to reconstruct images.

Proceeding Procedure:

noise is gradually added to the progression:

$$q(xt \mid x_{t-1}) = N(xt; \sqrt{1\text{-beta}_t}\, x_{t-1}, \text{beta}_t I)$$

Reverse Process:

$$L = E_{x_0, e, t}\, [\, \| e - e0(xt, t) \|2 \,]$$

It is an efficiency-based latent diffusion model that works in latent space through a convolutional autoencoder and minimizes computations without decreasing fidelity. The training details are shown in Table 1.

Table 1. Training details of diffusion model

Parameter	Value
Noise Steps	1000
Optimizer	AdamW
Learning Rate	1e-4
Batch Size	16
Loss Function	MSE
Epochs	200
Framework	PyTorch

The diffusion model was trained for 200 epochs based on empirical convergence observations. It was observed that the training and validation loss stabilized after approximately 180 epochs, indicating that the model had sufficiently learned the underlying data distribution without overfitting. Therefore, training was continued up to 200 epochs to ensure stable convergence.

The trained model is used to restore images that have been obscured or blurred and submit them to CoAtNet to get classified.

CoAtNet for Disease Classification

CoAtNet is a hybrid architecture that combines convolutional layers with attention mechanisms. The Training details of CoAtNet Disease Classification is shown in Table.2.

Architecture Components

1. Convolutional layers for local feature extraction
2. Attention blocks for global dependency modeling
3. Fully connected layers for classification

Table 2. Training details of CoAtNet disease classification

Parameter	Value
Model	CoAtNet-0
Optimizer	Adam
Learning Rate	1e-4

continued on following page

Table 2. Continued

Parameter	Value
Epochs	100
Augmentation	Random flips, rotations, jitter on brightness

Diffusion and CoAtNet Integration:
The proposed method integrates Diffusion to assist in the learning process.

Integration of Diffusion and CoAtNet

The integrated pipeline operates sequentially:

1. Input image is preprocessed
2. Diffusion model restores degraded features
3. Restored image is passed to CoAtNet
4. Disease label and confidence score are generated

Pseudocode:

```
def detect_disease(image):
img = preprocess(image)
restored_img = diffusion_model.restore(img)
label, confidence = coatnet.predict(restored_img)
return label, confidence5.6 Evaluation Metrics
```

GeoAI Integration

To incorporate spatial intelligence, the system integrates a GeoAI layer:

a. Images are tagged with GPS coordinates
b. Predictions are mapped onto GIS platforms
c. Spatial analysis enables disease hotspot detection

This allows:

a. Precision spraying
b. Field-level monitoring
c. Large-scale agricultural analytics

Additionally, inference time and FPS were measured for edge deployment evaluation.

Hardware and Software Configuration

Hardware:

a. NVIDIA RTX 3060 GPU
b. Intel i7 Processor
c. Raspberry Pi 4 (4GB RAM)

Software:

a. Python 3.10
b. PyTorch
c. OpenCV, NumPy
d. Ubuntu 22.04

APPLICATION AND TEST OUTCOMES

Experimental Setup

The Diffusion-CoAtNet framework proposed was trained with the PyTorch deep learning framework. The training process was carried out using the NVIDIA RTX 3060 GPU, whereas the deployment process was carried out using the Raspberry Pi 4 (4GB RAM) for testing the real-time performance.

Dataset Description

The experiments were carried out on the PlantVillage data. Although the original dataset contains 38 classes, this study focuses on a subset of 7 disease categories for experimental evaluation. The reported accuracy of 96.4% corresponds specifically to this selected subset and not the full dataset.

To simulate natural farming conditions, an enhanced dataset, called PlantVillage-O, is employed. Various transformations are used to obtain this dataset and they include:

Random Occlusion (overlapping of leaves)
Gaussian Blur (noise in the camera)
Shadow Overlay (variations in lighting)

Color Distortion

Data Splitting and Evaluation Protocol.
The dataset was split the following way:

a) Training Set: 70%
b) Validation Set: 15%
c) Test Set: 15%

The 96.4% accuracy reported is with regard to the performance on the held-out test set, where bias in evaluation does not exist.

To compare all the baseline models (CNN, GAN+CNN, Vision Transformer), all baseline models were trained and evaluated with the same data splits.

Implementation Details

Diffusion Model

a) Optimizer: AdamW
b) Learning Rate: 1e-4.
c) Batch Size: 16
d) Epochs: 200
e) Loss Function: Mean Squared Error.

The number of training epochs was selected based on convergence behavior observed during experimentation. Validation performance plateaued after approximately 180 epochs, and no significant overfitting was observed beyond this point.

CoAtNet Model

a) Model: CoAtNet-0 (ImageNet pretrained)
b) Optimizer: Adam
c) Learning Rate: 3 - 4e.
d) Batch Size: 32
e) Epochs: 100
f) Loss Function: Cross-Entropy.

The two modules have been combined to form a single inference pipeline in which an image is restored then classified.

Qualitative Results (Image Restoration)

The diffusion model has greatly enhanced the quality of images by reconstructing missing parts and damaged areas. Visual inspection confirmed:

a. Removal of shadows and noise
b. Restoration of parts of concealed leaf areas.
c. Better understanding of disease patterns.

This improvement directly led to a better accuracy of classification.

The Evaluation Metrics are considered for experimentation are shown in Table 3.

Table 3. Evaluation metrics

Metric	Formula	Description
Accuracy	$\frac{TP + TN}{TP + TN + FP + FN}$	Overall correctness
Precision	$\frac{TP}{(TP + FP)}$	Positive prediction quality
Recall	$\frac{TP}{(TP + FN)}$	Detection ability
F1-Score	$\frac{(2 * P * R)}{(P + R)}$	Balance of precision & recall
IoU	$\frac{TP}{(TP + FP + FN)}$	Overlap measure

GEOAI-BASED SPATIAL ANALYSIS

To validate the GeoAI component of the proposed framework, spatial analysis was performed by associating disease predictions with geographic coordinates. Each input image was assigned simulated GPS coordinates representing field locations.

The predicted disease labels were mapped onto a spatial grid using Geographic Information System (GIS) visualization techniques. This enabled the generation of disease distribution maps and hotspot regions.

The spatial visualization revealed that disease occurrences were clustered in specific regions, indicating potential hotspot zones. These hotspots can assist farmers in applying targeted treatments such as localized pesticide spraying, thereby reducing unnecessary chemical usage.

Furthermore, spatial mapping enables field-level monitoring and large-scale agricultural analytics, providing actionable insights for precision agriculture.

Figure 2. Disease distribution and hotspot visualization using GeoAI

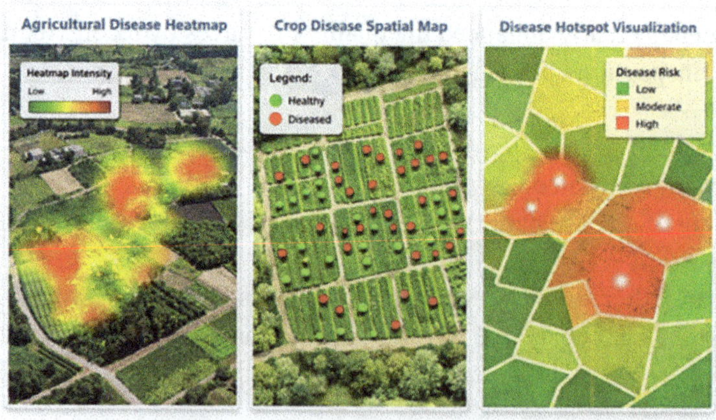

Figure 2 illustrates the spatial distribution of plant diseases using GeoAI-based analysis. The heat-map shows disease intensity across regions, while marked points represent healthy and diseased crops. High-risk zones (hotspots) are highlighted, demonstrating the potential of the proposed system for spatial decision-making in precision agriculture.

EXPERIMENTAL RESULTS

1. Visual Quality

The diffusion model significantly enhanced occluded images — restoring hidden regions, removing shadows and blur, and improving color uniformity for accurate classification.

The Quantitative Results are shown in Table 4.

Table 4. Quantitative results

Model	Dataset	Accuracy	Precision	Recall	F1
ResNet-50	Clean	98.1	97.8	98.0	97.9
ResNet-50	Occluded	83.4	80.9	82.1	81.5
GAN + CNN	Occluded	88.2	86.3	87.5	86.9
Vision Transformer	Occluded	91.7	90.8	90.9	90.8
Diffusion+CoAtNet (Proposed)	Occluded	96.4	95.7	96.0	95.8

The proposed hybrid improved occluded dataset accuracy from 83% (CNN) to 96.4%, outperforming all baselines.

Figure 3. Models comparison

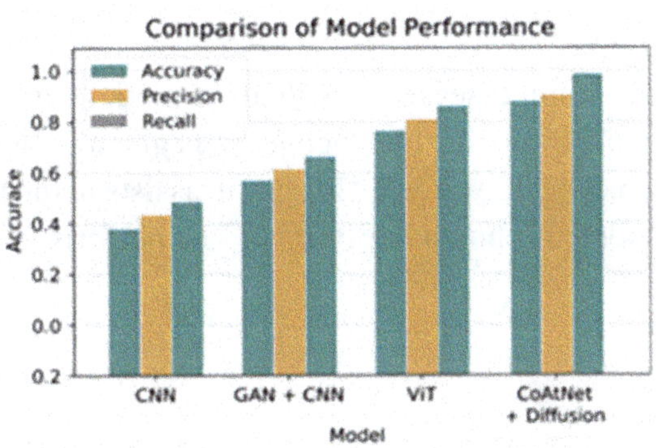

This bar chart compares the performance of four plant disease detection models: CNN, GAN + CNN, ViT, and CoAtNet + Diffusion. Accuracy, Precision, and Recall are shown for each model. The CoAtNet + Diffusion model achieves the highest scores across all metrics, demonstrating superior classification ability and robustness under real conditions are shown in Figure 3 and The Performance on edge devices are shown in Table 5.

Table 5. Performance on edge devices

Device	Inference Time	FPS	Power
NVIDIA RTX 3060	22 ms	45	160W
Raspberry Pi 4	420 ms	2.4	7W

Despite reduced compute, inference on Raspberry Pi achieved near real-time processing (~2.4 FPS), confirming its feasibility for field use.

Figure 4. Attention heatmap

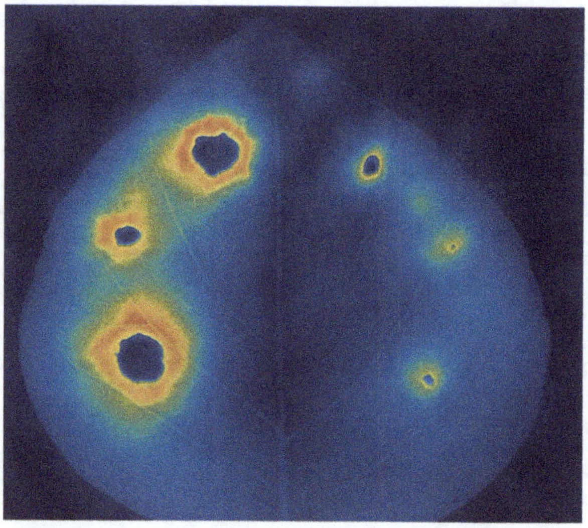

The heatmap shows the position of a diseased leaf that the model pays attention to when making a prediction. The red and yellow spots show that much concern is given to the disease outbreak areas like spots or lesions and blue spots depict healthy areas. This graph assists in the interpretation and validation of the decision-making of the model are shown in Figure 4. The performance comparison with existing work is shown in Table 6.

Table 6. Comparison with existing work

Study	Model	Limitation	Accuracy (%)
Mohanty et al., 2016	AlexNet	Poor lighting generalization	99.2 (clean), 76.3 (occluded)
Ferentinos,2018	VGG16	Overfitting	99.5, 80.2
Zhang et al., 2019	RNN-LSTM	Not image-suitable	85.7

continued on following page

Table 6. Continued

Study	Model	Limitation	Accuracy (%)
Lu et al., 2020	GAN-based	Training instability	88.1
Song et al., 2022	Transformer	High compute	91.2
Proposed (2025)	Diffusion + CoAtNet	Robust, efficient	96.4 (occluded)

The proposed system maintains high accuracy under occlusion, outperforming CNN, GAN, and Transformer-based approaches.

Visualization and Interpretability

Grad-CAM visualizations revealed model attention on disease-affected regions:

a. **Leaf Blight:** Brown lesion focus
b. **Mosaic Virus:** Discolored patches
c. **Powdery Mildew:** White fungal zones

This interpretability enhances trust and aids agronomists in validating predictions.

Figure 5. Confusion matrix heatmap

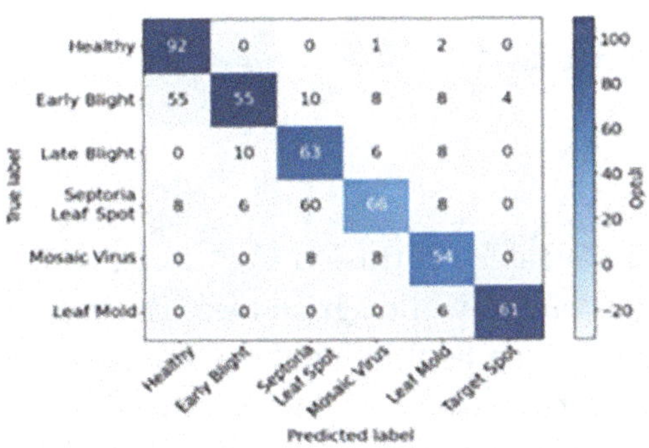

This confusion matrix shows classification results across the selected seven plant disease categories used in this study.. Dark diagonal cells represent correct predictions, while lighter off-diagonal cells show misclassifications. The model performs strongly on classes like Healthy, Leaf Mold, and Target Spot, with minor confusion between visually similar diseases are shown in Figure 5. The ablation studies are shown in Table 5.

Table 7. Ablation studies

Configuration	**Description**	**Accuracy (%)**
CoAtNet only	Without enhancement	90.5
Diffusion + CNN	Restoration + CNN	92.8
Diffusion + CoAtNet	Full hybrid	**96.4**

Findings support the assertion that diffusion restoration and hybrid attention provide a contribution to the overall performance. Model predictions were analyzed with the help of Grad-CAM visualizations. The results showed that:

1. The model targets the areas of the body affected by the disease like lesions and discolouration.
2. The irrelevant background areas are given minimal consideration.

This enhances transparency and trust on model predictions.

The suggested Diffusion–CoAtNet model showed great robustness in the actual real life situation. Diffusion model was able to restore the damaged image characteristics well whereas CoAtNet was able to capture local and global patterns to classify the image correctly.

The combination of these elements decreased the differences in performance between clean and occluded datasets and created a tradeoff between accuracy, efficiency, and deployability. The system is also compatible with edge devices, which allows it to be applied practically in agriculture.

CONCLUSION AND FUTURE SCOPE

Conclusion

This work has provided a GeoAI-enabled Diffusion-CoAtNet model of the powerful identification of plant diseases under the conditions of real-world agriculture. The proposed system has been developed to overcome practical challenges that can be faced in the existence of occlusions, noise, shadows, and changes in illumination unlike traditional methods which assume the availability of clean and controlled datasets.

The challenges of robust plant disease detection have been solved by the diffusion-based image restoration and CoAtNet-based classification that have helped the system to overcome its challenges. The experiment process has shown that the proposed system has achieved high accuracy of 96.4 percent in the cases of occluded datasets, which is much greater compared to conventional CNN, GAN-based, and even Transformer-based methods.

Besides enhancing the overall accuracy of the system, the introduction of GeoAI has also facilitated the addition of a new dimension of intelligence to the overall structure in plant disease detection. This has further helped in enhancing the overall applicability of the system in the domain of precision agriculture.

It was also demonstrated that the framework had effective execution on edge devices, and the framework was able to reach 2.4 FPS on the Raspberry Pi. Therefore, the suggested framework helps bridge the gap successfully between high accuracy in the laboratory setting and the real world application.

Key Contributions

The main contributions of the study can be summarized as follows. The research presents a new type of hybrid model, which integrates Diffusion and CoAtNet, to establish effective detection of plant diseases. It uses a diffusion-based image restoration method to address the issues like occlusions in the input images, thus enhancing overall model reliability. CoAtNet allows effective hybrid feature extraction, which improves the performance of classification. Moreover, the study introduces the GeoAI-driven spatial intelligence concept to facilitate disease mapping and enhance field-level analysis. The model is also designed in such a way that it can be deployed to the edge, which proves to be able to do real-time inference in low power devices. In addition, the paper has elaborate experimental validation on clean and occluded data to determine the effectiveness of the method.

Practical Implications

The suggested system has high practical relevance to contemporary agriculture, as it provides farmers with an opportunity to take pictures of the leaves on the crops with smartphones or other inexpensive cameras and get real-time predictions of the diseases. Such a possibility helps in the early detection of plant diseases and this is followed by early intervention and prevention. It also enables accuracy in application of pesticides, which will save on unnecessary pesticides and minimize environmental effects. The system can reduce the total cost of farming by optimizing the input resources, and also increase the productivity. Finally, it also leads to higher crop production and higher quality produce, thus it is a useful tool in efficient and sustainable farming methods.

Moreover, the GeoAI element is capable of integrating with GIS platforms, which makes it possible to monitor and make decisions based on data on a large scale in agriculture.

Limitations

Although the proposed study has good performance, it has some limitations which must be overcome. Diffusion models are relatively computationally intensive to train and thus can be less scalable and more accessible. Also, despite augmentation of the dataset, it might not be able to represent the broad environmental variations in the real-life environment. The system also has limitations in being able to support higher frame rates, and thus is not as viable in high-speed drone-based surveillance. Moreover, its functioning in extreme weather conditions is yet to be properly tested, which means that it requires additional testing and optimization.

Future Scope

This research has a number of good ways it can be extended in the future. The implementation of optimization methods, like model compression, pruning, and quantization, to enhance the efficiency and deployability of the system is one of the significant areas. The combination of the drone-based systems can also facilitate the large-scale, real-time monitoring of the crops. Moreover, adding multimodal data sources, including weather conditions, the nature of soils, and humidity, can help to make more context-dependent and accurate predictions. It is also possible to develop federated learning frameworks to support privacy-preserving model updates with distributed systems. Improving GeoAI with the addition of the spatiotemporal analysis would make disease prediction and forecasting more efficient. Lastly, the framework can be scaled to cover a wider scope of agricultural applications, including detection of pests, crop stress levels, and crop yield predictions.

The proposed Diffusion-CoAtNet approach shows how state-of-the-art deep learning methods can be combined with geospatial intelligence to solve pressing issues in agriculture. This study is a step towards intelligent, scalable and sustainable solutions to agriculture by incorporating image restoration, classification and spatial analysis into a single framework.

REFERENCES

Burks, T. F., Lee, W. S., & Pydipati, R. (2006). citrus disease detection based on discriminant analysis and color texture features. Agriculture Electronics and Computers. DOI: 10.13031/2013.21733

Dai, Z., Liu, H., Le, Q. V., & Tan, M. (2021). CoAtNet: Attention convolution with all data sizes. Innovations in Neural Information Processing Systems (NeurIPS). https://arxiv.org/abs/2106.04803

Dosovitskiy, A., Beyer, L., & Kolesnikov, A. (2021). An image is worth 16x16 words: Transformers for image recognition at scale. *International Conference on Learning Representations (ICLR 2021)*. DOI: 10.48550/arXiv.2010.11929

Ferentinos, K. P. (2018). Plant disease diagnosis and detection Deep learning model. *Computers and Electronics in Agriculture*. Advance online publication. DOI: 10.1016/j.compag.2018.01.009

Goodfellow, I., Pouget-Abadie, J., Mirza, M., Xu, B., Warde-Farley, D., Ozair, S., Courville, A., & Bengio, Y. (2020). Generative adversarial networks. *Communications of the ACM*, *63*(11), 139–144. DOI: 10.1145/3422622

He, K., Gkioxari, G., Dollár, P., & Girshick, R. (2017). Mask r-cnn. In *Proceedings of the IEEE international conference on computer vision* (pp. 2961-2969).

He, K., Zhang, X., Ren, S., & Sun, J. (2016). Deep residual learning for image recognition. In *Proceedings of the IEEE conference on computer vision and pattern recognition* (pp. 770-778).

Ho, J., Abbeel, P., & Jain, A. (2020). denoising probabilistic models of diffusion. Advancements in Neural Information Processing Systems (NeurIPS). https://doi.org//arXiv.2006.11239DOI: 10.48550

Huang, G., Liu, Z., Van Der Maaten, L., & Weinberger, K. Q. (2017). Densely connected convolutional networks. In *Proceedings of the IEEE conference on computer vision and pattern recognition* (pp. 4700-4708).

Isola, P., Zhu, J. Y., Zhou, T., & Efros, A. A. (2017). Image-to-image translation with conditional adversarial networks. In *Proceedings of the IEEE conference on computer vision and pattern recognition* (pp. 1125-1134).

Janowicz, K., Gao, S., McKenzie, G., Hu, Y., & Bhaduri, B. (2020). GeoAI: Spatially explicit artificial intelligence techniques for geographic knowledge discovery and beyond. *International Journal of Geographical Information Science*, *34*(4), 625–636. DOI: 10.1080/13658816.2019.1684500

Kamilaris, A., & Prenafeta-Boldú, F. (2018). Deep learning in agriculture. *Computers and Electronics in Agriculture*, *147*, 70–90. Advance online publication. DOI: 10.1016/j.compag.2018.02.016

Khan, S. H., Hayat, M., Naseer, M., & Khan, F. S. (2021). A vision survey of the transformers. Surveys of ACM Computing. DOI: 10.1145/3505244

Kingma, D. P., & Ba, J. (2014). Adam: A method for stochastic optimization. *arXiv preprint arXiv:1412.6980.*

Krizhevsky, A., Sutskever, I., & Hinton, G. E. (2012). Imagenet classification with deep convolutional neural networks. *Advances in Neural Information Processing Systems*, 25.

Li, W. (2020). *Deep learning in geospatial*. ISPRS Journal. DOI: 10.1016/j.isprsjprs.2020.04.012

Lin, T. Y., Goyal, P., Girshick, R., He, K., & Dollár, P. (2017). Focal loss for dense object detection. In *Proceedings of the IEEE international conference on computer vision* (pp. 2980-2988).

Lowe, D. (2004). *SIFT features*. IJCV. DOI: 10.1023/B:VISI.0000029664.99615.94

Lu, Y., Yi, S., Zeng, N., Liu, Y., & Zhang, Y. (2020). Data augmentation of the rice disease detection using CycleGAN. *Computers and Electronics in Agriculture*. Advance online publication. DOI: 10.1016/j.compag.2020.105434

Ma, L., Liu, Y., Zhang, X., Ye, Y., Yin, G., & Johnson, B. A. (2019). Deep learning in remote sensing applications: A meta-analysis and review. *ISPRS Journal of Photogrammetry and Remote Sensing*, *152*, 166–177. DOI: 10.1016/j.isprsjprs.2019.04.015

Mohanty, S. P., Salathé, M., & Hughes, D. P. (2016). Image-based deep learning plant disease diagnosis. *Frontiers in Plant Science*. Advance online publication. DOI: 10.3389/fpls.2016.01419 PMID: 27713752

Paszke, A., Gross, S., Massa, F., Lerer, A., Bradbury, J., Chanan, G., & Chintala, S. (2019). Pytorch: An imperative style, high-performance deep learning library. *Advances in Neural Information Processing Systems*, 32.

Redmon, J., Divvala, S., Girshick, R., & Farhadi, A. (2016). You only look once: Unified, real-time object detection. In *Proceedings of the IEEE conference on computer vision and pattern recognition* (pp. 779-788). DOI: 10.1109/CVPR.2016.91

Reichstein, M., Camps-Valls, G., Stevens, B., Jung, M., Denzler, J., Carvalhais, N., & Prabhat, F. (2019). Deep learning and process understanding for data-driven Earth system science. *Nature*, *566*(7743), 195–204. DOI: 10.1038/s41586-019-0912-1 PMID: 30760912

Shi, W., Cao, J., Zhang, Q., Li, Y., & Xu, L. (2016). Edge computing: Vision and challenges. *IEEE Internet of Things Journal*, *3*(5), 637–646. DOI: 10.1109/JIOT.2016.2579198

Shorten, C., & Khoshgoftaar, T. (2019). Data augmentation survey. *Journal of Big Data*. Advance online publication. DOI: 10.1186/s40537-019-0197-0

Simonyan, K., & Zisserman, A. (2015). VGG16. *ICLR*. arXiv.1409.1556

Song, J., Meng, C., & Ermon, S. (2022). Low-light image and obstructed picture restoration using denoising diffusion models. Image Processing Transactions of the IEEE. DOI: 10.1109/TIP.2022.3148289

Zhang, X., Cao, Z., & Dong, W. (2020). Overview of edge computing in the agricultural internet of things: Key technologies, applications, challenges. *IEEE Access : Practical Innovations, Open Solutions*, *8*, 141748–141761. DOI: 10.1109/ACCESS.2020.3013005

Zhang, Y., Wang, S., & Ji, G. (2019). *Time-series analysis RNN-based time-series analysis to monitor plant disease progression*. IEEE. DOI: 10.1109/ACCESS.2019.2931752

Chapter 8
Policy Frameworks and Governance for Data-Driven Agriculture

Marshall Nkosana Dzidzai Nhachi
http://orcid.org/0009-0009-2140-573X
Universitas Muhammadiyah Malang, Indonesia

M. P. Sutawi
Universitas Muhammadiyah Malang, Indonesia

Jabal Tarik Ibrahim
Universitas Muhammadiyah Malang, Indonesia

Rahayu Relawati
Universitas Muhammadiyah Malang, Indonesia

Nidhi Priya
Universitas Muhammadiyah Malang, Indonesia

ABSTRACT

This chapter examines how policy frameworks and governance structures shape the developmental outcomes of data-driven agriculture. It argues that digital technologies alone cannot ensure inclusive growth, as the benefits of agricultural data depend on institutional arrangements governing access, ownership, sharing, and commercialization. Using a data value chain perspective, the chapter analyzes how governance decisions from data collection to digital services influence value creation, market power, and farmer autonomy. It highlights two key tensions: openness versus enclosure and innovation versus justice. Applying the safeguards–enablers framework, the chapter shows how regulatory protections, interoperability standards, and participatory institutions promote trust, collaboration, and equitable benefit sharing. The chapter emphasizes the limits of privacy-focused regulation and the need for collective data rights and proposes an integrated governance roadmap to support interoperable, trustworthy, and socially just agricultural data ecosystems that advance sustainable development.

DOI: 10.4018/979-8-3373-8037-7.ch008

1. INTRODUCTION

Agriculture is moving towards being increasingly data-intensive as digital technologies progressively reorganize the sector (Rolandi et al., 2021). Farm machinery sensors, satellite and drone cameras, Internet-of-things (IoT) devices, farm management hubs and artificial intelligence (AI) applications have now mediated irrigation timing, pest management, logistics and market predictions decisions (Klerkx et al., 2019; Wolfert et al., 2017). As William and Kumar, (2024) observe, these infrastructures are progressively designing agricultural production, distribution, and risk management based on the flow of digital information. As a result, agricultural data is now a strategic asset that has significant implications on resource efficiency, productivity and climate resilience. Digital infrastructures are not only digital inputs in a GeoAI system but are geospatially enabled decision-support systems, which are based on satellite imagery, remote sensing, GIS layers and AI to generate spatially explicit predictions and recommendations. Hosseini et al., (2023) highlights that GeoAI and satellite-imagery fusion can be employed to predict soil physical properties to practice precision agriculture and manage land-resources, and Mai et al., (2025) describe GeoAI as the fusion of geospatial studies and AI. However, Melo-Velasco et al., (2025) points out that technological capability alone is not a measure of pervasive social benefits of digital transformation. Edler et al., (2023) goes on to highlight that the results of innovation depend on institutional settings and policy choices, which cautions against the use of technology. Similarly, OECD, (2019) notes that data-driven systems operate within organizational and regulatory structures to structure opportunities, risks, and incentives. Therefore, governance arrangements that control data generation, access, sharing and monetization mediate the effects of digitalization in agriculture (Duncan et al., 2026).

Bronson and Knezevic, (2016), argue that agricultural information is not presented in a value-free environment. Power and economic control is distributed by use of legal and commercial practices that bring up issues such as ownership, interoperability, privacy, licensing and reuse (Janssen et al., 2020). Birner et al. (2021) emphasize that the policy design determines the distribution of benefits and risks. Be that as it may, Bronson and Knezevic, (2016), contend that governance structures influence not only the adoption of technologies but also the distribution of power within food systems. Thus, the question as to whether data is a proprietary corporate resource, interoperable civic goods or surveillance instruments is an institutional one in essence (Jiang, 2025). This perspective is evidenced in other sectors. The existence of coordinated governance arrangements has facilitated inclusive innovation and broader creation of values, be that as it may data regimes that are poorly designed have been connected to concentration and exclusion across industries such as health, finance and urban systems (OECD, 2019). The amount of different data sources, including supply chain traceability, genomics, environmental monitoring, and production data at a farm level, and the number of actors involved also contribute to the problem of governance in agriculture (Wolfert et al., 2017).

Significant disparities in bargaining power and digital capacity, especially in smallholder-dominated and developing-country contexts, exacerbate these structural complexities. According to Carolan, (2018) farmers have little control over how data is used downstream, while big agribusiness companies and digital platforms frequently own data infrastructures, analytical tools, and contractual frameworks. In a similar vein, Rotz et al., (2019), show that rather than being equal participants in value chains, small-scale producers often serve as data providers. Because of this, technology companies and large purchasers typically benefit economically from datafication in a disproportionate way. Because of this, researchers now view agricultural data ecosystems as socio-technical systems where information flows and value creation interact with institutional structures and power dynamics (van der Burg et al., 2019).

These structural complexities are aggravated by significant differences in bargaining power and digital capacity, particularly in smallholder-led and developing-country situations. Carolan, (2018) notes that downstream data use is often not controlled by farmers, and that data infrastructures, tools of analysis, and contracts are often owned by large agribusiness corporations and online platforms. Similarly, Rotz et al., (2019), demonstrate that instead of being equal partners in value chains, small-scale producers tend to become providers of data. Due to this fact, datafication usually has a disproportionate economic advantage on technology firms and big buyers (van der Burg et al., 2019).

The existing debate on the governance of agricultural data revolves around two interrelated tensions. The former concerns the issue of enclosure versus interoperability. On one side, it is more likely to enhance the value of agricultural data by integrating, reusing and sharing datasets across systems, enabling collaborative innovation, early warning systems and climate-smart advisory services (Janssen et al., 2020). But dominant business models often encourage the use of closed platforms, proprietary standards and limited access, which increase vendor lock-in and market concentration (Bronson, 2019). Thus, policy interventions play a vital role in deciding whether enclosure or openness becomes the dominant organizing principle or not (Duncan et al., 2026). The second one is the battle between justice and innovation. Despite the fact that it is a normal belief that farmers will be benefiting out of data openness, empirical data shows the opposite (Fairbairn and Kish, 2023). As shown by Carolan, (2018), in the absence of adequate protections, institutional support and digital literacy, data-sharing agreements may undermine autonomy and escalate surveillance. Moreover, Canfield and Ntambirweki (2024) demonstrate that privacy-sensitive models are often unable to reflect group risks caused by profiling and aggregation. These challenges are particularly evident in Global South settings, where regulation may be weak.

Castellanos, (2023) emphasizes that, effective agricultural data governance requires integrated institutional arrangements operating across the entire data value chain of data collection and processing, sharing, data analysis, and commercialization. OECD, (2019), states that the balance between enabling circumstances and protective mechanisms should be achieved in governance structures. Clear data rights, privacy safeguards, cybersecurity standards and regulations controlling international data flows are examples of safeguards (Janssen et al., 2020).

By emphasizing governance over technological determinism, this chapter adds to the expanding corpus of literature that critically investigates techno-solutionist narratives in digital agriculture. Organizational capabilities, contractual arrangements, and regulatory design have a greater impact on the result of digital innovation than do technologies themselves (Bronson and Knezevic, 2016). These dynamics need to be understood in order to develop governance structures that would enhance equitable growth, environmental sustainability and rural livelihoods.

One of the crucial elements of the contemporary agrarian change is the regulation of agricultural data ecosystems (Devare et al., 2023). Digital infrastructures restructure power, market relations, and knowledge systems besides supporting production. Thus, in location-sensitive agricultural systems, GeoAI also presents governance issues with regard to explainability, privacy, fairness, and spatial bias. Although Gevaert, (2022) points out that explainable AI as an earth observation tool is already a significant regulatory and societal issue, Mai et al., (2025) cite fairness-aware, privacy-aware, and interpretable GeoAI as the future directions. These changes ultimately depend on the institutional structures that regulate data to create more resilient and equitable food systems or replicate existing inequalities (World Bank, 2019). This chapter looks at these governance frameworks and how they affect the development of data-driven agricultural systems that are reliable, fair, and interoperable.

2. CONCEPTUAL FOUNDATIONS: DATA-DRIVEN AGRICULTURE AS A GOVERNED VALUE SYSTEM

2.1 From "Data-Driven Agriculture" to Governed Data Ecosystems

The systematic use of digital data collected in farms, markets, landscapes, and biological systems to make decisions, automate processes, boost productivity and output in production and distribution is known as "data-driven agriculture" (Joshi et al., 2023). The sensors, satellite imagery, mobile platforms, and machine-learning algorithms are allowing agricultural actors to utilize data to control irrigation, fertilization, pest control, logistics, and participate in the market (Klerkx et al., 2019; Wolfert et al., 2017). These technologies are associated with increased productivity and sustainability which makes data one of the core economic and strategic assets in modern food systems (Al-Shaar et al., 2022; Wolfert et al., 2017). In a GeoAI framing, these infrastructures are geospatial intelligence systems where remote sensing, satellite imagery, GIS layer, and AI are integrated to generate spatially explicit predictions and suggestions. Mai et al. (2025) consider GeoAI as the combination of geospatial research and AI, whereas Song et al., (2023) demonstrate that GeoAI transforms the geospatial analysis and mapping of the complex human-natural systems. Hosseini et al., (2023) also show how GeoAI and satellite-imagery fusion could be employed to forecast soil physical characteristics to make accurate agricultural choices and land-resource management. However, the presence of data and analysis tools do not determine the presence of value creation and distribution (OECD, 2019). Different and even conflicting applications can be trained on the same datasets, such as farm management assistance, environmental surveillance, targeted input advertising, and insuring risk evaluation, credit rating, and educating proprietary artificial intelligence models (Carolan, 2018; Bronson, 2019).

This means that to comprehend data-driven agriculture, it is important to shift to an ecosystem perspective that recognizes the institutions and politics within which geospatially enabled data infrastructures are produced, distributed and commercialized (Combe and Camaréna, 2025). Remote sensing, satellite images, and GIS-based decision-support systems are also considered to be a part of infrastructures within a GeoAI framing, which integrate to establish data flows and economic outcomes (Van der Burg et al., 2022). These ecosystems comprise farmers, platform providers, agribusiness companies, financial institutions, government agencies, researchers, and development agencies all functioning under regulatory and market environments which arrange incentives and constraints. Klerkx et al. (2019) note that data infrastructures are integrated into social relations and power structures which define who has access, can set standards and can expropriate value.

The perspective of socio-technical governance views agricultural data systems as a collective output of institutional structures and technological object of the system. This approach combines theories of institutional economics, digital governance, and the theory of public value and focuses on the way data infrastructures are embedded into markets, law and organizational practices that determine how they operate (Janssen et al., 2020). Institutional economics contains an emphasis on the role of formal rules, informal norms and enforcement mechanisms in transaction costs and bargaining power and the public value theory emphasizes the relevance of assessing innovation in relation to the collective good and not on individual profitability. When applied to the field of digital agriculture, this view can attract attention to the ways in which data resources can be turned into social and economic value that can be mediated by policy regimes, contractual arrangements and administrative capabilities. Al-Shaar et al., (2022) emphasize that scenario-based spatial modeling allows simulating alternative developments that can be driven by different structural factors, and demonstrates how data infrastructures can be used to support future-oriented governance as opposed to less dynamic decision-making.

One of the main ideas of the given analysis is the concept of conversion factors, which is modified in accordance with the capability approach and the socio-technical transition theory. Conversion factors denote the social, economic and institutional circumstances under which technological resources can be converted into significant capabilities and outputs (Klerkx et al., 2019). These are digital literacy, existence of broadband infrastructure, access to devices, competitive market, regulatory power, and existence of farmer organizations or cooperatives in agricultural data ecosystems. In locations where such conversion factors are feeble, data infrastructures can be present in locations that are not providing physical returns to producers. On the other hand, favorable institutional conditions can help farmers use data to achieve better productivity, access to the market, and risk management.

This is the framework that is key to depth of analysis and policy relevance. Instead of introducing the concept of governance as a technical appendage, it aligns policy decisions as causal forces that determine interoperability, trust, adoption trends, market concentration, and distributions (Bronson and Knezevic, 2016). As a case in point, data portability and standardization choices affect a farmer who may or may not change platforms and get stuck within proprietary systems. The accessibility of datasets to public research and advisory services is influenced by regulations on data sharing. Competition policy determines the market structure and power of small manufacturers over the dominant platforms. In this regard, the best way of thinking about data-driven agriculture is a governed value system that institutions mediate the transformation of informational resources into economic and social outcomes.

This chapter does not follow a strictly descriptive overview of policy tools since the author frames the idea of the agricultural data embedded in the managed ecosystem. Rather, it prefigures the dynamic relationship between technology, institutions and power. In such way, a more subtle response to the question of why like digital tools can deliver dissimilar results in various regions and social groups and why governance reforms are the key to ensuring inclusive and sustainable digital transformation in agriculture can be given.

2.2 Data Governance as Authority, Control, and Accountability

An effective way of starting the analysis of agricultural data governance is the definition of governance as the execution of control and power over data assets management. Data governance includes formal and informal agreements with which actors establish rights, responsibilities, and decision-making processes concerning the data creation, access, use, and disposal (OECD, 2019; Janssen et al., 2020). It contains institutional procedures, which detail the individuals with the authority to gather information, the circumstances in which they can be processed or disclosed, and the resolution of conflicting situations and violations. In this respect, data governance is not a technical system only but a social order that determines the economic relations and power structures. (Weill & Ross, 2004; Khatri & Brown, 2010).

Governance structures are often made regarding how organizations roles are allocated, what is allowed and what is not, regarding what is expected of people, and how this will be enforced. Contractual conditions can be created by platform providers to limit the reuse of data, governments can create reporting requirements, and industry associations can market voluntary standards. Such structures are frequently explained by the improvement of data value, the minimization of the transaction expenses, and the alleviation of risks, including the misuse, breaches, or legal liability (OECD, 2019). Nonetheless, they are equally used as tools of control that determine who is allowed to join in data markets and under what conditions. To farmers, the association between government regimes and their freedom to control the information or not and to be mere data suppliers in vertically integrated structures (Carolan, 2018).

More importantly, the governance of agricultural data goes far beyond the traditional issues of privacy. Although the protection of personal data cannot be underestimated, particularly as the data about farms are tied to personal livelihood, the governance systems need to focus on a more comprehensive range of concerns. Ownership and de facto control is one of the main dimensions. The farmers can be the owners of the data in principle, but it is sometimes managed with the help of the contracts, platform architecture, and technical limitations, like proprietary formats, application programming interfaces (APIs) (Bronson, 2019). Such mechanisms may restrict the capacity of farmers to access, transfer or monetize their own data.

A second dimension that is important is standards and semantic interoperability. Interoperability allows datasets created on one device, platform, or institution to be integrated and re-used, and thus further enhanced in their social and economic value (Janssen et al., 2020). In the absence of common data format, metadata and terminology standards, information is fragmented and confined in silos. The contribution made by the policy of the government is very determinant in the advancement of open standards, reference architecture, and avoiding the domination aspect of the firms that can control competition by offering proprietary specifications.

A third pillar of agricultural data governance is security and cybersecurity. The more connected the farming systems are, the more vulnerable they become to data breaches, manipulation, and interference with infrastructure (Kshetri, 2021). Lack of strong security compromises the belief in online platforms and inclination towards implementation especially among risk-averse smallholders. Effective governance thus involves minimum levels of security, incident reporting processes as well as capacity building programs that enhance resilience within the data ecosystem.

A fourth governance issue is cross-border data flows, particularly in situations whereby cloud computing and multinational agribusiness companies control data processing. Agricultural data is often maintained and processed in other jurisdictions, but this creates concerns as to regulatory control, jurisdictional power, and adherence to national development aims (OECD, 2019). The data localization policies, transfer agreements, and international cooperation determine whether nations will be in a position to maintain strategic control of the vital agricultural information (Oguamanam et al., 2023).

Canfield and Ntambirweki (2024), note that benefit-sharing is central to fair data governance, especially when it comes to the smallholders, or in the context of discussions on digital sequence information and genetic resources. Unless it is mentioned that value is to be redistributed, data-intensive innovation may become another form of extractive models where local knowledge and biological resources become profits in other locations, Oguamanam et al., (2023) emphasises this point by showing that data governance must address not only access and control, but also the distribution of benefits across the value chain. There are several types of benefit-sharing, such as revenue-sharing agreement, preferential access to analytics service, capacity-building program, and collective data stewardship institution (Carroll et al., 2020).

Taken together, these dimensions show that governance of agricultural data concerns authority, control, and accountability mainly. It will decide who exercises authority, who bears risk, and who captures benefits of digital transformation. Effective governance should therefore balance efficiency and innovation with fairness, transparency and democratic oversight. (Haley et al., 2024).

3. GEOAI DATA VALUE CHAIN: WHERE GOVERNANCE DECISIONS CREATE OR DESTROY VALUE

3.1 Why the Value Chain Matters

Figure 1. Agricultural data value chain

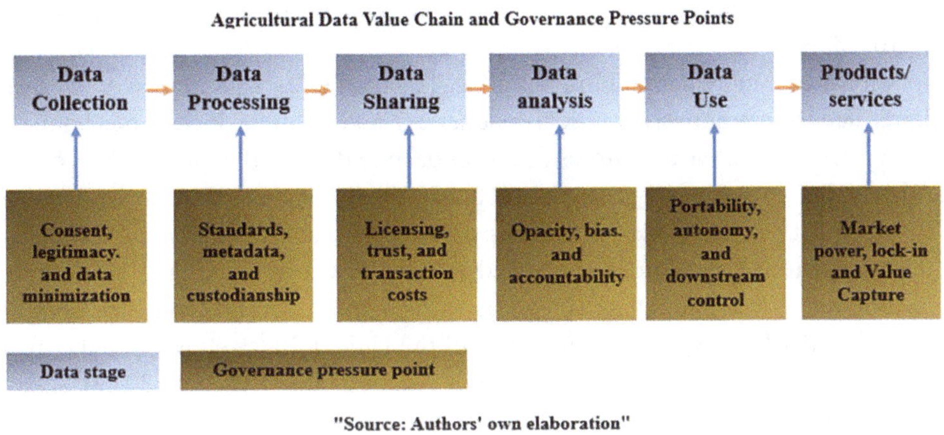

A value chain describes the sequential stages through which resources are transformed into socially and economically valuable outputs. Value chains are used in classical economic analysis to show how interconnected activities gradually refine, coordinate, and commercialize raw inputs (Helmold, 2020). According to empirical research, incorporating multi-dimensional datasets such as water, soil, and energy variables allows for more accurate and location-specific decision-making, which increases the value produced throughout the data value chain (Al-Shaar et al., 2022).

The value chain acts as a technical architecture and a governance map in the case of agricultural data system. Figure 3.1 represents the agricultural data value chain as a sequence of technical processes, but also as a group of governance "pressure points" affecting the production, administration and sharing of value. The six interconnected processes shown in Figure 3.1 data collection, processing, sharing, analysis, use, and products/services transform raw data into insights, which are useful and commercially viable. Notably, it emphasizes the fact that every stage is supported by certain governance concerns, such as consent, standards, licensing, algorithmic accountability, and market power. The figure highlights the significance of institutional arrangements that organize data flows, access, and control in the system by demonstrating that these parallel layers are equally significant to the value creation in digital agriculture as technological capability. Janssen et al. (2020) state that the rules, standards, and contractual agreements that are implemented at each stage define data quality, interoperability, portability, and accessibility. The governance failures can damage downstream innovation and public value in any of the stages, be it through opaque analytics, restrictive licensing, or poor consent procedures. Conversely, well-designed institutional structures have the capability to augment the impact of data infrastructures on sustainability and development.

Data collection, data processing, data sharing, data analysis, data use and data products or services are six analytically separate but interconnected stages in agricultural ecosystems (Castellanos, 2023). Each phase is a potential place of power concentration, value creation and risk concentration. Spatial modeling through scenarios enables policymakers to forecast possible value creation variations in varying market and climate circumstances, as well as map the existing data flows. By illustrating the ability of spatiotemporal predictive models to generate long-term predictions of land-use change and vulnerability, Al-Shaar et al., (2025) emphasize the importance of integrating forward-looking analytics into the governance design. As Carolan, (2018) notes, digital agriculture often alters the value chain of production to data-intensive intermediaries, and thus it is essential to examine how governance decisions influence such transfers.

This viewpoint has two more general implications. To begin with, governance is stage-specific. Processing standards, licensing regimes, analytical opaqueness, and technical lock-in mechanisms embedded in downstream services also serve as sources of power despite the fact that privacy and consent are most evident at the data collection point (Oguamanam et al., 2023). To begin with value capture is cumulative. Rotz et al. (2019) also state that farmers remain upstream sources of raw data with minimal bargaining power, and the largest rents are typically received in analytics, platforms, and subscription services. The lack of control over downstream value creation by farmers is complicated by the complex licensing structure and non-standard terms, as the experience of the United States shows that the management of agricultural data in the country mostly relies on private agreements (Bronson, 2019).

Thus, it is possible to systematically identify institutional bottlenecks and distributive asymmetries that influence the political economy of digital agriculture by comprehending the data value chain in terms of governance (Intellecap Advisory Services, 2024).

3.2 Governance Challenges at Each Stage of the Chain

Stage 1: Data Collection – Consent, Legitimacy, and Extraction

Yield maps, soil and nutrient profiles, machinery telemetry, animal health records, and satellite-based observations are among the increasingly detailed operational data that farms produce during the collection stage. According to Klerkx et al. (2019), this increase in data collection turns routine farming operations into ongoing monitoring and measurement locations. When consent procedures are formal rather than substantive, governance issues emerge. Farmers' capacity to make knowledgeable decisions about data use is frequently hampered by standardized contracts, low literacy rates, and unequal bargaining power (Carolan, 2018).

Canfield and Ntambirweki (2024) show that data extraction can cause collective harms through data pooling, profiling and market restructuring even in the presence of privacy frameworks. When the collection practices exceed functional necessity, data minimization principles are breached and secondary commercialization opportunities are enhanced. Lack of transparency also curtails accountability because there are numerous producers that do not know how their information is sold on platforms and supply chains. These dynamics indicate that governance on the collection level does not only revolve around personal privacy but distributive justice and structural power.

Stage 2: Data Processing – Quality, Standardization, and Custodianship

The way in which raw data is processed determines whether it is converted into reliable, understandable and reusable resources. Devare et al. (2023) report that the format of agricultural data is often not uniform, the metadata is often inadequate, and documentation practices are not very conducive to long-term usefulness. At this point, governance arrangements specify quality assurance procedures, assign stewardship duties, and control security procedures.

The importance of institutional support in the sustainable data management can be illustrated with the help of the FAIR (Findable, Accessible, Interoperable, Reusable) data movement. Wilkinson et al. (2016) argue that without sources of funding, professional incentives, and organizational cultures, in which long-term stewardship is given a high priority, technical standards do not suffice. Reuse and the creation of public value in agricultural research systems are still hampered by a lack of investment in data management capabilities. Moreover, inadequate custodianship increases the vulnerability of a system to cybersecurity threats, information loss and manipulation (Kshetri, 2021).

Stage 3: Data Sharing – Trust, Reciprocity, and Transaction Costs

The key coordination mechanism of an agricultural data ecosystem is sharing. According to OECD (2019), lack of data exchange between government agencies, research centers, and companies is one of the biggest limitations to innovation. Data hoarding is promoted by fear of competitive disadvantage, legal uncertainty, and reputational risks as opposed to cooperation.

According to Janssen et al. (2020), the key to successful sharing is standardized licensing, intermediaries that are trusted and reciprocity mechanisms that are institutionalized. In many agricultural contexts, these foundations are absent. Negotiating bilateral agreements repeatedly generates high transaction costs, while fragmented regulatory regimes discourage cross-sector cooperation. Farmers' distrust of corporate platforms and public agencies' concerns over commercial exploitation further undermine data circulation.

Stage 4: Data Analysis – Opacity, Bias, and Accountability

Analysis transforms processed data into predictive models, advisory systems, and automated decision tools that increasingly shape agricultural governance. However, according to Mittelstadt et al., (2016) algorithmic mediation is a cause of serious accountability issues when the decision-making process turns opaque. Black-box advisory systems in agriculture can give recommendations without disclosing their assumptions, limitations and uncertainties. Be that as it may, GeoAI is already contributing to agricultural analysis through spatial prediction and satellite-based decision support. Hosseini et al., (2023) show how Geospatial Artificial Intelligence and satellite-imagery fusion can be used to predict soil physical properties, supporting precision agriculture and land-resource management. Karmaoui et al., (2023) further show, through their bibliometric review, that geospatial analyses and artificial intelligence are an expanding research area in agriculture. Together, these studies indicate that spatial prediction is no longer a future possibility but an existing analytical practice that strengthens agricultural decision support. However, another significant issue is bias, particularly when models are trained mostly on data of large commercial farms and then implemented to the smallholder setting, which frequently results in the inappropriateness of their results and unequal benefits (Birner et al., 2021). Proprietary capture exacerbates these risks by having the private companies make commercial analytics products using the pooled datasets and sell them back to the same farmers who provided the data. Limited oversight over model validation and explainability further weakens farmers' ability to question or challenge harmful outputs.

Scenario-based spatial modeling extends data analysis beyond prediction toward policy-relevant decision support by comparing alternative futures across geographic space. Al-Shaar et al., (2022) show that scenario-based approaches can simulate land-use trajectories under different drivers, while Al-Shaar et al., (2025) demonstrate that deep neural network-enhanced Markov chain models can generate long-term spatial forecasts and climate vulnerability assessments under multiple scenarios. Brandt et al., (2017) further show that spatially explicit decision-support systems can integrate quantitative indicators with stakeholder knowledge to guide climate-smart interventions. Together, these approaches show that spatial modeling is not only an analytical instrument but also a governance tool for more transparent, place-sensitive, and adaptive decision-making.

Stage 5: Data Use – Rights, Portability, and Autonomy

Use represents the point at which data-derived insights intersect with everyday farming decisions. In practical terms, GeoAI-generated outputs can support context-sensitive decisions on input use, soil management, and land planning, showing how spatial prediction becomes operational at the point of use (Tamilkodi et al., 2025). Hosseini et al. (2023) demonstrate this through satellite-imagery-based soil prediction, while Karmaoui et al. (2023) confirm that geospatial analysis and AI are increasingly used across agricultural research for decision support. Similarly, Scenario-based spatial modeling strengthens data use by enabling decision-makers to test alternative intervention strategies before implementation. As shown by Al-Shaar et al., (2022) spatial simulation of different scenarios allows for more context-sensitive planning, particularly where environmental and economic conditions vary significantly across regions. Be that as it may, governance frameworks determine whether producers retain agency in applying recommendations or become dependent on platform-mediated guidance. In this context, data portability rights are crucial. While many other jurisdictions lack comparable protections, European Union regulations, for instance, encourage user control by permitting data transfer between services (OECD, 2019).

Carolan (2018) makes the connection between farmer autonomy and portability, contending that limited mobility stifles competitive innovation and strengthens vendor lock-in. It is equally important that downstream use governance. Farm data can be input price discriminated against, insured against, or scored on credit, without the express permission of producers. Interoperability standards and right-to-repair laws further determine the capacity of farmers to adjust, sustain and incorporate digital machines into different technological systems.

Stage 6: Data Products and Services – Market Power and Value Distribution

The last phase is about platforms, online marketplaces, analytics services and subscription-based advisory systems. Zuboff, (2019) states that data-intensive business models introduce new forms of economic concentration through the extraction of behavioral surplus in a systematic manner. The trend towards digital platforms that are more vertically integrated in the agricultural sector reflects such trends.

Ownership and governance studies demonstrate the means through which dominant businesses can manipulate the market conditions through controlling the access, interfaces, and pricing structures (Bronson, 2019). The bargaining positions of farmers are also undermined by asymmetries of knowledge between the providers and the users. Without regulatory intervention, benefit-sharing agreements are voluntary and unequal, particularly when genetic resources or smallholder production systems provide the data (Canfield & Ntambirweki, 2024).

Thus, the model of cooperative data stewardship, pro-competitive regulation and transparency requirements are needed to rebalance value. Potential institutional solutions to the issue of platform monopolization include public-interest data trusts, producer cooperatives, and open-access infrastructures, but these solutions have a scaling limit that requires continued policy backing (Janssen et al., 2020).

4. REGULATORY ARCHITECTURE: SAFEGUARDS AND ENABLERS AS A GOVERNANCE BLUEPRINT

4.1 The Safeguards–Enablers Distinction

Figure 2. Regulatory dimensions

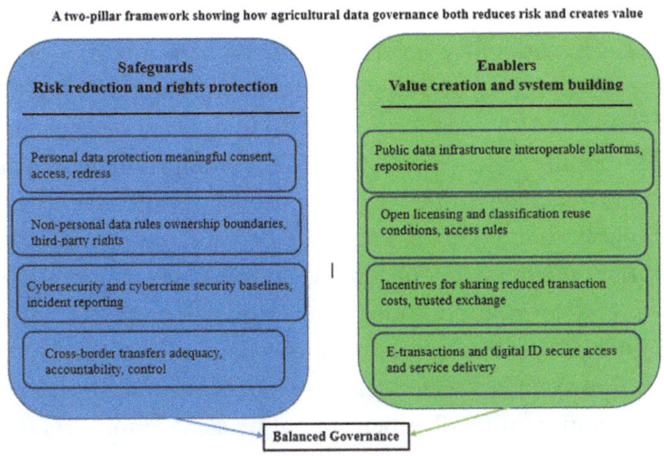

"Source: Authors' own elaboration"

Agricultural data governance requires regulatory frameworks that safeguard as well as empower productive utilization of data. One of the most common analytical frameworks is an analytical framework that conceptualizes this balance through two complementary pillars which are safeguards and enablers (OECD, 2019; Janssen et al., 2020). The safeguards minimize risk, avert abuse, and build trust, whereas the enablers provide institutional and technical environments that facilitate innovation, interoperability, and value creation (Chen, 2021). All these pillars represent a governance blueprint that determines the development of digital agriculture in practice.

The protective measures serve more as risk management and protection mechanisms. They include the minimum provisions of legal data processing, the responsibility in the requirements, and curtail detrimental practices. Protections are particularly necessary in the agricultural sector, which is becoming more and more interconnected between farm data and financial, insurance, and supply-chain systems, where abuse can have a major impact on livelihood (Carolan, 2018). Untrustworthy regulatory regimes that do not offer credible protection are likely to reduce the trust in farmers and lower their interest in digital ecosystems.

There are four key areas of regulation that are usually arranged to form safeguard systems. The former deals with the protection of personal data, its rights of access, correction, and erasure, and consent. Though the data of farms are commonly not considered as personal information, the connection to the person producer, household, and finances often places them under the jurisdiction of the privacy law (OECD, 2019). The second sphere is non-personal and governmentally owned data, where the rights of the third parties, the conditions of the license, and the need to ensure the interests of the people should be outlined. This region may facilitate unauthorized commercialization or exclusion in case of weak governance.

The third area is concerned with cybersecurity and cybercrime. Kshetri (2021) notes that digital agriculture systems are becoming more vulnerable to hacker attacks, data alterations, and system failures. Protective measures thus involve the criminalization of unauthorized access; mandatory security standards and reporting of incidents. In the absence of these protections, digital reliance may further increase systemic risk, and not resilience. The fourth area is related to cross-border data transfers. Cloud analytics and global agribusiness platforms are regularly transferring agricultural data between jurisdictions, which brings concerns of regulatory regulation and state possession. Mechanisms of adequacy, contractual protection, and accountability determine whether the international data flows will foster national development objectives or lead to unregulated data mining (OECD, 2019).

Whereas safeguards are directed toward risk reduction and rights protection on the other hand, enablers are oriented toward system-building and value creation. Enabling frameworks are the structures, incentives and institutional setups that enable the data to be useful, shareable and economically productive. These are regulations on public-interest data, exchange of data amongst the private sector and digital transactions. Policies on the collection, management and release of agricultural data to researchers, extension and early warning systems are outlined in the policies on public intent data. The regimes of private intent data determine the way that firms are able to pool, license, and reuse commercial datasets. E-commerce and e-transaction systems create legal status of electronic signatures, digital contracts, and authentication systems, facilitation of involvement in online service markets (UNCTAD, 2021).

Other enablers include interoperability standards, open licensing regimes and trusted intermediary institutions. Janssen et al. (2020) highlight that in the absence of common technical and legal frameworks, data will not be unified and will not be fully used. Interoperability lowers switching costs, enables competition and increases collaborative innovation. Licensing regimes make sense of reuse rights and low transaction costs. Data trusts, cooperatives, and public repositories can be neutral stewards allowing the balance between private incentives and the goals of public values.

Experience indicates that most regulatory systems focus excessively on protection systems, especially privacy protection, and inadequately in supporting infrastructures. Birner et al. (2021) note that policies on digital agriculture of developing countries mainly prioritize the security of data and consent without focusing on the lack of interoperability, poor extension connectivity, and data-sharing capacity. This vicious cycle limits creativity and solidifies reliance on mainstream channels. Equal control frameworks consequently demand the long-term investment in the safeguarding and empowering aspects.

4.2 Why "Personal vs Non-Personal" Is Not Enough in Agriculture

The fact that personal and non-personal data are different is a major aspect of conventional data protection regimes (Finck and Pallas, 2020). Whereas comparatively fewer regulations exist regarding non-personal data, personal data are safeguarded by strong legal provisions (GDPR Article 4(1)). This personal-data/non-personal-data dichotomy is convenient administratively, but it does not fit well into agricultural situations, where numerous datasets create collective and economically sensitive risks even though they do not necessarily qualify as formal privacy definitions. This focus on individual privacy, according to Canfield and Ntambirweki (2024), ignores such collective harms for smallholder farmers.

The non-personal information typically consists of environmental indicators, yield maps, land-use data, and production records. Nonetheless, this data may lead to aggregate disasters when aggregated, analyzed, and linked to market regimes. According to Carolan (2018), price discrimination, supply-chain restructuring, and market segmentation that hurt small producers are made possible by aggregated farm data. In the absence of explicit personal identifiers, production characteristic profiling can construct livelihoods through controlling access to contracts, insurance, and credit. These restrictions are demonstrated by the Kenyan case that Canfield and Ntambirweki (2024) looked at. The European-type models of privacy also had an impact on the regulatory frameworks, which focused on the rights of data subjects and individual consent. Whereas these mechanisms reduced the informational autonomy in some regards, it failed to mitigate the collective risks in informational asymmetrical contracting, platform concentration, and data aggregation. The farmers were still at the mercy of the exploitative market dynamics despite being granted data rights officially.

A more general conceptual issue is reflected in this gap. Agricultural data is often describing land, ecosystems, production systems, and community practices instead of describing particular individuals. The management of these information has distributive consequences and it is a shared resource. It distorts its impact on the territorial development, food security, and environmental governance when it is considered as a private property or a neutral commercial input (Bronson and Knezevic, 2016).

An improved regulatory policy recognizes group harms and group privacy. According to Taylor (2017), data governance needs to take into account how analytics impact communities as well as individuals. In the agricultural systems, collective risks are erosion of traditional knowledge systems, technological dependence and market exclusion. Structural inequality is not addressed by regulatory frameworks that disregard these dynamics. The political and economic sensitivity of data regarding land, production, and livelihoods must also be taken into consideration in agricultural data governance. Infrastructure development, land acquisition tactics, and investment patterns can all be impacted by land-use data. Trade negotiations and commodity pricing are impacted by production statistics. Eligibility for development programs, insurance, and subsidies is shaped by livelihood-related data. Producers are vulnerable to strategic manipulation by influential market players due to lax regulation (Birner et al., 2021).

Indigenous governance claims and traditional knowledge constitute another aspect that personal-data frameworks overlook. Genetic resources passed down through the generations and locally embedded ecological knowledge are incorporated into many farming systems. Digitization can disconnect knowledge from originating communities, as demonstrated by international discussions on digital sequence information and benefit-sharing (Canfield & Ntambirweki, 2024). Under digital conditions, governance regimes that disregard collective intellectual and cultural rights run the risk of replicating extractive models.

The individualistic regulation fails to work as evidenced by the fact that smallholder are more prone to digital market asymmetries. The lack of bargaining power, the inability to seek legal redress, and the dependence on integrated services impede the successful enforcement of rights. Even when there are formal protections in place, gaps in enforcement and differences in information make practical autonomy less likely (Carolan, 2018). There is a need to have collective institutions such as producer cooperatives and data stewardship organizations that will help in governance structures.

These challenges require the shift of the restrictive privacy-oriented regulation to the complete governance systems. These structures are group rights, protection of some sectors, competition rules, and data protection of people. They ensure that individual protections are compatible with systems that address market concentration, value extraction, and institutional dependency. In this manner, the regulatory frameworks will be in a better position to demonstrate how the political economy of agricultural data is varied.

Integrative Perspective

Regulatory architecture is one of the important aspects of agricultural data ecosystem development through the prism of the safeguards-enablers framework. Safeguards breed trust and reduce risk, whereas enablers provide the foundation of institutions that facilitate inclusive innovation. Simultaneously, by transcending the personal–non-personal dichotomy, governance regimes can address the cultural assertions, power relations, and susceptibility that are entrenched into agricultural information. Regulatory frameworks integrating these dimensions are better placed to enable sustainable, equitable, and developmental digital transformation in agriculture.

5. POLICY INSTRUMENTS AND GOVERNANCE MODELS FOR AGRICULTURAL DATA

Policy instruments transform the analytical issues into practical guidelines regarding how to handle agricultural data (Ehlers et al., 2022). Policy instruments shape behavior by setting up rights, duties, rewards, and ways to enforce them. In digital agriculture, they decide if data are controlled by soft law, hard law, contracts, interoperability standards, or public stewardship models (Castellanos, 2023). Haley et al., (2024) contend that it is essential to analyze how these instruments organize access, sharing, protection, and value distribution within agricultural data ecosystems, and how they can be integrated to facilitate reliable, inclusive, and equitable governance.

5.1 The Policy Instrument Toolkit: From Soft Law to Hard Law

5.1.1 Codes of Practice and Sectoral Norms: Soft Law with Coordination Power

Codes of practice (COPs) are one of the most important soft-law tools for managing agricultural data. They consist of a combination of industry groups, farmer groups, technology agencies, and government agencies that are attempting to devise some common concepts on how the ownership of data, transparency of the contracting process, and ethical use of the data should operate. According to Janssen et al, (2020)

COPs are coordination tools that reduce uncertainty, align expectations in fragmented ecosystems, and reduce uncertainty in such instances.

Castellanos, (2023) highlights the complexity of the connection between COPs and the idea of consent, disclosure, and transparency, making it a paradigm of creating trust without an official regulation. COPs accomplish this by introducing a common language to describe the key concepts like the originator of data, authorized user and secondary use and thus the bargain and dispute resolution. The number of those that contain the provisions of cybersecurity expectations, risk management, and audit procedures, the signs of the increased overlap of ethical norms and technical governance, is also significant.

This is also their major weakness since COPs are flexible and thus attractive. Compliance is voluntary and one-sided without any reasonable means to enforce the rules. Choudhury, (2018) notes that soft law has a tendency of favoring wealthy actors who are in a position to send signals without necessarily altering their actions. According to Castellanos (2023), the government recognition, accreditation systems and monitoring arrangements have significance in enhancing legitimacy and accountability.

An illustration of this type of mixed style of governance is the New Zealand Farm Data Code of Practice. The Code was developed through a lengthy procedure of consulting various groups of people. It establishes the lines on the manner in which data ought to be gathered, stored and shared and it is monitored by independent accreditation bodies. This combination of soft law, institutionalized regulation and buy in by the industry is a typology of scalable governance designs of fragmented agricultural sectors.

5.1.2 Data Protection and Privacy Legislation: Hard Law Data Protection through Structural Limits

Hard-law data protection laws define minimum rights and responsibilities that define digital markets. The Data Protection Act of 2019, which is based on the European Union General Data Protection Regulation (GDPR) as the template, enforces the issue of the spread of privacy standards on agriculture in Kenya (Canfield and Ntambirweki, 2024). The Act determines legal principles of processing, sets supervisory bodies, and develops enforcement.

This kind of legislation enhances trust and demystifies responsibility especially across the boundaries of farm data with financial and biometric data. According to OECD (2019), compliance standards are also facilitated in privacy regimes to allow cross-border cooperation. Nevertheless, Canfield and Ntambirweki (2024) note that privacy-oriented models are structurally weak in agricultural systems since a large portion of economically valuable information is not included in the definition of personal data.

This creates a regulatory asymmetry where individual identifiers are highly safeguarded, whereas aggregation production and land-use data are loosely controlled. Consequently, the privacy law is concerned with informational autonomy and does not challenge collective vulnerability, market profiling and exploitation of data to provide strategic advantages. Generic data protection statutes must be supplemented with sector-specific adaptation by using the services of the agricultural ministries, competition authorities, and development agencies.

5.1.3 Intellectual Property and Contractual Governance: De Facto Control over Agricultural Data

Berisha et al., (2026) argue that possession and control of agricultural data is exercised less through formal property titles than through stratified legal and technical systems. Oguamanam et al., (2023), also points out that control over data is often exercised through these embedded legal and technical infrastructures rather than through explicit ownership claims.

According to Bronson (2019), platform architectures frequently introduce access controls that override nominal claims to ownership. Farmers are free to legally own their data, but not to export, reuse or make use of it. Given the imbalance in bargaining power between farmers and platform providers, these contracts often entrench dependency and limit farmers' capacity to negotiate more equitable terms (Carolan, 2018).

This finding undermines the debate on policy that is limited to the data ownership. Practical governance relies more on template contract design, licensing templates, and collective negotiation capacity. Proper regulation would then need transparency, unfair contract terms, and aid to institutions of cooperative bargaining.

5.1.4 Open Data Laws and Public Data Infrastructures

Open data policies are key enabling instruments of agricultural governance (Fairbairn and Kish, 2023). These policies include access-to-information laws, data classification frameworks, interoperability standards, and open licensing regimes that facilitate the availability and reuse of data across sectors (Janssen et al., 2020). Chamorro-Padial et al., (2024), highlight that these infrastructures promote open research, extension services and early warning systems.

The safeguards–enablers framework recognizes public-interest data as a foundation for inclusive innovation and broad-based value creation (Rozenstein et al., 2024). This is because in cases where the meteorological, soil, and market data are openly available, the downstream actors are capable of creating localized advisory tools and risk management services. Nevertheless, OECD (2019) warns that openness, in the absence of governance, may enable private takeover, especially when public agencies do not have the ability to negotiate.

This means that differentiated licensing, rules on public-private partnerships and benefit-sharing schemes that do not erode the value of the people need to be the way to create a balance between access and strategic control.

5.2 Interoperability as Governance Standards, FAIR, and the Political Economy of Making Data Linkable

5.2.1 Why Interoperability Matters and Why Agriculture Lags

Interoperability refers to the ability of data systems, platforms, and devices to exchange, interpret, and reuse data across institutional and technical boundaries (Food and Agriculture Organization of the United Nations, 2021). In governance terms, it functions as both a technical standard and a policy instrument that shapes access, competition, and value creation within data ecosystems (Janssen et al., 2020). GeoAI systems further tighten these governance demands since they utilize spatially multiplied datasets and algorithmic decision-making activities that ought to be transparent, just, and responsible. According to Gevaert (2022), explainable AI in earth observation plays a key role in ensuring transparency and trust in geospatial decision-support systems. On the other hand, Oluoch (2024) emphasizes that GeoAI poses serious concerns on fairness, privacy, and ethical responsibility in geographically integrated data systems. This is further vital in agriculture since data sets which are distributed across farms, supply chain, research institutions and government agencies require interoperability (Atik & Urdu, 2025). Combining different types of data from different fields has been shown to improve decision-making, especially when water, soil, and energy data are combined to create localized solutions for sustainability (Al-Shaar et al., 2022).

Even though agriculture is important, it is behind fields like biomedicine when it comes to building interoperable data infrastructures. Traditionally, the agricultural systems were more focused on gathering data than arranging it, thus resulting in unequal formats, poor metadata, and minimal standardization. The FAIR principles indicate that data should be structured in a manner that enables it to get located, accessed, utilized with other data, and reused to generate value (Wilkinson et al., 2016). But weak institutional incentives and a lack of investment in data stewardship still make it hard for them to be put into action. Spatial modeling Scenario-based spatial modeling is most effective when it can access high-quality, interoperable datasets. Al-Shaar et al. (2025) demonstrated that predictive models require data streams, which integrate both spatial and temporal data. This shows how important it is to have infrastructures that follow FAIR principles in order to support advanced analytics and long-term forecasting.

Governance and market dynamics also influence interoperability challenges. Platform strategies and proprietary standards may limit data portability and strengthen market concentration (Bronson, 2019). According to the report by Intellecap Advisory Services, (2024), the lack of interoperability adds to the transaction costs and decreases the trust between the stakeholders. These challenges need to be resolved through policy intervention to advance open standards, enhance data stewardship, and make interoperability a public-good infrastructure that enables inclusive innovation (OECD, 2019).

5.2.2 FAIR, CARE, Sovereignty and Benefit Sharing

Figure 3. DSI, FAIR, and CARE

Integrating technical interoperability, ethical governance, and genetic data regulation in agricultural systems.

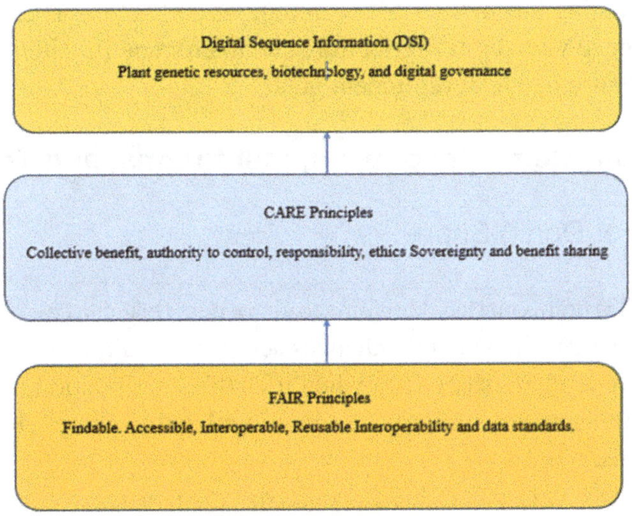

"Source: Authors' own elaboration"

Oguamanam et al., (2023) argue that FAIR principles should be interpreted alongside the CARE framework (Collective Benefit, Authority to Control, Responsibility, and Ethics) particularly in Indigenous data governance settings.This approach is important because openness alone does not guarantee fairness; as Carroll et al. (2020) caution, open data regimes can reproduce extractive dynamics when collective authority is not acknowledged. This concern is particularly relevant in GeoAI-enabled systems, where spatial data aggregation and algorithmic processing can reproduce or amplify existing inequalities if fairness and accountability are not explicitly governed (Oluoch, 2024). In this sense, governance must move beyond access alone and also safeguard community agency, ethical stewardship, and equitable participation in decisions about data use.

Digitalization further complicates these issues by introducing governance challenges related to data access, privacy, interoperability, and institutional capacity, all of which shape whether policy outcomes are equitable and effective (Ehlers et al., 2021). In Indigenous and smallholder systems, legitimacy depends on community consent, stewardship rights, and fair benefit-sharing (Wiseman et al., 2019). Accordingly, institutional arrangements must balance technical accessibility with social responsibility, ensuring that data systems support inclusive development rather than reinforcing asymmetries in control, value capture, and decision-making (Intellecap Advisory Services, 2024).

5.2.3 Digital Sequence Information (DSI) and Plant Genetic Resources: Governance Implications.

Digital genetic information extends agricultural data ecosystems from the farm plot to biodiversity, biotechnology and intellectual property management (Devare et al., 2023). Genomic data, or digital sequence information (DSI), introduces new questions about seed enclosure as value can be disaggregated from the people who developed or maintained plant genetic resources. This changes the way we access, use, and value biological resources in global data ecosystems.

DSI is not easily classified under existing regulatory frameworks as it bridges biodiversity governance and digital policy (Laird and Wynberg, 2018). If benefit-sharing frameworks are not updated, genomic databases may intensify existing inequalities between regions and actors. Therefore, the governance of biotechnology must be aligned with digital agriculture and interoperability policies to support equitable access, accountability and sustainable development goals.

5.3 Data Sharing Ecosystems: Incentives, Institutions, and Transaction Design

5.3.1 Why Actors Do (or Don't) Share

Oguamanam et al., (2023) point out that decisions to share data reflect a calculus of costs and benefits. Farmers are more likely to share data when it enhances access to credit, insurance and extension, while agribusiness firms are more likely to share data when it enables verification and traceability purposes (Smith, 2021). Government agencies, similarly, are more inclined to share when it enhances targeting and accountability functions.

However, data-hoarding is widespread due to perceived risks of misuse, legal liability and reputational risk (Wiseman et al., 2019). Be that as it may Rotz et al, (2019) argue that uneven risk exposure discourages small producers from participating in data markets dominated by powerful intermediaries.

5.3.2 Infrastructural Conditions for Data Sharing

There is evidence that institutional design can lower the cost of data sharing in well-designed data-sharing ecosystems (Janssen et al., 2020; OECD, 2019). Licensing templates for contracts and standardized agreements reduce the costs of renegotiation, while intermediaries particularly trusted ones, and cooperatives increase bargaining power and facilitate dispute resolution (OECD, 2019). Accreditation also improves compliance and legitimacy, while interoperable identity and authentication systems facilitate secure online transactions. Robust cybersecurity requirements further mitigate risks and foster trust (Janssen et al., 2020).

Overall, these practices convert the bilateral risk of data sharing into a standardized, collective practice. Research also shows that sharing and integrating multiple data sets in a coordinated fashion has a considerable impact on system efficiency and sustainability, further justifying the need for data-sharing ecosystems (Al-Shaar et al., 2022). Governance therefore not only provides structure but infrastructure for cooperation.

These findings are consistent with evidence that digitalization reshapes agricultural systems by improving information flows, reducing transaction costs, and enabling more adaptive and data-driven governance approaches (Ehlers et al., 2021). Soft law supports coordination, hard law defines boundaries, contractual arrangements organize power, and enabling infrastructures expand collective value creation. Trade-offs remain unavoidable, yet well-designed institutional frameworks can reduce asymmetries and align digital transformation with development objectives (OECD, 2019; Janssen et al., 2020).

5.4 Rights of Farmers, Participation, and Politics of Datafication: Lessons on Kenya and African Digital Agriculture

Canfield and Ntambirweki (2024), examination of how digital agriculture is transforming the agrarian power relations in the African continent represents one of the most empirically based accounts on the topic. They argue that with the growing size of data-driven systems, datafication restructures the control over agricultural knowledge and market access and decision-making power. Systems of governance that are strictly based on individual privacy cannot reflect such structural changes and collective vulnerabilities are not addressed. The analysis highlights the necessity to rebrand the rights and participatory governance of farmers to the center of agricultural data policy.

Instead of regarding digital agriculture as a neutral modernization process, the research places data infrastructures in historic political and economic inequalities among the smallholders, the agribusiness companies, and the transnational technology suppliers. It is demonstrated that data can serve both as productive resources and tools of surveillance, as well as, strategic assets. In the absence of institutional protection, these functions are likely to perpetuate instead of destabilizing unequal market relations. In this light, data governance emerges as a focal point where contest over autonomy, redistribution and recognition are brokered.

5.4.1 Kenya as a Laboratory of Governance—and Red Flag

Kenya is one of the most developed digital agriculture platforms in Sub-Saharan Africa, with an estimate of one hundred digital farming platforms, advisory apps, and data-driven services providers (Canfield and Ntambirweki, 2024). Such a heavy innovation climate has been backed by investments in mobile infrastructure, fintech integration, and e-government systems, early. Meanwhile, Kenya has implemented relatively elaborate data governance models, such as a GDPR-oriented Data Protection Act, and new industry-specific regulations.

This mix makes Kenya a governance laboratory that is where regulatory experimentation, privately led innovation and development goals overlap. The core of this path is the state-led the process of farmer registration and profiling focused on enhancing subsidy targeting, input allocation, and social protection. These are the systems behind the so-called e-incentive sites that connect biometric identification, production log, and financial services.

Nevertheless, Canfield and Ntambirweki (2024) show that such projects also produce high-value data, which is also of high interest to the private sector. Aggregated farmer profiles allow precise marketing, risk segmentation and optimization of supply-chain. Without a proper structure of governance over sharing of data and benefits delivery, the public infrastructures will be turned into upstream feeders to private value harvesting.

External influences on governance in Kenya are also determined by the geopolitical and regulatory forces. European regulatory diffusion is reflected in privacy legislation, regional harmonization efforts are policy convergence influenced by the African Union, and intellectual property and data related clauses are being increasingly included in bilateral trade negotiations. Such processes frequently happen without much involvement of farmers, which reinforces technocratic and externally driven policy processes.

The Kenyan example therefore demonstrates the development of digital agriculture regulation in multi-level forms of power. The autonomy of national policy is limited by international standards and market forces and farmers are not strongly represented in regulating negotiations. This structure introduces the risks of regulatory capture, disunity between accountability, and misalignment between development goals and commercial incentives.

5.4.2 The Importance of Participatory Governance

It is often encouraged that participatory governance can be a solution to power asymmetries in digital systems. Be that as it may, Canfield and Ntambirweki (2024) warn that formal mechanisms of inclusion are not enough. Consultative councils, stakeholders' forums, multi-stakeholder platforms etc. tend to perpetuate existing power dynamics in the case where farmers are not well versed in data governance issues.

The policies of digital agriculture discussion include very specialized knowledge on the subject of data structures, licensing frameworks, cybersecurity, and algorithmic responsibility. If farmer groups don't have their own technical expertise, the deliberative process is automatically taken over by the private sector. This is only participatory in name and not in substance, thereby legitimating decisions already affected by private interests.

This means that the participation is institutionalized. Training programs should go further than digital literacy and include legal education and data governance literacy and negotiation skills. Farmer organizations can be supported by the independent technical assistance organizations to review contracts, data sharing arrangements and regulatory consultations. The design of governance should also be such that the participatory bodies have real powers to make decisions as opposed to advisory powers.

Sondarjee, (2024) also notes that inclusive governance requires the redistribution of not just resources but also epistemic power. This would mean in agricultural data systems that the farmers are able to influence the definition of problems, the priorities of the policies, and the standards of evaluation. In the absence of this empowerment, participation will tend to become a mere formality that merely covers up structural exclusion.

5.4.3 Farmers' Data Rights and Governance

As a continuation of the historical battles of seed sovereignty and intellectual property, Canfield and Ntambirweki (2024) develop an idea of the rights of farmers in the digital context. Historically, the rights of farmers were the collective reaction to the enclosure in the form of the rights of plant breeders and patents, which stressed the right to access seeds, the sharing of benefits, and the acknowledgement of traditional knowledge. In the field of data, analogous processes of enclosure and appropriation are underway in the form of platform control and propriety analytics.

A digital farmers rights framework is not an individual consent but an extension that includes collective autonomy, economic participation and voice by an institution. The essential elements are fair access to digital infrastructures, security of old and locally ingrained knowledge, openness in information processing and commercialization and imposable accountability systems. The principles of data sovereignty focus on the right of farmers to control the manner in which the data collected and owned are stored, shared, and monetized.

The key to this framework is benefit-sharing arrangements. In areas where data create commercial value, producers must get proportional returns in the form of revenue sharing models, subsidized services or access to analytics. There should also be regulatory frameworks limiting exploitive downstream applications, including discriminatory pricing or making unilateral changes to the contract.

Notably, the rights of data of farmers cannot be reduced to property claims. They operate as rights to relations as institutionalized in the governance institutions. These rights are exercised in practice in terms of cooperative data trusts, collective licensing schemes, and producer-led stewardship bodies. They require effective legal acknowledgment and long-term financial support of the population.

5.5 Governance Typology: Four Practical Models and Their Trade-Offs

Continuing on the analysis above, the data governance arrangements in the agricultural sector can be categorized into four ideal-typical models depending on their distribution of control, risk, and value. Such models are very rarely found in pure form but serve an analytical purpose of gauging policy decisions.

Model 1: Market-Led Contractual Governance (Default Capitalism)

The market-led governance is heavily based on private ordering in form of contracts, platform terms of service, trade secrecy, and proprietary standards. Control is in the hands of the platform providers, agribusiness firms and farmers are hired as independent contractors. The model encourages speed in innovation and well-defined commercial incentives. Investment decisions are made on expected profits and therefore digital services are rapidly deployed. The risks are lock-in, contract complexities, weak market power and value capture. The case of U.S. agricultural data chains, as Castellanos, (2023) notes, governed by complex personal contracts is one such example. Without regulatory counterbalances, market-based data governance tends to mirror platform capitalism relationships, which prioritize the concentration and monetization of data over equity.

Model 2: Co-Regulation (Codes, Accreditation, and Selective Hard Law)

Co-regulatory systems are a combination of industry developed codes of practice and independent accreditation with government endorsement and minimum legal regulation. The public and private institutions have legitimacy and compliance is through reputational or market pressure. It is also dynamic and therefore able to adapt to the underlying technology. It can generate confidence in the industry and reduce regulatory uncertainty. The challenges are regulatory capture, uneven enforcement and voluntary compliance. This model is dependent on effective regulators and effective punishments.

Model 3: Public-Value Data Governance (Public Infrastructure and Open Stewardship)

The focus of public-value governance revolves around open data legislature, interoperability markets, community repositories, and stewardship organizations. Data are considered as strategic state resources to aid research, resilience planning and service delivery. This model improves the aspect of transparency, facilitates spillovers in innovation, and increases democratic checks and balances. But the absence of complementary equity would allow privatization of state resources to take place through openness. The institutional capacity should be strong in order to handle infrastructures, negotiate partnership, and implement licensing terms.

Model 4: Participatory Governance and Rights-Based (Justice-Oriented Stewardship)

The idea behind rights-based governance focuses on the rights of farmers, institutions of participation, sharing of benefits and principles of sovereignty in alignment with CARE. Power is shared by means of the collective governing bodies as well as by legally established rights. The model directly confronts the issue of structural power disparities and makes digital agriculture consistent with the goals of sustainable development. It enhances independence, appreciation and redistribution. Nonetheless, it is institutionally challenging and can face opposition by established business players. This needs long term political engagement and administrative strength.

Key Transition Argument

These models of governance are not counter-intuitive. Successful national policies usually have components of various policies. Markets and publics coexist, co-regulatory arrangements are complementary to rights regimes, and open data regimes co-exist with contractual regulations. It is however, the case that when the combinations occur automatically, they tend to reinforce asymmetries, instead of alleviating them.

The safeguards enablers framework is a design tool of building purposeful and balanced governance blends. Through a clear truncation of protective arrangements and enabling infrastructures and participatory institutions, policymakers can redirect digital agriculture into the inclusive and sustainable paths instead of falling into extractive arrangements.

6. A GOVERNANCE ROADMAP: DEVELOPING POLICY PACKS TO BE IMPLEMENTED

To transform the analytical knowledge into effective governance systems, policy designs that are institutionally realistic, politically viable, and administratively in sequence are needed. Reformers of agricultural data governance will not be effective by individual legal intervention or pilot projects. Instead, they need coordinated policy sets that would match regulatory instruments, institutional capabilities, and incentives of the stakeholders. An effective roadmap to governance thus organizes reform as a gradual process that gradually develops trust, capacity and responsibility throughout the data ecosystem.

6.1 Step 1: Mapping the National Agricultural Data Value Chain and Power Points

Good governance design starts with the systematic mapping of national agricultural data ecosystems. The six step value chain; collection, processing, sharing, analysis, use and products or services are used as an institutional diagnostic tool that indicates the concentration of authority, risk and value.

The mapping exercise will analyze major governance questions at every level: who gathers what types of data and under what consent regimes, in which technical and semantic standards are adopted or not, which data sets are contractually constrained or technical lock-ins, who holds control over analytical infrastructures and model ownership, and in which markets of inputs, financial services, insurance regimes, and interfaces economic rents are concentrated.

This kind of analysis reveals structural leverage areas of intervention. Scenario-based spatial modeling strengthens this diagnostic by showing how those leverage points may vary across locations under alternative climate, market, and infrastructure conditions. Similarly, In GeoAI-enabled agriculture, this makes value-chain mapping not only descriptive but also forward-looking, helping policymakers target interventions where they are most likely to produce measurable gains. In practice, a lack of portability at the use stage can be used as a justification of the regulatory requirements that allow data transfer between platforms. The national interoperability standards may be needed to facilitate fragmentation at the processing and sharing stages. Value concentration on the products and services level could require competition control or the system of sharing benefits.

Notably, value-chain mapping changes the mode of governance that is based on abstract regulation to institutional reform that is targeted. It also allows policymakers to focus interventions on the basis of empirical power relations, as opposed to normative assumptions, and enhance effectiveness and political feasibility.

6.2 Step 2: Creating Minimum Safeguards in the form of a Trust Foundation

Trust is the prerequisite of engaging in digital ecosystems. The absence of plausible securities makes farmers and social organizations unwilling to exchange information, restricting the value-generation in a system-wide context. Setting bare minimum protection is the next step of reform of governance thus a necessity.

Protections of personal data at the baseline are necessary, with conditions of meaningful consent required, as well as access and correction privileges, and independent redress. Nevertheless, protection should go beyond the traditional privacy models. Regulations on non-personal yet economically sensitive information are expected to demarcate the limits of intellectual property, conditional right of the third-party use, and circumstances of a derivative commercialization.

Provisions of cybersecurity and cybercrime are also central. Systemic resilience is increased by minimum security standards, certification schemes and mandatory breach reporting and reputational risk is minimized. Poor security compromises uptake and increases vulnerability especially in rural settings that are resource-starved.

There are well-defined systems in place that regulate cross-border data flows and enhance trust further. Adequacy mechanisms, contractual protection, and accountability standards facilitate cooperation on an international basis without damaging regulation. A combination of these measures creates an environment of legal predictability and portrays a sign of institutional responsibility in data management.

6.3 Step 3: Establishing Enablers Which Liberate Public Value

Security does not lead to developmental benefits. It should facilitate infrastructures and incentives to convert the protected data into a resource. The third stage therefore relates to the creation of systemic incentives for interoperability, innovation and access. Public data infrastructures are the underpinning of enabling governance. Common registries, interoperable government platforms and data banks are applied to coordinate across ministries, research centres and service providers. Infrastructures of this nature reduce duplication and enable policy responses.

The reuse conditions are further explained by data classification and open licensing regimes. The distinction between sensitive, restricted and publicly available datasets allows proportional governance. The open licensing of the proper public datasets helps the innovation in the private sector, monitoring by the civil society, and the development of localized services.

Both barrier-removal mechanisms and incentives tailored to induce voluntary information exchange among the private actors exist. These are tax breaks, government bids, regulatory sand boxes and risk-sharing. Such instruments cause incentives to be shifted towards collaboration as opposed to hoarding by minimizing coordination costs.

The other enablers are e-transaction frameworks and systems of digital identity. Trusted exchanges between platforms, financial institutions and government agencies are facilitated by legal recognition of electronic signatures, secure authentication systems and interoperable identity infrastructures. Combined, these actions legalize data circulation being part of the routine, instead of extraordinary cooperation.

6.4 Step 4: Capacity Building and Participation Institutionalization

In Kenya (and other similar settings) the empirical data shows that the mechanisms of formal participation do not work in the absence of a long-term investment in technical and organizational capacity. The governance councils, consultation platforms and stakeholder forums will not work when the representatives of the farmers do not have the expertise to communicate with multifaceted regulatory and contractual problems.

Participation has to be institutionalized, hence, relationships of multi-dimensional capacity-building. The independent technical assistance programs allow farmer organizations to consider platform contracts, licensing terms, and data-sharing agreements. Legal and economic training enhances negotiation skills and minimizes the informational asymmetries.

Government institutions need to be structured to have representation policies, accountability, and power of making choices. The capture risks are mitigated and legitimacy improves with rotational leadership, disclosure requirement, and conflict-of-interest policy. The participation should be incorporated in the regulation processes instead of ad hoc consultations.

Inclusive governance is also enhanced with the aid of public funding of FAIR data stewardship and standards adoption in research and extension systems. The metadata management, repository management and interoperability training investments enhances the knowledge infrastructures nationwide and lessens reliance on proprietary systems.

In these processes, participation turns out to be more of symbolic inclusion into institutionalized co-governance, where farmers can affect regulatory priorities and enforcement practices.

6.5 Step 5: Fair Value Distribution Mechanism Development

Efficient value dispersion is the final trial of the data governance of agriculture. Unless there are mechanisms that guarantee equal benefit distribution, there is a risk that digital transformation strengthens extractive market geometries. The last step of the road map is hence concerned with the institutional structures that match the use of data with distributive justice.

One of the tools includes model contractual clauses. The presence of standardized provisions that ensure portability of data, transparency of secondary-use, and termination endow bargaining power of farmers. Publicly approved templates minimize the asymmetries in negotiation and set a minimum standard of fairness.

Organizational vehicles of collective governance can be offered by cooperative data trusts and farmer-centric intermediaries. These institutions leverage the strength of data stewardship by combining data on a common basis and establishing negotiating leverage, providing pooled analytics, and developing revenue-sharing models. Their sustainability requires legal acceptance and financial resources.

The benefit-sharing models are especially decisive when data are based on smallholders or genetic resources or traditional knowledge systems. They can be royalty plans, priority in accessing services, infrastructure funds or community development funds. These arrangements make equity values work in the business ecosphere.

Regulatory limits on exploitative secondary uses represent an additional safeguard. Prohibitions on discriminatory pricing, opaque profiling, and unilateral contract modifications protect farmers from downstream manipulation. Competition authorities and sector regulators play complementary roles in monitoring market conduct.

Finally, transparency requirements enhance accountability in value distribution. Mandatory disclosure of revenue models, data monetization practices, and partnership arrangements enables public scrutiny and informed negotiation. Over time, such transparency supports norm formation around fair data practices.

Integrative Perspective

Taken together, these five steps constitute a sequenced governance roadmap that aligns institutional diagnosis, legal protection, infrastructural development, participatory capacity, and distributive justice.

Rather than prescribing uniform regulatory templates, the roadmap emphasizes adaptive policy packages tailored to national contexts and power structures.

By embedding safeguards within enabling infrastructures and participatory institutions, governance systems can transform agricultural data from sources of extraction into foundations for inclusive growth. Implementation remains politically and administratively demanding, but the cumulative design of these policy packages enhances resilience, legitimacy, and developmental impact in digital agriculture.

7. RECOMMENDATION: DATA-DRIVEN AGRICULTURE: TOWARDS INTEROPERABLE, TRUSTWORTHY, AND FAIR

This chapter has revealed that policy frameworks are not the peripheral structures to digital agriculture but they form the basis of its infrastructure. Governance structures define the extent to which data systems contribute to climate resilience, productivity increase, and inclusive innovation or whether they contribute to more and more enclosure, market concentration, and social inequality. Digital tools never work in the absence of institutional settings. Instead, their influences are mediated by legal regulations, organizational abilities, contractuality and relations of power that have formed the manner in which data has been produced, shared as well as monetized.

The agricultural data value chain analysis has shown that the governance decisions affect the value creation and distribution at all levels. Institutional arrangements predetermine the distribution of control of information resources and economic returns, starting at the consent mechanisms used during the collection of data to the licensing regimes and market structures of digital services. Asymmetries in the chain may be concentrated by accumulations in fragmented or poorly coordinated chains of governance with only partial upstream benefits and concentrated downstream rents. On the other hand, with coordinated measures to mitigate interoperability, accountability and market power, data infrastructures can become a source of shared value.

The safeguards-enablers framework presents an effective architecture on how to structure such interventions. The protections build trust and mitigate systemic risk as well as secure individual and communal rights. Enablers develop the technical and institutional bases of innovation, collaboration, and the creation of public value. Digital ecosystems do not work socially and economically well where these pillars are disproportionate, either by being too rigid in its regulatory form or because market liberalization is uncontrolled. Equal regulative structures are hence necessary to maintain both trust and dynamism.

Comparative evidence supports the boundaries of a privacy-driven governance. The case of Kenya shows that personal consent and rights over the data of a specific subject matter are important but cannot be applied in agriculture, where vulnerability is common, and the market is uneven. Privacy-based frameworks do not take into account structural risks due to aggregation, profiling, and contractual dependence as Canfield and Ntambirweki (2024) demonstrate. Without complementary institutions to manage bargaining power, benefits and citizen oversight datafication has the potential to amplify rural inequality.

The analysis also shows the key role of interoperability and data stewardship. Digitization investments yield little returns in a world where information is still isolated and unused beyond their organizational and sectoral boundaries. Data infrastructures, standards and professional guardianship institutions congruent to the FAIR principles thus cannot be regarded as technical accessories but as public goods. They enable early warning systems, climate change adaptation and scalable advisory services that are the backbone of resilient food systems. At the same time, the CARE principles and sovereignty arrangements will ensure that the openness is not pursued at the expense of the rights and authority of communities and local governance institutions.

Capacity-building mechanisms and institutions of participation become equally important. Nominal consultation is not sufficient for participation. It requires the redistribution of technical knowledge, legal knowledge and power. Farmer organizations, cooperatives, and data trusts represent key institutional vehicles for collective agency but the success of these entities depends on the continued support of the population and legal status. Without this support, the governance processes will remain in the hands of the business players with a superior resource and information advantage. Thus, the policy challenge is integrative and not incremental. Good governance processes should also at the same time implement certifiable safeguards, invest in interoperability and stewardship, build institutions for data sharing (trust) and implant equal distributive justice into market processes. These aspects are self-reinforcing. Sharing (trust) will trigger innovation; innovation will create values; and equality will create legitimacy. A governance system that does not account for all of these dimensions remains incomplete and lacks balance.

This co-integration is critical in terms of development, for smallholder-dominated systems. Digital agriculture will not necessarily lead to inclusive development. It may lead to a digital capture under a high-tech scenario otherwise. Data-driven transformation can be tailored to ensure food security, environmental sustainability and rural livelihoods through the rights-based and participatory governance principles.

Lastly, data-driven agriculture is not as technologically sophisticated as it could be and depends on the capacity of institutions to think and act differently. Long-term involvement of regulators, universities, private companies and producer groups is required to develop data ecosystems that are interoperable, trustful and fair. There is needs to move away from fragmented approaches to coherent policy packages in line with empirical evidence and normative certainty. Data-driven agriculture can be a resource for sustainable and equitable development if the governance system takes into account farmers as data mining units but also as co-governors and beneficiaries.

REFERENCES

Al-Shaar, W., Bonin, O., & de Gouvello, B. (2022). Scenario-based predictions of urban dynamics in Île-de-France region: A new combinatory methodologic approach of variance analysis and frequency ratio. *Sustainability (Basel)*, *14*(11), 6806. DOI: 10.3390/su14116806

Al-Shaar, W., Lehmann, X., Saad, N., Elmazi, G., Al-Shaar, M., & Tohme, C. (2025). Urban morphology and climate vulnerability assessment in Kuwait: A spatio-temporal predictive analysis utilizing deep neural network-enhanced Markov chain models for 2050 and 2100. *PLoS One*, *20*(8), e0318604. DOI: 10.1371/journal.pone.0318604 PMID: 40824951

Atik, C., & Urdu, D. (2025). *Bridging the Interoperability Gap in Digital Agriculture: Insights from Policy*. Practice, and Standards.

Berisha, F., Mooney, P., Pourzolfaghar, Z., & Helfert, M. (2026). Shared yet owned: The dual path of data ownership in agriculture: A systematic review. *Data & Policy*, *8*, e3. DOI: 10.1017/dap.2025.10050

Birner, R., Daum, T., & Pray, C. (2021). Who drives the digital revolution in agriculture? A review of supply-side trends, players and challenges. *Applied Economic Perspectives and Policy*, *43*(4), 1260–1285. DOI: 10.1002/aepp.13145

Brandt, P., Kvakić, M., Butterbach-Bahl, K., & Rufino, M. C. (2017). How to target climate-smart agriculture? Concept and application of the consensus-driven decision support framework "targetCSA.". *Agricultural Systems*, *151*, 234–245. DOI: 10.1016/j.agsy.2015.12.011

Bronson, K. (2019). Looking through a responsible innovation lens at uneven engagements with digital farming. *NJAS Wageningen Journal of Life Sciences*, *90–91*(1), 1–6. DOI: 10.1016/j.njas.2019.03.001

Bronson, K., & Knezevic, I. (2016). Big data in food and agriculture. *Big Data & Society*, *3*(1). Advance online publication. DOI: 10.1177/2053951716648174

Canfield, M., & Ntambirweki, B. (2024). Datafying African agriculture: From data governance to farmers' rights. *Development*, *67*(1-2), 5–13. DOI: 10.1057/s41301-024-00405-7

Carolan, M. (2018). "Smart" farming techniques as political ontology: Access, sovereignty and the performance of neoliberal and not-so-neoliberal worlds. *Sociologia Ruralis*, *58*(4), 745–764. DOI: 10.1111/soru.12202

Carroll, S. R., Garba, I., Figueroa-Rodríguez, O. L., Holbrook, J., Lovett, R., Materechera, S., Parsons, M., Raseroka, K., Rodriguez-Lonebear, D., Rowe, R., Sara, R., Walker, J. D., Anderson, J., & Hudson, M. (2020). The CARE principles for Indigenous data governance. *Data Science Journal*, *19*, 43. Advance online publication. DOI: 10.5334/dsj-2020-043

Castellanos, S. A. S. (2023). *A policy proposal for agricultural data governance* (Doctoral dissertation, Portland State University). DOI: 10.15760/etd.3567

Chamorro-Padial, J., García, R., & Gil, R. (2024). A systematic review of open data in agriculture. *Computers and Electronics in Agriculture*, *219*, 108775. DOI: 10.1016/j.compag.2024.108775

Chen, R. (2021). *Mapping data governance legal frameworks around the world: Findings from the global data regulation diagnostic*. https://ideas.repec.org/p/wbk/wbrwps/9615.html

Choudhury, B. (2018). Balancing soft and hard law for business and human rights. *The International and Comparative Law Quarterly*, *67*(4), 961–986. DOI: 10.1017/S0020589318000155

Devare, M., Arnaud, E., Antezana, E., & King, B. (2023). Governing agricultural data: Challenges and recommendations. In *Towards responsible plant data linkage: Data challenges for agricultural research and development* (pp. 201–222). Springer International Publishing. DOI: 10.1007/978-3-031-13276-6_11

Duncan, E., Ruder, S.-L., Kc, K. B., Wittman, H., & Fraser, E. (2026). The political economy of agricultural data governance: Farmer perspectives on trust in data sharing. *Journal of Rural Studies*, *123*, 104076. DOI: 10.1016/j.jrurstud.2026.104076

Edler, J., Blind, K., Kroll, H., & Schubert, T. (2023). Technology sovereignty as an emerging frame for innovation policy: Defining rationales, ends and means. *Research Policy*, *52*(5), 104765. DOI: 10.1016/j.respol.2023.104765

Ehlers, M.-H., Finger, R., El Benni, N., Gocht, A., Sørensen, C. A. G., Gusset, M., Pfeifer, C., Poppe, K., Regan, Á., Rose, D. C., Wolfert, S., & Huber, R. (2022). Scenarios for European agricultural policy-making in the era of digitalisation. *Agricultural Systems*, *196*, 103318. DOI: 10.1016/j.agsy.2021.103318

Ehlers, M.-H., Huber, R., & Finger, R. (2021). Agricultural policy in the era of digitalisation. *Food Policy*, *100*, 102019. DOI: 10.1016/j.foodpol.2020.102019

Fairbairn, M., & Kish, Z. (2023). Setting data free: The politics of open data for food and agriculture. *New Media & Society*, *25*(8), 1935–1959. DOI: 10.1177/14614448231174520

Finck, M., & Pallas, F. (2020). They who must not be identified—Distinguishing personal from non-personal data under the GDPR. *International Data Privacy Law*, *10*(1), 11–36. DOI: 10.1093/idpl/ipz026

Food and Agriculture Organization of the United Nations. (2021). *AGROVOC semantic data interoperability on food and agriculture*. DOI: 10.4060/cb2838en

Gans Combe, C., & Camaréna, S. (2025). Data sovereignty and valuation model for sustainable agriculture innovation and equity. *npj Sustainable Agriculture*, *3*, 61. DOI: 10.1038/s44264-025-00102-z

Gevaert, C. M. (2022). Explainable AI for earth observation: A review including societal and regulatory perspectives. *International Journal of Applied Earth Observation and Geoinformation*, *112*, 102869. DOI: 10.1016/j.jag.2022.102869

Haley, J., Conroy, D., & Errmann, A. (2024). *Shaping agricultural policy: Societal considerations of digitalization solutions. SSRN*. DOI: 10.2139/ssrn.4797293

Helmold, M. (2020). Lean management on the upstream (supply side). In *Supply management research* (pp. 91–106). Springer., DOI: 10.1007/978-3-030-46981-8_10

Hosseini, F. S., Seo, M. B., Razavi-Termeh, S. V., Sadeghi-Niaraki, A., Jamshidi, M., & Choi, S.-M. (2023). Geospatial artificial intelligence (GeoAI) and satellite imagery fusion for soil physical property predicting. *Sustainability (Basel)*, *15*(19), 14125. DOI: 10.3390/su151914125

Intellecap Advisory Services. (2024). *Unlocking data sharing in agriculture.* https://www.intellecap.com/publications/unlocking-data-sharing-in-agriculture/

Janssen, M., Brous, P., Estevez, E., Barbosa, L. S., & Janowski, T. (2020). Data governance: Organizing data for trustworthy artificial intelligence. *Government Information Quarterly, 37*(3), 101493. DOI: 10.1016/j.giq.2020.101493

Jiang, S. (2025). Big data sharing: A comprehensive survey. *Data, 10*(11), 182. DOI: 10.3390/data10110182

Joshi, K., Kumar, S., Bhatla, S., Kaushik, V., Anandaram, H., & Kumar, A. (2023). Automation intervention for monitoring agriculture farming scale: A review. In *2023 14th International Conference on Computing Communication and Networking Technologies (ICCCNT)* (pp. 1–6). IEEE. DOI: 10.1109/ICCCNT56998.2023.10307391

Karmaoui, A., El Jaafari, S., Chaachouay, H., & Hajji, L. (2023). A bibliometric review of geospatial analyses and artificial intelligence literature in agriculture. *GeoJournal, 88*(1), 343–360. DOI: 10.1007/s10708-023-10859-w

Khatri, V., & Brown, C. V. (2010). Designing data governance. *Communications of the ACM, 53*(1), 148–152. DOI: 10.1145/1629175.1629210

Klerkx, L., Jakku, E., & Labarthe, P. (2019). A review of social science on digital agriculture, smart farming and agriculture 4.0: New contributions and a future research agenda. *NJAS Wageningen Journal of Life Sciences, 90–91*(1), 100315. DOI: 10.1016/j.njas.2019.100315

Kshetri, N. (2021). Blockchain and sustainable supply chain management in developing countries. *International Journal of Information Management, 60*, 102376. DOI: 10.1016/j.ijinfomgt.2021.102376

Laird, S. A., & Wynberg, R. P. (2018). *Digital sequence information on genetic resources.* Convention on Biological Diversity. https://www.cbd.int/dsi-gr/2017-2018#peerreview

Mai, G., Xie, Y., Jia, X., Lao, N., Rao, J., Zhu, Q., Liu, Z., Chiang, Y.-Y., & Jiao, J. (2025). Towards the next generation of geospatial artificial intelligence. *International Journal of Applied Earth Observation and Geoinformation, 136*, 104368. DOI: 10.1016/j.jag.2025.104368

Melo-Velasco, J., Wilson, K., Hendrickson, M., & Valdivia, C. (2025). Understanding the social implications of digital agricultural technologies. *Journal of Agriculture, Food Systems, and Community Development*, 1–16. Advance online publication. DOI: 10.5304/jafscd.2025.143.010

Mittelstadt, B. D., Allo, P., Taddeo, M., Wachter, S., & Floridi, L. (2016). The ethics of algorithms: Mapping the debate. *Big Data & Society, 3*(2). Advance online publication. DOI: 10.1177/2053951716679679

OECD. (2019). *Enhancing access to and sharing of data: Reconciling risks and benefits for data re-use across societies.* OECD Publishing. DOI: 10.1787/276aaca8-

Oguamanam, C., de Beer, J., & Ubalijoro, É. (2023). *Ownership, control, and governance of the benefits of data for food and agriculture.* https://www.oguamanam.com/publications/2023-3-31/ownership-control-and-governance-of-the-benefits-of-data-for-food-and-agriculture

Oluoch, I. (2024). Crossing boundaries: The ethics of AI and geographic information technologies. *ISPRS International Journal of Geo-Information, 13*(3), 87. DOI: 10.3390/ijgi13030087

Regulation (EU) 2016/679 of the European Parliament and of the Council of 27 April 2016 on the protection of natural persons with regard to the processing of personal data and on the free movement of such data (General Data Protection Regulation). (2016). *Official Journal of the European Union.* https://eur-lex.europa.eu/eli/reg/2016/679/oj/eng

Rolandi, S., Brunori, G., Bacco, M., & Scotti, I. (2021). The digitalization of agriculture and rural areas: Towards a taxonomy of the impacts. *Sustainability (Basel), 13*(9), 5172. DOI: 10.3390/su13095172

Rotz, S., Duncan, E., Small, M., Botschner, J., Dara, R., Mosby, I., Reed, M., & Fraser, E. D. G. (2019). The politics of digital agricultural technologies: A preliminary review. *Sociologia Ruralis, 59*(2), 203–229. DOI: 10.1111/soru.12233

Rozenstein, O., Cohen, Y., Alchanatis, V., Behrendt, K., Bonfil, D. J., Eshel, G., Harari, A., Harris, W. E., Klapp, I., Laor, Y., Linker, R., Paz-Kagan, T., Peets, S., Rutter, S. M., Salzer, Y., & Lowenberg-DeBoer, J. (2024). Data-driven agriculture and sustainable farming: Friends or foes? *Precision Agriculture, 25*(2), 520–531. DOI: 10.1007/s11119-023-10061-5

Smith, G. (2021). *The power of farm data to enhance lending and crop insurance services.* U.S. Farmers & Ranchers in Action. https://usfarmersandranchers.org/stories/economic-sustainability/the-power-of-farm-data-to-enhance-lending-and-crop-insurance-services/

Sondarjee, M. (2024). Coloniality of epistemic power in international practices: NGO inclusion in World Bank policymaking. *Global Society, 38*(3), 328–350. DOI: 10.1080/13600826.2023.2236129

Song, Y., Kalacska, M., Gašparović, M., Yao, J., & Najibi, N. (2023). Advances in geocomputation and geospatial artificial intelligence (GeoAI) for mapping. *International Journal of Applied Earth Observation and Geoinformation, 120*, 103300. DOI: 10.1016/j.jag.2023.103300

Tamilkodi, R., Leelavathy, N., Sujatha, B., & Kumar, K. P. (2025). *Innovative AI approaches to enhance land use prediction and management.* https://doi.org/DOI: 10.21203/rs.3.rs-6352625/v1

Taylor, L. (2017). What is data justice? The case for connecting digital rights and freedoms globally. *Big Data & Society, 4*(2). DOI: 10.1177/2053951717736335

van der Burg, S., Bogaardt, M.-J., & Wolfert, S. (2019). Ethics of smart farming: Current questions and directions for responsible innovation towards the future. *NJAS Wageningen Journal of Life Sciences, 90–91*(1), 1–10. DOI: 10.1016/j.njas.2019.01.001

Weill, P., & Ross, J. W. (2004). *IT governance: How top performers manage IT decision rights for superior results.* Harvard Business Press.

Wilkinson, M. D., Dumontier, M., Aalbersberg, I. J., Appleton, G., Axton, M., Baak, A., Blomberg, N., Boiten, J.-W., da Silva Santos, L. B., Bourne, P. E., Bouwman, J., Brookes, A. J., Clark, T., Crosas, M., Dillo, I., Dumon, O., Edmunds, S., Evelo, C. T., Finkers, R., & Mons, B. (2016). The FAIR Guiding Principles for scientific data management and stewardship. *Scientific Data, 3*(1), 1–9. DOI: 10.1038/sdata.2016.18 PMID: 26978244

William & Kumar, S. (2024). The role of digital technologies in enhancing agricultural economic efficiency. *Stallion Journal for Multidisciplinary Associated Research Studies, 3*(5), 255–263. DOI: 10.55544/sjmars.3.5.19

Wiseman, L., Sanderson, J., Zhang, A., & Jakku, E. (2019). Farmers and their data: An examination of farmers' reluctance to share their data through the lens of the laws impacting smart farming. *NJAS Wageningen Journal of Life Sciences, 90–91*(1), 1–10. DOI: 10.1016/j.njas.2019.04.007

Wolfert, S., Ge, L., Verdouw, C., & Bogaardt, M.-J. (2017). Big data in smart farming: A review. *Agricultural Systems, 153*, 69–80. DOI: 10.1016/j.agsy.2017.01.023

World Bank. (2019). *Future of food: Harnessing digital technologies to improve food system outcomes*. https://www.worldbank.org/en/topic/agriculture/publication/future-of-food-harnessing-digital-technologies-to-improve-food-system-outcomes

Chapter 9
GeoAI-Enabled Multi-Model Agricultural Intelligence Framework for Plant Disease Detection and Crop Recommendation

G. Vennila
http://orcid.org/0000-0002-2205-518X
Mohan Babu University, India

N. Duraimutharasan
http://orcid.org/0009-0008-0569-7317
Reva University, India

A. Vijay
Kongu Engineering College, India

Dinesh Kumar Kumar K.
http://orcid.org/0000-0002-9912-3996
Vel Tech Rangarajan Dr. Sagunthala R&D Institute of Science and Technology, India

ABSTRACT

The current development of Geospatial Artificial Intelligence (GeoAI) is making possible the location-aware decision support system which can improve precision in agriculture by analysing spatial data and making intelligent predictions. This chapter introduces a GeoAI-based multi-model agricultural intelligence system that comprises of Vision Transformers (ViT) to detect early plant diseases, Temporal Fusion Transformers (TFT) to predict the climate and Graph Neural Networks (GNN) to provide spatially aware crop recommendations. The GNN component captures the spatial relationships among soil properties, environmental factors, and crop yields to come up with adaptive outputs. A chatbot interface serves as a layer of homogeneous interaction, allowing the farmer to obtain real-time information with

DOI: 10.4018/979-8-3373-8037-7.ch009

the lowest level of technical skills. The proposed GeoAI framework by combining spatial intelligence with multi-modal deep learning will be a step towards sustainable agriculture, lowered crop losses, and increased food security.

INTRODUCTION

Agriculture has remained one of the key pillars of the world economy, especially in the developing countries where agriculture is now a critical factor in securing food security, creation of jobs and rural socio-economic security. A large percentage of the people in these areas rely on agriculture as their major means of sustenance. Nevertheless, modern farming activities are being troubled by random weather changes, nutrient deficits in the soil, poor choice of crops and late detection of plant diseases. These problems are normally followed by low productivity, economical losses and poor utilisation of agricultural resources. The conventional farming methods which require extensive use of manual observation, experience and late intervention of experts in most cases are not sufficient in tackling the emerging environmental risks and biological threats because the changes occur very fast. Recent technological changes of Artificial Intelligence (AI), machine learning and sensor-based monitoring have created new opportunities of changing the methods of agricultural decision-making. Specifically, the Geospatial Artificial Intelligence (GeoAI) has become a relevant paradigm, integrating spatial information analytics, environmental sensors, time prediction and smart modelling to allow precision agriculture location-aware. With the application of geospatial context like weather dynamics in the region, soil diversity across cultivation areas and interactions between crop development and the environmental conditions, GeoAI-fueled systems have the potential to help farmers to make prompt and data-driven judgments. These features are particularly important in remote and resource-constrained areas in which access to agronomic skills and real-time monitoring applications is sparse.

The developments have inspired the creation of more and more pressure on integrated agriculture intelligence systems that do not just deal with a single task automation, but instead offer decision support throughout the crop lifecycle. The current solutions usually target individual functions like the identification of plant diseases, climate forecasting or crop advisory without sufficiently representing the interrelationship of agricultural ecosystems. This constraint has brought out the need to come up with multi-modelling frameworks that are effective in analysing the visual, temporal and spatial agricultural data in an integrated way.

To address this knowledge gap, this chapter presents a GeoAI-enabled multi-model agricultural intelligence system where Vision Transformers were used to detect early plant disease, Temporal Fusion Transformers was used to predict the climate and Graph Neural Networks was used to recommend crops spatially. The proposed system will optimize productivity of the farm, reduces uncertainty in agricultural planning and sustainable farming practices by providing real time and positioning information in the form of easy to use conversational interface.

Problem Statement

In most developing nations agriculture is a major source of livelihood and yet the modern systems of farming have remained at a serious operational and ecological difficulty owing to a lack of access to information-powered decision support. In rural and resource-intensive areas, the farmers are usually left to rely on traditional knowledge and manual observation methods which have been inherited over the years. Although these practices are still useful in some settings, they often fall short in responding to the emerging agricultural uncertainties like fast changing plant diseases, spatial inconsistency in soil health and growing climate insecurity. This is due to the fact that the absence of timely technological intervention can result in the suboptimal crop planning, late disease control and financial losses.

Early detection of plant disease is one of the problems in agricultural management. Manual inspection is a method that relies on the expertise and visual determination of symptoms, and may not be noticed until the disease has taken over vast areas of the farmland. Such procrastination in action may drastically decrease the yield of crops and raise reliance on the use of chemicals. On the same note, poor measurement of soil characteristics like pH, nutrient content and moisture content leads to cyclic crop rotation and eventual soil erosion. The farmers are not aware of the fact that their cultivation practices are shifting them to a lower yield and high cost production unless it is systematically monitored and spatially analyzed to determine the soil conditions.

There is an amplification of these issues by climate variability. The occurrence of irregular rain patterns, change in temperatures and extreme weather conditions distort the crop growth cycles and bring uncertainty in the decision on sowing or harvesting. Conventional forecasting techniques usually give generalized forecasts that do not reflect the environment location dynamics. These drawbacks show the necessity of intelligent farm systems that can combine spatial, temporal and visual information to provide an overall farm management. Although recent developments in Artificial Intelligence have brought a variety of solutions, the current solutions address individual tasks, including disease classification, yield prediction, but not the interconnections and geospatial characteristics of agricultural ecosystems. This gap in the research highlights the need to build GeoAI-enabled multi-modal frameworks that can be useful in location-aware crop recommendation, real-time climate forecasting and early disease diagnosis. These issues can be successfully addressed by using integrated spatial intelligence and be of great help to enhance agricultural resilience, sustainability, and decision-making efficiency in various farming settings.

Objectives

The main aim of the study is to design, develop and test an intelligent GeoAI based multi-modal agricultural decision support system capable of integrating the diagnosis of plant diseases, climatic risk forecasting, soil-conscious crop recommendation and real time field monitoring into a single system. The idea behind the system is to assist farmers particularly in rural areas by giving them the correct and timely advice using an easy-to-use conversational interface.

The targeted objectives of the proposed study are the following:

- To come up with a coherent Artificial Intelligence-driven agricultural system that will integrate plant disease detection, climate prediction and crop advice features into a single decision support system.
- To build and train a Vision Transformer (ViT) model that can be used to detect plant diseases in early stages relying on the field-collected leaf images and predict the severity rates to allow preventive measures to be taken in time.

- To create the Temporal Fusion Transformer (TFT) based climate prediction module capable of creating multi-horizon based predictions of environmental factors including temperature, rainfall and humidity to help farmers reduce the risks caused by weather.
- To develop a Graph Neural Network (GNN) that forecasts crops that is conditioned on the properties of soil nutrient and geographic associations and climate projections to determine the appropriate crop option to maximize yield.
- To combine IoT-enabled sensing tools to collect continuous environmental measurements of the field such as soil moisture, nutrient content and atmospheric conditions to enhance the accuracy of model prediction.
- In order to create an easy to use conversational chatbot interface that allows farmers to communicate with the system through queries in a natural language and multimodal inputs, including pictures, thus leaving less reliance on expert assistance.
- To improve the efficiency of agricultural decision-making through the ability to make crop planning decisions with data and proactive climate risks management, early disease detection, and prevention.

Scope

The area of the suggested system includes the creation and implementation of an intelligent and scalable Agri-decision system that could be applied in a small and large-scale farming setting. The model combines the computer vision, temporal analytics, spatial recommendation modelling, and IoT-enabled monitoring and conversational interaction mechanisms, to offer full farm management support.

The scope of the system is the important part of the system as it is:

- Identification and classification of frequent plant leaf diseases based on visual inputs that the farmer obtains in the field, as well as the evaluation of severity and recommendations.
- Both seasonal and long term predictive models of environmental conditions (temperature variations, rainfall patterns and humidity trend etc.) to facilitate sound agricultural planning.
- Prescription of appropriate crops depending on the characteristics of the soil such as the level of pH, nutrient ratio and moisture content in addition to the future prediction of the climate to enhance the productivity of farms in the long run.
- Introduction of easy to use conversational interface which supports text and picture queries, so that farmers who lack technical expertise can use intelligent advisory services.
- Multilingual interaction: As a support to ensure that usability occurs in the various geographical locations and farming communities.
- Implementation of an architectural design that supports the addition of other features at a later stage like yield estimation, market price analytics, fertiliser recommendation and increasing environmental sensing inputs.
- Deployable to various regions of the agricultural system at the cloud and edge computing infrastructure with minimal or no changes to the basic architecture of the system.
- Alternation of unfriendly farming approaches by cutting down on the volume of waste on crops with respect to prompt illness awareness to facilitate forewarning danger management and reduce the unnecessary application of chemical pesticides.

Significance

The fast adoption of digital technologies and Artificial Intelligence in the agricultural sector is transforming the conventional methods of farming, as it allows making decisions based on the data and enhances the management of resources. Geospatial Artificial Intelligence (GeoAI) in this regard has become a paradigm shift that integrates spatial data analytics, environmental sensing, and predictive modelling and intelligent automation to assist precision agriculture. The suggested multi-model agricultural intelligence framework is an important one since it shows that several AI methods can be combined into a GeoAI-based ecosystem to tackle some of the most crucial issues in the agricultural field in an integrated way.

The primary contribution that can be made when it comes to this framework is that it can facilitate the detection of early plant diseases with the help of some advanced deep learning models. The system can prevent the damage of great numbers of crops and the unnecessary use of pesticides and chemical treatment by means of timely detection of the signs of the diseases basing on the imagery of the plants. This does not only enhance crop output and economic performance of farmers but also enhances sustainable agricultural methods and conservation of long term soil health. Planting of crops that are spatially informed also enhances the agricultural planning through analysing the nutrient variability of soils, the climatic conditions and the regional patterns of suitability of specific crops. These mechanisms of intelligent crop selection have the potential to increase stability of yields as well as optimising the use of natural resources.

The second notable feature of the suggested framework is that it has the capacity to forecast the weather, which allows farmers to have predictive data on how the weather risks like changes in rainfall, temperature changes and extreme weather conditions would impact the same. These predictions allow reduction of uncertainty in the timing of sowing, irrigation planning and harvesting choices by facilitating proactive management strategies used in the farming. This comes in handy especially to farmers working in agro-climatically sensitive areas whose hiding weather patterns can translate to a significant loss of funds.

Accessibility and usability are also the attributes that are highlighted in the framework, as its conversational chatbot interface will simplify communicating with the complex AI-driven analytics. This is the design to be considered, which makes the system easy to implement in remote rural settings where the level of digital literacy can be different and there is not always access to expert agricultural advice. Moreover, the proposed system has a modular architecture, which allows it to be deployed to scale on edge and cloud computing platforms, and in low-connectivity conditions, it is operationally viable. In addition to the direct practical value, the study helps the overall vision of smart agricultural eco-systems by showing that spatial intelligence, temporal prediction and visual analytics may be connected in one decision support system. The framework offers the basis of future extensions about the integration of satellite imagery, market predictive models, market intelligence analytics and explainable AI methods. Altogether, the suggested GeoAI-enpowered multi-model solution is instrumental to the reduction of the gap between the conventional experience-based agriculture and the current sustainable practice in the agricultural environment, contributing to enhanced productivity, sustainability, and food security in various agricultural settings.

Challenges

Although there is an increasing utilization of Artificial Intelligence and the digital technologies in the field, the real implementation of the GeoAI-enabled multi-model decision support systems still remains under a number of technical, operational and socio-economic issues. In its nature, agriculture is dynamic and due to the complicated interplay of environmental factors, soil properties, crop behaviour and disease patterns. Creation of a uniform intelligent platform that can provide consistent predictions in various agro-climatic zones hence continues to be a major research and engineering problem.

Among the key constraints is associated with the data quality and availability. GeoAI systems are principally dependent on massive spatial, temporal and visual datasets such as satellite imagery, sensor measurements and annotated plant disease imagery. Nonetheless, these datasets are regularly disjointed, area-specific or labelled inconsistently. The inconsistency in imaging conditions, soil measurements standards and climatic report practices may provide noise and bias in model training and, consequently, the performance of generalisation. Besides, plant diseases might appear differently in geographical regions based on the varying differences in environmental stress factors, which make classification and provision of recommendations of diseases quite hard to rely on.

There are also infrastructure limitations that are significant in the real deployment. Constant check-in using IoTs needs stable connectivity, and good power supply, and hardware maintenance design that are usually low in remote rural farm conditions. Poor internet infrastructure may limit the real-time transfer of data and cloud-based processing, which requires the formulation of edge-optimised AI models that balance between computation and prediction. Combining several AI components like Vision Transformers, Temporal Fusion Transformers and Graph Neural Networks into one unified system architecture increases the computational complexity and requires the careful organization and control of resources by the model. The other burning issue is with respect to the use of technology in the agricultural communities. The farmers who are not very digitally literate or those who have encountered intelligent systems previously might struggle to properly use AI-based recommendations. Even technically sound system can bring no significant change without proper training, usability, and trust-building mechanisms. Adoption is also affected by economic aspects because the initial cost of implementation of sensor networks, intelligent devices, and supporting infrastructure can be costly to small and marginal farmers.

Moreover, it is essential that GeoAI systems are updated on a regular basis in order to retain their relevance to the real world in the long run, this is achieved by updating datasets, retraining models and incorporating new agricultural knowledge. The changing weather patterns, changing pest behaviour and changes in the soil conditions require adaptive learning strategies to guarantee performance sustainability. To overcome such issues with scalable architecture design, explainable AI methods, farmer-oriented interface design and better data-sharing ecosystems is the way forward to allow the consistent and extensive deployment of intelligent agricultural decision support systems in all farming environments around the world.

LITERATURE SURVEY AND COMPARATIVE STUDIES

The use of Artificial Intelligence has been a major revolution in the modern precision agriculture, allowing smart crop surveillance, automatic diseases diagnosis, decision support and resource optimization based on climate awareness. Among the initial real-world applications of AI-based management of plant diseases, Reddy et al (2025) also have shown the use of a deep learning-based mobile solution that should be used to identify leaf diseases. Their article brought to the limelight the possibility of implementing AI-based solutions directly in actual farming conditions, using user-friendly interfaces and minimizing the need for manual inspection and consultation with an expert.

The next study was aimed at enhancing the accuracy of classification through sophisticated convolutional neural network models. Liang and Jiang (2023) suggested a ResNet-based disease recognition model of tomato crops that performed better in feature representation and accuracy of disease classification. In the same manner, Ali et al. (2025) used EfficientNet models to determine apple leaf disease and showed its superiority in computational efficiency and prediction. Kotwal et al. (2024) also investigated hybrid segmentation-classification models and introduced U-Net segmentation with optimized CNN backbones, which leads to better localization and classification.

Although they are effective, CNN-based approaches tend to fail to generalize to the field conditions by failure to handle the variations in the illumination, the background complexity and environmental noise. In order to overcome such challenges, the idea of transformer-based architectures has been presented in agricultural image analytics. The multitask Vision Transformer created by Hemalatha and Jayachandran(2024) can be used to locate and classify disease simultaneously, which enhances learning in contextual features. Salman et al.(2025) also established the strength of transformer models in real field setting through the mechanism of mixture-of-experts, which implies that there is an increasing trend towards attention-based visual intelligence in precision agriculture.

In addition to the detection of diseases, crop recommendation systems have developed beyond rule-based and statistical models to graph-based machine learning models that obtain spatial and environmental data. According to Barvin and Sampradeepraj (2023), the use of Graph Convolutional Networks in predicting the nutrient interactions and environmental conditions of soil in order to facilitate intelligent choice of crops has been discussed as significant. Gupta and Singh(2023) developed this idea by coming up with a GraphSAGE-based agricultural prediction model that incorporates both genotypic and environmental data to improve the yield prediction effectiveness.

The issue of climate variability has been one of the most important aspects of agricultural planning that has a considerable impact on crop productivity and risk management. In the recent development of deep learning, there are transformer-based models that can model long-term temporal dependencies in environmental data. Li et al. (2020) presented the CNN Transformer hybrid model that made use of multitemporal remote sensing data to classify crops in their study and managed to effectively learn the phenological patterns to predict better. In addition, transformer-based models have demonstrated high potential in converting spatial and time characteristics to predict agriculture. GeoAI also improves this by integrating geospatial and remote sensing with machine learning and allowing scaleable agricultural intelligence. Boutayeb et al. (2025) proposed a concise overview of geoAI applications and the significance of spatial analytics to enhance agricultural sustainability and decision support. Kussul et al. (2017) also showed remote sensing-based deep learning methods in crop monitoring and classification, which confirms the application of spatial intelligence in the present-day agricultural systems. Also, smart agriculture systems have been considered using IoT to realize real-time soil and environmental monitoring. Farooq et al.(2019) addressed the role of the IoT technologies in facilitating the use of automated agricultural decision support and El Jarroudi et al. (2024) emphasized the opportunities of the edge AI systems in facilitating sustainable agriculture by implementing localized intelligence.

Most of the current research has been on single problem environments as opposed to integrated agricultural intelligence system, despite the fact that significant advances have been recorded in individual areas like detection of diseases, crop advising and climate prediction. This shortcoming highlights the importance of combined GeoAI-oriented systems that have the capacity to concurrently process visual, temporal and spatial agricultural data. The proposed multi-modal architecture fills this gap by integrating the use of Vision Transformers, Temporal Fusion Transformers and Graph Neural Networks into a conversational decision support system that has to be used to manage precision agriculture as a whole.

Existing Methods

The fast adoption of the Artificial Intelligence and the data-driven technologies resulted in the creation of multiple intelligent approaches that can be used to enhance agricultural monitoring, crop planning, and environmental risk management. These applications are usually focused on particular parts of the farm decision process including plant disease diagnostics, crop prescription and weather prediction. Even though these approaches have played a significant role in precision agriculture, the majority of current systems are independent and do not provide the ability of making space-related decisions.

A. Convolutional Neural Network Based-Plant Disease Classification

Convolutional Neural Networks (CNNs) have been extensively used in automated plant disease identification with the capacity to identify hierarchical visual features of leaf images. The features that are studied through these models include colour changes, texture patterns, and lesion patterns to categorise crops as healthy or infected. The application of CNN-based methods has minimized the need to use manual inspection and expert human intervention because discriminative representations are learned directly using image datasets. Nevertheless, their work might be worse among real-life conditions when the lighting conditions vary widely, the background noise and image angles can significantly vary. Their generalisation ability in field deployment situations is constrained by such domain shift challenges.

B. Detection of Diseases Using Vision Transformer

Transformer-based architectures are new in comparison to the classical convolutional models of agricultural image analysis. The mechanism of attention used by Vision Transformers help to take into consideration all the relationships of the global context of plant imagery, thus being able to be more robust in challenging environment conditions. These models examine complete patches of images as opposed to local feature extractors, thus assisting the more precise diagnosis of diseases even in heterogeneous or noisy field circumstances. Also, multitask learning is made possible by the transformer structures, which can localise and classify the disease simultaneously in single architectures. Regardless of these strengths, transformer models generally need substantial data volume and increased computational power to be trained and deployed successfully.

C. Processing of Images to Estimate the Severity

In addition to the disease classification, a number of papers have investigated the severity estimation by applying classical computer vision methods like image segmentation and region-based analysis. The purpose of these methods is to measure the degree of infection through separating diseased areas of the leaf images and measuring proportions of the affected areas. This type of severity assessment assists the economic decision-making process by informing the level of treatment and allocation of resources. Nonetheless, conventional segmentation methods can be affected by mixed background interference and different morphology of crops which restricts its consistency in different agricultural environments.

D. GNN-Based Crop Recommendation

Graph-based learning has continued to be extensively used in crop recommendation systems to represent relationships between soil nutrients, climatic conditions and crop yield patterns. In such methods, those agricultural factors that include pH levels, moisture content and nutrient availability are modeled as interrelated nodes in a graph form. Graphics Neural Networks allow reasoning in space by allowing them to catch up on the relationship between neighbourhood fields and environmental factors, which facilitates more context-sensitive crop selection plans. However, most of the current solutions are still limited in terms of real-time flexibility and a lack of unification with predictive climate analytics.

E. Machine Learning-Based Climate Forecasting

Agricultural weather forecasting has also been conducted using temporal machine learning models where the forecasting ability is to predict the pattern of rainfall, changes in temperature and seasonal changes. These models can assist in decision-making at the farm level, as they can give information about the ideal time to plan the sowing and irrigation decisions. Although methods of making a prediction based on neural networks have better predictive potential than the traditional statistical models, most of the systems that are used on the ground can still provide rough or globalised output that may not capture the local dynamics in the environment.

In general, the current intelligent agricultural procedures offer useful solutions to the individual work but frequently do not offer integrated systems that can work together to process visual, temporal and spatial information. It is this shortcoming that underlines the importance of GeoAI-enabled multi-mod systems that are able to combine disease detection, climate forecasting and crop recommendation in to one scalable decision support architecture.

PROPOSED METHODOLOGY

Managing and controlling agriculture is an extremely complicated process because of the interplay of plant health statuses, weather changes and soil nutrient cycle and the area crop suitability trends. The traditional intelligent agricultural systems would usually implement these challenges separately with a single-task model like disease classifier or yield predictor. Nevertheless, the actual farming conditions demand combined analytics, which take into account spatial, temporal, and biological relationships together. In order to overcome this weakness, the presented study presents a GeoAI-powered multi-model based agricultural intelligence framework that incorporates plant diseases detection, climate predictions and spatial crop recommendation into a single precision agriculture structure. The framework takes advantage of geospatial background by taking advantage of geographically specific soil sensing with the use of IoT sensors, time based environmental modifications like rain and temperatures as well as relating modelling of crops based on the surrounding agricultural lands. The architecture allows the integration of three deep learning paradigms that are complementary to each other: Vision Transformers to examine visual disease, Temporal Fusion Transformers to forecast the environment, and Graph Neural Networks to recommend the spatial crop. These models communicate predictive representations between themselves to give an intelligence of the farm as a whole.

System Architecture

The proposed system architecture is aimed at being a single GeoAI-equipped intelligent agricultural decision support system through which farmers can receive real-time information regarding plant health monitoring, weather prediction and crop appropriateness analysis. Supported by mechanisms of conversational interaction, inference engine of deep learning, environment sensing in space and scalable data management blocks, the architecture will guarantee the delivery of agricultural intelligence to be reliable and accessible.

The system has a layered architecture made up of five components that are interconnected with each other i.e., the User Interaction Layer, Orchestration Layer, AI Model Layer, IoT Data Integration Layer and Data Storage and Deployment Layer. These layers work together to promote a smooth flow of the data between the field-level sensing and user query and actionable agricultural recommendations.

1. User Interaction Layer

The User Interaction Layer is used as the main access point of communicating with the intelligent agricultural system by the farmers. This interface is achieved by applying a chatbot application in the form of a conversation that can be used both in a mobile application and a web portal. The chatbot allows the farmers to communicate with natural language queries aided by regional languages and thus saves them the time of having knowledge about the technical side of the chat. Farmers would be able to post plant leaf pictures to be diagnosed with diseases, asked to receive crop recommendations, depending on the soil conditions in the specific region, and receive weather-related advisories. The conversational interface accepts multimodal inputs such as textual query and uploading of images. This layer facilitates the easier usability of the systems by simplifying the complex outputs of analytical systems into conversational replies; this would enhance the adoption of the systems in rural farming communities.

2. Orchestration Layer

The Orchestration Layer is an intelligent middleware controller which is used to regulate communication between user interface and the backend analytical models. This component uses natural language understanding techniques to detect the intent of the user request and classifies the user request as one of disease detection, climate forecasting or crop recommendation tasks. Depending on the perceived intent, the orchestration engine formulates the request to the respective AI module whilst preserving the continuity of the conversation session. It also integrates the outputs of several models where necessary thus facilitating integrated decision support. This layer guarantees the systematic control of workflow, validation of response and generation of messages in context, which serves as the coordinating point of the architecture.

3. AI Model Layer

The AI Model Layer is the analytical depth of the suggested framework and is composed of three specialised deep learning modules that can solve various agricultural decision tasks. Vision Transformer module identifies plant diseases based on the imagery of the leaf. Through modeling the global spatial relationships among patterns of the disease, the model identifies the type of infections and approximates the severity level so that early intervention strategies can be made. Temporal Fusion Transformer is an AI module that is able to make multi-horizon climate forecasts based on environmental time-series data (temperature, rainfall and humidity). Seasonal trends and unexpected changes in the environment are captured by this module to help in proactive planning of agriculture. The Graph Neural Network module engages in crop recommendation by modeling of the spatial relationships among the soil nutrients and climatic factors as well as patterns of crop suitability. The model determines crops that have greater potential of yielding in particular environment by spreading contextual information to the neighbouring agricultural areas. The three modules used in conjunction comprise a single engine of agricultural intelligence that assists farmers on the crop lifecycle and assist farmers in planning, managing the crop, mitigating of risks and maximising their yields.

4. IoT Data Integration Layer

The IoT Data Integration Layer will be charged with the responsibility of acquiring real-time observations of the environment at agricultural fields. Sensor nodes in farms measure the levels of nutrients in the soil, pH levels, water content and the environmental atmospheric condition. This level carries out preprocessing of data, authentication and transmission of data to the central analytics node. The use of continuous data acquisition is important in that the predictions made in the model are made using dynamic field conditions instead of using the historical data that is not dynamic. This layer has made climate forecasting and spatial crop recommendation modules more accurate by allowing them to be made more accurate by knowing real-time the environmental awareness.

5. Data Storage and Deployment Layer

The Data storage and deployment layer offers scalable infrastructure to the management of large volumes of multimodal agricultural data such as image data, environmental time-series measurements and interactions between farmers. Cloud-based storage solutions make it easier to store long-term data and retrain the model and edge deployment strategies allow the inference to take place with low latency in regions with rural connectivity. Model monitoring services provide a measure of performance and system reliability and make sure that predictive accuracy is constantly improved. This layer forms the backbone of the architecture as it supports distributed deployment, protection of data and optimal use of computational resources that are needed to support real-life precision agriculture applications.

These are the five major layers which are being used in the AI based multi model system architecture.

Figure 1. System architecture of the proposed model

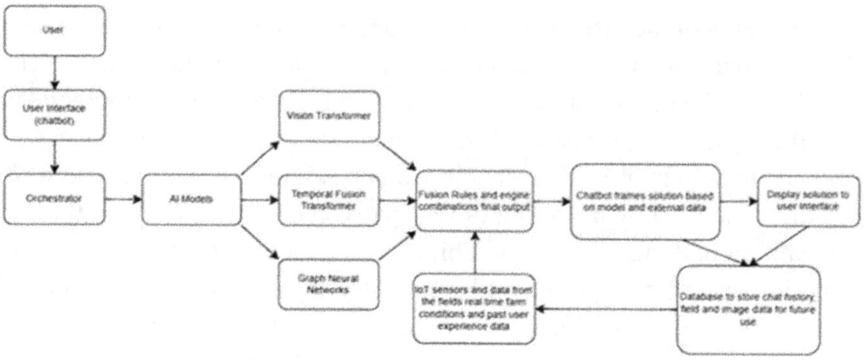

A. Dataset Description and Pre-Processing

The New Plant Diseases dataset is used to train the disease detection module and it has around 87, 000 RGB images that represent 38 types of diseases in 15 different crops: apple, corn, grape, tomato, soybean and strawberry. The data is split into training and validation subsets of 80 and 20 respectively.

Where the input image is modeled to be $\mathbf{X} \in \mathbb{R}^{\mathbf{H} \times \mathbf{W} \times \mathbf{C}}$ H, W are the dimensions of space and C = 3 is RGB channels.

Image Resizing

There is a variability in the spatial resolution of images obtained in diverse sources. Because transformer architectures have fixed-dimensional inputs, inputs, every image is transformed by interpolation-based up-sampling.

$$X \in \boldsymbol{R}^{H \times W \times C} \tag{1}$$

where R is resizing operator and $X_r \in \boldsymbol{R}(224 \times 224 \times 3)$

This is done to assure the dimensional uniformity but retain disease texture features needed to classify the disease accurately.

Pixel Normalization

The pixel intensities of the RGB images are usually 0-255. When such values are directly fed into deep learning models they can cause unstable gradients and slow convergence.

Thus, an interpolation of pixels is done using the following formula,

$$X_n = \frac{X_r}{255} \tag{2}$$

Mean-standard deviation normalisation is used to further stabilise learning by the formula below,

$$x_{norm} = \frac{(X_n - \mu)}{\sigma} \quad (3)$$

where μ and σ are mean and standard deviation of a dataset respectively.

Such transformation enhances stability of training and increases convergence.

Patch Extraction for Vision Transformer

Vision Transformers work with sequences of image patches unlike convolutional architectures which operate on local receptive fields. Every downsized image is broken into P x P patches which are not overlapping.

The amount of patches produced as $N = \frac{HW}{P^2}$ *(4)*

The sequence of flattened patches is denoted by $X_p \in R^{N\times(P^2C)}$ *(5)*

This representation transforms data of spatial images into token series, which can be processed by transformers.

B. Notation and Symbol Definitions

Multi-model GeoAI systems rely on the idea of having similar interpretations of symbols in order to achieve mathematical clarity and reproducibility. The suggested framework incorporates visual, temporal and spatial learning modules; thus, a common notation is presented to describe feature representations, graph representations, the ordering of times and prediction. Where $X \in R^{H\times W\times C}$ is an input image as a scalar value or a 2D or 3D image.

Transformer patch embedding sequence is modeled as $Z_0 \in R^{N\times D}$

The sequence of temporal environment is represented as $X_t = \{x_1, x_2, \ldots, x_T\}$

Agricultural entity representation in the form of a graph is $G = (V, E)$

Hidden node at layer l is $H^{(l)} \in R^{|V|\times d}$

Prediction outputs of the final models are: $\widehat{y_c}, \widehat{y_s}, \widehat{y_{t+k}}, \widehat{y_{rec}}$ indicating disease type, disease severity index, climate prediction and crop advisory respectively. Plant Disease Detection Vision Transformer.

C. Vision Transformer for Plant Disease Detection

Vision Transformers record global contextual view of relationship of disease-affected areas of the plant leaves. Patch tokens are encoded in a latent embedding space as each is projected into the latent embedding space with a learnable projection matrix E and positional encoding vector E pos so that spatial sorting is maintained.

$$Z_0 = [x_1 E; x_2 E; \ldots\ldots; x_n E] + E_{pos} \quad (6)$$

Relationships between patch tokens are computed by the transformer encoder by use of scaled dot-product self-attention.

$$\text{Attention}(\boldsymbol{Q},\boldsymbol{K},\boldsymbol{V}) = \text{softmax}\left(\frac{\boldsymbol{QK}^T}{\sqrt{\boldsymbol{d}_k}}\right)\boldsymbol{V} \tag{7}$$

The model allows attention to disease-specific visual patterns like lesions, discoloration, and vein distortions, which is facilitated by this mechanism. Probability distribution of the final disease classification is acquired with the help of.

$$\hat{y}_c = \text{softmax}\left(\boldsymbol{W}_o\boldsymbol{h}_{\text{cls}}\right) \tag{8}$$

Besides, the task of estimation of the severity of the disease is structured as a regression task. $\hat{y}_s = \boldsymbol{W}_s\boldsymbol{h}_{\text{cls}} + \boldsymbol{b}_s$ *(9)*

This will enable the system to measure the level of infection and give actionable information about the early treatment.

D. Temporal Fusion Transformer for Climate Forecasting

Time climatic patterns such as the variable rainfall, variation in humidity and temperature have strong effects on crop productivity. In order to model it, the system uses a Temporal Fusion Transformer that is able to perform multi-horizon forecasting. Observations of the environment are given as:

$$\boldsymbol{X}_t = [\boldsymbol{x}_1,\boldsymbol{x}_2,\ldots,\boldsymbol{x}_T] \tag{10}$$

The Feature fusion combines both the non dynamic spatial features like the soil type and the geographic location with the dynamical climatic signals.

$$\boldsymbol{Z}_t = \boldsymbol{f}(\boldsymbol{S},\boldsymbol{X}_t,\boldsymbol{X}_t^{\text{known}}) \tag{11}$$

Gated recurrent encoding has been computed to calculate temporal hidden representations.

$$\boldsymbol{h}_t = \text{LSTM}(\boldsymbol{x}_t,\boldsymbol{h}_{t-1}) \tag{12}$$

Attention-based weighting is used to predict the future environmental conditions.

$$\widehat{y_{t+k}} = \sum_{i=t-L}^{t} \alpha_i \boldsymbol{h}_i \tag{13}$$

This predictive tool can be used to do preventive agricultural planning and risk aversion.

E. Graph Neural Network for Spatial Crop Recommendation

Spatial relationships between nutrient of the soil, environmental conditions and crop yield potential are the factors of crop recommendation. A graph is used to model these relationships.

$$G = (V, E) \tag{14}$$

where nodes are agricultural entities and edges are environmental similarity.
Adjacency normalization is applied by

$$\widetilde{A} = D^{-1/2} A D^{-1/2} \tag{15}$$

Graph convolution propagates contextual information by

$$H^{(l+1)} = \sigma\left(\widetilde{A} H^{(l)} W^{(l)}\right) \tag{16}$$

Node update is computed as

$$h_v^{(l+1)} = \sigma\left(\sum_{u \in N(v)} \frac{W^{(l)} h_u^{(l)}}{\sqrt{d_u d_v}}\right) \tag{17}$$

The recommended crop is obtained using

$$\hat{y} = \underset{c \in C}{\operatorname{argmax}} f(h_v) \tag{18}$$

This spatial reasoning facilitates adaptive planning of crops to suit regionally where there is variation in soil and weather conditions.

F. Training Objective Functions

In order to provide the best performance with a heterogeneous task prediction, the proposed multi-model framework utilises task-specific loss functions that collectively inform the network parameter optimisation.

Disease Classification Loss

Categorical cross-entropy loss is used in training multi-class disease prediction.

$$\mathcal{L}_{cls} = -\sum_{i=1}^{N} \sum_{k=1}^{K} y_{ik} \log\left(\widehat{y_{ik}}\right) \tag{19}$$

In which y_{ik} means ground-truth class label and $\hat{y}_{ik}$ means predicted probability.

Disease Severity Regression Loss

The estimation of the severity of leaf infections is conceived as a regression problem where Mean Absolute Error is used.

$$\mathcal{L}_{sev} = \frac{1}{N}\sum_{i=1}^{N}|y_i - \hat{y}_i| \quad (20)$$

This loss guarantees an effective learning in noisy visual field field conditions.

Climate Forecasting Loss

Root Mean Square Error objective is used to optimise the temporal prediction accuracy.

$$\mathcal{L}_{tft} = \sqrt{\frac{1}{T}\sum_{t=1}^{T}(y_t - \hat{y}_t)^2} \quad (21)$$

This produces huge forecasting deviations which can have negative impacts on agricultural planning.

Crop Recommendation Ranking Loss

Graph recommender learning is optimised through a pairwise ranking optimisation.

$$\mathcal{L}_{rec} = -\sum_{(u,v)} \log\sigma(\hat{y}_u - \hat{y}_v) \quad (22)$$

where sigmoid activation, denoted as σ, is used to guarantee preference ranking in candidate crop activation.

Joint Multi-Task Objective

The general training goal incorporates every loss of tasks.

$$\mathcal{L}_{total} = \lambda_1\mathcal{L}_{cls} + \lambda_2\mathcal{L}_{sev} + \lambda_3\mathcal{L}_{tft} + \lambda_4\mathcal{L}_{rec} \quad (23)$$

Where weighting coefficients balancing task accomplishments are λ_i.

G. GeoAI Multi-Model Decision Fusion

Visual, temporal and spatial intelligence modules have to be combined to provide agricultural decision support. In order to facilitate combined reasoning, the suggested framework uses feature-level fusion and decision-level optimisation.

F_v, F_t, F_g can be used to represent learned representations of ViT, TFT and GNN respectively.

The integrated agricultural intelligence representation is

$$F_{fusion} = \Phi(w_1F_v + w_2F_t + w_3F_g) \quad (24)$$

where w_i are learnable fusion weights and $\phi(\cdot)$ denotes nonlinear activation.

The final precision agriculture recommendation score is $\hat{Y} = \arg\max f(F_{fusion})$ **(25)**

The role of this fusion mechanism is to provide coordinated control of the diseases, reduction of the climate risks and crop selection.

Evaluation Metrics

The disease classification performance is measured with the help of

$$\text{Accuracy} = \frac{TP + TN}{TP + FP + TN + FN} \tag{26}$$

The performance of climate forecasting is measured with the help of

$$\text{RMSE} = \sqrt{\frac{1}{N}\sum_{i=1}^{N}(y_i - \hat{y}_i)^2} \tag{27}$$

The relative forecasting error is calculated with the help of

$$\text{MAPE} = \frac{100}{N}\sum_{i=1}^{N}\left|\frac{y_i - \hat{y}_i}{y_i}\right| \tag{28}$$

All of these metrics compare the success of the suggested GeoAI multi-model platform.

RESULTS

The proposed GeoAI-based multi-modal agricultural decision support system was realized based on the modular deep learning model that incorporates visual analytics, time prediction and space recommendation modelling. Vision Transformer (ViT), Temporal Fusion Transformer (TFT) and Graph Neural Network (GNN) are used to generate system architecture that is coordinated by a chatbot interface to offer a single farm-level intelligence.

The Python language was used to develop the models and do experiments with deep learning packages such as PyTorch and TensorFlow. The environs of an NVIDIA Tesla T4 were used to run training and inference tasks to facilitate the effective processing of large-dimensional agricultural datasets. The chatbot orchestration layer was done with Rasa and Fast API services and simulated and real-time sensor streams with MQTT communication protocols. ONNX conversion was used to optimise model deployment to make it compatible with the edge computing environments.

1. Plant Disease Detection Module

Kaggle New Plant Diseases Dataset was used to train the ViT-based disease classifier and consisted of over 87,000 images that have been pre-processed. The images were uniformly reduced to 224x224 in order to have consistent input and computational efficiency. The model was able to recognize and estimate the severity of the diseases even when the background conditions and lighting varied.

Table 1. Results using vision transformer

Model	Accuracy	Precision	Recall	F1-Score
Vision Transformer	98.42%	97.85%	98.01%	97.92%

According to the confusion matrix in Figure 2, the plant disease detection system produces accurate predictions for all four characteristics of Leaf Spot, Early Blight, and Late Blight (healthy), with most entries being located in the diagonal area of the matrix (exact matching classifications). The accuracy of the healthy and Leafspot cases is demonstrated by the fact that the total amount of incorrect predictions associated with these categories is very small. Additionally, Early Blight and Late Blight classifications are very high, but there is some confusion between those two categories, which is not uncommon given the visual similarities associated with both of those diseases. Based on the results presented in this confusion matrix, it can be concluded that the plant disease detection procedure is reliable as an approach to detecting plant diseases and provides the user with only minor issues related to distinguishing between closely related plant diseases.

Figure 2. Confusion matrix

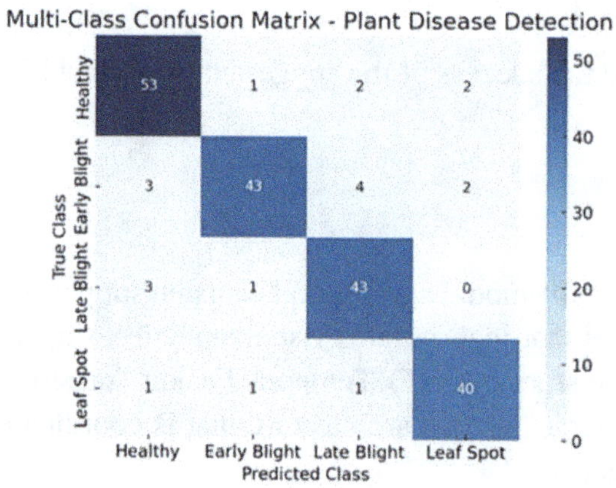

Figure 3 is an ROC Curve that depicts the ability of a disease detection model to separate positive cases from negative cases (also called True Positive and False Positive cases) at many different threshold levels for decision-making.

Figure 3. ROC curve disease detection

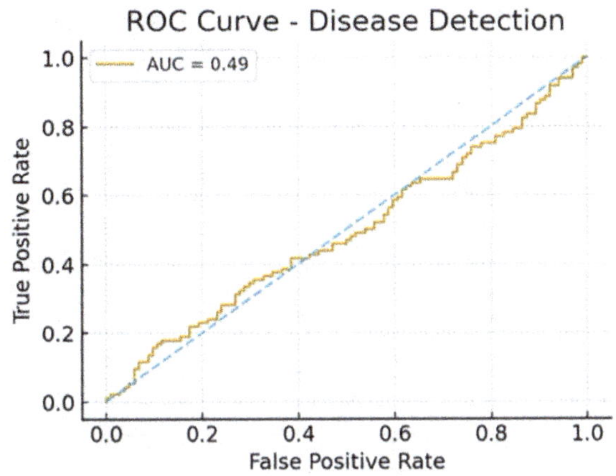

2. Climate Prediction Module

To predict temperature, humidity and precipitation, TFT model was trained on actual and simulated weather data. The model achieved high results in terms of time-series forecasting with low Root Mean Square Error (RMSE), indicating its reliability in the proactive decision support.

Table 2. Results using TFT module

Model	Rainfall	Temperature	Humidity	Forecast Duration
TFT (RMSE values)	0.72	0.48	0.61	7 days ahead

According to the RMSE trending plot in Figure 4, the amount of error associated with the prediction of the climate model has changed throughout its training periods. The lower the range of RMSEs indicates a higher-level accuracy of prediction.

Figure 4. RMSE trend

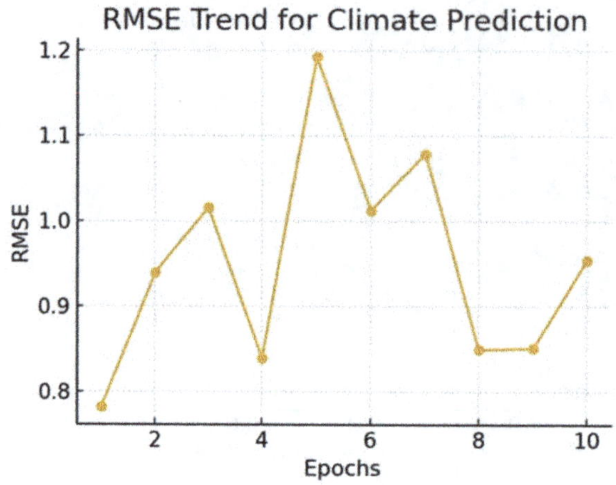

The plot in Figure 5 illustrates a comparison of observed and predicted precipitation during a one-week forecasting period. Both precipitation curves appear to be relatively consistent, with an overall pattern of decreasing precipitation through days 1-4, a dramatic drop-off on day 5, then increasing precipitation again through days 6-7.

Figure 5. Climate forecast

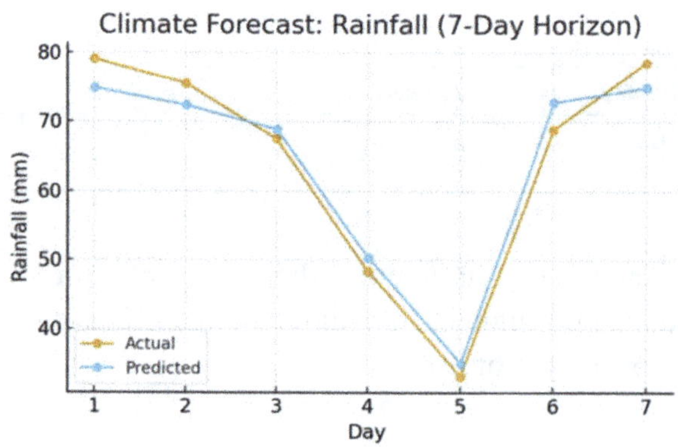

3. Crop Recommendation Module

The GNN-based recommendation system was validated using soil nutrient (N, P, and K), pH, moisture content, and weather data generated by the TFT model. It suggested the finest crops on the basis of forecasted consistency of yield and appropriateness.

Table 3. Results using GNN module

Model	Recommendation Accuracy	Top-3 Recommendation Success
GNN	93.16%	97.8%

The crop suitability score heatmap presented in Figure 6 demonstrates how suitable various crops are across the four soil plots, with darker shades of green indicating greater suitability of each crop to grow in each soil plot. Among the crops evaluated, groundnut exhibits the highest crop suitability across all the plots and receives a perfect suitability score of one hundred percent for Plot 4, demonstrating that it is the most adaptable crop between all soil types across the four regions.

Figure 6. Heat map

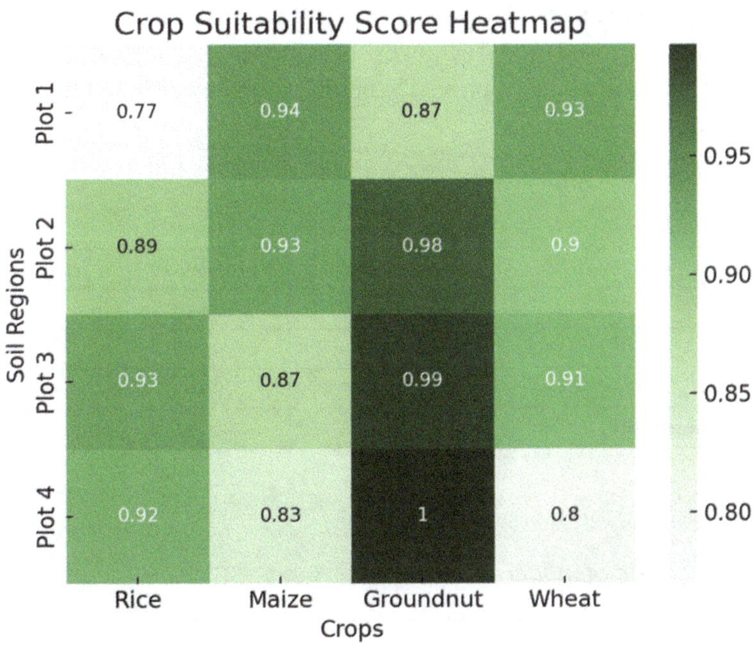

4. Chatbot Performance Evaluation

The chatbot interface was tested using 100 type of input questions by text and by the use of images. The findings showed that the intent detection and user satisfaction was highly accurate, which ensured smooth running in both technical and non-technical users.

Table 4. Results for chatbot performance

Evaluation Parameter	Score
Query Understanding Accuracy	92.8%
Response Accuracy	95.2%
Average Reply Time	1.8 seconds

Comparative Performance Evaluation

To validate the effectiveness of the proposed multi-model AI framework, a comparative analysis is conducted against three established baseline algorithms: Traditional CNN, ResNet-50, and EfficientNet-B0. These models are widely used in agricultural image classification and serve as standard benchmarks. All models were evaluated on the same dataset using identical experimental settings to ensure fairness.

As shown in Table 5, the proposed model achieves superior performance across all evaluation metrics. The higher recall and F1-score indicate improved generalization and reduced false negatives, which are essential for early plant disease detection.

Table 5. Comparative results

Model	Accuracy (%)	Precision (%)	Recall (%)	F1-Score (%)
Traditional CNN	92.15	91.30	90.85	91.07
ResNet-50	95.62	95.10	94.88	94.99
EfficientNet-B0	96.48	96.05	95.92	95.98
Proposed Multi-Model Framework (ViT)	98.42	97.85	98.01	97.92

Figure 7. Accuracy comparison of models

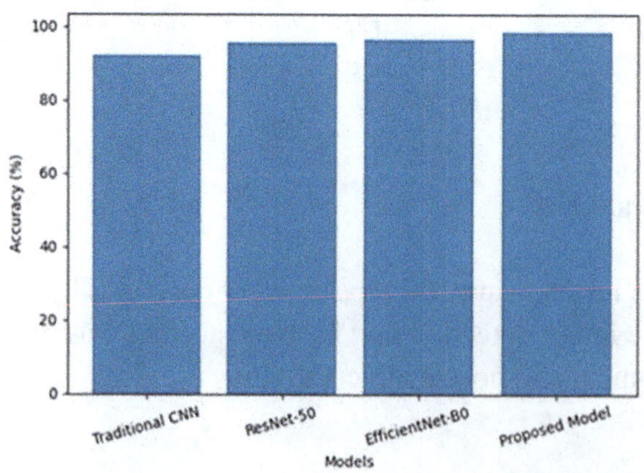

CONCLUSION AND RESEARCH DIRECTIONS

Conclusion

Agriculture has remained the mainstay of livelihoods in most of the developing areas but the farmers have always been faced with constant problems with regards to late detection of diseases, failure to select their crops and the escalating climate uncertainties. This study introduced a multi-modal agricultural decision support system, where plant disease diagnosis, climatic forecast and soil-conscious crop suggestion interoperability are integrated within a single precision agriculture platform based on GeoAI.

The experimental analysis has shown that the Vision Transformer model had high disease classification performance with an accuracy of 98.42 which can support trustworthy early-stage detection in the field conditions of variability. Temporal Fusion Transformer offered predictable multi-horizon environment predictions with reduced prediction errors, useful in proactive planning in response to rain variability, temperature changes and crop risks due to humidity. Moreover, the Graph Neural Network was useful in modeling the spatial associations among soil features and climatic predictors with a recommendation accuracy of 93.16 and a Top-3 success rate of 97.8 to enhance crop suitability decision-making.

The combination of IoT-based sensing made sure that the contextual data is constantly obtained, which improves the reliability of the model and increases its flexibility to the real-life agricultural conditions. Moreover, conversational chatbot interface lowered the technological barrier among farmers since it allowed them to interact through natural language queries and image-based inputs in a way that was easy to use. All in all, the offered multi-modal framework proves that the integration of visual intelligence, temporal analytics and spatial reasoning can contribute greatly to the farming productivity, minimization of relying on too much chemicals and promotion of sustainable approaches to the farm management. The paper validates the practicality of implementing intelligent decision support systems that will bridge the gap between the conventional farming knowledge and precision agriculture based on data.

Research Directions

Even though the suggested framework delivered high results in various agricultural intelligence tasks, there are a number of ways in which the framework can be expanded to increase its functionality and real-world influence.

- Fertiliser and Treatment Advice: The future work can be done by using agronomic advisory modules that will create accurate amounts of fertilizers and pesticides treatment plans using an estimation of the disease severity, and the soil nutrient profile.
- Market Intelligence and Price Forecasting: Economic indicators of crop planning and timing of harvesting operations can be assisted by agricultural market analytics and price forecast models.
- Edge AI Deployment: Lightweight transformer and graph learning architectures will be developed to deploy on low-power edge devices and can enhance access to the system in low-internet access locations.
- Explainable Artificial Intelligence (XAI): The inclusion of interpretability (via attention visualisation, feature importance mapping and confidence scoring) can promote the transparency of the model and increase trust in the model by farmers.

- Dataset Expansion and Generalisation: It will be possible to expand training datasets with more crop species, disease types and various agro-climatic regions to increase the robustness and scale of models.
- Aerial Monitoring Integration: The integration of drone-based imaging with satellite remote sensing can provide the ability to monitor large fields and identify the outbreak of epidemics at geographically separated farms.
- Multilingual Conversational Intelligence: The language adaptability will increase the usability of farmers with different lingual backgrounds and digital literacy levels.
- Feedback-Driven Learning Systems: Feedback loop through feedback mechanisms on farmers and outcome tracking mechanisms can be used to maintain an ever refining model as well as maintain accuracy in adaptive recommendations.
- Secure Agricultural Data Management: Investigating the concept of smart agriculture ecosystems based on the idea of blockchain-enabled data storage can guarantee transparency, data integrity, and privacy protection.

Finally, this study forms a platform of move towards smart, scalable and sustainable precision farming solutions. The framework can help in meeting long-term objectives like optimising food security, sustainable farming and enhanced socio-economic resilience in agricultural communities by incorporating GeoAI-driven perception, prediction and recommendation capabilities.

REFERENCES

Ali, H., Shifa, N., Benlamri, R., Farooque, A. A., & Yaqub, R. (2025). A fine tuned EfficientNet-B0 convolutional neural network for accurate and efficient classification of apple leaf diseases. *Scientific Reports*, *15*(1), 25732. DOI: 10.1038/s41598-025-04479-2 PMID: 40670396

Ayesha Barvin, P., & Sampradeepraj, T. (2023). Crop recommendation systems based on soil and environmental factors using graph convolution neural network: A systematic literature review. *Engineering Proceedings*, *58*(1), 97. DOI: 10.3390/ecsa-10-16010

Boutayeb, A., Lahsen-Cherif, I., & El Khadimi, A. (2025). When Machine Learning Meets Geospatial Data: A Comprehensive GeoAI Review. *IEEE Journal of Selected Topics in Applied Earth Observations and Remote Sensing*, *18*, 13135–13191. Advance online publication. DOI: 10.1109/JSTARS.2025.3568715

Cartolano, A., Cuzzocrea, A., & Pilato, G. (2024). Analyzing and assessing explainable AI models for smart agriculture environments. *Multimedia Tools and Applications*, *83*(12), 37225–37246. DOI: 10.1007/s11042-023-17978-z

Dosovitskiy, A., Beyer, L., Kolesnikov, A., Weissenborn, D., Zhai, X., Unterthiner, T., . . . Houlsby, N. (2020). An image is worth 16x16 words: Transformers for image recognition at scale. *arXiv preprint arXiv:2010.11929*

El Jarroudi, M., Kouadio, L., Delfosse, P., Bock, C. H., Mahlein, A. K., Fettweis, X., Mercatoris, B., Adams, F., Lenné, J. M., & Hamdioui, S. (2024). Leveraging edge artificial intelligence for sustainable agriculture. *Nature Sustainability*, *7*(7), 846–854. DOI: 10.1038/s41893-024-01352-4

Farooq, M. S., Riaz, S., Abid, A., Abid, K., & Naeem, M. A. (2019). A Survey on the Role of IoT in Agriculture for the Implementation of Smart Farming. *IEEE Access : Practical Innovations, Open Solutions*, *7*, 156237–156271. DOI: 10.1109/ACCESS.2019.2949703

Fei, S., Hassan, M. A., Xiao, Y., Su, X., Chen, Z., Cheng, Q., Duan, F., Chen, R., & Ma, Y. (2023). UAV-based multi-sensor data fusion and machine learning algorithm for yield prediction in wheat. *Precision Agriculture*, *24*(1), 187–212. DOI: 10.1007/s11119-022-09938-8 PMID: 35967193

Gupta, A., & Singh, A. (2023). Agri-gnn: A novel genotypic-topological graph neural network framework built on graphsage for optimized yield prediction. *arXiv preprint arXiv:2310.13037*

Hemalatha, S., & Jayachandran, J. J. B. (2024). A multitask learning-based vision transformer for plant disease localization and classification. *International Journal of Computational Intelligence Systems*, *17*(1), 188. DOI: 10.1007/s44196-024-00597-3

Kotwal, J., Kashyap, R., Shafi, P. M., & Kimbahune, V. (2024). Enhanced leaf disease detection: UNet for segmentation and optimized EfficientNet for disease classification. *Software Impacts*, *22*, 100701. DOI: 10.1016/j.simpa.2024.100701

Kussul, N., Lavreniuk, M., Skakun, S., & Shelestov, A. (2017). Deep learning classification of land cover and crop types using remote sensing data. *IEEE Geoscience and Remote Sensing Letters*, *14*(5), 778–782. DOI: 10.1109/LGRS.2017.2681128

Li, Z., Chen, G., & Zhang, T. (2020). A CNN-transformer hybrid approach for crop classification using multitemporal multisensor images. *IEEE Journal of Selected Topics in Applied Earth Observations and Remote Sensing*, *13*, 847–858. DOI: 10.1109/JSTARS.2020.2971763

Liang, J., & Jiang, W. (2023). A ResNet50-DPA model for tomato leaf disease identification. *Frontiers in Plant Science*, *14*, 1258658. DOI: 10.3389/fpls.2023.1258658 PMID: 37908831

Lim, B., Arık, S. Ö., Loeff, N., & Pfister, T. (2021). Temporal fusion transformers for interpretable multi-horizon time series forecasting. *International Journal of Forecasting*, *37*(4), 1748–1764. DOI: 10.1016/j.ijforecast.2021.03.012

Liu, X., Chen, J., & Wen, Q. (2023). A survey on graph classification and link prediction based on gnn. *arXiv preprint arXiv:2307.00865*

Mohanty, S. P., Hughes, D. P., & Salathé, M. (2016). Using deep learning for image-based plant disease detection. *Frontiers in Plant Science*, *7*, 1419. DOI: 10.3389/fpls.2016.01419 PMID: 27713752

Noyan, M. A. (2022). Uncovering bias in the PlantVillage dataset. *arXiv preprint arXiv:2206.04374*

Reddy, B. R., Kalnoor, G., Devashish, M., & Reddy, P. S. K. (2025). Deep Learning Based Mobile Application for Automated Plant Disease Detection. *IEEE Access : Practical Innovations, Open Solutions*, *13*, 107917–107925. Advance online publication. DOI: 10.1109/ACCESS.2025.3581099

Salman, A. G., Kanigoro, B., & Heryadi, Y. (2015, October). Weather forecasting using deep learning techniques. In *2015 international conference on advanced computer science and information systems (ICACSIS)* (pp. 281-285). IEEE. DOI: 10.1109/ICACSIS.2015.7415154

Salman, Z., Muhammad, A., & Han, D. (2025). Plant disease classification in the wild using vision transformers and mixture of experts. *Frontiers in Plant Science*, *16*, 1522985. DOI: 10.3389/fpls.2025.1522985 PMID: 40606487

Chapter 10
Remote Sensing and GIS Applications in Assessing Climate Change Impacts on Land Use and Land Cover Changes

Walid Al-Shaar
http://orcid.org/0000-0001-6822-8245
Novum Research and Innovation, Lebanon

Hessah Hassan
Novum Research and Innovation, Lebanon

May Al-Shaar
Novum Research and Innovation, Lebanon

Fatima Al-Juwaisri
Novum Research and Innovation, Lebanon

Fatima Al-Yousefi
Novum Research and Innovation, Lebanon

ABSTRACT

This chapter examines how remote sensing and GIS assess climate change impacts on LULC. These technologies enable real-time monitoring across vast scales through enhanced Earth observation data, time-series analysis, multi-source integration (optical, radar, LiDAR), and machine learning. Cloud platforms have democratized global analyses. Applications span urban areas (heat islands, expansion), agriculture (crop monitoring, irrigation), forests (deforestation, carbon storage, species shifts), and coasts (sea-level rise, erosion). These tools support evidence-based climate decisions despite challenges in data integration, computational demands, and uncertainty quantification. Emerging innovations—AI, IoT sensors, hyperspectral imaging, advanced LiDAR, and small satellites—expand capabilities. These

DOI: 10.4018/979-8-3373-8037-7.ch010

technologies shape policy in climate action, urban planning, agriculture, conservation, and disaster reduction. Integrating technological advances into climate-land use research is essential for building resilience in natural and human-modified environments facing global change.

1. INTRODUCTION

Climate change has emerged as one of the most pressing global challenges of the 21st century, with far-reaching implications for ecosystems, human societies, and economies worldwide. Climate change is reshaping many aspects of the Earth's systems, but its effects on land use and land cover (LULC) are especially noteworthy. Landscapes play a vital role in maintaining climate balance, preserving biodiversity, providing food resources, and supporting human communities. With global temperatures continuing to rise as a result of human-driven greenhouse gas emissions, there is a growing need to closely track and understand changes in land use and land cover. This understanding has become increasingly urgent for planning and adaptation efforts. This makes accurate, up-to-date, and comprehensive monitoring of LULC dynamics not just valuable, but essential.

Remote sensing (RS) and Geographic Information Systems (GIS) have changed how researchers study land use and land cover. Instead of relying only on field surveys or limited local observations, it is now possible to track changes across very large regions and follow how these changes unfold over many years. These tools make it much easier to see patterns and shifts that would otherwise remain unnoticed.

These technologies offer unparalleled capabilities for tracking changes in vegetation cover, urban expansion, agricultural transitions, and myriad other land surface processes that are both influenced by and contribute to climate change. Remote sensing, as it is known today, began to take shape with the first Earth-observation satellites in the 1970s (Wulder et al. 2022). Early missions like Landsat provided images that were somewhat coarse, roughly 80 meters for one pixel, meaning that only broad patterns on the landscape could be seen. The technology was still valuable, but it offered more of a general impression than a detailed picture of what was happening on the ground. Today, sub-meter commercial satellites and advanced radar platforms offer far more detailed and varied views of the Earth (Everett et al. 2002). Alongside improvements in the sensors themselves, there have also been major advances in how RS data are processed and analyzed. Increased computing power and new analytical approaches now allow researchers to draw much more precise and meaningful insights about changes taking place on the Earth's surface.

GIS has developed alongside remote sensing, giving researchers practical ways to combine, examine, and visualize different types of spatial data. Together, these tools provide a framework for studying how LULC respond to climate change. They are widely used to track forest cover loss, monitor urban heat, evaluate changes in agricultural production, follow coastal erosion, and investigate many other processes that reveal how landscapes and climate influence each other. This combination makes it possible to integrate climate data, socioeconomic information, and field observations within a single analytical environment. The range of applications continues to grow. RS and GIS have been used to map global forest loss, monitor the spread of urban heat islands, examine changes in agricultural productivity, track coastal erosion, and support many other efforts aimed at understanding how climate and landscapes interact. In addition, these tools are essential for guiding evidence-based decisions on climate change mitigation and adaptation. Understanding how climate, land use, environmental impacts, and management strategies influence one another requires a systems perspective.

Figure 1 presents the conceptual framework that guides the discussion in this chapter. It shows how climate change can drive changes in land use and land cover, which may then lead to environmental consequences such as reductions in biodiversity and shifts in ecosystem services. These environmental changes can, in turn, influence local and regional climate conditions, creating feedback loops that further complicate the relationship between climate and landscape dynamics. RS and GIS technologies play a central role in this system by monitoring these changes, analyzing the relationships between different components, and providing the information needed for effective policy and management responses. The framework also shows how well-informed policies and management strategies can help mitigate climate change impacts and manage LU changes more sustainably.

Figure 1. Conceptual framework illustrating the interconnected relationships between climate change, LULC changes, policies and management and the role of RS and GIS technologies in monitoring and managing these interactions

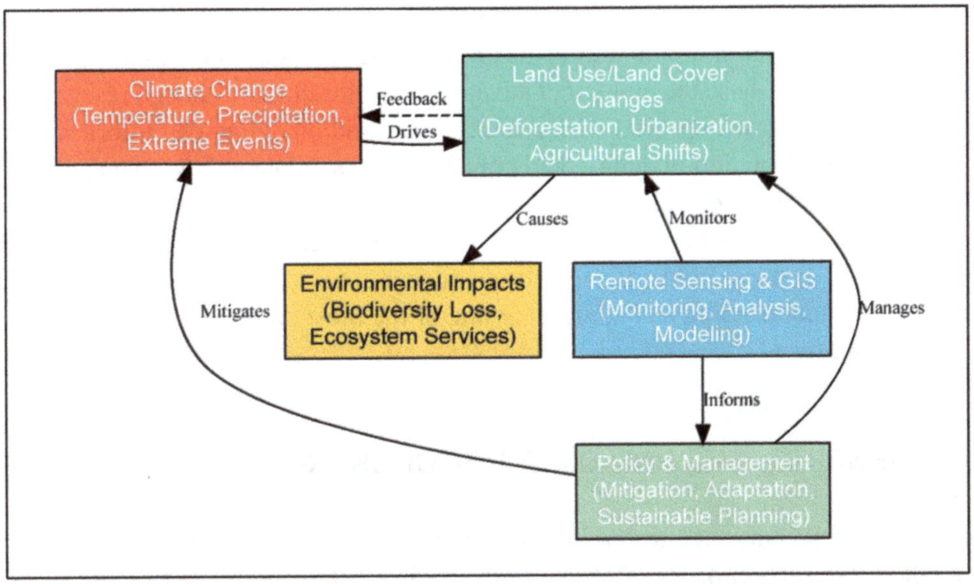

This integrated perspective emphasizes that RS and GIS are not merely technical tools for data collection and analysis but are essential components of a comprehensive approach to understanding and addressing one of the most pressing environmental challenges of our time. The following sections examine each component of this framework in detail, demonstrating how RS and GIS technologies enable more effective monitoring, analysis, and management of climate-land interactions across various scales and ecosystems.

As climate change accelerates and its impacts on landscapes become more pronounced, the role of RS and GIS in monitoring and analyzing these changes is becoming increasingly critical. These technologies not only provide a means to document current and historical LULC patterns but also offer valuable tools for projecting future scenarios, enabling policymakers and land managers to anticipate and plan for potential changes.

This chapter provides a comprehensive overview of the applications of RS and GIS in assessing climate change impacts on LULC. It begins by examining the fundamental principles and recent advancements in RS and GIS technologies relevant to LULC studies. The discussion then delves into the specific methodologies employed for detecting, quantifying, and analyzing climate driven LULC changes across various ecosystems and LU sectors. A significant portion of the chapter is dedicated to exploring case studies and applications in key areas, including urban environments, agricultural lands, forests, and coastal zones. These examples illustrate the practical implementation of RS and GIS techniques in real-world contexts and highlight their importance in addressing critical environmental challenges.

The chapter also addresses the current challenges faced in applying RS and GIS to climate-LULC studies, including issues of data integration, scale, and uncertainty quantification. It explores emerging technologies and methodologies that promise to overcome these challenges and push the boundaries of what is possible in Earth observation and analysis. Looking to the future, the chapter discusses upcoming satellite missions, novel sensor technologies, and advanced analytical techniques that are poised to further enhance the capacity for monitoring and understanding climate induced LULC changes.

It considers the potential of big data analytics, artificial intelligence, and cloud computing to revolutionize the processing and interpretation of Earth observation data. Finally, the chapter reflects on the broader implications of these technological advancements for climate change research, policy, and management. It considers how improved LULC monitoring, and modeling can inform climate mitigation strategies, support adaptation planning, and contribute to sustainable land management practices. By providing a comprehensive examination of the intersection between RS, GIS, climate change, and LULC dynamics, this chapter aims to serve as a valuable resource for researchers, practitioners, and policymakers working in these fields. It seeks to demonstrate the crucial role that Earth observation technologies play in understanding and addressing one of the most complex and urgent environmental challenges of current time.

2. REMOTE SENSING, GIS, AND DATA PROCESSING

The application of RS and GIS technologies, for instance for LULC change assessment, follows systematic methodological approaches that integrate multiple data sources, processing techniques, and analytical workflows. While modern implementations increasingly leverage advanced technologies such as artificial intelligence, cloud computing, and multi-sensor fusion, understanding the fundamental workflow provides essential context for the sophisticated techniques discussed throughout this section. Figure 2 illustrates the core methodological framework that underlies most RS based LULC studies. This traditional workflow serves as the foundation upon which contemporary approaches build, incorporating the advanced processing techniques, machine learning algorithms, and integration strategies examined in the following subsections.

Figure 2. Fundamental RS workflow for LULC change assessment. This framework illustrates the essential methodological steps underlying RS based LULC studies. The reference feedback loop (dashed line) represents the validation process using ground truth data

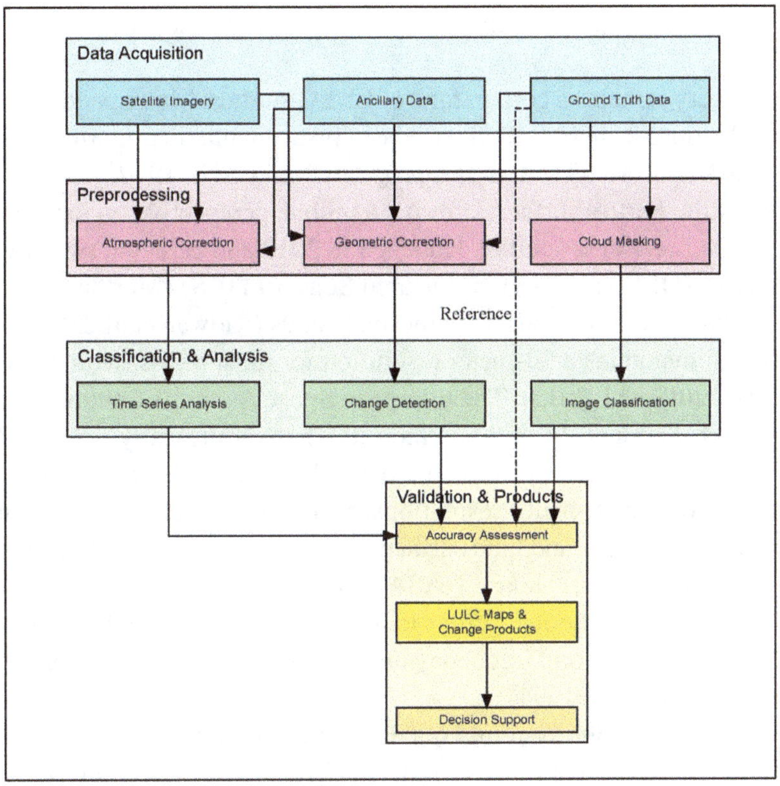

This foundational workflow emphasizes several key principles that guide effective LULC change assessment: the critical importance of integrating multiple data sources, the necessity of rigorous preprocessing to ensure data quality, and the iterative nature of validation that improves product reliability. However, contemporary RS applications extend far beyond this linear progression. Modern LULC change assessment increasingly employs sophisticated approaches that transcend traditional workflow boundaries. These include machine learning algorithms that automate feature extraction and classification, big data analytics platforms that process petabyte-scale archives, multi-sensor fusion techniques that leverage complementary sensor capabilities, and AI-driven systems that enable near real-time monitoring and change detection. The following sections examine each component of this framework in detail, with particular emphasis on how emerging technologies are revolutionizing traditional approaches.

The integration of these advanced capabilities has transformed RS from a primarily manual, expert-driven process to an increasingly automated, scalable, and precise scientific discipline capable of addressing the complex challenges posed by global environmental change.

2.1 Remote Sensing Data Sources

The field of RS encompasses a diverse array of data sources, each offering unique capabilities for observing and analyzing the Earth's surface. These data sources can be broadly categorized based on the type of sensor, the platform on which they are mounted, and the specific characteristics of the data they provide.

Optical satellite imagery remains a cornerstone of LULC studies. Multispectral sensors aboard satellites such as Landsat, Sentinel-2, and MODIS have been instrumental in providing consistent, long-term observations of the Earth's surface. The Landsat program, initiated in 1972, represents the longest continuous global record of the Earth's surface, offering a unique perspective on long-term LULC changes (Wulder et al. 2022). The Landsat 8 satellite, launched in 2013 and carrying two advanced sensors: the Operational Land Imager (OLI) and Thermal Infrared Sensor (TIRS) (Markham et al. 2015), provides data in nine shortwave bands and two longwave thermal bands (Ridwan et al. 2018). The satellite offers a 16-day revisit time and maintains a 30-meter resolution for most bands, with 15-meter resolution for panchromatic imagery (Lulla et al. 2013). The European Space Agency's Sentinel-2 mission, consisting of two identical satellites (Sentinel-2A and Sentinel-2B), has significantly enhanced the temporal and spectral resolution of freely available multispectral data. With a five-day revisit time at the equator and 13 spectral bands at 10, 20, and 60-meter resolutions, Sentinel-2 data has proven particularly valuable for monitoring vegetation dynamics and land surface changes (Gascon, 2018). MODIS (Moderate Resolution Imaging Spectroradiometer), aboard NASA's Terra and Aqua satellites, provides daily global coverage at moderate spatial resolutions (250m to 1km). While its spatial resolution is coarser than Landsat or Sentinel-2, its high temporal frequency makes it invaluable for monitoring rapid changes and phenological cycles (Running et al. 1994).

The diversity of available satellite systems creates a complex landscape of trade-offs between spatial resolution, temporal resolution, and data accessibility. Understanding these trade-offs is crucial for selecting appropriate data sources for specific LULC change and climate impact studies. Figure 3 illustrates the positioning of major satellite systems discussed in this chapter across the spatial-temporal resolution spectrum, while also indicating their data accessibility policies.

Figure 3. Satellite data characteristics - spatial vs temporal resolution trade-offs. The figure displays major satellite systems used in climate change and LULC studies, showing the fundamental trade-offs between spatial resolution (y-axis, meters) and temporal resolution (x-axis, days). Point size indicates spatial resolution categories, while colors represent data accessibility: green (free/open access), red (commercial), and orange (mixed/limited access)

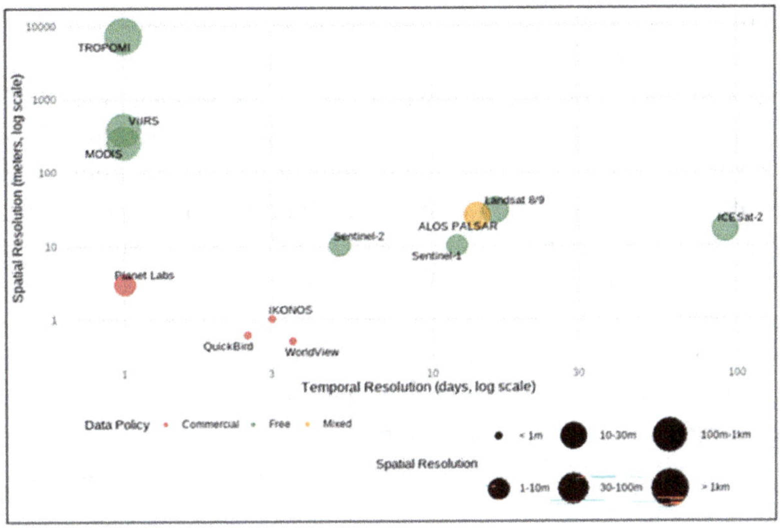

As illustrated in Figure 3, satellite systems generally exhibit an inverse relationship between spatial and temporal resolution, reflecting fundamental constraints in sensor design, data storage, and downlink capabilities. Daily global coverage systems like MODIS and VIIRS operate at moderate to coarse spatial resolutions (250m-7km), making them ideal for monitoring broad-scale phenomena such as vegetation phenology, fire detection, and atmospheric composition. In contrast, commercial high-resolution satellites (WorldView, IKONOS, QuickBird) provide sub-meter to 4 meters-scale detail but with revisit times of several days, making them suitable for detailed urban studies and precision agriculture applications. The intermediate zone occupied by Landsat and Sentinel-2 represents an optimal balance for many LULC applications, providing sufficient spatial detail (10-30m) for landscape-scale analysis while maintaining reasonable temporal frequency (5-16 days) for change detection studies. SAR systems like Sentinel-1 and ALOS PALSAR offer unique capabilities by penetrating clouds and operating day/night, though they typically provide structural rather than spectral information. Data accessibility policies significantly influence the practical application of these systems. The shift toward free and open data policies by major space agencies (ESA's Sentinels, USGS Landsat archive, NASA missions) has democratized access to Earth observation data, enabling global research and operational applications. However, the highest spatial resolution capabilities remain largely commercial, creating cost barriers for some applications while driving innovation in the private Earth observation sector.

High-resolution commercial satellites, such as WorldView, QuickBird, and IKONOS, offer imagery at sub-meter to 4-meter resolutions. These data sources are particularly useful for detailed urban studies, precision agriculture, and fine-scale ecosystem mapping (Yang, 2018; Blaschke, 2010). However, their high cost and limited spatial coverage often restrict their use to localized studies or for validating analyses based on medium-resolution imagery.

Synthetic Aperture Radar (SAR) sensors have gained prominence in LULC studies due to their ability to penetrate clouds and operate in day or night conditions (Reiche et al. 2016). Similarly, the European Space Agency's Sentinel-1 mission provides C-band SAR data globally, (Huang et al. 2018). JAXA's ALOS PALSAR and PALSAR-2 offer L-band SAR data, which is superior to C-band for forest monitoring, is particularly useful due to its ability to penetrate vegetation canopies (Mitchell et al. 2012).

LiDAR (Light Detection and Ranging) systems, both airborne and spaceborne, have revolutionized the ability to measure three-dimensional vegetation structure and terrain. Airborne LiDAR provides high-resolution elevation data crucial for urban modeling, forest structure analysis, and detailed terrain mapping. The Global Ecosystem Dynamics Investigation (GEDI) mission, launched in 2018, provided the first global, high-resolution LiDAR observations of vertical forest structure (Dubayah et al. 2020).

Hyperspectral sensors, which capture data in hundreds of narrow, contiguous spectral bands, offer unprecedented capabilities for detailed surface material characterization. While currently limited primarily to airborne platforms, upcoming satellite missions like NASA's Surface Biology and Geology (SBG) mission promise to provide global hyperspectral observations (Sen, 2023; Transon et al. 2018).

Thermal infrared sensors, such as those aboard Landsat and MODIS satellites, measure land surface temperature, which is crucial for studying urban heat islands, monitoring drought conditions, and assessing evapotranspiration (Al-Shaar et al. 2022). In that context, the European Space Agency's Land Surface Temperature Monitoring (LSTM) mission, part of the Copernicus programme expansion, aims to provide high spatio-temporal resolution land surface temperature and evapotranspiration data globally (Bernard et al. 2023).

Passive microwave sensors, while offering coarser spatial resolution, provide valuable information on soil moisture, snow cover, and sea ice extent. For instance, NASA's Soil Moisture Active Passive (SMAP) mission, initiated in 2015, provides global measurements of soil moisture state (Sabbir et al. 2025). The integration of data from these diverse sources, each with its own strengths and limitations, allows for a more comprehensive understanding of LULC dynamics and their interactions with climate change. Multi-sensor approaches have become increasingly common, leveraging the complementary nature of different data types to overcome the limitations of individual sensors.

2.2 GIS Data Integration

GIS serve as a crucial framework for integrating RS data with other spatial datasets, enabling comprehensive analyses of LULC changes in the context of climate, topography, and human activities. The ability to overlay and analyze multiple layers of spatial information is fundamental to understanding the complex interactions between different environmental and anthropogenic factors driving LULC changes.

Digital Elevation Models (DEMs) form a critical component of many LULC analyses, providing essential information on terrain characteristics that influence LU patterns, vegetation distribution, and hydrological processes. Global DEMs, such as the Shuttle Radar Topography Mission (SRTM) data at 30-meter resolution and the ASTER Global Digital Elevation Model (GDEM) at 30-meter resolution, have been widely used in LULC studies (Tadono et al. 2014). More recently, the TanDEM-X mission has provided a global DEM at 12-meter resolution, offering unprecedented detail for large-scale topographic analyses (Zink et al. 2014).

Climate data integration is fundamental to understanding the relationships between LULC changes and climate variability. Gridded climate datasets, such as WorldClim and the Climatic Research Unit Time Series (CRUTS), provide long-term average climate variables and time series of climate data at various spatial resolutions (Harris et al. 2020; Fick and Hijmans, 2017). These datasets typically include variables such as temperature, precipitation, solar radiation, and wind speed, which are crucial for analyzing climate driven LULC changes and for parameterizing land surface models.

Soil data provides essential information on the physical and chemical properties of soils that influence vegetation growth, agricultural productivity, and LU suitability. Global soil databases such as the Harmonized World Soil Database (HWSD) and SoilGrids offer spatially explicit soil information at various depths (Hengl et al. 2017). These datasets typically include properties such as soil texture, pH, organic carbon content, and water holding capacity, which are crucial for understanding the potential impacts of climate change on soil resources and land productivity.

Socioeconomic data integration is essential for understanding the human dimensions of LULC changes. Datasets on population density, economic indicators, infrastructure, and land ownership provide context for observed LULC patterns and help in modeling future scenarios. The integration of remotely sensed LULC data with socioeconomic information allows for more nuanced analyses of the drivers and impacts of LU changes (Al-Shaar, 2024).

Historical maps and atlases serve as valuable resources for extending LULC change analyses beyond the era of satellite observations. Digitized historical maps can be georeferenced and integrated into GIS databases, allowing for long-term assessments of landscape changes (Yang et al. 2014). This integration of historical and contemporary data sources is particularly valuable for understanding long-term trends and cycles in LULC dynamics.

Field survey data plays a crucial role in calibrating and validating RS based LULC assessments. Ground truth information collected through field surveys provides essential reference data for training classification algorithms and assessing the accuracy of LULC maps derived from satellite imagery (Al-Shaar and Bonin, 2021). The integration of field data with RS observations in a GIS environment allows for the development of more robust and accurate LULC products.

The effective integration of these diverse data sources in a GIS framework presents both opportunities and challenges. Issues of data harmonization, including reconciling differences in spatial and temporal resolutions, coordinate systems, and data formats, must be carefully addressed. Moreover, the integration of qualitative and quantitative data, often with varying levels of uncertainty, requires sophisticated analytical approaches to ensure meaningful results.

Advanced GIS techniques, such as spatial statistics, geostatistics, and multi-criteria decision analysis, are increasingly being employed to analyze the complex relationships between LULC patterns and multiple environmental and socioeconomic variables. These techniques allow for the identification of spatial patterns, trends, and correlations that may not be apparent from visual inspection alone.

The advent of cloud-based GIS platforms and big data analytics has further expanded the capabilities for large-scale, multi-source data integration and analysis. Platforms such as Google Earth Engine provide access to petabytes of satellite imagery and geospatial datasets, along with the computational power to process this data at global scales (Gorelick et al. 2017). These advancements have democratized access to Earth observation data and analytics, enabling researchers worldwide to conduct sophisticated LULC analyses. As the volume and diversity of available geospatial data continue to grow, the development of efficient data management systems and metadata standards becomes increasingly important. Initiatives such as the Group on Earth Observations (GEO) and the Committee on Earth Observation Satellites (CEOS) are working towards improving data interoperability and accessibility, facilitating more comprehensive and integrated approaches to studying global environmental changes.

2.3 Data Processing and Analysis Techniques

The processing and analysis of RS data for LULC studies involve a wide range of techniques, from basic image preprocessing to advanced machine learning algorithms. These methods have evolved rapidly in recent years, driven by advancements in computing power, algorithm development, and the increasing volume and complexity of available Earth observation data.

Preprocessing of satellite imagery is a crucial first step in any RS based LULC analysis.

Atmospheric correction is a critical preprocessing step in RS, aiming to remove the effects of atmospheric scattering and absorption, ensuring that the derived surface reflectance values accurately represent ground conditions. As described by Hagolle et al. (2015), this process can involve a range of techniques, from simpler approaches such as dark object subtraction to more sophisticated radiative transfer model-based approaches.

Geometric correction and orthorectification are necessary to ensure that images are accurately aligned with their true positions on the Earth's surface. This process involves correcting for distortions caused by sensor geometry, satellite orbit, and terrain variations. Modern satellite data products often come with predefined geometric corrections, but additional processing may be required for historical imagery or when combining data from multiple sensors (Toutin, 2004).

Cloud masking is another critical preprocessing step, particularly for optical satellite imagery. Various algorithms have been developed to detect and mask cloud and cloud shadow pixels, ranging from simple thresholding approaches to more complex machine learning methods (Zhu et al. 2015). The choice of cloud masking algorithm can significantly impact subsequent analyses, especially in regions with persistent cloud cover.

Image classification techniques form the backbone of many LULC mapping efforts. Traditional pixel-based classification methods, including supervised and unsupervised approaches, have been widely used for decades. Supervised classification algorithms, such as Maximum Likelihood and Minimum Distance, rely on training data to categorize pixels into predefined LULC classes. Unsupervised methods, like K-means clustering and ISODATA, group pixels with similar spectral characteristics without the need for training data (Al-Shaar and Bonin, 2024).

Object-Based Image Analysis (OBIA) has gained prominence in recent years, particularly for high-resolution imagery. OBIA approaches first segment the image into homogeneous objects and then classify these objects based on their spectral, spatial, and contextual properties. This method can often produce more accurate and visually appealing LULC maps compared to pixel-based approaches, especially in heterogeneous landscapes (Blaschke, 2010).

Machine learning algorithms have become increasingly popular for LULC classification due to their ability to handle complex, non-linear relationships in high-dimensional data. Random Forests, Support Vector Machines (SVM), and ensemble methods have shown excellent performance across a wide range of LULC classification tasks (Maxwell et al. 2018). These algorithms can effectively integrate multiple data sources, including spectral bands, vegetation indices, texture measures, and ancillary data.

Deep learning techniques, particularly Convolutional Neural Networks (CNNs), have demonstrated state-of-the-art performance in many RS tasks, including LULC classification (Castelluccio et al. 2015). CNNs can automatically learn hierarchical features from raw image data, often outperforming traditional machine learning approaches. However, they typically require large amounts of training data and significant computational resources. Despite their remarkable effectiveness, CNNs typically require large amounts of labeled training data and considerable computational resources, which can limit their applicability in domains with scarce annotated datasets.

More, transfer learning approaches have gained traction in recent years, allowing models trained on one dataset to be fine-tuned for use on a different but related dataset. This technique has proven particularly valuable in RS applications, where labeled training data may be scarce for certain regions or LULC types (Xie et al. 2016).

Change detection is a critical component of LULC studies, especially in the context of climate change impacts. Post-classification comparison remains a widely used method, where independently classified images from different time points are compared to identify changes. While straightforward, this method can propagate errors from individual classifications (Singh, 1989).

More sophisticated change detection techniques include image differencing, change vector analysis, and spectral mixture analysis. These methods can detect subtle changes in land cover that may not result in a change of categorical class but could indicate important shifts in ecosystem structure or function (Coppin et al. 2004).

Time series analysis has become increasingly important with the availability of long-term satellite records and high-frequency observations. Methods such as BFAST (Breaks For Additive Season and Trend) and LandTrendr (Landsat-based Detection of Trends in Disturbance and Recovery) enable the detection of both abrupt and gradual changes in LULC over time (Verbesselt et al. 2010; Kennedy et al. 2010). These approaches are particularly valuable for understanding the trajectory of landscape changes in response to climate variability and long-term climate trends.

Continuous change detection and classification (CCDC) algorithms have been developed to leverage the entire time series of available imagery, rather than relying on bi-temporal comparisons. These methods can provide more robust change detection results and allow for near real-time monitoring of LULC changes (Zhu and Woodcock, 2014).

Biophysical parameter estimation is crucial for quantifying the impacts of climate change on vegetation and ecosystems. Vegetation indices, such as the Normalized Difference Vegetation Index (NDVI) and Enhanced Vegetation Index (EVI), have been widely used to assess vegetation greenness, productivity, and health (Huete et al. 2002).

More advanced techniques, including retrieval of terrestrial vegetation methods, allow for the estimation of specific vegetation properties such as leaf area index (Verrelst et al. 2015).

Spectral unmixing techniques have been developed to address the issue of mixed pixels, which is particularly relevant in medium to coarse resolution imagery. Linear spectral unmixing assumes that the reflectance of a pixel is a linear combination of the reflectances of its constituent cover types, weighted by their fractional abundances. More complex non-linear unmixing models have also been developed to account for multiple scattering effects in certain environments (Keshava and Mustard, 2002).

Data fusion techniques play an increasingly important role in LULC studies, allowing for the integration of data from multiple sensors to overcome the limitations of individual data sources. Optical-radar fusion, for example, can combine the spectral information from optical sensors with the structural information from SAR data to improve LULC classification accuracy, especially in areas prone to cloud cover (Joshi et al. 2016).

Multi-scale analysis approaches have been developed to address the issue of scale dependency in LULC studies. Techniques such as wavelet transforms and object-based multi-resolution segmentation allow for the analysis of LULC patterns across multiple spatial scales, providing insights into the hierarchical organization of landscapes and the scale-dependent relationships between LULC and environmental variables (Wu, 2004).

Geostatistical methods, including kriging and its variants, have been applied to interpolate and analyze spatial patterns in LULC data. These techniques can be particularly useful for filling gaps in satellite observations and for understanding the spatial structure of landscape heterogeneity (Oliver and Webster, 2015).

Landscape metrics, derived from classified LULC maps, provide quantitative measures of landscape composition and configuration. Metrics such as patch size, shape, connectivity, and diversity can offer valuable insights into the fragmentation and heterogeneity of landscapes, which are often indicative of anthropogenic impacts and climate-induced changes (Lausch et al. 2015).

The integration of RS data with process-based models has become an important approach for understanding and projecting LULC changes. LU change models, such as CLUE (Conversion of LU and its Effects) and LTM (Land Transformation Model), incorporate RS-derived LULC information along with socioeconomic and biophysical variables to simulate future LU patterns under different scenarios (Verburg et al. 2002).

Ecosystem models, ranging from simple regression-based approaches to complex dynamic global vegetation models (DGVMs), increasingly rely on RS data for parameterization, calibration, and validation. These models can simulate the responses of vegetation to climate variability and change, providing insights into potential future shifts in ecosystem structure and function (Poulter et al. 2015).

The assimilation of RS data into climate models has improved the representation of land surface processes in these models. Techniques such as ensemble Kalman filtering allow for the optimal combination of model predictions with observational data, leading to more accurate simulations of land-atmosphere interactions (Quaife et al. 2008).

Uncertainty quantification and error propagation analysis are critical aspects of LULC studies based on RS data. Uncertainty quantification in LULC classification has been addressed using various methods. For instance, Valle et al. (2023) employed conformal statistics approach to quantify uncertainty in land cover classification, demonstrating superior computational efficiency compared to traditional resampling methods. Cheng et al. (2021) utilized stochastic simulation approaches to assess classification accuracy uncertainty, accounting for variability in training sample selection. Bayesian approaches have gained prominence, with Cockx et al. (2014) proposing a Monte Carlo simulation framework combined with Bayesian approach for urban land-use mapping uncertainty.

As the volume of available Earth observation data continues to grow exponentially, big data analytics and cloud computing platforms have become essential for processing and analyzing these massive datasets. Platforms such as Google Earth Engine provide researchers with access to petabytes of satellite imagery and the computational power to analyze this data at global scales (Gorelick et al. 2017).

The development of near real-time monitoring systems for LULC changes has been facilitated by advances in both data availability and processing techniques. Systems such as Global Forest Watch utilize daily satellite observations and automated processing chains to provide alerts on forest cover changes within days of their occurrence (Hansen et al. 2016).

The integration of citizen science and crowdsourcing approaches with RS has opened new possibilities for LULC mapping and validation. Platforms such as Geo-Wiki have demonstrated the potential of leveraging human interpretation to improve global land cover maps and to provide reference data for algorithm training and validation (Fritz et al. 2017).

As RS technology continues to advance rapidly, innovative analytical techniques are emerging. For instance, deep reinforcement learning has shown promise in optimizing the scheduling and acquisition of satellite imagery (Hadj-Salah et al. 2019). Meanwhile, quantum computing still in its early stages, holds potential to revolutionize the processing of large RS datasets in the future (Zaidenberg et al. 2021).

3. DISCUSSION

3.1 Advancements in Multi-temporal Satellite Imagery Analysis

The analysis of multi-temporal satellite imagery has been fundamental to understanding the dynamics of LULC changes in the context of climate change. Recent advancements in this field have significantly enhanced the ability to detect, quantify, and characterize both abrupt and gradual changes in landscapes over time. One of the most significant developments has been the improvement in the spatial and temporal resolution of satellite imagery. The launch of Sentinel-2 in 2015 marked a significant advancement in Earth observation, providing global coverage every 5 days at 10-20 meter resolution (Gascon, 2014). This high-frequency, high-resolution data has enabled more detailed and timely monitoring of LULC changes, particularly in rapidly changing environments such as urban areas and agricultural regions. The long-term continuity of satellite missions, particularly the Landsat program, has been crucial for understanding decadal-scale changes in LULC.

The Landsat archive, with its consistent record dating back to 1972, provides an unparalleled resource for analyzing long-term trends in landscape change (Wulder et al. 2012). The recent release of the entire Landsat archive as analysis-ready data has greatly facilitated large-scale, long-term studies of LULC dynamics. Advancements in time series analysis techniques have allowed researchers to extract more information from these long-term datasets. The development of algorithms such as LandTrendr (Kennedy et al. 2010) and BFAST (Verbesselt et al. 2010) has enabled the detection of both abrupt changes (e.g., deforestation events) and gradual trends (e.g., vegetation decline due to climate stress) within a single analytical framework. These methods have been particularly valuable for understanding the complex dynamics of forest ecosystems under changing climatic conditions.

The integration of data from multiple sensors has become increasingly common, allowing for more robust and comprehensive analyses. Fusion techniques that combine optical imagery with SAR data, for example, have improved the ability to monitor LULC changes in areas prone to persistent cloud cover, such as tropical rainforests (Reiche et al. 2016). The synergistic use of sensors with different characteristics has enhanced the ability to discriminate between LULC classes and to detect subtle changes in landscape properties.

Machine learning approaches, particularly deep learning techniques, have revolutionized the analysis of multi-temporal satellite imagery. Convolutional Neural Networks (CNNs) and Recurrent Neural Networks (RNNs) have demonstrated superior performance in tasks such as LULC classification and change detection, often surpassing traditional methods in both accuracy and efficiency. These techniques are particularly well-suited to handling the high-dimensional, temporally dense data produced by modern satellite systems.

The development of global, high-resolution LULC products has been another significant advancement. Initiatives such as the Copernicus Global Land Service have produced consistent, annually updated land cover maps, providing a valuable resource for monitoring global LULC trends. These products leverage the latest advancements in machine learning and big data processing to generate accurate, timely information on global land cover dynamics.

Near real-time monitoring systems have emerged as powerful tools for detecting and responding to rapid LULC changes. Systems such as Global Forest Watch utilize daily satellite observations and automated processing chains to provide alerts on forest cover changes within days of their occurrence (Hansen et al. 2016). These systems have become invaluable for forest management, conservation efforts, and policy enforcement in many regions.

The analysis of land surface phenology using time series of vegetation indices has provided important insights into the impacts of climate change on ecosystem functioning. Studies have revealed shifts in the timing of vegetation green-up and senescence, changes in growing season length, and alterations in productivity patterns across various biomes (Garonna et al. 2016). These phenological changes serve as sensitive indicators of ecosystem responses to climate variability and long-term climate trends.

Advances have been made in the characterization of the dynamics of land surface temperature (LST) through the use of thermal infrared data, leading to a better understanding of urban heat island effects and their links to LULC patterns. Complex interactions between urban expansion, vegetation changes, and local climate modifications have been revealed through time series analyses of LST data (Peng et al. 2012). The development of automated change detection and classification systems has greatly enhanced the ability to monitor LULC changes over large areas. The Continuous Change Detection and Classification (CCDC) algorithm, for example, leverages all available Landsat imagery to provide updated land cover information and to detect changes as they occur (Zhu and Woodcock, 2014). Such systems are particularly valuable for maintaining up-to-date LULC information in rapidly changing environments.

Despite recent technological advances, working with multi-temporal satellite imagery is still not easy. Modern satellites produce huge amounts of data, and handling all of it can be slow and demanding without smart algorithms or cloud-based tools. Combining data from different sensors, each with its own resolution and timing, also requires careful handling to ensure results are accurate and consistent. On top of that, maintaining continuous datasets can be difficult as older satellites reach the end of their service. Ensuring seamless transitions between satellite missions and maintaining long-term consistency in Earth observation records is crucial for ongoing monitoring of climate-induced LULC changes.

Looking ahead, advances in satellite technology are likely to make tracking LU dynamics over time even more effective. Hyperspectral missions such as EnMAP and PRISMA, for example, will provide far more detailed spectral information, helping researchers better understand the characteristics of Earth's surface (Guanter et al. 2015). Likewise, new SAR missions with enhanced capabilities will improve the ability to monitor structural changes in vegetation and urban environments.

The integration of satellite-based Earth observation with in-situ sensor networks and unmanned aerial vehicles (UAVs) is likely to become increasingly important, providing a multi-scale perspective on LULC dynamics. This integration will be particularly valuable for validating satellite-based analyses and for understanding fine-scale processes that influence broader landscape patterns.

As climate change continues to strongly impact ecosystems and human land use, analyzing multi-temporal satellite imagery will become increasingly important for tracking, understanding, and anticipating these changes. The ongoing advancements in this field promise to provide ever more detailed and timely information to support scientific research, policy-making, and land management decisions in the face of rapid environmental change.

3.2 Integration of Remote Sensing With Climate Data

The integration of RS data with climate datasets has emerged as a powerful approach for understanding the complex interactions between climate variability, climate change, and LULC dynamics. This integration allows for a more comprehensive analysis of how climatic factors drive changes in vegetation patterns, agricultural practices, urban development, and other aspects of LU.

Climate data sources used in conjunction with RS observations include ground-based weather station records, gridded climate datasets, reanalysis products, and outputs from climate models. Global datasets such as WorldClim and the CRUTS (Climatic Research Unit Time Series) provide long-term average climate variables and time series of climate data at various spatial resolutions, which can be readily integrated with satellite derived LULC information (Harris et al. 2020; Fick and Hijmans, 2017).

One of the primary applications of this integrated approach has been in the analysis of vegetation responses to climate variability and change. Time series of vegetation indices derived from satellites, such as NDVI from MODIS or LandSat, have been extensively used in combination with climate data to assess the sensitivity of different ecosystems to climate fluctuations.

For instance, Piao et al. (2014) used this approach to analyze vegetation greening trends and their relationship to temperature and precipitation changes.

The integration of climate data with RS observations has been particularly valuable for understanding and predicting agricultural responses to climate change. By combining crop-specific land cover maps derived from high-resolution satellite imagery with climate data, researchers have been able to assess the impacts of temperature and precipitation changes on crop yields and distribution patterns. Lobell et al. (2011) demonstrated this approach by analyzing the sensitivity of crop yields to climate trends in different regions, providing important insights for agricultural adaptation strategies.

In urban environments, the integration of land surface temperature data derived from thermal infrared sensors with climate records has improved the understanding of urban heat island effects and their relationship to urban growth patterns. Researchers have used this integrated approach to analyze how urbanization modifies local climate conditions and to assess the effectiveness of various urban planning strategies in mitigating heat stress (Li et al. 2019).

The fusion of RS data with climate model outputs has become an important tool for projecting future LULC changes under different climate scenarios. For example, Sohl et al. (2016) integrated land-use models with climate projections to simulate potential changes in LU patterns across the United States under different emission scenarios. Such integrated modeling approaches provide valuable information for long-term LU planning and climate change adaptation strategies.

In forestry applications, the combination of RS-derived forest cover maps with climate data has enhanced the ability to model and predict forest disturbance patterns, such as wildfire risk and insect outbreaks. These integrated analyses have revealed complex relationships between climate variables, forest structure, and disturbance regimes, informing forest management practices in the context of climate change (Seidl et al. 2017).

The assessment of climate change impacts on biodiversity has also benefited from the integration of RS and climate data. Species distribution models that incorporate both climate variables and remotely sensed habitat characteristics can improve predictions of potential range shifts under future climate scenarios.

In coastal areas, satellite-derived land cover maps can be combined with projections of sea level rise and storm surges to identify regions prone to this risk. Such information is used by planners and local communities to guide the protection of these vulnerable areas. For example, Kulp and Strauss (2019) used improved elevation models in combination with sea-level rise scenarios to reveal that coastal exposure to flooding is much higher than previously estimated, highlighting the urgency for targeted adaptation strategies. The use of RS data has also played a critical role in improving the spatial resolution and accuracy of climate datasets themselves. Satellite-derived products, including precipitation estimates from the Global Precipitation Measurement (GPM) mission and land surface temperature data, have been used to improve the spatial detail of gridded climate datasets, especially in areas where ground-based observations are limited (Huffman et al. 2019). This improved spatial characterization of climate variables has, in turn, enabled more detailed analyses of climate-LULC interactions at regional and local scales. The integration of RS and climate data has been particularly valuable in understanding and monitoring climate extremes and their impacts on landscapes. For example, drought indices calculated from a combination of satellite-based vegetation health indicators and climate data, such as the Vegetation Health Index (VHI), have been used to assess the impacts of drought on agriculture and natural ecosystems (Kogan et al. 2019). For example, drought indices calculated from a combination of satellite-based vegetation health indicators and climate data, such as the Vegetation Health Index (VHI), have been used to assess the impacts of drought on agriculture and natural ecosystems (Kogan et al. 2019). These integrated drought monitoring approaches offer advantages over traditional climate-based indices by capturing both the climatic drivers and vegetation responses to water stress. In the field of phenology, the combination of satellite-derived phenological metrics with climate data has revealed important insights into the impacts of climate change on ecosystem functioning. Using this integrated approach, changes in the timing of spring green-up, autumn senescence, and overall growing season length have been observed across several biomes. These shifts have been associated with long-term patterns in temperature and precipitation (Piao et al. 2019). These studies have highlighted the sensitivity of different ecosystems to climate change and have implications for carbon cycling, wildlife habitat, and agricultural management.

The assimilation of RS data into climate models represents another important area of integration. Satellite-derived land surface parameters, such as albedo, leaf area index, and soil moisture, have been increasingly used to improve the representation of land surface processes in climate models (Boussetta et al. 2015). This integration has led to more accurate simulations of land-atmosphere interactions and has improved the ability to project future climate conditions and their potential impacts on LULC.

In the field of carbon cycle studies, the integration of RS-derived biomass estimates with climate data has enhanced understanding of the terrestrial carbon sink and its response to climate variability.

For example, Liu et al. (2017) combined satellite-based estimates of aboveground biomass with climate data to analyze the sensitivity of tropical forests to drought, providing insights into the vulnerability of these important carbon stores to climate change.

The use of integrated RS and climate data approaches has also been valuable in attribution studies, which seek to disentangle the relative influences of climate change and other factors on observed LULC changes. For instance, Zhu et al. (2016) used an integrated analysis of satellite imagery, climate data, and atmospheric CO2 measurements to attribute global vegetation greening trends to various drivers, including climate change, CO2 fertilization, and LU practices.

In coastal environments, the combination of RS-derived shoreline change data with climate and oceanographic datasets has improved understanding of coastal dynamics under changing climate conditions. These integrated analyses have revealed complex relationships between sea-level rise, storm patterns, sediment transport, and coastal LU, informing coastal management and adaptation strategies (Mentaschi et al. 2018).

The integration of RS and climate data has also played a crucial role in validating and improving climate model projections. Satellite observations of LULC changes and their associated impacts on surface properties (e.g., albedo, evapotranspiration) have been used to evaluate the performance of climate models in simulating land-atmosphere interactions and to constrain projections of future climate conditions (Feddema et al. 2005).

In the field of ecosystem service assessment, the combination of RS-derived LULC maps with climate data has enabled more comprehensive evaluations of how climate change may impact the provision of various ecosystem services. For example, Schirpke et al. (2017) integrated land cover maps, topographic data, and climate projections to assess potential changes in multiple ecosystem services in mountain regions under different climate scenarios.

The development of data cube technologies, which allow for the efficient storage and analysis of multi-dimensional Earth observation data, has further facilitated the integration of RS and climate datasets. Platforms such as the Australian Geoscience Data Cube provide researchers with the ability to analyze long time series of combined satellite imagery and climate data across large geographic areas, enabling new insights into climate-LULC interactions (Lewis et al. 2017). Despite the significant advances in integrating RS and climate data, several challenges remain. These include issues of scale mismatch between RS observations and climate data, uncertainties in both datasets, and the complexity of disentangling climate signals from other drivers of LULC change. Additionally, the high computational demands of processing and analyzing these large, integrated datasets pose challenges for many researchers and institutions.

Looking to the future, several emerging trends are likely to shape the continued integration of RS and climate data in LULC studies:

1. The increasing availability of high-resolution, globally consistent climate datasets, such as ERA5-Land, will enable more detailed analyses of climate-LULC interactions at local to regional scales (Muñoz-Sabater et al. 2021).
2. Advances in machine learning and artificial intelligence techniques are expected to improve the ability to extract meaningful patterns and relationships from the complex, high-dimensional datasets resulting from the integration of RS and climate data.
3. The launch of new satellite missions, such as the NASA Surface Biology and Geology (SBG) mission, will provide unprecedented observations of Earth system processes, allowing for more comprehensive integration with climate data (Sen, 2023).
4. Improvements in Earth system models and their ability to assimilate RS observations are likely to lead to more accurate and spatially detailed projections of future climate conditions and their impacts on LULC.
5. The continued development of cloud computing platforms and big data technologies will facilitate the processing and analysis of integrated RS and climate datasets at global scales and over long time periods.

3.3 Applications in Key Sectors

The integration of RS and GIS technologies has found widespread application in assessing climate change impacts on LULC across various sectors. This section explores the specific applications in key areas, highlighting how these technologies have enhanced understanding and management of climate-related challenges.

3.3.1 Urban Environments

Urban areas are particularly vulnerable to climate change impacts due to their high population densities, concentration of infrastructure, and modified local climates. RS and GIS have played crucial roles in understanding and addressing these challenges:

Urban Heat Island (UHI) Effect:

Satellite-derived land surface temperature (LST) data, primarily from thermal infrared sensors, have been instrumental in mapping and analyzing urban heat island effects. Studies have utilized long-term time series of LST data to quantify the intensification of UHI effects in relation to urban expansion and climate change.

Chakraborty and Lee (2019) used MODIS-derived land surface temperature (LST) data and a simplified urban-extent algorithm to analyze surface urban heat island (UHI) intensity across 9483 urban clusters. Their study revealed consistent increases in UHI intensity over time, with significant spatial and temporal variability influenced by climate zones and differences in vegetation cover between urban and rural areas.

High-resolution thermal mapping using airborne sensors and UAVs has enabled detailed analysis of intra-urban temperature variations, informing targeted mitigation strategies. Gaitani et al. (2011) demonstrated the use of high-resolution thermal imagery to identify urban hot spots and evaluate the cooling effects of green spaces in Athens, Greece.

Urban Growth Monitoring:

Multi-temporal analysis of high and medium resolution satellite imagery has been widely used to track urban expansion and its impacts on surrounding ecosystems. Long-term studies using consistent data sources like Landsat have revealed global trends in urban growth and their relationship to climate vulnerabilities. For example, Liu et al. (2020) used a 40-year time series of Landsat images to analyze global urban expansion patterns, providing insights into the environmental consequences of rapid urbanization.

Object-based image analysis (OBIA) techniques have improved the accuracy of urban land cover classification, particularly in complex urban environments. Berger et al. (2013) demonstrated the effectiveness of OBIA in mapping detailed urban structures and materials using high-resolution satellite imagery, which is crucial for understanding urban climate dynamics.

Green Space Dynamics:

RS has been vital in monitoring changes in urban vegetation cover, which plays a critical role in mitigating climate change impacts in cities. Time series analysis of vegetation indices, such as NDVI, has revealed trends in urban green space dynamics. For instance, Kabisch et al. (2019) used Landsat-derived NDVI time series to analyze changes in urban green space in Leipzig, Germany, linking these changes to climate adaptation policies.

LiDAR data has revolutionized the three-dimensional characterization of urban vegetation, allowing for accurate quantification of urban forest structure and biomass. Alonzo et al. (2016) demonstrated the use of airborne LiDAR data for mapping and quantifying urban tree species diversity and structure, providing valuable information for urban forest management and climate resilience planning.

Urban Water Management:

SAR technology has proven particularly useful for flood mapping and monitoring in urban areas, due to its ability to penetrate cloud cover and detect standing water. Chini et al. (2019) showcased the use of Sentinel-1 SAR data for high-resolution flood mapping in urban environments, demonstrating its potential for near real-time flood monitoring and risk assessment.

RS-based evapotranspiration (ET) modeling has improved understanding of urban water dynamics and their response to climate change. Nouri et al. (2013) reviewed various approaches for estimating urban ET using RS data, highlighting its importance for urban water management in a changing climate.

Air Quality Assessment:

Satellite-based air quality monitoring has significantly enhanced the ability to assess and manage urban air pollution, which is often exacerbated by climate change and urban growth. Sensors such as MODIS and TROPOMI provide data on aerosol optical depth (AOD) and trace gas concentrations, allowing for the analysis of long-term trends and spatial patterns in urban air quality (Duncan et al. 2014).

The integration of satellite-derived air quality data with LU information has revealed important relationships between urban form, climate, and air pollution patterns.

Climate Change Vulnerability Assessment:

GIS-based multi-criteria analysis techniques have been widely applied to assess urban vulnerability to climate change impacts, integrating RS-derived land cover data with socioeconomic and climate information. Pramanik et al. (2021) utilized such an approach to develop a climate change vulnerability index for coastal urban areas, demonstrating the power of integrated spatial analysis in identifying high-risk zones and informing adaptation strategies.

Future Urban Scenario Modeling:

The integration of RS data with urban growth models has enhanced the ability to project future urban landscapes under different climate and development scenarios. Al-Shaar et al. (2024) coupled a Multilayer-Perceptron Cellular Automata model with climate change scenarios to simulate future urban growth patterns in Lebanon, providing valuable insights for sustainable urban planning.

3.3.2 Agricultural Lands

Agriculture is one of the sectors most directly affected by climate change, with significant implications for global food security. RS and GIS have become indispensable tools for monitoring and managing agricultural systems in the face of climatic uncertainties:

Crop Monitoring and Yield Estimation:

Time series analysis of vegetation indices derived from satellites like MODIS and Sentinel-2 has enabled near real-time monitoring of crop health and productivity at regional to global scales. Becker-Reshef et al. (2010) demonstrated the use of MODIS data for global crop yield forecasting, providing valuable information for food security assessments and agricultural policy decisions.

Machine learning approaches integrating satellite imagery with climate data have improved the accuracy of crop yield predictions. You et al. (2017) developed a deep learning model combining RS data for county-level crop yield prediction in the United States, showcasing the potential of integrated data approaches in agricultural monitoring.

Agricultural Land Use Changes:

Multi-temporal analysis of satellite imagery has revealed significant shifts in agricultural LU patterns in response to climate change. Yin et al. (2020) used the full Landsat time series to map cropland abandonment in different countries, providing insights into how changing climate conditions are influencing agricultural LU decisions. The integration of RS data with climate projections can allow the modeling of potential future shifts in crop suitability zones.

Irrigation Mapping and Water Use Efficiency:

SAR technology has proven particularly effective for mapping irrigated areas and monitoring irrigation practices, especially in regions where cloud cover limits the use of optical sensors. Bégué et al. (2018) reviewed various RS approaches for irrigation mapping, highlighting the potential of RS data for monitoring irrigation dynamics in the context of climate change and water scarcity.

Thermal infrared data has been used to assess crop water stress and irrigation efficiency. Anderson et al. (2012) developed the Atmosphere-Land Exchange Inverse (ALEXI) model, which uses Landsat data to estimate evapotranspiration and drought stress, providing valuable information for irrigation management under changing climate conditions.

Soil Health Monitoring:

Monitoring soil health, by including factors such as organic matter content, salinity, and erosion, has been improved through the combination of multispectral and hyperspectral remote sensing data. Applications of visible and near-infrared spectroscopy for soil analysis have been examined by Ge et al. (2011), and its usefulness for large-scale soil assessments has been demonstrated. More, monitoring soil moisture has been carried out using SAR technology, due to its critical influence on agricultural productivity and susceptibility to climate change. Estimation techniques for soil moisture based on SAR data have been explored by Kornelsen and Coulibaly (2013), and the relevance of these methods for drought monitoring as well as for planning the use of agricultural water resources has been emphasized.

Crop Disease and Pest Monitoring:

High-resolution multispectral and hyperspectral imagery has improved the ability to detect and monitor crop diseases and pest infestations, which are expected to be exacerbated by climate change. Mahlein et al. (2012) demonstrated the use of hyperspectral imaging for early detection of plant diseases, showcasing its potential for improving crop management strategies in a changing climate. The integration of RS data with climate information and pest models has enhanced the ability to predict and manage pest outbreaks. Jaramillo et al. (2011) combined climate modeling and ecological modeling to predict shifts in the potential spread of the coffee berry borer under future climate change scenarios.

Climate-Smart Agriculture:

RS has played a crucial role in supporting the implementation and monitoring of climate-smart agricultural practices. Lipper et al. (2014) discussed the potential of Earth observation technologies in supporting climate-smart agriculture, highlighting applications in crop monitoring, soil health assessment, and water management.

Precision agriculture techniques, heavily reliant on RS data, have been increasingly adopted to optimize resource use and increase resilience to climate variability. Weiss et al. (2020) reviewed the applications of RS in precision agriculture, emphasizing its role in supporting adaptive management strategies in the face of climate change.

Crop Phenology Analysis:

Time series analysis of vegetation indices has revealed significant shifts in crop phenology in response to changing climate conditions. Bolton et al. (2013) used MODIS data to analyze trends in crop phenology across the U.S., demonstrating the sensitivity of agricultural systems to climate variability and change.

The integration of satellite-derived phenological metrics with climate data has improved understanding of crop-climate interactions. Zeng et al. (2016) combined MODIS data with climate records to analyze the impacts of temperature changes on crop phenology, providing insights into potential adaptation strategies.

Agricultural Drought Monitoring:

The development of RS-based drought indices has significantly enhanced the ability to monitor and assess agricultural drought conditions.

The integration of multiple RS data sources has improved the accuracy and reliability of agricultural drought assessments. For instance, agricultural drought can be effectively monitored using multi sensor approaches that integrate optical and thermal RS data (Rhee et al. 2010).

Greenhouse Gas Emissions:

RS has contributed to the monitoring and estimation of greenhouse gas emissions from agricultural activities. Zhang et al. (2016) used a combination of satellite imagery and biogeochemical modeling to estimate methane emissions from rice paddies, demonstrating the potential of RS in supporting climate change mitigation efforts in agriculture.

Land Degradation Assessment:

Multi-temporal analysis of RS data has been crucial in assessing and monitoring land degradation processes in agricultural areas. For instance, Prăvălie et al. (2021) used long-term NDVI time series to analyze global trends in land degradation, providing insights into the impacts of climate change and land management practices on soil health and productivity.

The integration of RS data with socioeconomic information and climate projections has enhanced the ability to model future land degradation risks.

The integration of multiple data sources, including optical, thermal, and microwave RS, along with climate data and in-situ observations, has significantly enhanced the ability to comprehensively assess and monitor agricultural systems. This multi-sensor, multi-scale approach is particularly valuable in the context of climate change, where impacts can manifest across various aspects of agricultural systems and at different spatial and temporal scales.

Looking ahead, emerging technologies such as hyperspectral imaging from space, high-resolution SAR systems, and Internet of Things (IoT) sensors promise to further enhance the capabilities for agricultural monitoring and management. The integration of these advanced Earth observation technologies with big data analytics, artificial intelligence, and cloud computing platforms is likely to revolutionize agricultural practices, enabling more precise, timely, and effective responses to the challenges posed by climate change.

3.3.3 Forests and Biodiversity

Forests play a crucial role in global climate regulation and biodiversity conservation. RS and GIS technologies have become indispensable tools for monitoring and managing forest ecosystems in the face of climate change:

Deforestation Monitoring:

The development of global forest monitoring systems based on satellite data has revolutionized the ability to track deforestation in near real-time. Hansen et al. (2013) used Landsat data to produce global maps of 21st-century forest cover change, providing unprecedented insights into the patterns and drivers of deforestation worldwide.

Time series analysis techniques, such as the Breaks For Additive Season and Trend (BFAST) algorithm, have enhanced the ability to detect subtle changes in forest cover and to differentiate between various types of forest disturbances. DeVries et al. (2015) demonstrated the application of BFAST for near real-time deforestation monitoring in tropical forests, highlighting its potential for supporting REDD+ initiatives.

The integration of optical and radar data has improved forest monitoring capabilities, particularly in regions prone to persistent cloud cover. Reiche et al. (2016) developed a multi-sensor approach combining Landsat and Sentinel data for near real-time deforestation detection in tropical forests, demonstrating significant improvements in monitoring frequency and accuracy.

Forest Degradation Assessment:

LiDAR technology has proven particularly valuable for assessing forest structure and detecting subtle changes associated with forest degradation. Dubayah et al. (2020) reviewed the applications of space-borne LiDAR, particularly the Global Ecosystem Dynamics Investigation (GEDI) mission, in mapping forest structure and biomass globally.

The combination of LiDAR with hyperspectral data has enabled more comprehensive assessments of forest health and biodiversity. Asner et al. (2017) used airborne LiDAR and imaging spectroscopy to map forest functional diversity in the Amazon, demonstrating the potential of these technologies for monitoring climate change impacts on forest ecosystems.

Carbon Stock Estimation:

RS-based approaches have significantly improved the ability to estimate and monitor forest carbon stocks at large scales. Baccini et al. (2012) used a combination of satellite LiDAR and radar data to produce pan-tropical maps of forest carbon stocks, providing crucial information for climate change mitigation efforts.

The integration of RS data with forest inventory plots and ecological models has enhanced the accuracy of carbon stock estimates.

Biodiversity Assessment:

High-resolution satellite imagery and advanced classification techniques have improved the ability to map and monitor habitat types, a crucial aspect of biodiversity conservation. Pettorelli et al. (2014) reviewed the applications of satellite RS in biodiversity monitoring, highlighting its potential for tracking changes in habitat extent and quality.

The use of imaging spectroscopy has enabled more detailed characterization of plant biodiversity from space. Jetz et al. (2016) discussed the potential of forthcoming global biodiversity monitoring systems based on imaging spectroscopy, emphasizing their importance for understanding climate change impacts on biodiversity.

Species Distribution Modeling:

The integration of RS-derived environmental variables with species occurrence data has enhanced the ability to model and predict species distributions under changing climate conditions. He et al. (2015) reviewed the use of RS in species distribution modeling, highlighting its potential for improving predictions of climate-driven range shifts.

Time series analysis of vegetation indices has revealed phenological shifts that can impact species interactions and ecosystem functioning. Buitenwerf et al. (2015) used long-term satellite observations to detect global changes in vegetation phenology, providing insights into the ecological consequences of climate change.

Fire Risk Mapping and Monitoring:

RS has played a crucial role in assessing and monitoring wildfire risk, which is expected to increase in many regions due to climate change.

Thermal infrared sensors have enabled near real-time detection and monitoring of active fires globally. Schroeder et al. (2014) described the fire detection algorithm used in the VIIRS (Visible Infrared Imaging Radiometer Suite) system, demonstrating its capabilities for global fire monitoring.

The analysis of long-term satellite records has revealed changing patterns in fire frequency and severity related to climate change. Andela et al. (2017) used MODIS burned area data to analyze global trends in fire activity, revealing complex relationships between climate, human activities, and fire regimes.

Forest Resilience and Recovery:

Time series analysis of satellite imagery has provided insights into forest resilience and recovery processes following disturbances. Kennedy et al. (2010) developed the LandTrendr algorithm for analyzing forest disturbance and recovery trajectories using Landsat time series, enabling assessments of long-term forest dynamics. The integration of RS data with climate information and ecological models has improved understanding of forest vulnerability to climate change.

Invasive Species Monitoring:

High-resolution multispectral and hyperspectral imagery has enhanced the ability to detect and map invasive plant species, which can be favored by changing climate conditions.

The combination of RS data with species distribution models has improved the ability to predict the potential spread of invasive species under future climate scenarios.

Protected Area Effectiveness:

Satellites are now a key tool for seeing how well protected areas keep forests and biodiversity intact. Nagendra et al. (2013) showed that remote sensing can track biodiversity and help manage these areas, offering useful insights for conservation planning.

Combining RS data with socioeconomic information has further helped researchers understand the complex factors that influence protected area effectiveness. In one study, Andam et al. (2008) applied matching methods alongside RS data to assess the impact of protected areas on deforestation in Costa Rica, highlighting the importance of considering confounding factors when evaluating conservation outcomes.

Combining multiple remote sensing technologies, such as optical, radar, and LiDAR systems, with advanced data analysis and ecological modeling has greatly improved recent ability to monitor and understand forest ecosystems at different scales. This integrated approach is especially important under climate change framework, where impacts can affect forest structure, composition, and function in complex ways.

Looking ahead, new technologies like hyperspectral satellite missions, high-resolution SAR constellations, and advanced airborne and terrestrial LiDAR systems are expected to further strengthen forest and biodiversity monitoring. Combining these Earth observation tools with AI, cloud computing, and field-based sensors could change the way ecological research and conservation practices are done, making it possible to respond to climate change impacts more quickly and effectively.

3.3.4 Coastal and Marine Environments

Coastal and marine ecosystems are particularly vulnerable to climate change impacts, including sea-level rise, ocean acidification, and changing temperature and circulation patterns. RS and GIS technologies have become crucial tools for monitoring and managing these dynamic environments:

Sea Level Rise Impact Assessment:

Satellite altimetry has revolutionized the ability to measure global sea level changes with high precision. Nerem et al. (2018) used 25 years of satellite altimeter data to assess global mean sea level rise, providing crucial insights into the acceleration of sea level rise in recent decades.

The integration of satellite-derived digital elevation models (DEMs) with sea level projections has enhanced the ability to assess coastal vulnerability to inundation. Kulp and Strauss (2019) used a new

DEM and probabilistic sea level projections to produce improved global estimates of population exposure to sea level rise and coastal flooding.

Interferometric Synthetic Aperture Radar (InSAR) techniques have enabled the detection of land subsidence in coastal areas, which can exacerbate the impacts of sea level rise. Shirzaei and Bürgmann (2018) used InSAR data to map land subsidence rates in the coastal areas along San Francisco Bay, demonstrating its importance for refining local sea level rise projections.

Coastal Erosion Monitoring:

Multi-temporal analysis of high-resolution satellite imagery has improved the ability to track shoreline changes and coastal erosion patterns. Vos et al. (2019) developed an automated shoreline detection algorithm using Landsat imagery, enabling global-scale analysis of shoreline dynamics. The integration of optical imagery with LiDAR and SAR data has enhanced coastal topographic mapping capabilities.

Time series analysis of satellite imagery has revealed long-term trends in coastal morphology related to climate change. Luijendijk et al. (2018) used 33 years of satellite imagery to assess global sandy beach erosion rates, providing a comprehensive picture of coastal change worldwide.

Mangrove Ecosystem Dynamics:

High-resolution satellite imagery and advanced classification techniques have improved the ability to map and monitor mangrove forests globally. Hamilton and Casey (2016) used Landsat data to produce a global map of mangrove forest extent and change, revealing significant losses due to both natural and anthropogenic factors.

The integration of optical and radar data has enhanced the ability to assess mangrove forest structure and biomass. Simard et al. (2019) combined multiple RS datasets to map global mangrove forest height and biomass, providing crucial information for blue carbon assessments.

Time series analysis of vegetation indices has revealed impacts of climate change on mangrove phenology and productivity. Songsom et al. (2019) used MODIS data to analyze long-term trends in mangrove greenness in Southern Thailand, demonstrating the sensitivity of these ecosystems to climate variability.

Coral Reef Monitoring:

High-resolution multispectral imagery has enabled large-scale mapping and monitoring of coral reef ecosystems. Thermal infrared sensors have played a crucial role in detecting and monitoring coral bleaching events related to ocean warming. Skirving et al. (2019) described the Coral Reef Watch program, which uses satellite-derived sea surface temperature data to provide near real-time monitoring and alerts for coral bleaching risk globally.

The integration of RS data with ecological models has improved the ability to assess coral reef vulnerability to climate change. Kumagai et al. (2018) combined satellite-derived environmental data with species distribution models to project future habitat suitability for corals under different climate scenarios.

Ocean Productivity and Ecosystem Changes:

Ocean color RS has provided unprecedented insights into global patterns of marine primary productivity and their response to climate variability. Behrenfeld et al. (2006) used satellite ocean color data to analyze global ocean primary production, revealing significant interannual variations related to climate oscillations.

Time series analysis of ocean color data has revealed long-term trends in marine ecosystem structure related to climate change. Boyce et al. (2010) used a century-long dataset of ocean transparency measurements, combined with recent satellite data, to detect global phytoplankton declines, highlighting the sensitivity of marine ecosystems to climate warming.

The integration of multiple satellite datasets has improved the ability to monitor and understand complex ocean processes. Siegel et al. (2014) demonstrated the use of satellite data to assess global ocean carbon export, combining observations of ocean color and sea surface temperature.

Coastal Urban Development:

High-resolution satellite imagery has enabled detailed mapping and monitoring of urban development in coastal areas.

For instance, the integration of RS data with socioeconomic information and climate projections has enhanced the ability to assess future risks to coastal urban areas. Vousdoukas et al. (2018) combined satellite-derived urban extent data with projections of sea-level rise and socioeconomic development to assess future flood risk in coastal cities globally.

More, thermal infrared RS has provided insights into the urban heat island effect in coastal cities, which can be exacerbated by climate change. Azevedo et al. (2016) used Landsat thermal data to analyze the relationship between urban expansion and land surface temperature in coastal metropolitan areas.

Marine Pollution Detection:

Synthetic Aperture Radar (SAR) has proven particularly effective for detecting oil spills and other forms of marine pollution. Topouzelis and Singha (2017) reviewed SAR applications in oil spill detection, highlighting recent advances in algorithms and sensor technologies.

Ocean color RS has enabled the detection and monitoring of harmful algal blooms, which may be influenced by climate change. Blondeau-Patissier et al. (2014) reviewed RS techniques for detecting and monitoring algal blooms in coastal and oceanic waters.

Sea Ice Monitoring:

Microwave RS has played a crucial role in monitoring global sea ice extent and thickness, revealing dramatic declines related to Arctic warming. Comiso et al. (2017) used long-term passive microwave data to analyze trends in Arctic Sea ice extent, demonstrating significant losses in recent decades.

The integration of multiple satellite datasets has improved the ability to assess sea ice thickness and volume. Kwok and Cunningham (2015) combined data from ICESat and CryoSat-2 satellite altimeters to estimate Arctic sea ice thickness and volume changes, providing a more complete picture of sea ice decline.

High-resolution optical and SAR imagery has enabled detailed mapping of sea ice features and dynamics. Scheuchl et al. (2016) demonstrated the use of Sentinel-1 SAR data for mapping sea ice types and tracking ice movement, highlighting its potential for supporting Arctic navigation and climate studies.

The integration of multiple RS technologies, including optical, thermal, radar, and altimetry systems, along with in-situ observations and numerical models, has significantly enhanced the ability to monitor and understand complex coastal and marine processes. This multi-faceted approach is particularly valuable in the context of climate change, where impacts can manifest across various spatial and temporal scales and involve complex interactions between physical, biological, and human systems.

Looking ahead, emerging technologies such as satellite-derived bathymetry, high-resolution ocean color sensors, and advanced SAR systems promise to further enhance capabilities for coastal and marine monitoring. The integration of these Earth observation technologies with artificial intelligence, big data analytics, and underwater vehicles is likely to revolutionize oceanographic research and coastal management practices, enabling more comprehensive, timely, and effective responses to the challenges posed by climate change in these critical environments.

4 CHALLENGES AND OPPORTUNITIES

While RS and GIS have greatly enhanced the ability to assess climate change impacts on LULC, several challenges remain. This section explores these challenges and discusses emerging opportunities for addressing them.

4.1 Data Integration and Harmonization

One of the primary challenges in using RS for LULC change and climate impact assessment is the integration and harmonization of diverse data sources. These sources often vary in spatial and temporal resolution, spectral characteristics, and data formats. Advancements in data fusion techniques and the development of standardized data formats offer promising solutions. For instance, the Harmonized Landsat Sentinel-2 (HLS) project (Claverie et al. 2018) demonstrates the potential for creating consistent, multi-sensor time series for land monitoring. Furthermore, the development of Analysis-Ready Data (ARD) products, such as those provided by the USGS for Landsat data, significantly reduces preprocessing requirements and facilitates multi-sensor integration. Machine learning approaches, particularly deep learning techniques, show great promise for data fusion and harmonization.

4.2 Computational Requirements

The increasing volume and complexity of RS data pose significant computational challenges for LULC change and climate impact assessment. Processing and analyzing petabyte-scale satellite data archives require substantial computational resources, often beyond the capacity of individual researchers or institutions. Cloud computing platforms and distributed processing technologies offer solutions to these computational challenges. Platforms like Google Earth Engine provide researchers with access to vast satellite imagery archives and the computational power to analyze them at global scales.

These platforms democratize access to big data analytics in Earth observation, enabling researchers worldwide to conduct sophisticated LULC analyses. The development of efficient algorithms optimized for big data processing is another area of opportunity.

4.3 Uncertainty Quantification

Quantifying and communicating uncertainties in RS-based LULC change assessments is crucial for informed decision-making. However, characterizing uncertainties across complex, multi-step analysis workflows remains challenging.

Advancements in statistical methods and ensemble modeling approaches offer improved frameworks for uncertainty quantification. Bayesian hierarchical models, for instance, provide a robust approach for propagating uncertainties through complex analytical workflows (Cressie et al. 2009).

The development of standardized protocols for accuracy assessment and uncertainty reporting in LULC products is another area of opportunity. Olofsson et al. (2014) proposed good practice recommendations for estimating area and assessing classification accuracy of land change, which could serve as a foundation for more comprehensive uncertainty quantification frameworks.

4.4 Temporal Frequency and Data Continuity

While the temporal resolution of satellite observations has improved significantly, gaps in coverage due to cloud cover or satellite revisit times can still limit the ability to detect rapid LULC changes or short-term climate impacts.

The launch of satellite constellations and the integration of data from multiple sensors offer opportunities to increase the effective temporal resolution of Earth observation data. For example, the combination of Landsat and Sentinel-2 data can provide 2-3 day revisit times for many parts of the world (Li and Roy, 2017). Data interpolation and fusion techniques also show promise for filling temporal gaps in satellite observations.

4.5 Attribution of Changes

Distinguishing climate driven LULC changes from those caused by other factors (e.g., economic drivers, policy changes) remains a significant challenge. This attribution is crucial for understanding the specific impacts of climate change and designing appropriate mitigation and adaptation strategies.

The integration of RS data with climate models, socioeconomic information, and process-based ecosystem models offers a more comprehensive approach to attribution. Zhu et al. (2016) demonstrated this approach by combining satellite observations with climate and atmospheric CO2 data to attribute global vegetation changes to various drivers.

4.6 Scale Issues

Bridging the gap between the spatial scales of RS observations and the scales at which climate processes and LU decisions operate poses ongoing challenges. This includes issues of upscaling fine-resolution data to regional or global assessments and downscaling coarse climate data to landscape scales.

Multi-scale analysis approaches and the development of scale-aware models offer potential solutions to these scaling issues. Wu and Li (2006) reviewed hierarchical patch dynamics as a conceptual framework for integrating pattern and process across scales in landscape ecology, which could be adapted for RS based LULC analyses.

The increasing availability of high-resolution satellite data, combined with advanced machine learning techniques, also offers opportunities for developing more sophisticated upscaling and downscaling methods. For instance, Senanayake et al. (2024) demonstrated the use of deep learning for downscaling climate data to finer spatial resolutions.

4.7 Validation and Ground Truth Data

Obtaining adequate ground truth data for validating RS based LULC classifications and change detections, particularly in remote or inaccessible areas, remains a significant challenge. The integration of citizen science approaches and crowdsourcing platforms with RS offers innovative solutions for collecting ground truth data. Projects like Geo-Wiki (Fritz et al. 2017) have demonstrated the potential of crowdsourcing for validating and improving global land cover maps. Advances in high-resolution satellite imagery and unmanned aerial vehicles (UAVs) also provide new opportunities for collecting detailed validation data.

4.8 Data Accessibility and Standardization

Despite the increasing availability of RS data, issues of data accessibility, interoperability, and standardization can still hinder large-scale or long-term LULC studies.

Opportunity:

The development of open data policies and standardized data formats is improving data accessibility and interoperability. Initiatives like the Committee on Earth Observation Satellites (CEOS) Analysis Ready Data (ARD) are working towards standardizing satellite data products for easier use in time series analyses and multi-sensor applications (Dwyer et al. 2018).

Cloud-based platforms and data cubes offer new paradigms for storing, accessing, and analyzing large volumes of Earth observation data. The Australian Geoscience Data Cube initiative, for instance, provides an open-source framework for organizing and analyzing large satellite data collections (Lewis et al. 2017). These challenges and opportunities highlight the dynamic nature of RS and GIS applications in climate change and LULC studies. As technology continues to advance and new analytical approaches are developed, the field is poised for significant progress in addressing these challenges and enhancing the ability to monitor, understand, and respond to climate induced LULC changes.

5. EMERGING TECHNOLOGIES AND FUTURE DIRECTIONS

The field of RS and GIS is rapidly evolving, with new technologies and methodologies continually emerging. Figure 4 illustrates the integrated technological ecosystem that encompasses both established foundation technologies and emerging innovations, showing how cloud computing and big data analytics serve as the central integration hub connecting these diverse technologies to support multiple application domains.

Figure 4. Technological ecosystem for current and emerging RS applications

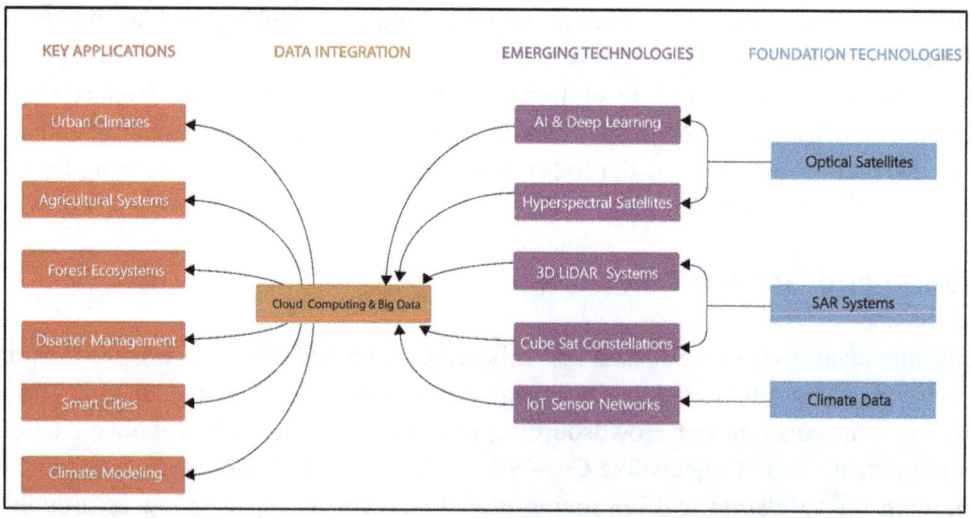

The emerging technologies highlighted in this ecosystem like the artificial intelligence, hyperspectral satellites, 3D LiDAR systems, CubeSat constellations, and IoT sensor networks are poised to revolutionize Earth observation capabilities and expand the scope of climate change and LULC applications. This section explores some of the most promising developments that are likely to shape the future of climate change and LULC studies.

5.1 Artificial Intelligence and Deep Learning

Artificial Intelligence (AI), particularly deep learning, is revolutionizing the analysis of RS data, offering new possibilities for extracting information from complex and large-scale datasets.

Advanced Image Classification and Feature Extraction:

Convolutional Neural Networks (CNNs) have demonstrated state-of-the-art performance in land cover classification tasks. For instance, Zhu et al. (2017) investigated deep learning framework for large-scale land cover mapping across diverse landscapes.

Transfer learning approaches are enabling the adaptation of models trained on one dataset to be applied to different geographical regions or sensors. This is particularly valuable in areas where training data may be scarce. Xie et al. (2016) demonstrated the effectiveness of transfer learning for mapping, using deep learning, across different zones.

Object-based image analysis (OBIA) integrated with deep learning is improving feature extraction and classification in high-resolution imagery. Zaabar et al. (2021) combined OBIA with CNNs for detailed urban LU classification, showing significant improvements over traditional methods.

Time Series Analysis and Change Detection:

Recurrent Neural Networks (RNNs) and Long Short-Term Memory (LSTM) networks are being applied to analyze satellite image time series, enabling more sophisticated change detection and trend analysis. Rußwurm and Körner (2018) used LSTMs for crop classification from multi-temporal satellite imagery, demonstrating their ability to capture complex temporal patterns.

Attention mechanisms in deep learning models are improving the ability to focus on relevant temporal patterns in LULC change detection. Zaabar et al. (2021) incorporated attention mechanisms in a deep learning framework for land cover change detection, enhancing the model's ability to identify significant changes while ignoring irrelevant variations.

Theory-guided data science approaches, including the integration of physical understanding with neural network architectures, show promise for improving the generalization and robustness of AI models in scientific applications (Karpatne et al. 2017).

5.2 Internet of Things (IoT) and Sensor Networks

The integration of IoT technologies and sensor networks with RS is opening new avenues for comprehensive environmental monitoring and LULC change assessment.

In-situ Sensor Networks:

The deployment of wireless sensor networks for continuous ground-based monitoring is enhancing the ability to validate and calibrate satellite observations. For example, Vanitha et al. (2021) demonstrated the use of IoT-based soil moisture sensor networks for validating satellite-derived soil moisture products in agricultural settings.

Edge Computing:

The development of edge computing capabilities is enabling real-time processing and integration of sensor data with satellite observations. This is particularly valuable for applications requiring rapid response, such as disaster monitoring or precision agriculture.

Self-organizing and adaptive sensor networks that optimize data collection based on satellite observations and AI-driven analytics could dynamically adjust their sampling strategies to focus on areas or phenomena of particular interest or uncertainty.

The integration of blockchain technologies for secure and transparent management of distributed sensor networks and data streams is another promising direction, particularly for applications involving multiple stakeholders or requiring high data integrity (Leng et al. 2020).

5.3 Hyperspectral Remote Sensing

Hyperspectral RS, which captures data in hundreds of narrow spectral bands, offers unprecedented capabilities for detailed characterization of Earth surface properties.

Material Identification and Mapping:

Hyperspectral data enables more accurate identification and mapping of surface materials, including specific mineral types, vegetation species, and urban materials. Transon et al. (2018) reviewed the state-of-the-art in hyperspectral RS for land cover mapping, highlighting its potential for improved discrimination of land cover types.

Vegetation Health and Stress Monitoring:

The fine spectral resolution of hyperspectral data allows for the detection of subtle changes in vegetation properties, enabling early detection of plant stress and disease. Zarco-Tejada et al. (2018) employed hyperspectral indices for early detection of olive tree disease, showcasing the potential for improved vegetation health monitoring.

The launch of new satellite-based hyperspectral missions, such as EnMAP and PRISMA, will provide global hyperspectral coverage, opening up new possibilities for large-scale, detailed surface characterization. Guanter et al. (2015) discussed the potential applications of the EnMAP mission, highlighting its importance for vegetation monitoring and ecosystem research.

The development of advanced machine learning techniques specifically tailored for hyperspectral data analysis is an important area of future research. Deep learning models that can effectively leverage the high-dimensional information in hyperspectral data while dealing with limited training samples are particularly promising (Audebert et al. 2019).

5.4 LiDAR and 3D Remote Sensing

LiDAR (Light Detection and Ranging) technology is revolutionizing the ability to characterize the three-dimensional structure of landscapes, with important implications for LULC and climate change studies.

Forest Structure and Biomass Estimation:

Airborne and spaceborne LiDAR systems are enabling more accurate estimation of forest structure and biomass, crucial for understanding the role of forests in the carbon cycle. Dubayah et al. (2020) reviewed the potential of the Global Ecosystem Dynamics Investigation (GEDI) mission, a spaceborne LiDAR system, for global forest biomass mapping.

Urban 3D Modeling:

LiDAR is enhancing the ability to create detailed 3D models of urban environments, important for urban climate studies and planning. Yan et al. (2015) demonstrated the use of airborne LiDAR for 3D building reconstruction and urban vegetation mapping, highlighting its potential for comprehensive urban environmental analysis.

When LiDAR is combined with hyperspectral and SAR data, it opens up new possibilities for understanding ecosystems in detail. By bringing these different types of data together, researchers can gain clearer insights into both the physical structure of landscapes and the variety of species and vegetation they support. Advances such as single-photon LiDAR, especially if deployed on satellites, could greatly expand the scope and resolution of global 3D Earth observations. This would allow more frequent and detailed monitoring of forest structure and biomass changes linked to climate and LU (Dubayah et al. 2020).

5.5 Small Satellites and CubeSats

The proliferation of small satellites and CubeSats is dramatically increasing the temporal resolution of Earth observation data, enabling near-continuous monitoring of LULC changes.

High-Frequency Earth Observation:

Constellations of small satellites, like Planet's Dove fleet, now provide daily global imagery at moderate resolution. This frequent coverage makes it possible to detect rapid changes in LULC, as well as short-term environmental events. McCabe et al. (2017) reviewed the potential of CubeSats for Earth observation, highlighting their ability to provide high spatiotemporal resolution data for environmental monitoring.

Specialized Earth Observation Missions:

CubeSats are also enabling specialized Earth observation missions focused on specific environmental variables or phenomena. For example, NASA's CYGNSS mission uses a constellation of small satellites for hurricane forecasting and global soil moisture monitoring (Ruf et al. 2018).

The development of interconnected, autonomous CubeSat constellations that can dynamically adjust their observation strategies based on detected changes or events, provides targeted, high-resolution observations of rapidly evolving LULC changes or environmental disasters.

Adding edge computing to small satellites could allow data to be processed directly on board, cutting down the need to transmit large volumes of data and allowing for quicker responses in time-sensitive applications (Santos et al. 2022).

These emerging technologies and future directions highlight the dynamic and rapidly evolving nature of RS and GIS applications in climate change and LULC studies.

As these technologies advance and new analytical methods emerge, the capacity to track, understand, and respond to climate-driven changes in LULC will keep improving, offering vital support for environmental management and policy-making in a rapidly changing world.

6. CONCLUSION AND FUTURE PERSPECTIVES

Remote sensing (RS) and geographic information systems (GIS) have become increasingly central to assessing how climate change influences land use and land cover (LULC). Their use has grown rapidly in recent years, supported by advances in sensor technology, data analytics, and computational power. Together, these tools now play a crucial role in observing, interpreting, and responding to the complex and evolving relationships between climate dynamics and land-use patterns across different regions and ecosystems. Recent progress in this field has been shaped by several important technological developments. First, satellite data now offer higher spatial and temporal resolution than in the past, which allows researchers to observe land use and land cover changes with greater detail and frequency. Another important advancement is depicted by the refinement of time-series methods which allows the identification both sudden shifts and slower, long-term changes in landscapes. Research has also benefited from combining different types of remote sensing data, including optical images, radar signals, and LiDAR measurements, to build a more complete picture of surface conditions. Alongside this, machine learning and artificial intelligence tools have improved how data are processed, classified, and interpreted. Finally, cloud-based platforms have made it easier to handle and analyze the large volumes of data now produced through Earth observation programs.

These advancements have significantly enhanced the ability to: (i) Detect and quantify climate-induced changes in vegetation patterns, agricultural systems, and urban environments; (ii) Assess the effectiveness of conservation strategies and protected areas in the face of climate change, (iii) Monitor and project the impacts of sea-level rise on coastal ecosystems and urban areas; (iv) Evaluate the carbon sequestration potential of different LU strategies; (v) Provide early warning for climate-related disasters and support disaster risk reduction efforts.

Looking to the future, several key areas are likely to drive further advancements in the field:

1. Artificial Intelligence and Deep Learning: The continued development of AI techniques tailored for Earth observation data is likely to revolutionize LULC change detection and prediction. This includes the development of explainable AI models that can provide insights into the drivers of observed changes.
2. Internet of Things and Sensor Networks: The integration of satellite observations with ground-based sensor networks and IoT devices will enable more comprehensive and real-time monitoring of environmental changes.
3. Hyperspectral RS: The launch of new hyperspectral satellite missions will provide unprecedented capabilities for detailed surface characterization, enhancing the ability to detect subtle changes in ecosystem composition and health.
4. Small Satellites and CubeSats: The proliferation of small satellite constellations will dramatically increase the temporal resolution of Earth observation data, enabling near-continuous monitoring of LULC changes.
5. Data Fusion and Assimilation: Advanced techniques for integrating multi-source Earth observation data with climate models and in-situ measurements will improve the ability to understand and project climate-LULC interactions.
6. Quantum Computing: While still in its early stages, quantum computing has the potential to revolutionize the processing and analysis of large Earth observation datasets, enabling more complex and sophisticated analyses.

These technological advances, together with a deeper understanding of how climate interacts with LULC, will be essential for tackling the complex challenges of global environmental change. At the same time, fully harnessing the potential of these tools requires addressing several key issues:

1. Integrating and harmonizing data from different sensors and across multiple scales.
2. Developing reliable methods to quantify uncertainty in RS-based LULC assessments.
3. Filling the gap between Earth observation research and policy-making to ensure that RS insights inform practical decisions.
4. Strengthening capacity building and technology transfer to provide more equitable access to Earth observation tools and data worldwide.
5. Considering ethical aspects of using high-resolution Earth observation data, including privacy concerns and the risk of misuse.

REFERENCES

Al-Shaar, W. (2024). Analyzing the effect size of urban growth driving factors: Application of multilayer-perceptron Markov-chain model for the Riyadh city. *Modeling Earth Systems and Environment, 10*(1), 303–312. DOI: 10.1007/s40808-023-01781-8

Al-Shaar, W., & Bonin, O. (2021). Factors behind the dynamics of land use evolution: Case of Lebanon. *SN Applied Sciences, 3*(6), 677. DOI: 10.1007/s42452-021-04639-9

Al-Shaar, W., & Bonin, O. (2024). A comparative study of land use dynamics in urban and peri-urban areas of Greater Beirut Agglomeration and Greater Paris Region: A geospatial approach. In Sahana, M. (Ed.), *Modern cartography series* (Vol. 11, pp. 89–115). Academic Press. DOI: 10.1016/B978-0-443-15832-2.00005-8

Al-Shaar, W., Bonin, O., Faour, G., Zeidan, N., & Al-Shaar, M. (2022). Spatial analysis of land surface temperature distribution: Case of the Greater Beirut Area. *Euro-Mediterranean Journal for Environmental Integration, 7*(4), 483–495. DOI: 10.1007/s41207-022-00330-6

Al-Shaar, W., Bonin, O., Seregina, T., Faour, G. (2024). Morphological dynamics, landscape fragmentation and climate change vulnerability of actual and predicted urban areas of Lebanon: Application of Multilayer-Perceptron Markov-Chain model. Journal of Urban Management.

Alfieri, L., Salamon, P., Pappenberger, F., Wetterhall, F., & Thielen, J. (2012). Operational early warning systems for water-related hazards in Europe. *Environmental Science & Policy, 21*, 35–49. DOI: 10.1016/j.envsci.2012.01.008

Alonzo, M., McFadden, J. P., Nowak, D. J., & Roberts, D. A. (2016). Mapping urban forest structure and function using hyperspectral imagery and lidar data. *Urban Forestry & Urban Greening, 17*, 135–147. DOI: 10.1016/j.ufug.2016.04.003

Andam, K. S., Ferraro, P. J., Pfaff, A., Sánchez-Azofeifa, G. A., & Robalino, J. (2008). Measuring the effectiveness of protected area networks in reducing deforestation. *Proceedings of the National Academy of Sciences of the United States of America, 105*(42), 16089–16094. DOI: 10.1073/pnas.0800437105 PMID: 18854414

Andela, N., Andela, N., Morton, D. C., Giglio, L., Chen, Y., Werf, G. V., Kasibhatla, P., DeFries, R. S., Collatz, G. J., Hantson, S., Kloster, S., Bachelet, D., Forrest, M., Lasslop, G., Li, F., Mangeon, S., Melton, J. R., Yue, C., & Randerson, J. T. (2017). A human-driven decline in global burned area. *Science, 356*(6345), 1356–1362. DOI: 10.1126/science.aal4108 PMID: 28663495

Anderson, M. C., Kustas, W. P., Alfieri, J. G., Gao, F., Hain, C. R., Prueger, J. H., Evett, S. R., Colaizzi, P. D., Howell, T., & Chávez, J. L. (2012). Mapping daily evapotranspiration at Landsat spatial scales during the BEAREX'08 field campaign. *Advances in Water Resources, 50*, 162–177. DOI: 10.1016/j.advwatres.2012.06.005

Asner, G. P., Martin, R. E., Knapp, D. E., Tupayachi, R., Anderson, C. B., Sinca, F., Vaughn, N. R., & Llactayo, W. (2017). Airborne laser-guided imaging spectroscopy to map forest trait diversity and guide conservation. *Science, 355*(6323), 385–389. DOI: 10.1126/science.aaj1987 PMID: 28126815

Audebert, N., Le Saux, B., & Lefèvre, S. (2019). Deep learning for classification of hyperspectral data: A comparative review. *IEEE Geoscience and Remote Sensing Magazine*, *7*(2), 159–173. DOI: 10.1109/MGRS.2019.2912563

Azevedo, J. A., Chapman, L., & Muller, C. L. (2016). Quantifying the daytime and night-time urban heat island in Birmingham, UK: A comparison of satellite derived land surface temperature and high-resolution air temperature observations. *Remote Sensing (Basel)*, *8*(2), 153. DOI: 10.3390/rs8020153

Baccini, A., Goetz, S. J., Walker, W. S., Laporte, N. T., Sun, M., Sulla-Menashe, D., Hackler, J., Beck, P. S. A., Dubayah, R., Friedl, M. A., Samanta, S., & Houghton, R. A. (2012). Estimated carbon dioxide emissions from tropical deforestation improved by carbon-density maps. *Nature Climate Change*, *2*(3), 182–185. DOI: 10.1038/nclimate1354

Bastin, J., Finegold, Y., Garcia, C. A., Mollicone, D., Rezende, M., Routh, D., Zohner, C. M., & Crowther, T. W. (2019). The global tree restoration potential. *Science*, *365*(6448), 76–79. DOI: 10.1126/science.aax0848 PMID: 31273120

Becker-Reshef, I., Justice, C., Barker, B., Humber, M.L., Rembold, F., Bonifaçio, R., Zappacosta, M., Budde, M.E., Magadzire, T., Shitote, C., Pound, J., Constantino, A., Nakalembe, C., Mwangi, K., Sobue, S., Newby, T.S., Whitcraft, A.K., Jarvis, I.A., Verdin, J.P. (2020). Strengthening agricultural decisions in countries at risk of food insecurity: The GEOGLAM Crop Monitor for Early Warning. Remote Sensing of Environment.

Becker-Reshef, I., Vermote, E. F., Lindeman, M. A., & Justice, C. O. (2010). A generalized regression-based model for forecasting winter wheat yields in Kansas and Ukraine using MODIS data. *Remote Sensing of Environment*, *114*(6), 1312–1323. DOI: 10.1016/j.rse.2010.01.010

Bégué, A., Arvor, D., Bellon, B., Betbeder, J., De Abelleyra, D., Ferraz, R. P., Lebourgeois, V., Lelong, C., Simões, M., & Verón, S. R. (2018). Remote sensing and cropping practices: A review. *Remote Sensing*, *10*(1), 99. DOI: 10.3390/rs10010099

Behrenfeld, M. J., O'Malley, R. T., Siegel, D. A., McClain, C. R., Sarmiento, J. L., Feldman, G. C., Milligan, A. J., Falkowski, P. G., Letelier, R. M., & Boss, E. S. (2006). Climate-driven trends in contemporary ocean productivity. *Nature*, *444*(7120), 752–755. DOI: 10.1038/nature05317 PMID: 17151666

Berger, C., Voltersen, M., Hese, S., Walde, I., & Schmullius, C. (2013). Robust Extraction of Urban Land Cover Information From HSR Multi-Spectral and LiDAR Data. *IEEE Journal of Selected Topics in Applied Earth Observations and Remote Sensing*, *6*(5), 1–16. DOI: 10.1109/JSTARS.2013.2252329

Bernard, F., Manolis, I., Barat, I., Bolea Alamañac, A., Such Taboada, M., Mingorance, P., Ciapponi, A., Cardone, T., Fernandez Nunez, I., Garcia, A., Hallibert, P., Hammar, A., Henriot, P., Merodio Codinachs, D., Patti, S., Prado Montes, P., Skrzypek, P., Steenari, D., Weixler, S., & Gandullo Holgueras, D. (2023). The Copernicus land surface temperature monitoring (LSTM) mission: Design, technology and status. *Remote Sensing.*

Bhunia, G. S., & Shit, P. K. (2021). *Geospatial Technology for Multi-hazard Risk Assessment*. Advances in Geographic Information Science.

Blaschke, T. (2010). Object based image analysis for remote sensing. *ISPRS Journal of Photogrammetry and Remote Sensing*, *65*(1), 2–16. DOI: 10.1016/j.isprsjprs.2009.06.004

Blondeau-Patissier, D., Gower, J. F., Dekker, A. G., Phinn, S. R., & Brando, V. E. (2014). A review of ocean color remote sensing methods and statistical techniques for the detection, mapping and analysis of phytoplankton blooms in coastal and open oceans. *Progress in Oceanography*, *123*, 123–144. DOI: 10.1016/j.pocean.2013.12.008

Bolton, D. K., & Friedl, M. A. (2013). Forecasting crop yield using remotely sensed vegetation indices and crop phenology metrics. *Agricultural and Forest Meteorology*, *173*, 74–84. DOI: 10.1016/j.agrformet.2013.01.007

Boussetta, S., Balsamo, G., Dutra, E., Beljaars, A., & Albergel, C. (2015). Assimilation of surface albedo and vegetation states from satellite observations and their impact on numerical weather prediction. *Remote Sensing of Environment*, *163*, 111–126. DOI: 10.1016/j.rse.2015.03.009

Boyce, D. G., Lewis, M. R., & Worm, B. (2010). Global phytoplankton decline over the past century. *Nature*, *466*(7306), 591–596. DOI: 10.1038/nature09268 PMID: 20671703

Bradley, B. A. (2013). Remote detection of invasive plants: A review of spectral, textural and phenological approaches. *Biological Invasions*. Advance online publication. DOI: 10.1007/s10530-013-0578-9

Brito, M. M., Evers, M., & Almoradie, A. (2017). Participatory flood vulnerability assessment: A multi-criteria approach. *Hydrology and Earth System Sciences*, *22*(1), 373–390. DOI: 10.5194/hess-22-373-2018

Buitenwerf, R., Rose, L., & Higgins, S. I. (2015). Three decades of multi-dimensional change in global leaf phenology. *Nature Climate Change*, *5*(4), 364–368. DOI: 10.1038/nclimate2533

Calvino-Cancela, M., Mendez-Rial, R., Reguera-Salgado, J., & Martin-Herrero, J. (2014). Alien plant monitoring with ultralight airborne imaging spectroscopy. *PLoS One*, *9*(7), e102381. DOI: 10.1371/journal.pone.0102381 PMID: 25010601

Castelluccio, M., Poggi, G., Sansone, C., & Verdoliva, L. (2015). Land use classification in remote sensing images by convolutional neural networks. https://arxiv.org/abs/1508.00092

Chakraborty, T., & Lee, X. (2019). A simplified urban-extent algorithm to characterize surface urban heat islands on a global scale and examine vegetation control on their spatiotemporal variability. *International Journal of Applied Earth Observation and Geoinformation*, *74*, 269–280. DOI: 10.1016/j.jag.2018.09.015

Cheng, K. S., Ling, J., Lin, T., Liu, Y., Shen, Y., & Kono, Y. (2021). Quantifying Uncertainty in Land-Use/Land-Cover Classification Accuracy: A Stochastic Simulation Approach. *Frontiers in Environmental Science*, *9*, 628214. Advance online publication. DOI: 10.3389/fenvs.2021.628214

Chini, M., Pelich, R., Pulvirenti, L., Pierdicca, N., Hostache, R., & Matgen, P. (2019). Sentinel-1 InSAR coherence to detect floodwater in urban areas: Houston and Hurricane Harvey as a test case. *Remote Sensing (Basel)*, *11*(2), 107. DOI: 10.3390/rs11020107

Claverie, M., Ju, J., Masek, J. G., Dungan, J. L., Vermote, E. F., Roger, J. C., Skakun, S. V., & Justice, C. (2018). The harmonized Landsat and Sentinel-2 surface reflectance data set. *Remote Sensing of Environment*, *219*, 145–161. DOI: 10.1016/j.rse.2018.09.002

Cockx, K., Voorde, T. V., & Canters, F. (2014). Quantifying uncertainty in remote sensing-based urban land-use mapping. *International Journal of Applied Earth Observation and Geoinformation, 31*, 154–166. DOI: 10.1016/j.jag.2014.03.016

Comiso, J. C., Meier, W. N., & Gersten, R. (2017). Variability and trends in the Arctic sea ice cover: Results from different techniques. *Journal of Geophysical Research. Oceans, 122*(8), 6883–6900. Advance online publication. DOI: 10.1002/2017JC012768

Coppin, P., Jonckheere, I., Nackaerts, K., Muys, B., & Lambin, E. (2004). Review ArticleDigital change detection methods in ecosystem monitoring: A review. *International Journal of Remote Sensing, 25*(9), 1565–1596. DOI: 10.1080/0143116031000101675

Cressie, N., Calder, C. A., Clark, J. S., Hoef, J. M. V., & Wikle, C. K. (2009). Accounting for uncertainty in ecological analysis: The strengths and limitations of hierarchical statistical modeling. *Ecological Applications : A Publication of the Ecological Society of America, 19*(3), 553–570. DOI: 10.1890/07-0744.1 PMID: 19425416

DeVries, B., Decuyper, M., Verbesselt, J., Zeileis, A., Herold, M., & Joseph, S. (2015). Tracking disturbance-regrowth dynamics in tropical forests using structural change detection and Landsat time series. *Remote Sensing of Environment, 169*, 320–334. DOI: 10.1016/j.rse.2015.08.020

. Dubayah, R.O., Blair, J.B., Goetz, S., Fatoyinbo, L., Hansen, M., Healey, S., Hofton, M.A., Hurtt, G.C., Kellner, J., Luthcke, S.B., Armston, J., Tang, H., Duncanson, L.I., Hancock, S., Jantz, P.A., Marselis, S.M., Patterson, P.L., Qi, W., Silva, C. (2020). The Global Ecosystem Dynamics Investigation: High-resolution laser ranging of the Earth's forests and topography. Science of Remote Sensing.

Duncan, B. N., Prados, A. I., Lamsal, L. N., Liu, Y., Streets, D. G., Gupta, P., Hilsenrath, E., Kahn, R. A., Nielsen, J. E., Beyersdorf, A. J., Burton, S. P., Fiore, A. M., Fishman, J., Henze, D. K., Hostetler, C. A., Krotkov, N. A., Lee, P., Lin, M., Pawson, S., & Ziemba, L. D. (2014). Satellite Data of Atmospheric Pollution for U.S. Air Quality Applications: Examples of Applications, Summary of Data End-User Resources, Answers to FAQs, and Common Mistakes to Avoid. *Atmospheric Environment, 94*, 647–662. DOI: 10.1016/j.atmosenv.2014.05.061

Dwyer, J. L., Roy, D. P., Sauer, B., Jenkerson, C. B., Zhang, H. K., & Lymburner, L. (2018). Analysis ready data: Enabling analysis of the Landsat archive. *Remote Sensing (Basel), 10*(9), 1363. DOI: 10.3390/rs10091363

Estoque, R. C., Murayama, Y., & Myint, S. W. (2017). Effects of landscape composition and pattern on land surface temperature: An urban heat island study in the megacities of Southeast Asia. *The Science of the Total Environment, 577*, 349–359. Advance online publication. DOI: 10.1016/j.scitotenv.2016.10.195 PMID: 27832866

Everett, J. R., Staskowski, R. J., & Jengo, C. M. (2002). Remote sensing and GIS enable future exploration success.

Feddema, J. J., Oleson, K. W., Bonan, G. B., Mearns, L. O., Buja, L. E., Meehl, G. A., & Washington, W. M. (2005). The importance of land-cover change in simulating future climates. *Science, 310*(5754), 1674–1678. DOI: 10.1126/science.1118160 PMID: 16339443

Fick, S. E., & Hijmans, R. J. (2017). WorldClim 2: New 1-km spatial resolution climate surfaces for global land areas. *International Journal of Climatology*, *37*(12), 4302–4315. Advance online publication. DOI: 10.1002/joc.5086

Fritz, S., See, L., Perger, C., McCallum, I., Schill, C., Schepaschenko, D., Duerauer, M., Karner, M., Dresel, C., Laso-Bayas, J.-C., Lesiv, M., Moorthy, I., Salk, C. F., Danylo, O., Sturn, T., Albrecht, F., You, L., Kraxner, F., & Obersteiner, M. (2017). A global dataset of crowdsourced land cover and land use reference data. *Scientific Data*, *4*(1), 170075. DOI: 10.1038/sdata.2017.75 PMID: 28608851

Gaitani, N., Spanou, A., Saliari, M., Synnefa, A., Vassilakopoulou, K., Papadopoulou, K., Pavlou, K., Santamouris, M., Papaioannou, M., & Lagoudaki, A. (2011). Improving the microclimate in urban areas: A case study in the centre of Athens. *Building Services Engineering Research and Technology*, *32*(1), 53–71. DOI: 10.1177/0143624410394518

Garonna, I., Jong, R., & Schaepman, M. E. (2016). Variability and evolution of global land surface phenology over the past three decades (1982–2012). *Global Change Biology*, *22*(4), 1456–1468. DOI: 10.1111/gcb.13168 PMID: 26924776

Gascon, F. (2018). Sentinel-2 for Agricultural Monitoring. IGARSS 2018 - 2018 IEEE International Geoscience and Remote Sensing Symposium, 8166-8168, DOI: 10.1109/IGARSS.2018.8517942

. Gascon, F., Cadau, E., Colin, O., Hoersch, B., Isola, C., Fernández, B.L., Martimort, P. (2014). Copernicus Sentinel-2 mission: products, algorithms and Cal/Val. Optics and Photonics - Optical Engineering + Applications.

Ge, Y., Thomasson, J. A., & Sui, R. (2011). Remote sensing of soil properties in precision agriculture: A review. *Frontiers of Earth Science*, *5*(3), 229–238. DOI: 10.1007/s11707-011-0175-0

Geldmann, J., Manica, A., Burgess, N. D., Coad, L., & Balmford, A. (2019). A global-level assessment of the effectiveness of protected areas at resisting anthropogenic pressures. *Proceedings of the National Academy of Sciences of the United States of America*, *116*(46), 23209–23215. DOI: 10.1073/pnas.1908221116 PMID: 31659036

Gorelick, N., Hancher, M., Dixon, M., Ilyushchenko, S., Thau, D., & Moore, R. (2017). Google Earth Engine: Planetary-scale geospatial analysis for everyone. *Remote Sensing of Environment*, *202*, 18–27. DOI: 10.1016/j.rse.2017.06.031

Griscom, B. W., Adams, J., Ellis, P. W., Houghton, R. A., Lomax, G., Miteva, D. A., Schlesinger, W. H., Shoch, D. T., Siikamäki, J., Smith, P., Woodbury, P. B., Zganjar, C., Blackman, A., Campari, J. S., Conant, R. T., Delgado, C., Elias, P., Gopalakrishna, T., Hamsik, M. R., & Fargione, J. E. (2017). Natural climate solutions. *Proceedings of the National Academy of Sciences of the United States of America*, *114*(44), 11645–11650. DOI: 10.1073/pnas.1710465114 PMID: 29078344

Guanter, L., Kaufmann, H., Segl, K., Foerster, S., Rogass, C., Chabrillat, S., Küster, T., Hollstein, A., Rossner, G., Chlebek, C., Straif, C., Fischer, S., Schrader, S., Storch, T., Heiden, U., Müller, A., Bachmann, M., Mühle, H., Müller, R., & Sang, B. (2015). The EnMAP Spaceborne Imaging Spectroscopy Mission for Earth Observation. *Remote Sensing (Basel)*, *7*(7), 8830–8857. DOI: 10.3390/rs70708830

Hadj-Salah, A., Verdier, R., Caron, C., Picard, M., & Capelle, M. (2019). Schedule Earth observation satellites with deep reinforcement learning. arXiv preprint arXiv:1911.05696. https://arxiv.org/abs/1911.05696

Hagolle, O., Huc, M., Villa Pascual, D., & Dedieu, G. (2015). A multi-temporal and multi-spectral method to estimate aerosol optical thickness over land, for the atmospheric correction of FormoSat-2, Landsat, VENμS and Sentinel-2 images. *Remote Sensing (Basel)*, *7*(3), 2668–2691. DOI: 10.3390/rs70302668

Hamilton, S. E., & Casey, D. (2016). Creation of a high spatio-temporal resolution global database of continuous mangrove forest cover for the 21st century (CGMFC-21). *Global Ecology and Biogeography*, *25*(6), 729–738. DOI: 10.1111/geb.12449

Hansen, M. C., Krylov, A., Tyukavina, A., Potapov, P. V., Turubanova, S., Zutta, B., Ifo, S., Margono, B., Stolle, F., & Moore, R. (2016). Humid tropical forest disturbance alerts using Landsat data. *Environmental Research Letters*, *11*(3), 034008. DOI: 10.1088/1748-9326/11/3/034008

Hansen, M. C., Potapov, P. V., Moore, R., Hancher, M., Turubanova, S. A., Tyukavina, A., Thau, D., Stehman, S. V., Goetz, S. J., Loveland, T. R., Kommareddy, A., Egorov, A., Chini, L., Justice, C. O., & Townshend, J. R. G. (2013). High-resolution global maps of 21st-century forest cover change. *Science*, *342*(6160), 850–853. DOI: 10.1126/science.1244693 PMID: 24233722

Harris, I. C., Osborn, T. J., Jones, P. D., & Lister, D. H. (2020). Version 4 of the CRU TS monthly high-resolution gridded multivariate climate dataset. *Scientific Data*, *7*(1), 7. DOI: 10.1038/s41597-020-0453-3 PMID: 32246091

He, K. S., Bradley, B. A., Cord, A. F., Rocchini, D., Tuanmu, M., Schmidtlein, S., Turner, W., Wegmann, M., & Pettorelli, N. (2015). Will remote sensing shape the next generation of species distribution models? *Remote Sensing in Ecology and Conservation*, *1*(1), 4–18. Advance online publication. DOI: 10.1002/rse2.7

Hengl, T., Mendes de Jesus, J., Heuvelink, G. B., Ruiperez González, M., Kilibarda, M., Blagotić, A., Wei, S., Wright, M. N., Geng, X., Bauer-Marschallinger, B., Guevara, M., Vargas, R., MacMillan, R. A., Batjes, N. H., Leenaars, J. G., Ribeiro, E., Wheeler, I., Mantel, S., & Kempen, B. (2017). SoilGrids250m: Global gridded soil information based on machine learning. *PLoS One*, *12*(2), 12. DOI: 10.1371/journal.pone.0169748 PMID: 28207752

Hislop, S., Jones, S., Soto-Berelov, M., Skidmore, A., Haywood, A., & Nguyen, T. H. (2018). Using Landsat Spectral Indices in Time-Series to Assess Wildfire Disturbance and Recovery. *Remote Sensing (Basel)*, *10*(3), 460. DOI: 10.3390/rs10030460

Huang, X., Ziniti, B., Torbick, N., & Ducey, M. J. (2018). Assessment of Forest above Ground Biomass Estimation Using Multi-Temporal C-band Sentinel-1 and Polarimetric L-band PALSAR-2 Data. *Remote Sensing (Basel)*, *10*(9), 1424. DOI: 10.3390/rs10091424

Huete, A., Didan, K., Miura, T., Rodriguez, E. P., Gao, X., & Ferreira, L. G. (2002). Overview of the radiometric and biophysical performance of the MODIS vegetation indices. *Remote Sensing of Environment*, *83*(1–2), 195–213. DOI: 10.1016/S0034-4257(02)00096-2

Huffman, G. J., Bolvin, D. T., Braithwaite, D., Hsu, K., Joyce, R., Kidd, C., Nelkin, E. J., Sorooshian, S., Tan, J., & Xie, P. (2019). Algorithm theoretical basis document (ATBD) version 06 NASA global precipitation measurement (GPM) integrated multi-satellitE retrievals for GPM (IMERG). National Aeronautics and Space Administration. https://pmm.nasa.gov/sites/default/files/document_files/IMERG_ATBD_V06.pdf

Ines, A. V., Das, N. N., Hansen, J. W., & Njoku, E. G. (2013). Assimilation of remotely sensed soil moisture and vegetation with a crop simulation model for maize yield prediction. *Remote Sensing of Environment, 138*, 149–164. DOI: 10.1016/j.rse.2013.07.018

Jaramillo, J., Muchugu, E., Vega, F. E., Davis, A. P., Borgemeister, C., & Chabi-Olaye, A. (2011). Some Like It Hot: The Influence and Implications of Climate Change on Coffee Berry Borer (Hypothenemus hampei) and Coffee Production in East Africa. *PLoS One*, *6*(9), 6. DOI: 10.1371/journal.pone.0024528 PMID: 21935419

Jetz, W., Cavender-Bares, J. M., Pavlick, R. P., Schimel, D. S., Davis, F. W., Asner, G. P., Guralnick, R. P., Kattge, J., Latimer, A. M., Moorcroft, P. R., Schaepman, M. E., Schildhauer, M., Schneider, F. D., Schrodt, F., Stahl, U., & Ustin, S. L. (2016). Monitoring plant functional diversity from space. *Nature Plants*, 2. PMID: 27249357

Joshi, N., Baumann, M., Ehammer, A., Fensholt, R., Grogan, K., Hostert, P., Jepsen, M. R., Kuemmerle, T., Meyfroidt, P., Mitchard, E. T. A., Reiche, J., Ryan, C. M., & Waske, B. (2016). A Review of the Application of Optical and Radar Remote Sensing Data Fusion to Land Use Mapping and Monitoring. *Remote Sensing (Basel)*, *8*(1), 70. DOI: 10.3390/rs8010070

Kabisch, N., Selsam, P., Kirsten, T., Lausch, A., & Bumberger, J. (2019). A multi-sensor and multi-temporal remote sensing approach to detect land cover change dynamics in heterogeneous urban land-scapes. *Ecological Indicators*, *99*, 273–282. DOI: 10.1016/j.ecolind.2018.12.033

Kaku, K., Aso, N., & Takiguchi, F. (2015). Space-based response to the 2011 Great East Japan Earth-quake: Lessons learnt from JAXA's support using earth observation satellites. *International Journal of Disaster Risk Reduction*, *12*, 134–153. DOI: 10.1016/j.ijdrr.2014.12.009

Karpatne, A., Atluri, G., Faghmous, J. H., Steinbach, M., Banerjee, A., Ganguly, A., Shekhar, S., Samatova, N., & Kumar, V. (2017). Theory-guided data science: A new paradigm for scientific discovery from data. *IEEE Transactions on Knowledge and Data Engineering*, *29*(10), 2318–2331. Advance online publication. DOI: 10.1109/TKDE.2017.2720168

Kennedy, R. E., Yang, Z., & Cohen, W. B. (2010). Detecting trends in forest disturbance and recovery using yearly Landsat time series: 1. LandTrendr—Temporal segmentation algorithms. *Remote Sensing of Environment*, *114*(12), 2897–2910. DOI: 10.1016/j.rse.2010.07.008

Keshava, N., & Mustard, J. F. (2002). Spectral unmixing. *IEEE Signal Processing Magazine*, *19*(1), 44–57. DOI: 10.1109/79.974727

Kogan, F., Guo, W., & Yang, W. (2019). Drought and food security prediction from NOAA new generation of operational satellites. *Geomatics, Natural Hazards & Risk*, *10*(1), 651–666. DOI: 10.1080/19475705.2018.1541257

Kornelsen, K. C., & Coulibaly, P. (2013). Advances in soil moisture retrieval from synthetic aperture radar and hydrological applications. *Journal of Hydrology (Amsterdam)*, *476*, 460–489. DOI: 10.1016/j.jhydrol.2012.10.044

Kulp, S. A., & Strauss, B. H. (2019). New elevation data triple estimates of global vulnerability to sea-level rise and coastal flooding. *Nature Communications*, *10*(1), 4844. DOI: 10.1038/s41467-019-12808-z PMID: 31664024

Kumagai, N. H., Molinos, J. G., Yamano, H., Takao, S., Fujii, M., & Yamanaka, Y. (2018). Ocean currents and herbivory drive macroalgae-to-coral community shift under climate warming. *Proceedings of the National Academy of Sciences of the United States of America*, *115*(36), 8990–8995. DOI: 10.1073/pnas.1716826115 PMID: 30126981

Kwok, R., & Cunningham, G. F. (2015). Variability of Arctic sea ice thickness and volume from CryoSat-2. *Philosophical Transactions. Series A, Mathematical, Physical, and Engineering Sciences*, *373*(2045), 20140157. Advance online publication. DOI: 10.1098/rsta.2014.0157 PMID: 26032317

Lausch, A., Blaschke, T., Haase, D., Herzog, F., Syrbe, R.-U., Tischendorf, L., & Walz, U. (2015). Understanding and quantifying landscape structure—A review on relevant process characteristics, data models and landscape metrics. *Ecological Modelling*, *295*, 31–41. DOI: 10.1016/j.ecolmodel.2014.08.018

Lefsky, M. A., Cohen, W. B., Parker, G. G., & Harding, D. J. (2002). Lidar remote sensing for ecosystem studies: Lidar, an emerging remote sensing technology that directly measures the three-dimensional distribution of plant canopies, can accurately estimate vegetation structural attributes and should be of particular interest to forest, landscape, and global ecologists. *Bioscience*, *52*(1), 19–30. DOI: 10.1641/0006-3568(2002)052[0019:LRSFES]2.0.CO;2

Leng, J., Jiang, P., & Xu, K. (2020). Blockchain-empowered sustainable manufacturing and product lifecycle management in Industry 4.0: A survey. *Renewable & Sustainable Energy Reviews*, *132*, 110112. DOI: 10.1016/j.rser.2020.110112

Lewis, A., Oliver, S., Lymburner, L., Evans, B. J., Wyborn, L., Mueller, N., Raevksi, G., Hooke, J., Woodcock, R., Sixsmith, J., Wu, W., Tan, P., Li, F., Killough, B. D., Minchin, S., Roberts, D. O., Ayers, D., Bala, B., Dwyer, J. L., & Wang, L. (2017). The Australian Geoscience Data Cube - foundations and lessons learned. *Remote Sensing of Environment*, *202*, 276–292. DOI: 10.1016/j.rse.2017.03.015

Li, D., Liao, W., Rigden, A. J., Liu, X., Wang, D., Malyshev, S., & Shevliakova, E. (2019). Urban heat island: Aerodynamics or imperviousness? *Science Advances*, *5*(4), eaau4299. Advance online publication. DOI: 10.1126/sciadv.aau4299 PMID: 30949572

Li, J., & Roy, D. P. (2017). A global analysis of Sentinel-2A, Sentinel-2B and Landsat-8 data revisit intervals and implications for terrestrial monitoring. *Remote Sensing (Basel)*, *9*(9), 902. DOI: 10.3390/rs9090902

Lipper, L., Thornton, P. K., Campbell, B. M., Baedeker, T., Braimoh, A. K., Bwalya, M., Caron, P., Cattaneo, A., Garrity, D. P., Henry, K., Hottle, R. D., Jackson, L. E., Jarvis, A., Kossam, F., Mann, W., McCarthy, N., Meybeck, A., Neufeldt, H., Remington, T., & Torquebiau, E. F. (2014). Climate-smart agriculture for food security. *Nature Climate Change*, *4*(12), 1068–1072. DOI: 10.1038/nclimate2437

Littlefield, C. E., Mcrae, B. H., Michalak, J. L., Lawler, J. J., & Carroll, C. (2017). Connecting today's climates to future climate analogs to facilitate movement of species under climate change. *Conservation Biology : The Journal of the Society for Conservation Biology*, *31*(6), 31. DOI: 10.1111/cobi.12938 PMID: 28339121

Liu, J., Bowman, K. W., Schimel, D. S., Parazoo, N. C., Jiang, Z., Lee, M., Bloom, A. A., Wunch, D., Frankenberg, C., Sun, Y., O'Dell, C. W., Gurney, K. R., Menemenlis, D., Gierach, M., Crisp, D., & Eldering, A. (2017). Contrasting carbon cycle responses of the tropical continents to the 2015–2016 El Niño. *Science*, *358*(6360), eaam5690. Advance online publication. DOI: 10.1126/science.aam5690 PMID: 29026011

Liu, X., Huang, Y., Xu, X., Li, X., Li, X., Ciais, P., Lin, P., Gong, K., Ziegler, A. D., Chen, A., Gong, P., Chen, J., Hu, G., Chen, Y., Wang, S., Wu, Q., Huang, K., Estes, L. D., & Zeng, Z. (2020). High-spatiotemporal-resolution mapping of global urban change from 1985 to 2015. *Nature Sustainability*, *3*(7), 1–7. DOI: 10.1038/s41893-020-0521-x

Lobell, D., Schlenker, W., & Costa-Roberts, J. (2011). Climate Trends and Global Crop Production Since 1980. *Science*, *333*(6042), 616–620. DOI: 10.1126/science.1204531 PMID: 21551030

Loosvelt, L., Peters, J., Skriver, H., Lievens, H., Van Coillie, F. M. B., De Baets, B., & Verhoest, N. E. C. (2012). Random forests as a tool for estimating uncertainty at pixel-level in SAR image classification. *International Journal of Applied Earth Observation and Geoinformation*, *19*, 173–184. DOI: 10.1016/j.jag.2012.05.011

Luijendijk, A., Hagenaars, G., Ranasinghe, R., Baart, F., Donchyts, G., & Aarninkhof, S. (2018). The state of the world's beaches. *Scientific Reports*, •••, 8.

Lulla, K. P., Duane Nellis, M., & Rundquist, B. C. (2013). The Landsat 8 is ready for geospatial science and technology researchers and practitioners. *Geocarto International*, *28*(3), 191–191. DOI: 10.1080/10106049.2013.812346

Mahlein, A. K., Steiner, U., Hillnhütter, C., Dehne, H. W., & Oerke, E. C. (2012). Hyperspectral imaging for small-scale analysis of symptoms caused by different sugar beet diseases. *Plant Methods*, *8*(1), 3. DOI: 10.1186/1746-4811-8-3 PMID: 22273513

Markham, B. L., Storey, J. C., & Morfitt, R. A. (2015). Landsat-8 Sensor Characterization and Calibration. *Remote Sensing (Basel)*, *7*(3), 2279–2282. DOI: 10.3390/rs70302279

Maxwell, A. E., Warner, T. A., & Fang, F. (2018). Implementation of machine-learning classification in remote sensing: An applied review. *International Journal of Remote Sensing*, *39*(9), 2784–2817. DOI: 10.1080/01431161.2018.1433343

Mccabe, M. F., Rodell, M., Alsdorf, D. E., Miralles, D. G., Uijlenhoet, R., Wagner, W., Lucieer, A., Houborg, R., Verhoest, N. E., Franz, T. E., Shi, J., Gao, H., & Wood, E. F. (2017). The Future of Earth Observation in Hydrology. *Hydrology and Earth System Sciences*, *21*(7), 3879–3914. DOI: 10.5194/hess-21-3879-2017 PMID: 30233123

Mekasha, A., Gérard, B., Tesfaye, K., Nigatu, L., & Duncan, A. J. (2014). Inter-connection between land use/land cover change and herders'/farmers' livestock feed resource management strategies: A case study from three Ethiopian eco-environments. *Agriculture, Ecosystems & Environment*, *188*, 150–162. DOI: 10.1016/j.agee.2014.02.022

Mentaschi, L., Vousdoukas, M. I., Pekel, J. F., Voukouvalas, E., & Feyen, L. (2018). Global long-term observations of coastal erosion and accretion. *Scientific Reports*, *8*(1), 12876. DOI: 10.1038/s41598-018-30904-w PMID: 30150698

Mitchell, A. L., Williams, M. L., Tapley, I., & Milne, A. K. (2012). Interoperability of multi-frequency SAR data for forest information extraction in support of national MRV systems. *2012 IEEE International Geoscience and Remote Sensing Symposium*, 3166-3169. DOI: 10.1109/IGARSS.2012.6350752

Montesino Pouzols, F., Toivonen, T., Di Minin, E., Kukkala, A. S., Kullberg, P., Kuusterä, J., Lehtomäki, J., Tenkanen, H., Verburg, P. H., & Moilanen, A. (2014). Global protected area expansion is compromised by projected land-use and parochialism. *Nature*, *516*(7531), 383–386. DOI: 10.1038/nature14032 PMID: 25494203

Muñoz-Sabater, J., Dutra, E., Agustí-Panareda, A., Albergel, C., Arduini, G., Balsamo, G., Boussetta, S., Choulga, M., Harrigan, S., Hersbach, H., Martens, B., Miralles, D. G., Piles, M., Rodríguez-Fernández, N. J., Zsoter, E., Buontempo, C., & Thépaut, J.-N. (2021). ERA5-Land: A state-of-the-art global reanalysis dataset for land applications. *Earth System Science Data*, *13*(9), 4349–4383. DOI: 10.5194/essd-13-4349-2021

Nagendra, H., Lucas, R. M., Honrado, J. P., Jongman, R. H., Tarantino, C., Adamo, M., & Mairota, P. (2013). Remote sensing for conservation monitoring: Assessing protected areas, habitat extent, habitat condition, species diversity, and threats. *Ecological Indicators*, *33*, 45–59. DOI: 10.1016/j.ecolind.2012.09.014

National Academies of Sciences, Engineering, and Medicine. (2018). Thriving on our changing planet: A decadal strategy for Earth observation from space. The National Academies Press. DOI: 10.17226/24938

Nerem, R. S., Beckley, B., Fasullo, J. T., Hamlington, B. D., Masters, D., & Mitchum, G. T. (2018). Climate-change–driven accelerated sea-level rise detected in the altimeter era. *Proceedings of the National Academy of Sciences of the United States of America*, *115*(9), 2022–2025. DOI: 10.1073/pnas.1717312115 PMID: 29440401

Ng, A. H., Ge, L., Li, X., & Zhang, K. (2012). Monitoring ground deformation in Beijing, China with persistent scatterer SAR interferometry. *Journal of Geodesy*, *86*(6), 375–392. DOI: 10.1007/s00190-011-0525-4

Nouri, H., Beecham, S., Kazemi, F., & Hassanli, A. (2013). A review of ET measurement techniques for estimating the water requirements of urban landscape vegetation. *Urban Water Journal*, *10*(4), 247–259. DOI: 10.1080/1573062X.2012.726360

Oliver, M. A., & Webster, R. (2015). *Basic steps in geostatistics: The variogram and kriging*. Springer., DOI: 10.1007/978-3-319-15865-5

Olofsson, P., Foody, G. M., Herold, M., Stehman, S. V., Woodcock, C. E., & Wulder, M. A. (2014). Good practices for estimating area and assessing accuracy of land change. *Remote Sensing of Environment*, *148*, 42–57. DOI: 10.1016/j.rse.2014.02.015

Peng, S., Piao, S., Ciais, P., Friedlingstein, P., Ottle, C., Breon, F., Nan, H., Zhou, L., & Myneni, R. B. (2012). Surface Urban Heat Island Across 419 Global Big Cities. *Environmental Science & Technology*, *46*(2), 696–703. DOI: 10.1021/es2030438 PMID: 22142232

Pettorelli, N., Laurance, W. F., O'Brien, T. G., Wegmann, M., Nagendra, H., & Turner, W. (2014). Satellite remote sensing for applied ecologists: Opportunities and challenges. *Journal of Applied Ecology*, *51*(4), 839–848. DOI: 10.1111/1365-2664.12261

Piao, S., Liu, Q., Chen, A., Janssens, I. A., Fu, Y., Dai, J., Liu, L., Lian, X., Shen, M., & Zhu, X. (2019). Plant phenology and global climate change: Current progresses and challenges. *Global Change Biology*, *25*(6), 1922–1940. DOI: 10.1111/gcb.14619 PMID: 30884039

Piao, S., Nan, H., Huntingford, C., Ciais, P., Friedlingstein, P., Sitch, S. A., Peng, S., Ahlström, A., Canadell, J. G., Cong, N., Levis, S., Levy, P. E., Liu, L., Lomas, M. R., Mao, J., Myneni, R. B., Peylin, P., Poulter, B. I., Shi, X., & Chen, A. (2014). Evidence for a weakening relationship between interannual temperature variability and northern vegetation activity. *Nature Communications*, *5*(1), 5018. Advance online publication. DOI: 10.1038/ncomms6018 PMID: 25318638

Poulter, B., MacBean, N., Hartley, A., Khlystova, I., Arino, O., Betts, R., Bontemps, S., Boettcher, M., Brockmann, C., Defourny, P., Hagemann, S., Herold, M., Kirches, G., Lamarche, C., Lederer, D., Ottlé, C., Peters, M., & Peylin, P. (2015). Plant functional type classification for earth system models: Results from the European Space Agency's Land Cover Climate Change Initiative. *Geoscientific Model Development*, *8*(7), 2315–2328. DOI: 10.5194/gmd-8-2315-2015

Pramanik, M. K., Dash, P., & Behal, D. (2021). Improving outcomes for socioeconomic variables with coastal vulnerability index under significant sea-level rise: An approach from Mumbai coasts. *Environment, Development and Sustainability*, *23*(9), 13819–13853. DOI: 10.1007/s10668-021-01239-w

Prăvălie, R. (2021). Exploring the multiple land degradation pathways across the planet. *Earth-Science Reviews*, *220*, 103689. DOI: 10.1016/j.earscirev.2021.103689

Quaife, T., Lewis, P., De Kauwe, M., Williams, M., Law, B. E., Disney, M., & Bowyer, P. (2008). Assimilating canopy reflectance data into an ecosystem model with an Ensemble Kalman Filter. *Remote Sensing of Environment*, *112*(4), 1347–1364. DOI: 10.1016/j.rse.2007.05.020

Reiche, J., Lucas, R., Mitchell, A. L., Verbesselt, J., Hoekman, D. H., Haarpaintner, J., Kellndorfer, J. M., Rosenqvist, A., Lehmann, E. A., Woodcock, C. E., Seifert, F. M., & Herold, M. (2016). Combining satellite data for better tropical forest monitoring. *Nature Climate Change*, *6*(2), 120–122. DOI: 10.1038/nclimate2919

Rhee, J., Im, J., & Carbone, G. J. (2010). Monitoring agricultural drought for arid and humid regions using multi sensor remote sensing data. *Remote Sensing of Environment*, *114*(12), 2875–2887. DOI: 10.1016/j.rse.2010.07.005

Ridwan, M.A., Radzi, N.A., Ahmad, W.S., Ahmad, W.S., Mustafa, I.S., Mustafa, I.S., Din, N.M., Jalil, Y.E., Jalil, Y.E., Isa, A.M., Isa, A.M., Othman, N.S., Zaki, W.M. (2018). Applications of Landsat-8 Data: a Survey. International Journal of Engineering and Technology.

Ruf, C. S., Chew, C., Lang, T., Morris, M. G., Nave, K., Ridley, A., & Balasubramaniam, R. (2018). A new paradigm in Earth environmental monitoring with the CYGNSS small satellite constellation. *Scientific Reports*, *8*(1), 8782. DOI: 10.1038/s41598-018-27127-4 PMID: 29884899

Running, S. W., Justice, C. O., Salomonson, V. V., Hall, D. K., Barker, J. L., Kaufmann, Y. J., Strahler, A. H., Huete, A. R., Muller, J., Vanderbilt, V. C., Wan, Z., Teillet, P. M., & Carneggie, D. M. (1994). Terrestrial remote sensing science and algorithms planned for EOS/MODIS. *International Journal of Remote Sensing*, *15*(17), 3587–3620. DOI: 10.1080/01431169408954346

Rußwurm, M., & Körner, M. (2018). Multi-temporal land cover classification with sequential recurrent encoders. *ISPRS International Journal of Geo-Information*, *7*(4), 129. DOI: 10.3390/ijgi7040129

Sabbir, M., Ruman, A., Azim, N.A., & Mahmud, S., Parvez, Miah, T. (2025). Soil Moisture Retrieval from Soil Moisture Active Passive (SMAP) Satellite: A Validation and Error Analysis Using Ground-Based Observation Over Bangladesh. *European Journal of Applied Sciences*.

Santos, R., Bertolami, O., Castanho, E., Silva, P., Costa, A., Guerra, A. G. C., Arantes, M., Martin, M., Figueiredo, P., Cecilio, C. M., Castelão, I., Azevedo, L. F., Faria, J., Silva, H., Fontes, J., Prendergast, S., Tieppo, M., Pereira, E., Miranda, T., . . . Payne, C. (2022). The AEROS ocean observation mission and its CubeSat pathfinder. The 4S Symposium 2022.

Scheuchl, B., Mouginot, J., Rignot, E., Morlighem, M., & Khazendar, A. (2016). Grounding line retreat of Pope, Smith, and Kohler Glaciers, West Antarctica, measured with Sentinel-1a radar interferometry data. *Geophysical Research Letters*, *43*(16), 8572–8579. DOI: 10.1002/2016GL069287

Schirpke, U., Kohler, M., Leitinger, G., Fontana, V., Tasser, E., & Tappeiner, U. (2017). Future impacts of changing land-use and climate on ecosystem services of mountain grassland and their resilience. *Ecosystem Services*, *26*, 79–94. DOI: 10.1016/j.ecoser.2017.06.008 PMID: 31110934

Schroeder, W., Oliva, P., Giglio, L., & Csiszar, I. (2014). The New VIIRS 375 m active fire detection data product: Algorithm description and initial assessment. *Remote Sensing of Environment*, *143*, 85–96. DOI: 10.1016/j.rse.2013.12.008

Seidl, R., Thom, D., Kautz, M., Martin-Benito, D., Peltoniemi, M., Vacchiano, G., Wild, J., Ascoli, D., Petr, M., Honkaniemi, J., Lexer, M. J., Trotsiuk, V., Mairota, P., Svoboda, M., Fabrika, M., Nagel, T. A., & Reyer, C. P. (2017). Forest disturbances under climate change. *Nature Climate Change*, *7*(6), 395–402. DOI: 10.1038/nclimate3303 PMID: 28861124

Sen, A. (2023). *Overview of the imaging spectrometer instrument for NASA's Surface Biology and Geology Mission*. Sensors, Systems, and Next-Generation Satellites XXVII.

Senanayake, I. P., Pathira Arachchilage, K. R. L., Yeo, I.-Y., Khaki, M., Han, S.-C., & Dahlhaus, P. G. (2024). Spatial Downscaling of Satellite-Based Soil Moisture Products Using Machine Learning Techniques: A Review. *Remote Sensing (Basel)*, *16*(12), 2067. DOI: 10.3390/rs16122067

Shirzaei, M., & Bürgmann, R. (2018). Global climate change and local land subsidence exacerbate inundation risk to the San Francisco Bay Area. *Science Advances*, *4*(3), eaap9234. Advance online publication. DOI: 10.1126/sciadv.aap9234 PMID: 29536042

Siegel, D. A., Buesseler, K. O., Doney, S. C., Sailley, S. F., Behrenfeld, M. J., & Boyd, P. W. (2014). Global assessment of ocean carbon export by combining satellite observations and food-web models. *Global Biogeochemical Cycles*, *28*(3), 181–196. DOI: 10.1002/2013GB004743

Simard, M., Fatoyinbo, L., Smetanka, C., Rivera-Monroy, V. H., Castañeda-Moya, E., Thomas, N., & Van der Stocken, T. (2019). Mangrove canopy height globally related to precipitation, temperature and cyclone frequency. *Nature Geoscience*, *12*(1), 40–45. DOI: 10.1038/s41561-018-0279-1

Singh, A. (1989). Digital change detection techniques using remotely sensed data. *International Journal of Remote Sensing*, *10*(6), 989–1003. DOI: 10.1080/01431168908903939

Skirving, W. J., Heron, S. F., Marsh, B. L., Liu, G., De La Cour, J. L., Geiger, E. F., & Eakin, C. M. (2019). The relentless march of mass coral bleaching: A global perspective of changing heat stress. *Coral Reefs*, *38*(4), 547–557. DOI: 10.1007/s00338-019-01799-4

Sohl, T. L., Wimberly, M. C., Radeloff, V. C., Theobald, D. M., & Sleeter, B. M. (2016). Divergent projections of future land use in the United States arising from different models and scenarios. *Ecological Modelling*, *337*, 281–297. DOI: 10.1016/j.ecolmodel.2016.07.016

Songsom, V., Koedsin, W., Ritchie, R. J., & Huete, A. (2019). Mangrove phenology and environmental drivers derived from remote sensing in southern Thailand. *Remote Sensing (Basel)*, *11*(8), 955. DOI: 10.3390/rs11080955

Symeonakis, E., Caccetta, P., Koukoulas, S., Furby, S., & Karathanasis, N. (2012). Multi-temporal land-cover classification and change analysis with conditional probability networks: The case of Lesvos Island (Greece). *International Journal of Remote Sensing*, *33*(13), 4075–4093. DOI: 10.1080/01431161.2011.640961

Tadono, T., Ishida, H., Oda, F., Naito, S., Minakawa, K., & Iwamoto, H. (2014). Precise global DEM generation by ALOS PRISM. *ISPRS Annals of the Photogrammetry, Remote Sensing and Spatial Information Sciences*, *II-4*, 71–76. DOI: 10.5194/isprsannals-II-4-71-2014

Tellman, B., Sullivan, J. A., Kuhn, C., Kettner, A. J., Doyle, C. S., Brakenridge, G. R., Erickson, T. A., & Slayback, D. A. (2021). Satellite imaging reveals increased proportion of population exposed to floods. *Nature*, *596*(7870), 80–86. DOI: 10.1038/s41586-021-03695-w PMID: 34349288

Topouzelis, K., & Singha, S. (2017). Oil spill detection using space-borne Sentinel-1 SAR imagery. *Oil spill science and technology*, 387-394.

Toutin, T. (2004). Review article: Geometric processing of remote sensing images: Models, algorithms and methods. *International Journal of Remote Sensing*, *25*(10), 1893–1924. DOI: 10.1080/0143116031000101611

Transon, J., d'Andrimont, R., Maugnard, A., & Defourny, P. (2018). Survey of Hyperspectral Earth Observation Applications from Space in the Sentinel-2 Context. *Remote Sensing (Basel)*, *10*(2), 157. DOI: 10.3390/rs10020157

Ustin, S. L., DiPietro, D., Olmstead, K. S., Underwood, E. C., & Scheer, G. J. (2002). Hyperspectral remote sensing for invasive species detection and mapping. *IEEE International Geoscience and Remote Sensing Symposium*, 3, 1658-1660. DOI: 10.1109/IGARSS.2002.1026212

Valle, D. R., Izbicki, R., & Leite, R. V. (2023). Quantifying uncertainty in land-use land-cover classification using conformal statistics. *Remote Sensing of Environment, 295*, 113682. DOI: 10.1016/j.rse.2023.113682

Vanitha, V., Dwibedi, R. K., Painadthu, B., Venugopal, K., & Thomas, A. (2021). IOT based Soil Moisture Measuring System for Indian Agriculture. *Journal of Physics: Conference Series, 1964*(6), 062054. Advance online publication. DOI: 10.1088/1742-6596/1964/6/062054

Verbesselt, J., Hyndman, R., Newnham, G., & Culvenor, D. (2010). Detecting trend and seasonal changes in satellite image time series. *Remote Sensing of Environment, 114*(1), 106–115. DOI: 10.1016/j.rse.2009.08.014

Verburg, P., Soepboer, W., Veldkamp, A., Limpiada, R., Espaldon, V., & Mastura, S. S. A. (2002). Modeling the spatial dynamics of regional land use: The CLUE-S model. *Environmental Management, 30*(3), 391–405. DOI: 10.1007/s00267-002-2630-x PMID: 12148073

Verrelst, J., Camps-Valls, G., Muñoz-Marí, J., Rivera, J. P., Veroustraete, F., Clevers, J. G. P. W., & Moreno, J. (2015). Optical remote sensing and the retrieval of terrestrial vegetation bio-geophysical properties – A review. *ISPRS Journal of Photogrammetry and Remote Sensing, 108*, 273–290. DOI: 10.1016/j.isprsjprs.2015.05.005

Vos, K., Splinter, K. D., Harley, M. D., Simmons, J. A., & Turner, I. L. (2019). CoastSat: A Google Earth Engine-enabled Python toolkit to extract shorelines from publicly available satellite imagery. *Environmental Modelling & Software, 122*, 122. DOI: 10.1016/j.envsoft.2019.104528

Vousdoukas, M. I., Mentaschi, L., Voukouvalas, E., Verlaan, M., Jevrejeva, S., Jackson, L. P., & Feyen, L. (2018). Global probabilistic projections of extreme sea levels show intensification of coastal flood hazard. *Nature Communications, 9*(1), 2360. DOI: 10.1038/s41467-018-04692-w PMID: 29915265

Weiss, M., Jacob, F., & Duveiller, G. (2020). Remote sensing for agricultural applications: A meta-review. *Remote Sensing of Environment, 236*, 111402. DOI: 10.1016/j.rse.2019.111402

Wu, J. (2004). Effects of changing scale on landscape pattern analysis: Scaling relations. *Landscape Ecology, 19*(2), 125–138. DOI: 10.1023/B:LAND.0000021711.40074.ae

Wu, J., & Li, H. (2006). Concepts of scale and scaling. In Wu, J., Jones, K. B., Li, H., & Loucks, O. L. (Eds.), *Scaling and uncertainty analysis in ecology: Methods and applications* (pp. 3–15). Springer. DOI: 10.1007/1-4020-4663-4_1

Wulder, M. A., Masek, J. G., Cohen, W. B., Loveland, T. R., & Woodcock, C. E. (2012). Opening the archive: How free data has enabled the science and monitoring promise of Landsat. *Remote Sensing of Environment, 122*, 2–10. DOI: 10.1016/j.rse.2012.01.010

Wulder, M. A., Roy, D. P., Radeloff, V. C., Loveland, T. R., Anderson, M. C., Johnson, D. M., Healey, S. P., Zhu, Z., Scambos, T. A., Pahlevan, N., Hansen, M. C., Gorelick, N., Crawford, C. J., Masek, J. G., Hermosilla, T., White, J. C., Belward, A. S., Schaaf, C., Woodcock, C. E., & Cook, B. D. (2022). Fifty years of Landsat science and impacts. *Remote Sensing of Environment, 280*, 113195. Advance online publication. DOI: 10.1016/j.rse.2022.113195

Xie, M., Jean, N., Burke, M., Lobell, D., & Ermon, S. (2016). Transfer learning from deep features for remote sensing and poverty mapping. In *Proceedings of the Thirtieth AAAI Conference on Artificial Intelligence (AAAI-16)* (pp. 3929–3935). AAAI Press. DOI: 10.1609/aaai.v30i1.9906

Yan, W. Y., Shaker, A., & El-Ashmawy, N. (2015). Urban land cover classification using airborne LiDAR data: A review. *Remote Sensing of Environment, 158*, 295–310. DOI: 10.1016/j.rse.2014.11.001

Yang, C. (2018). High resolution satellite imaging sensors for precision agriculture. *Frontiers of Agricultural Science and Engineering, 5*(0), 393–405. DOI: 10.15302/J-FASE-2018226

Yang, Y., Zhang, S., Yang, J., Chang, L., Bu, K., & Xing, X. (2014). A review of historical reconstruction methods of land use/land cover. *Journal of Geographical Sciences, 24*(4), 746–766. DOI: 10.1007/s11442-014-1117-z

Yin, H., Brandão, A.Jr, Buchner, J., Helmers, D., Iuliano, B., Kimambo, N. E., Lewińska, K. E., Razenkova, E., Rizayeva, A., Rogova, N., Spawn, S. A., Xie, Y., & Radeloff, V. C. (2020). Monitoring cropland abandonment with Landsat time series. *Remote Sensing of Environment, 246*, 111873. DOI: 10.1016/j.rse.2020.111873

You, J., Li, X., Low, M., Lobell, D., & Ermon, S. (2017, February). Deep gaussian process for crop yield prediction based on remote sensing data. *Proceedings of the AAAI Conference on Artificial Intelligence, 31*(1). DOI: 10.1609/aaai.v31i1.11172

Zaabar, N., Niculescu, S., Mihoubi, M.K. (2021). Assessment of combining convolutional neural networks and object-based image analysis to land cover classification using sentinel 2 satellite imagery (TENES REGION, ALGERIA).

Zaidenberg, D. A., Sebastianelli, A., Spiller, D., & Ullo, S. L. (2021). Advantages and Bottlenecks of Quantum Machine Learning for Remote Sensing. 2021 IEEE International Geoscience and Remote Sensing Symposium IGARSS, 5680-5683.

Zarco-Tejada, P. J., Camino, C., Beck, P. S. A., Calderon, R., Hornero, A., Hernández-Clemente, R., Kattenborn, T., Montes-Borrego, M., Susca, L., Morelli, M., Gonzalez-Dugo, V., North, P. R. J., Landa, B. B., Boscia, D., Saponari, M., & Navas-Cortes, J. A. (2018). Previsual symptoms of Xylella fastidiosa infection revealed in spectral plant-trait alterations. *Nature Plants, 4*(7), 432–439. DOI: 10.1038/s41477-018-0189-7 PMID: 29942047

Zeng, L., Wardlow, B. D., Wang, R., Shan, J., Tadesse, T., Hayes, M. J., & Li, D. (2016). A hybrid approach for detecting corn and soybean phenology with time-series MODIS data. *Remote Sensing of Environment, 181*, 237–250. DOI: 10.1016/j.rse.2016.03.039

Zhang, B., Tian, H., Ren, W., Tao, B., Lu, C., Yang, J., Banger, K., & Pan, S. (2016). Methane emissions from global rice fields: Magnitude, spatiotemporal patterns, and environmental controls. *Global Biogeochemical Cycles*, *30*(9), 1246–1263. DOI: 10.1002/2016GB005381

Zhu, X. X., Tuia, D., Mou, L., Xia, G.-S., Zhang, L., Xu, F., & Fraundorfer, F. (2017). Deep learning in remote sensing: A comprehensive review and list of resources. *IEEE Geoscience and Remote Sensing Magazine*, *5*(4), 8–36. DOI: 10.1109/MGRS.2017.2762307

Zhu, Z., Piao, S., Myneni, R. B., Huang, M., Zeng, Z., Canadell, J. G., Ciais, P., Ciais, P., Sitch, S. A., Friedlingstein, P., Arneth, A., Cao, C., Cheng, L., Kato, E., Koven, C. D., Li, Y., Lian, X., Liu, Y., Liu, R., & Zeng, N. (2016). Greening of the Earth and its drivers. *Nature Climate Change*, *6*(8), 791–795. DOI: 10.1038/nclimate3004

Zhu, Z., Wang, S., & Woodcock, C. E. (2015). Improvement and expansion of the Fmask algorithm: Cloud, cloud shadow, and snow detection for Landsats 4–7, 8, and Sentinel-2 images. *Remote Sensing of Environment*, *159*, 269–277. DOI: 10.1016/j.rse.2014.12.014

Zhu, Z., & Woodcock, C. E. (2014). Continuous change detection and classification of land cover using all available Landsat data. *Remote Sensing of Environment*, *144*, 152–171. DOI: 10.1016/j.rse.2014.01.011

Zink, M., Bachmann, M., Brautigam, B., Fritz, T., Hajnsek, I., Moreira, A., Wessel, B., & Krieger, G. (2014). TanDEM-X: The New Global DEM Takes Shape. *IEEE Geoscience and Remote Sensing Magazine*, *2*(2), 8–23. DOI: 10.1109/MGRS.2014.2318895

Chapter 11
Precision Agriculture and Crop Monitoring With Geospatial AI

Debika Chaudhuri
http://orcid.org/0000-0002-9026-3191
Department of Electronics and Communication Engineering, School of Engineering and Technology, IFTM University, Moradabad, India

Atanu Nag
http://orcid.org/0000-0003-0561-1946
Department of Physics, School of Sciences, IFTM University, Moradabad, India

ABSTRACT

Precision agriculture is fundamentally transforming the management, monitoring, and optimization of agricultural systems through data-driven technologies that enhance productivity, resource efficiency, and environmental sustainability. By quantifying spatial variability in soil properties, crop development, and climatic conditions, it enables site-specific application of inputs, reducing waste and environmental impacts while maximizing yields. Central to this transformation is Geospatial Artificial Intelligence (GeoAI), which integrates remote sensing, GIS, GNSS, UAVs, and AI-driven analytics into a cohesive decision-support framework. Machine learning and deep learning methods extract meaningful patterns from complex geospatial datasets, supporting crop health monitoring, pest and disease detection, yield forecasting, and optimized resource management. Despite challenges related to data quality, infrastructure, and equitable access, GeoAI represents a vital pathway toward resilient, climate-smart, and sustainable agricultural systems capable of addressing growing global food demands.

1. INTRODUCTION

Precision agriculture (PA) has developed in recent decades as a fundamental practice of sustainable and knowledge-based agricultural production systems. The fundamental idea of PA is to find a balance between the dual and often contradictory demands of raising agricultural production to meet the increasing global food requirement and at the same time limiting the environmental impact of intensive agricultural production. PA fundamentally differs from traditional agricultural production in the fact that it views agricultural fields not as homogeneous entities but rather recognizes that spatial and temporal variability

DOI: 10.4018/979-8-3373-8037-7.ch011

is an inherent feature of agricultural production systems. For example, differences in soil texture, nutrient availability, topography, microclimate, crop growth, and biotic stress factors such as pests and diseases may occur at different scales ranging from a few meters to the whole field or even the whole landscape (McBratney et al., 2005). Managing this heterogeneity using uniform management practices is likely to result in inefficiencies in agricultural production, where fertilizers and pesticides may be applied in excess in some areas and in deficit in others, resulting in yield penalties and wastage of resources and environmental degradation. PA provides solutions to this problem by ensuring that inputs are applied at the right rate, at the right place, and at the right time (Pedersen & Lind, 2017). Traditionally, the evolution of PA has been associated with the advancement of positioning, sensing, and mechanization technologies. The introduction of Global Positioning System (GPS)-based guidance, yield monitors in harvester machines, variable rate technology in fertilizers and pesticides, and field sensors revolutionized farming by shifting focus from generalized management strategies to data-driven management. These technologies facilitated the acquisition of data on crop performance and management using spatial references, thereby enabling informed management strategies (Auernhammer, 2001; Pandit et al., 2026). For example, yield maps were mostly used to gain retrospective knowledge about crop performance, whereas soil grids were used in the management of fertilizers. Although these methods were fundamental in the development of PA, they were largely viewed as passive and narrow in scope (Pierce & Nowak, 1999). The data acquisition process in traditional PA has been regarded as irregular and conducted at low temporal resolution. Moreover, traditional PA has been regarded as primarily focused on machine optimization rather than continuous monitoring of crop and environmental conditions. The growing intricacy of the system and the rise in issues such as climate change, land degradation, water scarcity, and labor shortages have made it obvious that the conventional paradigms of PA are insufficient (Getahun et al., 2024). For instance, climate change has caused crop responses to become even more unpredictable. The conventional approach to decision-making in modern agriculture is based on historical averages. However, climate change has made crop responses unpredictable. Weather conditions such as droughts, floods, and heatwaves can suddenly change the conditions of crops in just days or even hours (Chen et al., 2025).

Under such circumstances, the need for timely and spatially specific information becomes essential. Farmers and agronomists need tools that not only recognize the changes but also interpret them in order to provide the required information in a timely manner rather than waiting until the harvest is over (Han et al., 2024). This has further given rise to the interest in the application of advanced geospatial technologies and data-driven analytical techniques that can provide effective support in the real-time or near-real-time monitoring of crops. The role of geospatial technologies is critical in addressing the emerging issues and challenges in the PA domain. Satellite remote sensing has revolutionized the way crops are observed and monitored in the agricultural sector. By providing the ability to repeatedly cover large areas from space at increasingly finer resolutions, satellite remote sensing has greatly impacted the PA domain (Mulla, 2013). The modern Earth observation systems offer a range of data types including multi-spectral, hyper-spectral, thermal, and radar data that can be used to measure and assess vegetation conditions, soil moisture levels, evapotranspiration rates, and land surface changes. Unmanned aerial vehicles (UAVs), also known as drones, offer ultra-high-resolution data and offer flexible data collection times, making them useful in detecting changes in fields and early warning signs of diseases in plants (Zhang & Kovacs, 2012). Geographic Information Systems (GIS) offer an umbrella system that integrates all these diverse spatial data types and provides an effective framework for management zone development and spatial pattern analysis (Pierce & Nowak, 1999). All these geospatial technologies have significantly enhanced the observational capabilities of PA. The full potential of these geospatial technologies is achieved when they are linked with artificial intelligence (AI) technologies. The volume, velocity, and variety of data obtained using geospatial technologies including satellite imaging, drones, and ground sensors pose significant analytical problems that cannot be solved using traditional data analysis techniques. AI technologies such as machine learning (ML) and deep learning (DL) offer powerful means to discover significant patterns in complex, high-dimensional data sets. The potential of AI technologies to learn directly from data and discover complex relationships offers new opportunities for gaining valuable insights, which would otherwise be unattainable through conventional analysis or human thought (Zhu et al., 2017; Zhao et al., 2024). The fusion of geospatial technologies with AI technologies has thus introduced Geospatial Artificial Intelligence (GeoAI) as a new paradigm in the analysis and use of spatial data in agriculture. The conventional application of geospatial technologies in agriculture was based on descriptive analysis. GeoAI today allows for the progression from descriptive analytics to predictive and prescriptive analytics. For instance, ML models have been successfully utilized for the classification of crops, estimation of biomass, and detection of pest and diseases, utilizing features such as spectral and textural information from remotely sensed imagery. Convolutional neural network (CNN) models, for example, have shown remarkable performance for image-based applications, such as crop stress analysis and weed detection, as they are capable of learning features from raw input images. In addition, recurrent neural networks, and their extensions, are especially useful for dealing with time-series data, making them useful for yield prediction and phenology analysis (Kamilaris & Prenafeta-Boldú, 2018; Kussul et al., 2017; Albahar, 2023). The inclusion of these AI models within a GeoAI environment allows for the generation of results that are location-specific.

GeoAI platforms have numerous advantages compared to conventional methods of crop monitoring. Crop growth and health are influenced by a complex array of factors, such as nutrient supply, water scarcity, diseases, and weather patterns. In most cases, these factors affect crops subtly, for example, through alterations in chlorophyll and canopy temperature. Such effects can be easily identified through spectral and canopy temperature indices (Weiss et al., 2020). GeoAI models can incorporate diverse information sources, including satellite and drone images, soil, weather, and historical yield information, to help identify early signs of crop stress. Such a data integration capacity not only helps identify anomalies but also helps differentiate between the sources of stress, making the information more accurate and relevant for decision-making (Pantazi et al., 2016). This is especially useful for precision interventions, since it helps to address issues before yield loss is too great to reverse. Finally, continuous monitoring is possible over spatial scales ranging from local to regional and national levels, which makes GeoAI relevant not just for farm management, but also for agricultural policy, food security, and supply chain issues (Jin et al., 2016). However, the increasing relevance of GeoAI in agriculture has to be viewed in the context of global issues. For one, the world's population is still growing, while land and water resources are finite. On the other hand, agriculture is at once a victim and a cause of climate change, since it is responsible for a considerable percentage of greenhouse gas emissions and is very susceptible to extreme weather. Sustainable intensification, or growing more food from existing land while reducing its environmental footprint, has become one of the major goals in agricultural development (Amritha et al., 2025; Godfray et al., 2010; Foley et al., 2011). PA, with the power of GeoAI, resonates well with this goal, as it promotes efficient and effective resource utilization, minimizes input losses, and encourages climate-responsible practices. These aspects of GeoAI resonate well with the sustainable development paradigm advocated in global agendas such as the Sustainable Development Goals (SDGs) presented by the United Nations, including aspects of zero hunger, climate change, and sustainable consumption and production patterns

(Barroso-Barroso et al., 2026). Though GeoAI in PA is a potentially effective practice, it faces some limitations in its implementation. The first and foremost issue in the implementation of GeoAI in PA is the availability of data. The power of GeoAI is highly dependent on the availability of data. The availability of data in cloud cover and sensor noise, and the differences in spatial and temporal resolutions of satellite and UAV imagery, may act as limitations in the implementation of GeoAI in PA (Zhang & Kovacs, 2012). The availability of ground truth data is also a major issue in the implementation of GeoAI in PA. The ground truth data are the most critical data in the implementation of GeoAI in PA, and their unavailability may act as a limitation in the implementation of GeoAI in PA (Mulla, 2013). The availability of computing power may also act as a limitation in the implementation of GeoAI in PA. Another barrier is the computational requirements, especially in areas where access to high-performance computing is limited. Moreover, the black box problem of some AI techniques has sparked concerns over interpretability and trust, especially when AI model output is used to make critical decisions in agriculture (Ryo, 2022; Yenkikar et al., 2025). Apart from these technical considerations, socioeconomic and institutional aspects also play an important role in the adoption of GeoAI-based solutions (Burabe et al., 2024; Kroupová et al., 2025). Large farming enterprises are in a better position to adopt advanced technologies and reap the benefits of data-based agriculture, compared to small-scale farmers, who are a significant portion of the world's farming community (Kroupová et al., 2025; Chen, 2025). In addition, data ownership, data privacy, and data governance are some of the emerging concerns in the context of GeoAI-based solutions, considering the economic and strategic importance of agricultural data in digital environments. Therefore, it is imperative to ensure that GeoAI-based solutions are developed and deployed in an inclusive, transparent, and ethical manner (Dara et al., 2022; Sullivan et al., 2024).

Figure 1 presents the conceptual framework of the crop monitoring system with the integration of GeoAI in the context of PA.

Figure 1. Conceptual framework of a GeoAI-enabled crop monitoring system for PA

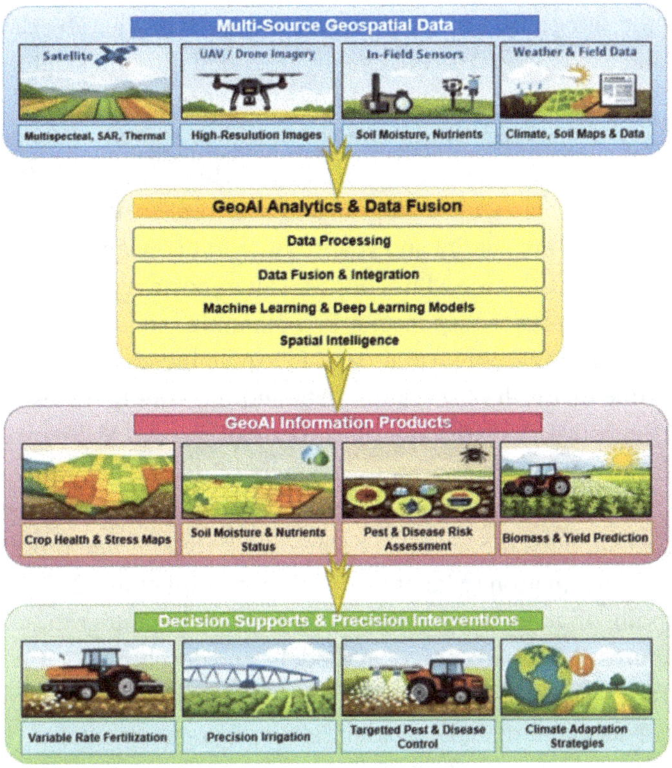

This chapter is designed to offer an overarching, balanced, and comprehensive overview of PA and crop monitoring with reference to the application of GeoAI. In addition, it builds on the foundations of PA and geospatial science, with reference to the application of AI-enhanced analytical frameworks with regard to the observation, interpretation, and management of agricultural variability. The discussion will cover the relevant data sources, including satellite remote sensing, UAVs, and sensor networks, with reference to the associated workflow processes with regard to the application of GeoAI. The pertinent applications are also addressed in the context of crop monitoring, yield prediction, and optimization in various agro-ecological conditions. In the process, the chapter attempts to bridge the gap that usually exists between theory and practice by providing information that is pertinent to researchers working on methodological advancements and practitioners interested in the application of GeoAI in agriculture. In this regard, this chapter critically evaluates the current status, challenges, and prospects of using GeoAI in PA, emphasizing the transformative role of GeoAI in PA, as well as the challenges that need to be overcome for PA to reap the full benefits of GeoAI. The integration of GeoAI with the advanced features of PA provides a fundamental route toward the development of stronger and more sustainable agricultural systems that can address the issues of the twenty-first century (Getahun et al., 2024; Ruder et al., 2026).

2. GEOAI: CONCEPTS AND RELEVANCE TO CROP MONITORING

GeoAI is defined as the integration of AI and geospatial data, whereby data is specifically referenced in terms of space and time. Such data are usually obtained through satellite imagery, UAVs, GIS, Global Navigation Satellite Systems (GNSS), etc. In most AI applications, the data used are considered independent and identically distributed. In this regard, specific spatial concepts like location, spatial dependence, scale, and temporal effects are specifically factored into the analytical process. These concepts are particularly significant in understanding agricultural systems, whereby crop growth, soil moisture patterns, and crop pests are spatial and spatio-temporal concepts (Pierdicca & Paolanti, 2022; VoPham et al., 2018). In recent years, GeoAI has become an important paradigm in many fields, including environmental monitoring, disaster response, urban planning, and resource management, due to its ability to provide valuable insights from spatial data. GeoAI's increasing importance in PA is largely due to the increasing volume of Earth observation data and the simultaneous progress in ML and DL techniques that can effectively process large amounts of complex data (VoPham et al., 2018; Ajith et al., 2025). Agricultural landscapes produce enormous amounts of diverse data types, including multi- and hyper-spectral images obtained using satellite and UAV platforms, soil maps, terrain models, and time-series weather data. GeoAI supplies the necessary framework to integrate these diverse data types and transform them into useful information (Mai et al., 2025; Hosseini et al., 2023).

From the conceptual point of view, GeoAI could be defined as the integration of three main elements: geospatial data infrastructures, AI-based analytical models, and domain-specific agricultural knowledge (Bordogna et al., 2016; Tan et al., 2024; Aderele et al., 2025). Geospatial data infrastructures provide the ability to geocode, standardize, and manage spatial data in GIS environments (Bordogna et al., 2016). At the same time, AI-based models, which include random forests, support vector machines, gradient boosting models, and deep neural network (DNN) models, provide support for pattern recognition, classification, regression, and prediction (Tan et al., 2024). Domain-specific agricultural knowledge offers the required context for the interpretation of the results of AI models, ensuring that the results are meaningfully aligned with agricultural processes (Aderele et al., 2025). The combination of these three components sets GeoAI apart from conventional remote sensing and GIS-based approaches, which often heavily rely on manual feature engineering and expert decision-making (Mai et al., 2025; Choi, 2023). GeoAI is especially significant for crop monitoring, as it allows for continuous and multi-scale evaluation of crops throughout the growth season (Dong et al., 2025). Crop monitoring involves a range of interconnected activities, including crop type determination, evaluation of crop vigor and health, detection of stress conditions, biomass estimation, and final yield prediction. Traditionally, crop monitoring has used methods such as scouting crops in the field, which is a tedious process (Omia et al., 2023; Wu et al., 2023). With the use of remote sensing, it was possible to monitor crops on a larger scale. Despite this, there have been challenges in interpreting the results, which have affected the use of the method (Wu et al., 2023).

GeoAI helps to avoid many of these limitations by using data to extract features and identify complex relationships. Satellite images are considered the main source of data for the GeoAI-based crop monitoring system, given their wide spatial coverage and short revisit times (Omia et al., 2023; Sishodia et al., 2020). Multispectral imagery can be exploited to compute vegetation indices such as those associated with chlorophyll content, canopy architecture, and photosynthetic processes, while radar imagery can be used to extract information under cloudy conditions and crop structural characteristics (Sishodia et al., 2020; Hu et al., 2024). UAVs can be used as supporting sources of data given their capacity to provide data with very high spatial resolution and their flexible deployment times (Tsouros et al., 2019; Bazrafkan et al., 2025). Moreover, ground sensors and IoT devices can provide in-situ data, such as soil moisture, temperature, and nutrient status, which can further enhance the data ecosystem and support more accurate crop monitoring (Islam et al., 2023; Kumari et al., 2025). The GeoAI models combine this multi-source information through data fusion techniques, making it possible for the GeoAI models to leverage the spatial extent of satellite data and the high spatial detail of UAV and in situ data. The ML algorithms are the heart of this information transformation. The supervised ML algorithms are commonly employed for crop classification, disease detection, and soil property estimation, where the required information is readily available. Among supervised models of ML, ensemble models such as random forest and gradient boosting are used in crop monitoring due to their robustness and ability to recognize intricate and non-linear relationships (Ngugi et al., 2024; Shawon et al., 2025). DL models, especially CNNs, have shown promising results in image-based crop monitoring by automatically detecting spatial features such as texture, shape, and spectral properties of crops (Li, K. et al., 2023; Lee et al., 2024). Adding temporal data further improves analysis by facilitating the identification of phenology and growth trends of crops, which is useful in early yield prediction and timely detection of stress in crops (Meghraoui et al., 2024; Muruganantham et al., 2022).

The importance of GeoAI in crop monitoring is further enhanced in relation to its predictive and prescriptive capabilities. In addition to descriptive analysis, GeoAI also offers predictive analysis through the integration of past yield data, weather forecasts, and soil properties to predict future crop performance. Such predictive analytics will play an important role in proactive management decisions in relation to timely interventions in crop management practices such as precision irrigation, fertilization, and pest management (Budah et al., 2025; Jagan Mohan et al., 2025; Saha et al., 2025). The predictive and prescriptive analytics capabilities of GeoAI in crop monitoring also enhance its application in PA because it provides not only predictive but also prescriptive analytics capabilities, shifting from reactive to proactive management (Yange et al., 2020). In relation to its application in sustainability, GeoAI in crop monitoring directly contributes to resource and environmental sustainability (Getahun et al., 2024; Aijaz et al., 2025). Further, identifying the spatial variability in crop demand will allow for the application of inputs in a more precise way, reducing leaching, water waste, and chemical runoff. It will also help in increasing the ability to respond to climate change through early warning systems for droughts, heat stress, and pest occurrence (Schuster et al., 2023; Shen et al., 2025). These applications are very relevant to global sustainability goals and visions, as advocated by the Food and Agriculture Organization (FAO, 2013; FAO, 2018), and the United Nations (UN, 2015), among other bodies. Though large-scale commercial farms are the first adopters of GeoAI technologies due to their financial and technological potential, the applicability of GeoAI is just as relevant to small-scale farming systems (Khaspuria et al., 2024), and its application is being made possible through the free access to satellite images and other technologies. Thus, when GeoAI is leveraged together with other approaches like participation and local knowledge, it holds promise for empowering smallholders through information provision that is timely and location-specific (Mendes et al., 2020; Adla et al., 2025).

In summary, GeoAI is a highly influential concept and application for promoting the advancement of crop monitoring through PA (Gupta & Kumar Pal, 2025). By combining spatially explicit data with advanced AI technologies, GeoAI facilitates the development of precise, timely, and predictive crop conditions (Espinel et al., 2024). The relevance of GeoAI is not just about the development of new technologies but also lies in the fact that it is critical in promoting sustainable, efficient, and resilient agricultural systems across various scales.

Figure 2 presents an overall workflow for GeoAI that encompasses the integration of various geospatial data sources such as satellite data, UAV data, GIS data, and in-situ data under one umbrella. By utilizing ML and DL techniques, spatial and spatio-temporal patterns are captured, allowing for continuous monitoring, multi-scale crop monitoring, yield prediction, and stress detection. These outputs facilitate PA and decision-making processes, allowing predictive and prescriptive management to be carried out in an efficient manner while ensuring sustainability and resiliency.

Figure 2. Conceptual framework of GeoAI in crop monitoring for PA

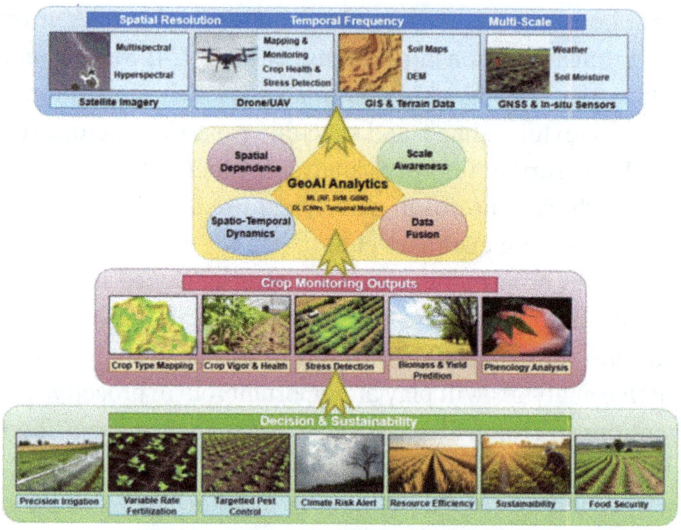

3. DATA SOURCES AND SENSING TECHNOLOGIES IN CROP MONITORING

For effective crop monitoring in the context of GeoAI, it is imperative that various data sources and sensing technologies be synergistically used. This is because no single data source is able to capture the full complexities of agricultural systems, characterized by significant spatial heterogeneity and dynamic changes in time (Karmakar, 2024). GeoAI in crop monitoring integrates data sources from satellite imagery, UAVs, field sensors, and geospatial infrastructure layers in an effective manner while utilizing the strength of multiple data sources in an effective manner (Xing et al., 2026; Yang et al., 2025). The data sources and sensing technologies used in crop monitoring will significantly affect the performance and accuracy of GeoAI models.

Satellite remote sensing: It is considered the backbone of large-scale crop monitoring owing to its large spatial extent and repetitive temporal cycles. Modern satellite platforms of the Earth observation system offer multispectral and even hyperspectral images in visible, near-infrared (NIR), and short-wave infrared (SWIR) regions (Sishodia et al., 2020; Atzberger, 2013). Such spectral bands are appropriate for generating vegetation indices such as the Normalized Difference Vegetation Index (NDVI) and Normalized Difference Red Edge Index (NDRE), which have been used as surrogates for crop vigor, chlorophyll content, and photosynthesis (Vidican et al., 2023). The modern Earth observation satellites like Sentinel-2, which provide systematic acquisition of multispectral data with high revisit time and free data access, have provided dense time series data that can be used for continuous monitoring of crop phenology. The time series data provided by these satellites can be used by GeoAI models to detect the anomaly in crop growth caused by water stress, nutrient starvation, or disease development. The free data policy of the Sentinel satellites has greatly reduced the adoption barrier of satellite-based crop monitoring in both developed and developing agricultural areas (Drusch et al., 2012). However, the cloud interference in the data and the moderate spatial resolution of the data provided by these satellites make it necessary to use additional data sources.

UAVs: Commonly referred to as drones, the role of UAVs as an intermediate form of remote sensing, linking satellite imagery of coarse resolution to ground-based observation, cannot be overstated. The fact that they are able to fly low and collect imagery of RGB, multispectral, hyperspectral, and thermal bands, allowing for imagery of high spatial resolution of a few centimeters, helps in the detection of fine-scale variations within a field that may not be apparent to satellite imagery. UAVs are of significant importance in crop monitoring and PA in GeoAI systems due to their ability to collect ultra-high-resolution imagery (Tsouros et al., 2019). High-resolution data is especially useful in monitoring row crops, specialty crops, and crops in early growth stages, in which precise spatial data is required to recognize minor stress signals in crops (Sankaran et al., 2015). The thermal cameras carried by UAVs make it possible to spot anomalies in crop canopy temperature, which is strongly associated with water stress and transpiration rates (Lacerda et al., 2022; Messina & Modica, 2020). UAVs also provide flexibility in PA applications due to their ability to be flown on demand to match specific crop growth stages or areas of concern identified through satellite monitoring. Moreover, the use of UAV and satellite images with strong GeoAI models can significantly improve the spatial resolution and accuracy of crop monitoring and diagnosis at various scales. The use of images from various sources can improve the accurate detection of crop stress, growth, and management needs (Xing et al., 2026; Maimaitijiang et al., 2020).

Field- and Ground-Based Sensors: Field sensors are important for offering in-situ data, which can be used for the validation of remotely sensed data as well as information obtained from AI models. Soil sensors, weather stations, and nutrient sensors can be used for the real-time monitoring of environmental factors affecting crop growth, which can improve the reliability of GeoAI models (Zhang et al., 2024). For example, yield monitors used in harvesting equipment can generate spatially explicit yield maps, which can be used as training and validation data sets for GeoAI models used in yield prediction tasks (Pathak et al., 2023). Proximal sensing technologies, which are usually attached to tractors or irrigation systems, can be used to collect data on crop growth factors such as plant height and biomass. The technologies are important in providing data with high accuracy under real-field conditions (Pallottino et al., 2019; Thorp et al., 2015). These data sets can be used to reduce uncertainty in AI models by grounding them in physical measurements.

Geospatial infrastructure and GIS layers: Static layers such as soil properties, digital elevation models (DEMs), slope, drainage, and historical management zones carry information about biophysical constraints that affect crop growth and development and provide the necessary context for the interpretation of dynamic sensor data and the implementation of site-specific PA strategies (Pedersen & Lind, 2017; Auernhammer, 2001). The use of historical yield maps, as well as records of management by farmers, can add to this context by incorporating the legacy effects of previous management (Maestrini & Basso, 2021). Spatial layers, in association with data from remote sensing, can facilitate the use of GeoAI by pointing to the variability of crops within the field. GIS can be used for the integration, visualization, and mapping of specific zones, which can be used for precision interventions in accordance with spatial heterogeneity (Weiss et al., 2020). Aside from that, these geospatial layers have also been used in the formulation of geospatial risk assessment frameworks where terrain variables such as slope, height, and drainage patterns could be used in modeling geospatial hazards or risks, such as soil erosion or flooding, and their impacts on agricultural productivity. This is in line with quantitative spatial risk modeling methodologies that make use of terrain-based process simulation with exposure analysis, as in recent studies that used 3D geospatial modeling in assessing hazards or risks (Al-Shaar et al., 2024). The use of similar risk-based approaches in GeoAI-based crop monitoring can improve the utility of such an approach for food security planning and resilience analysis, especially in environmentally sensitive agricultural landscapes. For instance, the use of multi-source sensing data within a geospatial framework of risk assessment enables GeoAI systems to not only monitor crops but also predict spatial risks to agricultural productivity, thus enabling pro-active agricultural decisions.

Temporal resolution: Regular observations are vital in crop monitoring, considering that crop processes are constantly changing. Temporal sensing can be used in tracking phenological stages, detecting stress, and assessing management interventions, which can be used in making appropriate decisions for optimal crop performance (Yu et al., 2022). Satellite imagery offers temporal and regional scale data, whereas UAVs and ground sensors provide high-frequency data at the field scale. Through the integration of multi-temporal data from these heterogeneous sources, more accurate, timely, and spatially precise crop conditions can be provided by GeoAI systems (Maimaitijiang et al., 2020; Allu & Mesapam, 2025). Time-series analysis and spatio-temporal modeling are also capable of differentiating between transient and persistent trends.

However, despite all these advantages, there are certain issues that crop up with the integration of heterogeneous data. Specifically, with regard to the huge amounts of data, especially from UAV images, there is a need for substantial storage facilities for effective management (Chi et al., 2016). Spatial and temporal data alignment from various sources require significant preprocessing, including geometric correction, radiometric correction, and normalization. Moreover, optical remote sensing is affected by clouds, whereas connectivity issues affect the transmission of data in rural areas (Xiao et al., 2023; Khanal et al., 2020). Cost factors also affect the selection and application of sensors in crop monitoring and PA, especially in the case of small-scale farmers. However, recent technological developments have greatly reduced the challenges faced in crop monitoring and PA. These include the use of cloud-based geospatial platforms, automated preprocessing tools, edge computing devices in UAVs and sensors, and data standards in GeoAI applications (Wang, X et al., 2025; Ampatzidis et al., 2020). These technologies have greatly improved the effectiveness of crop monitoring systems in near-real-time, which can fully capitalize on the potential of different sensors in PA. Collectively, these data sources lay the foundation for PA through the use of GeoAI, which facilitates more efficient, data-driven, and effective decisions in managing crops (Sishodia et al., 2020; Sangeetha et al., 2024). The various data sources used in crop monitoring are summarized in Table 1, which describes their main characteristics, such as sensors, spatial and temporal resolutions, as well as their applications, advantages, and limitations.

Table 1. Comparison of major geospatial data sources/ sensing platforms used in crop monitoring/ PA

Data Source / Platform	Typical Sensors	Spatial Resolution	Temporal Resolution	Key Crop Monitoring Applications	Strengths	Limitations	References
Satellite remote sensing (e.g., Sentinel-2, Landsat)	Multispectral, hyperspectral	10–30 m	5–16 days	Crop vigor (NDVI, NDRE), phenology, stress detection, regional yield estimation	Large spatial coverage, consistent time series, free data availability	Cloud contamination, moderate spatial resolution	(Omia et al., 2023; Budah et al., 2025)
Field-based sensors	Soil moisture probes, nutrient sensors, weather stations	Point-based	Continuous / real-time	Soil–plant–atmosphere monitoring, irrigation scheduling, model calibration	High accuracy, real-time data	Sparse spatial coverage, installation and maintenance costs	(Brinkhoff et al., 2018; Hardie, 2020)
UAVs (drones)	RGB, multispectral, hyperspectral, thermal (and optional LiDAR)	Ultra-high (often a few cm/pixel)	On-demand/ high	Vegetation growth and health monitoring, yield estimation, weed mapping, disease/ stress detection, irrigation management	Very high spatial detail; flexible, rapid deployment; field-scale insights	Limited coverage per flight; complex data processing; need for standardized workflows; operational constraints	(Tsouros et al., 2019)
Proximal sensors (tractor-/ irrigation-mounted)	LiDAR, optical reflectance, biomass sensors	Sub-meter	High-frequency (operational)	Biomass estimation, canopy structure, variable-rate applications	High precision under field conditions	Limited spatial extent, equipment dependency	(Farhan et al., 2024; Jennewein et al., 2025)
Yield monitors	Mass flow, moisture sensors	Sub-field	Seasonal	Yield mapping, model training and validation	Direct productivity measurement	Available only at harvest	(Liu et al., 2022)
GIS & geospatial layers	Soil maps, DEMs, slope, aspect, management zones	Static / multi-scale	Low change frequency	Contextualization of crop response, management zone delineation	Encodes biophysical constraints	Often coarse or outdated	(Nenkam et al., 2024; Sangeetha et al., 2024)

Figure 3. Schematic of the multi-source sensing ecosystem and GeoAI-based workflow for crop monitoring. The satellite, UAV, field sensors, and geospatial infrastructure layers provide spatial, spectral, and temporal information for crop monitoring. The GeoAI-based workflow involves the integration of multi-source data, where ML and DL are applied to explicitly consider spatial dependence, scale, and temporal dynamics for generating spatially explicit crop intelligence, including crop vigor, stress, phenology, and yield

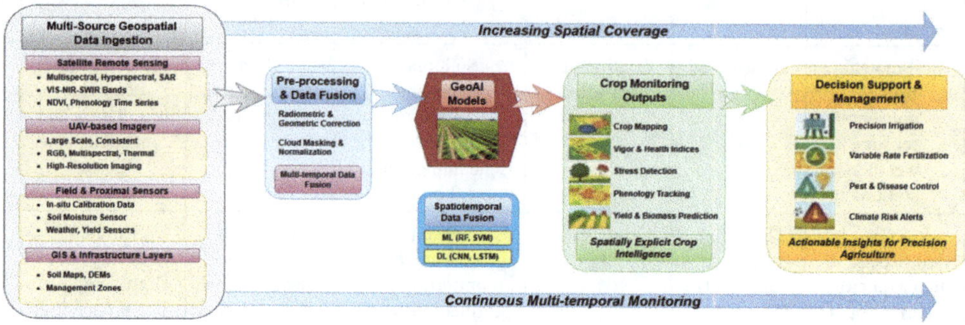

Figure 3 presents a holistic GeoAI-based crop monitoring system where there is a systematic integration of multi-source geospatial information from satellite imagery, UAV, field sensors, and GIS layers, and where ML and DL are applied for analysis. The workflow enables heterogeneous and multi-temporal information from various sources to be converted into spatially explicit crop intelligence for PA decision-making and sustainable crop management.

4. INTEGRATED ANALYTICAL FRAMEWORK FOR GEOAI-ENABLED PA

With the establishment of a holistic data system based on satellite imagery, UAV observations, in situ sensor networks, and geospatial infrastructure layers, the next challenge in the context of GeoAI-enabled PA is the transformation of the vast and heterogeneous spatial data streams into reliable and actionable intelligence (Mai et al., 2025; Dritsas & Trigka, 2025). This section proposes an integrated analytical framework, covering both conceptual and operational aspects of GeoAI systems. Contrary to viewing individual analytical components as separate, discrete, and sequential entities, this framework views them as interconnected and dynamic functions, which change over time during the crop growth period. The framework is based on an operational cycle, from defining the baseline to continuous monitoring, modeling, decision-making, and refinement. Each component contributes incrementally to the analytical value chain, with flexibility to adjust to crop types, environmental conditions, and management needs.

4.1 Objective Definition and Baseline Spatial Context

The GeoAI workflow starts with the formulation of agricultural objectives, then the development of baseline spatial data sets. This involves demarcating boundaries in the fields and creating management zones based on past yield maps, soil surveys, and terrain data (Sishodia et al., 2020; Javadi et al., 2022;

Breunig et al., 2020). This phase embodies the concept of spatial heterogeneity as it creates a reference point for assessing crop dynamics.

4.2 Continuous Data Acquisition and Monitoring

After establishing the crops, continuous monitoring is necessary to observe the temporal changes in the crop conditions. This involves the integration of multi-source data streams. These data streams include periodic satellite data, on-demand UAV data, and sensor data. This occurs throughout the growth season (Bazrafkan et al., 2025; Maimaitijiang et al., 2020). This phase provides temporal data richness. More importantly, data acquisition occurs in close association with analytical activities.

4.3 Data Pre-processing and Harmonization

Pre-processing is the fundamental step for assuring data reliability, data interoperability, and data analysis consistency. This involves radiometric correction and atmospheric correction for mitigating sensor and atmospheric noise effects (Zhang et al., 2022). Moreover, geometric correction and geo-referencing are necessary for spatial data alignment within GIS frameworks (Bazrafkan et al., 2025). For data collected via UAVs, mosaicking and ortho-rectification are necessary for spatial data alignment (Adão et al., 2017). However, one of the most important aspects within pre-processing is data harmonization from various sources, as data have different spatial resolution, spatial reference systems, and temporal frequencies. This facilitates data integration. After that, important agronomic parameters like vegetation indices (e.g., NDVI, NDRE), Leaf Area Index (LAI), and canopy temperature are calculated (Xu et al., 2020; Moravec et al., 2021). Although the preprocessing stage is defined as a specific stage, in reality, the preprocessing stage is part of each data acquisition cycle.

4.4 Management Zone Delineation

The process of management zone delineation converts the continuous spatial variability into discrete units, thus facilitating site-specific management (Filintas et al., 2023; Zhang et al., 2010). By using GIS-based spatial analysis tools such as geostatistics, variography, and kriging, the agricultural fields are segmented based on parameters such as soil, moisture levels, status of vegetation, and yield history (Bhagwan et al., 2025). This is perhaps the most important aspect of the entire spatial analytics process because it is the link between spatial analytics and management interventions. Furthermore, the zones are dynamic, i.e., they change over time based on the availability of new data, which is true because agro-ecosystems are dynamic.

4.5 Feature Extraction and Multi-Source Data Fusion

In feature extraction, the raw data is converted into useful variables that represent the conditions of the crops or the environment. It includes the extraction of canopy temperature using thermal imaging, texture information using high-resolution UAV data, or terrain information such as slope or aspect using DEMs (Li et al., 2022). Following this process, data fusion combines different types of information obtained from remote sensing, soil information, weather information, or sensor data into a single dataset that is used for analysis (Allu & Mesapam, 2025; Mena et al., 2025). From a theoretical perspective, the process of data fusion is used to reduce uncertainty or improve the quality of the data by using different types of information available in the data set, which is the basis of GeoAI modeling.

4.6 ML and DL Modeling

The foundation of the GeoAI framework is ML and DL modeling, which facilitates the recognition of patterns, classification, and predictive analysis (Allu & Mesapam, 2025; Joshi et al., 2023). Supervised ML algorithms are best suited when labeled datasets are accessible, while unsupervised ML algorithms allow the detection of spatial patterns in the data. DL modeling, especially CNNs, has found significant application in image analysis at high resolutions, thus recognizing spatial relationships (Jiang et al., 2025). This helps in recognizing relationships between environmental parameters and crop response, thus aiding in stress recognition, yield prediction, and anomaly recognition. This is in line with the broader application of the predictive paradigm in understanding spatial and temporal land change modeling scenarios, wherein scenario-based modeling is used for understanding environmental parameters (Al-Shaar et al., 2022). In terms of implementation, the model is run iteratively and continuously updated as and when fresh data becomes available.

4.7 Decision Support and Prescription Mapping

This decision support system helps in the interpretation of model outputs. Spatially explicit prescription maps are created to facilitate the application of irrigation water, fertilizers, and pest control agents in variable rates (Nikolić et al., 2025; Romero & Heenkenda, 2024). The outputs of the model are presented in an easily interpretable manner to the users through the help of dashboards and mobile apps (Hu et al., 2025; Barbosa Júnior et al., 2024). This phase reflects the integration of analytics with the real world.

4.8 Feedback Mechanisms and Iterative Learning

Another key aspect of the GeoAI system is the feature of adaptive learning. The feedback mechanisms involve ground truth validation and post-season evaluation to enhance the models as well as the management strategies (Li et al., 2007; Zheng et al., 2025). With the acquisition of more data over the seasons, the models become more efficient in their predictions. GeoAI is thus seen as an evolving system that can be used in the management of agricultural sustainability.

Table 2 highlights the different relationships between the data sources, methods of analysis, features extracted, and the AI techniques in the unified GeoAI system.

The proposed unified analytical framework for GeoAI in PA covers both the theory and practice of GeoAI. This framework illustrates how data from various sources can be transformed into intelligence in a systematic way through a cyclical and adaptive approach.

Table 2. Relationship between data sources, analytical methods, and AI techniques in GeoAI workflows

Data Source	GIS / Remote Sensing Analytics	Extracted Features	AI Techniques Applied	Typical Applications	References
Satellite imagery	Atmospheric correction, time-series analysis	Vegetation indices, phenology metrics	RF, SVM, LSTM	Crop type mapping, yield forecasting	(Tufail et al., 2022; Joshi et al., 2023)
UAV imagery	Orthomosaics, texture analysis	Canopy structure, row patterns	CNNs	Disease detection, vigor assessment	(Kamilaris & Prenafeta-Boldú, 2018; Mohanty et al., 2016)
Soil & terrain layers	Spatial interpolation, terrain analysis	Texture, slope, aspect	RF, ensemble models	Management zone delineation	(Jena et al., 2022)
In situ sensors	Temporal aggregation, spatial linking	Soil moisture, temperature	Regression, hybrid ML	Irrigation scheduling	(Tefera et al., 2026; Togneri et al., 2022)

Figure 4 depicts an integrated season-long GeoAI workflow in crop monitoring, which illustrates the transformation of multi-source remote sensing and in-situ data into actionable and site-specific PA management decisions through geospatial processing and AI analytics, as well as the feedback loop of ground truth validation and post-season evaluation in continuously refining the PA management strategy.

Figure 4. Overall GeoAI process for the continuous and long-season crop monitoring process with the integration of multi-source remote sensing and in-situ data, geo-processing, AI-based data analytics, decision support services, and feedback mechanisms for improving the models and supporting the adaptive PA management process

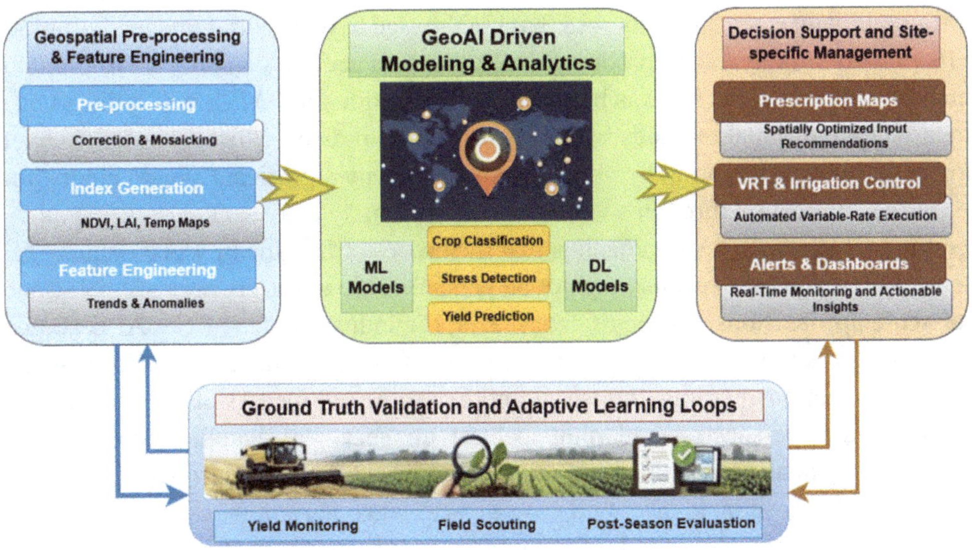

5. APPLICATIONS

GeoAI is now being recognized as having tremendous potential in revolutionizing the efficiency, sustainability, and productivity of modern farming systems. However, its applications show significant differences in terms of spatial scales—from farm-level decision support systems to regional-scale monitoring systems and national-scale policies—and these differences are significant in understanding the capabilities and limitations of GeoAI in PA management.

Farm-scale applications: One of the most notable applications of GeoAI is the fusion of UAV and satellite data for crop monitoring. UAV technology provides ultra-high spatial resolution data for the intra-field crop analysis process, whereas satellite technology provides consistent temporal coverage at the regional level (Bazrafkan et al., 2025; Liu et al., 2025). The integration of these data sources with the application of ML models helps in improving the estimation of nitrogen levels, biomass production, and yield prediction (Maimaitijiang et al., 2020). Specifically, UAV-based hyperspectral and multispectral data enable the detection of intra-field nutrient levels and crop stress detection. AI-based fusion of both datasets is more effective in providing accurate predictions than individual datasets. This is becoming increasingly popular in commercial farms, especially for the optimal use of nitrogen and reduction of losses without affecting crop yield (Xing et al., 2026). Empirical studies have proven that the application of the holistic approach of GeoAI can bring significant agronomic benefits. For example, the integration of UAV hyperspectral data with satellite time series data has been proven to increase the accuracy of yield prediction by 10-20% and enable savings of up to 15-30% in fertilizers. The benefits of such improvements are particularly important in high-input systems because even small improvements in efficiency can bring about significant benefits in terms of the environment and the economy. Moreover, a specific case study (Maimaitijiang et al., 2020) showed that UAV-based hyperspectral images, combined with ML algorithms like random forest, could produce high prediction accuracy in crop yield prediction, where the coefficient of determination (R^2) was higher than 0.80, while the RMSE was reduced by about 15% to 25% compared to traditional methods based on vegetation indices. Moreover, the study emphasized the reduction of nitrogen application by about 20% without any significant yield penalty, thereby confirming the GeoAI application through specific quantification.

However, the problem with the approach is that the process of acquiring UAV data is expensive and labor-intensive. In addition, the results may not be generalizable to different crops, soils, and climates.

Regional-scale applications: At the regional level, the industrial use of GeoAI is showing its efficiency and effectiveness in supporting crop monitoring and farm management (Hosseini et al., 2023). In the latest PA applications, various industrial and experimental GeoAI platforms have been able to integrate UAV-based hyperspectral and multispectral imagery with satellite-based time-series data and ML and AI-based analytics. This data fusion capability helps agronomists identify water stress, nutrient deficiencies, disease incidence, and crop growth stage variations at an earlier stage and with higher accuracy than traditional methods. Research has confirmed that the integration of UAV-based high-resolution imagery with satellite-based spectral data can improve the accuracy of yield prediction, biomass and nitrogen estimation, and optimize the application of water and fertilizers in an effective and reduced manner without compromising productivity (Xing et al., 2026; Bazrafkan et al., 2025). In addition, Schauberger et al. (2020) presented a large-scale case study where ML models using satellite-based information on vegetation indices, weather conditions, and historical crop yields were implemented in different regions to attain high accuracy in crop yield predictions with R^2 values ranging from 0.70

to 0.85. Their proposed framework facilitated the early season forecasting of crop yields with errors reduced by 10-20% using conventional statistical methods.

The above applications clearly indicate the operational benefits of GeoAI-based applications. For regional-scale monitoring, there is an increasing trend in using GeoAI-based systems for monitoring crop conditions in all agro-ecological zones. This is helpful in creating early warning systems for droughts, pest infestations, or yield anomalies. However, there is also an increase in issues associated with data heterogeneity, more uncertainty in the system, and the need for harmonization in calibrating diverse farming systems.

National and policy-scale applications: The use of GeoAI in PA is also being applied in agricultural mapping and spatial intelligence initiatives. GIS, remote sensing, and ML are commonly used for crop type classification, yield estimation, and land use monitoring (Budah et al., 2025; Espinel et al., 2024). Crop type classifiers using AI can be used to accurately classify crops using multi-temporal satellite images, even in complex landscapes (Li, Q et al., 2023). Yield estimation models use yield history maps, weather, and vegetation to support early season forecasting and logistics (Zhu et al., 2021; Schauberger et al., 2020). The optimization of resources, such as those provided by the use of resource optimization maps for irrigation demand zones or fertilizer requirements, can be conducted in a much more efficient manner, thus reducing waste to a minimum.

At the national level, GeoAI-based applications have come to be recognized as essential tools in the development of agricultural policies, subsidies, and climate resilience strategies. It is through such systems that governments, as well as international bodies, have sought to make use of them for purposes of large-scale yield forecasting, drought monitoring, and disaster preparedness. The advantage of scalability and consistency of satellite imagery can be leveraged for these applications, though they are often carried out at a lower spatial resolution, which may not consider the fine-scale variations that are important for decision-making. In addition, there are inherent uncertainties associated with the model outputs, and this may affect the policy decisions.

Societal Impact: The role of GeoAI frameworks can be observed in both local and regional settings (Budah et al., 2025; Saha et al., 2025). While they can be used to improve the quality of decisions made at the local level, they can also be used to improve regional agricultural decisions. Through the development of integrated initiatives that make use of satellite images, UAV images, AI methodologies, and cloud computing, it has become possible to monitor crops in their various zones in real-time. The initiatives have been recognized for their suitability in smallholder-dominated agricultural systems due to their scalability and access to satellite data (Xing et al., 2026; Maimaitijiang et al., 2020; Demissie et al., 2026).

However, the benefits of these applications to society are not evenly spread. For instance, the barriers to the adoption of these applications by smallholder farmers might be costly. In this regard, the application might even worsen the inequalities that already exist. In most of the reported application outcomes, the successes are highlighted with little emphasis on the failures and uncertainties that might be experienced. Thus, the evaluation of the GeoAI application needs to be critical.

GeoAI has also been found to aid in the early diagnosis of stress, precise input allocation, yield forecasting, and sustainability in different research, industrial, and governmental initiatives. As these technologies continue to develop and advance, their role in the integration of PA will be instrumental in creating more resilient, efficient, and data-driven food systems (Gupta & Kumar Pal, 2025; Demissie et al., 2026; Aslan et al., 2024). The applications of GeoAI in PA, which describe the integration of remote sensing, GIS, and AI in different scales and decision contexts, are listed in Table 3.

Table 3. Key applications of GeoAI in PA

Application Domain	Primary Data Sources	GeoAI Techniques	Key Outputs	Reported Benefits	References
UAV–Satellite Data Fusion for Crop Monitoring	UAV multispectral/ hyperspectral imagery, satellite time-series (NDVI, EVI), weather data	ML regression, data fusion models, DL	Nitrogen status, biomass, yield potential maps	Improved nutrient management, reduced fertilizer losses, higher yield stability	(Maimaitijiang et al. 2020; Morales-Ona et al., 2025)
Automated Stress and Anomaly Detection	Satellite multispectral data, UAV thermal imagery	CNN-based image analysis, anomaly detection algorithms	Early detection of water stress, disease, uneven growth	Faster detection, reduced water use, improved yield forecasting accuracy	(Guimarães et al., 2024; Gao et al., 2025)
Crop Type Mapping and Classification	Multi-temporal satellite imagery (optical, SAR)	Supervised ML classifiers (RF, SVM, DL)	Crop type maps, cropping pattern analysis	High classification accuracy, regional-scale monitoring	(Tufail et al., 2022; Hamidi et al., 2024)
Yield Estimation and Forecasting	Historical yield maps, vegetation indices, climate data	ML, time-series modeling	Early-season yield forecasts	Improved logistics, market planning, food security assessment	(Zhu et al., 2021; Li et al., 2024)
Village- and Regional-Scale Crop Monitoring	Satellite data, UAV observations, GIS layers, cloud platforms	AI-driven analytics, big data architectures	Crop condition, acreage, productivity indicators	Scalable monitoring, policy support, smallholder relevance	(Xing et al., 2026; Gokool et al., 2024)

6. CHALLENGES, LIMITATIONS, AND ENABLING FACTORS

Despite its potential, the large-scale application of GeoAI in PA is subject to a number of technical, economic, and institutional obstacles. These obstacles are not necessarily uniform in nature; rather, they can be broadly categorized into technical constraints and socio-institutional barriers. It is important to differentiate between the two categories, as they call for fundamentally different intervention approaches.

Data quality and availability: Data availability and quality are significant challenges that are faced in GeoAI studies. The acquisition of high-resolution images, especially through the use of UAV-based platforms, may be costly, weather-dependent, and subject to regulatory constraints. It may also require specialized knowledge for operating and working with the equipment. Satellite images may not have enough resolution and may be affected by cloud conditions.

The process of data fusion and data calibration also introduces complexity in GeoAI research, as mentioned in various recent research papers (Bazrafkan et al., 2025; Allu & Mesapam, 2025; Agrawal & Arafat, 2024). Apart from data availability, issues of data inconsistency, noise, and missing data, and temporal incompatibility may also affect the performance of GeoAI models. In addition, differences in sensor calibration and data acquisition protocols may affect the generation of standard data sets, which may affect the transfer and scalability of GeoAI-based solutions for different regions and cropping systems.

Infrastructure and connectivity challenges: The limitations in digital connectivity and infrastructural facilities, especially in rural regions, may create barriers for the efficient operation of GeoAI-based solutions. In addition, the high costs of acquiring necessary sensors, UAV platforms, and software, as well as specialized knowledge, may create barriers for the adoption of GeoAI-based solutions, especially for small-scale farmers, unless the associated economic benefits are well demonstrated (Khaspuria et al., 2024; Orellana et al., 2025; Nxumalo & Chauke, 2025).

In addition, the lack of access to electricity and cloud computing facilities and data storage is a challenge, particularly in developing regions. Infrastructural challenges not only influence the effective and efficient functioning of the system but also contribute to the digital divide between large-scale and small-scale farming systems. In addition, the ability of the AI model to generalize is a challenge, particularly in regions where the model was not initially developed or where the climate is not the same (Zhang et al., 2025).

Economic barriers and adoption constraints: Apart from costs related to infrastructure, there is also an element of economic viability with regards to GeoAI systems, particularly among many small-scale farmers whose financial conditions may not be favorable enough to accommodate risks related to return on investment (ROI). Moreover, there is also an issue related to skewed cost structures that tend to favor capital-intensive options over other forms of adoption.

Model interpretability and explainability: Interpretability of AI output is critical, and farmers need actionable insights, not just the output of the AI model. However, the black-box nature of most DL models makes them less trustworthy and increases the difficulty in decision-making for critical applications. Recently, expandable AI (XAI) techniques, such as feature attribution, saliency map, and rule-based model approximation, are providing new ways to interpret AI output and make it more transparent and interpretable. However, the challenge is to integrate XAI techniques into GeoAI systems, and the complexity, accuracy, and interpretability of the model need to be traded off. This is important not only for the accuracy of the results but also for the interpretability and usability of the results generated by the AI. However, scaling pilot systems is challenging, especially in terms of workflow, data, and privacy issues in drone imagery (Dara et al., 2022; Barbedo et al., 2026).

Governance, data privacy, and institutional framework: As GeoAI systems increasingly require high-resolution spatial and temporal data, issues related to data ownership, data privacy, and data governance become more critical. Drone and IoT technologies used in monitoring generate critical data at the farm level, and issues related to data access and benefits become more significant. There is a lack of proper institutional framework in many developing countries related to data usage, data sharing, and data security in agriculture. Moreover, there is also a lack of institutional support in terms of extension and training in GeoAI systems.

However, for the adoption to be successful, specific enabling factors need to be developed, which cover both technical and socio-institutional issues. For example, effective cloud and edge computing workflows, inter-disciplinary approaches, user-centric dashboards, extension services, policies, rural connectivity, and scalable platforms need to be accompanied by investments in digital literacy, data governance, and participatory approaches that allow the involvement of farmers during system development.

Recent research has also proven the effectiveness of the cloud-edge integrated architecture in the reduction of latency and bandwidth requirements in the implementation of a real-time crop monitoring system, thus enhancing the responsiveness of decisions (Yu et al., 2025). Additionally, the adoption and trust in the implementation of AI-based advisory systems have been proven to be improved by the participatory approach in the design and implementation in the context of smallholder farming (Kroupová et al., 2025). In addition, the success stories of pilot projects that have integrated IoT-based sensing with mobile-based advisory service delivery have demonstrated the value of user-centric dashboards and extension service delivery in addressing usability challenges (Orellana et al., 2025). Besides, the initiatives and projects undertaken by the government in facilitating the development of rural connectivity and open data initiatives have been instrumental in addressing infrastructure challenges and facilitating the deployment of GeoAI in the context of the developing world (Nxumalo & Chauke, 2025). These instances, therefore, indicate that the enabling factors are not just facilitators but also the driving forces in addressing the challenges associated with the deployment and use of the technology. This type of approach is significant to ensure that the PA system, enabled by GeoAI, not only becomes technologically sound but also socially inclusive, economically viable, and environmentally sustainable (Kroupová et al. 2025; Yu et al., 2025).

Figure 5 emphasizes that the successful adoption of GeoAI is not merely a technological issue but rather a socio-technical system issue that requires simultaneous progress in data, infrastructure, economics, governance, and human factors.

While there are challenges to the adoption and scalability of GeoAI in PA, such as those concerning data quality and availability, infrastructural limitations, financial feasibility, model generalization, and governance, there are significant opportunities for the scalability and sustainability of GeoAI-based PA, considering the progress being achieved in cloud-edge computing, adaptive AI models, interpretable analytics, and supportive institutional frameworks.

Figure 5. Conceptual framework showing major barriers and facilitating factors influencing the adoption of GeoAI in PA

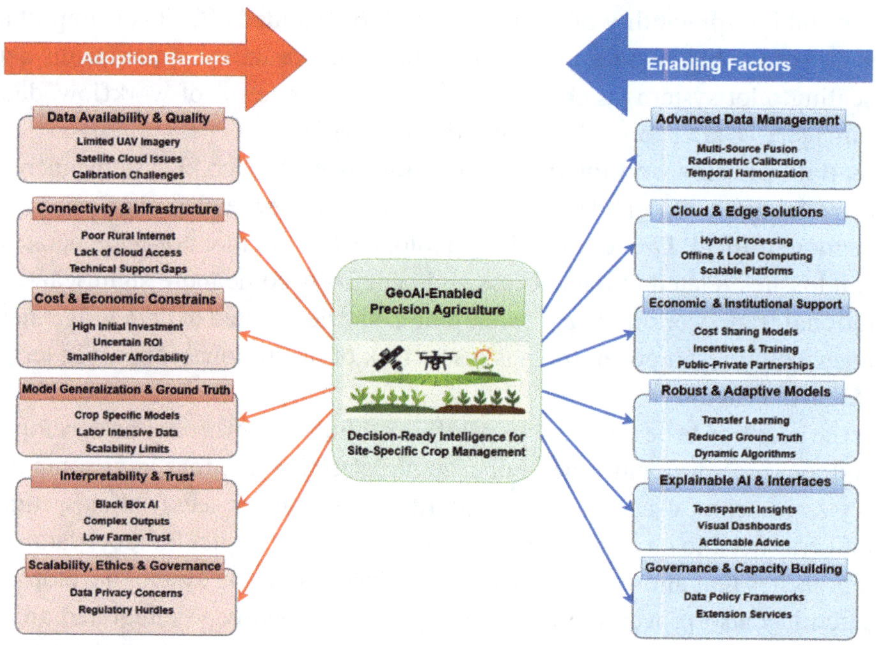

7. FUTURE DIRECTIONS

The future of crop monitoring in PA appears to be bright and promising, considering the rapid progress being made in the fields of GeoAI, edge computing, and sensor technologies. These technological innovations have empowered smart, quick, and environmentally friendly farming techniques that facilitate informed decision-making in near real-time. Future crop monitoring systems in PA have significant potential to maximize productivity while ensuring environmental sustainability (Miller et al., 2025; Sudha & Loret, 2026). It is expected that future progress in agricultural decision-support systems will result in a significant increase in speed, accuracy, and scalability. One of the most significant aspects of future progress in agricultural decision-support systems, especially in the field of crop monitoring, is that it will shift from descriptive and diagnostic analytics to structured geospatial risk assessment. In this regard, it is expected that GeoAI will incorporate concepts of hazard, exposure, and vulnerability in agricultural decision-support systems. This will not only elevate the role of crop monitoring but also transform it into a strategic part of food security planning, especially in environmentally vulnerable and hazard-prone regions of the world.

Key trends and developments include:

- **AI on device for near real-time decision-making:** The integration of AI into devices such as drones, tractors, and other smart agricultural machines will play a significant role in agricultural decision-support systems. The use of AI on devices will help in the early detection of diseases, nutrient deficiencies, and other problems, especially in near real-time, without relying on connectivity (Guebsi et al., 2024; Sapna, 2026).
- **Merging multiple sensing modalities:** Using multispectral, hyperspectral, LiDAR, SAR, thermal, or IoT sensors will allow for an integrated view of crop-soil-atmosphere interactions, thereby improving crop stress detection or yield forecasting under different environmental conditions (Karmakar, 2024; Sudha & Loret, 2026).
- **GeoAI- based integration of geospatial risk assessment:** With the integration of data from various sources of sensing and terrain attributes such as slope, elevation, and hydrological flow paths, it is possible to model the risks associated with agricultural activities such as soil erosion, flooding, and land degradation. This is possible by using quantitative spatial risk assessment methodologies that incorporate process models and exposure maps, similar to the 3D hazard propagation models that are applicable in other domains of geospatial science (Al-Shaar et al., 2024).
- **Generative AI and data augmentation:** The use of generative AI and data augmentation techniques is possible to overcome the challenges associated with the availability of data and improve the generality of models in dealing with various crops and regions (Mirzaei et al., 2023; Wang, C et al., 2025).
- **Scalable platforms:** With the integration of scalable platforms such as cloud platforms, it is possible to overcome the challenges associated with the adoption of data analytics and visualization by farmers and provide them with the necessary information and advice in managing their farms (Sudha & Loret, 2026; Vissarapu, 2025).
- **Localized AI models:** These models, which are specific to certain soils, crop phenology, and climates, will be used to send real-time alerts for pest, nutrient, and irrigation needs, going beyond post-season analysis (Sudha & Loret, 2026; Peeters et al., 2024).
- **Sustainability-centric GeoAI systems:** The future of GeoAI systems is expected to go beyond yield-based evaluations by incorporating various factors such as soil quality, nutrient runoff, biodiversity, and carbon footprint, thus focusing on sustainable agricultural practices (Getahun et al., 2024; Ruder et al., 2026).

- **Risk integration for sustainability analytics:** The future of GeoAI systems will incorporate spatial risk factors such as erosion risk, flood risk, and land degradation, along with sustainability factors, for evaluating the sustainability of agricultural practices, especially in the face of environmental risks.
- **Farmer- centricity and ease of access:** It is critical to develop tools and applications that are accessible and usable by farmers, especially small-scale farmers, to ensure effective and efficient use of the GeoAI systems (Cerjak et al., 2025; Muromba et al., 2025).
- **Deep Learning-Based Spatio-Temporal Scenario Forecasting:** Recent advances in scenario-based spatio-temporal forecasting are increasingly leveraging the capabilities of deep learning-based models, such as DNNs, in conjunction with other probabilistic models such as Markov chains.

For example, in a recent publication (Al-Shaar et al., 2025), it is evident how DNN-based Markov chain models can be used to project scenarios of urbanization and its corresponding risks of climate change for 2050 and 2100. This is a potential application for integrating GeoAI-based crop monitoring systems with climate change adaptation. This can hypothetically allow PA systems to shift from being purely diagnostic to anticipatory and resilient in nature.

- **Scenario-based geospatial risk forecasting:** Future PA systems are expected to build upon these spatio-temporal scenario-based models to include scenario-based geospatial risk forecasting models. These models can potentially assess the risks of various natural hazards such as flooding, erosion, and climate change, which can potentially impact agricultural productivity in the future.

These methods have the potential to allow PA systems to transcend the domain of immediate diagnostics and reach the domain of anticipatory and climate-resilient decision-making.

8. CONCLUSION

GeoAI is clearly an innovative technology with tremendous possibilities for the advancement of crop monitoring in the context of PA. The applications that have been presented in the various parts of this chapter demonstrate the various benefits that can be obtained with the implementation of GeoAI. However, in order to reach the true potential of GeoAI, the various challenges that are currently being experienced need to be addressed. While the level of technical innovation is significant, it is not the only factor.

Successful application will depend on the level of interdisciplinarity, user-friendliness, policy support, and capacity development. Scaling from demonstration to operational application is still an objective. In conclusion, it is evident that with further innovation in geospatial technologies and AI, intelligent integration is an opportunity for achieving resilient, efficient, and sustainable agriculture. While it is necessary to transform GeoAI from an emerging innovation to an accepted practice in agriculture, it is an imperative for achieving the twin challenges of sustainable agriculture, environmental protection, and climate change mitigation.

REFERENCES

Adão, T., Hruška, J., Pádua, L., Bessa, J., Peres, E., Morais, R., & Sousa, J. J. (2017). Hyperspectral imaging: A review on UAV-based sensors, data processing and applications for agriculture and forestry. *Remote Sensing (Basel)*, *9*(11), 1110. DOI: 10.3390/rs9111110

Aderele, M. O., Srivastava, A. K., Butterbach-Bahl, K., & Rahimi, J. (2025). Integrating Machine Learning with Agroecosystem Modelling: Current State and Future Challenges. *European Journal of Agronomy*, *168*, 127610. DOI: 10.1016/j.eja.2025.127610

Adla, S., Aravindakshan, A., Tyagi, A., Guntha, R., Ponce-Pacheco, M. A., Nagi, A., Pastore, P., & Pande, S. (2025). Participatory development of mobile agricultural advisory driven by behavioural determinants of adoption. *Journal of Environmental Management*, *374*, 124140. DOI: 10.1016/j.jenvman.2025.124140 PMID: 39826370

Agrawal, J., & Arafat, M. Y. (2024). Transforming farming: A review of AI-powered UAV technologies in precision agriculture. *Drones (Basel)*, *8*(11), 664. DOI: 10.3390/drones8110664

Aijaz, N., Lan, H., Raza, T., Yaqub, M., Iqbal, R., & Pathan, M. S. (2025). Artificial intelligence in agriculture: Advancing crop productivity and sustainability. *Journal of Agriculture and Food Research*, *20*, 101762. DOI: 10.1016/j.jafr.2025.101762

Ajith, S., Vijayakumar, S., & Elakkiya, N. (2025). Yield prediction, pest and disease diagnosis, soil fertility mapping, precision irrigation scheduling, and food quality assessment using machine learning and deep learning algorithms. *Discover Food*, *5*(1), 67. DOI: 10.1007/s44187-025-00338-1

Al-Shaar, M., Gerard, P.-C., Faour, G., Al-Shaar, W., & Adjizian-Gérard, J. (2024). A comprehensive approach to quantitative risk assessment of rockfalls on buildings using 3D model of rockfall runout. J, 7(2), 183–203. DOI: 10.3390/j7020011

Al-Shaar, W., Bonin, O., & de Gouvello, B. (2022). Scenario-based predictions of urban dynamics in Île-de-France region: A new combinatory methodologic approach of variance analysis and frequency ratio. *Sustainability (Basel)*, *14*(11), 6806. DOI: 10.3390/su14116806

Al-Shaar, W., Lehmann, X., Saad, N., Elmazi, G., Al-Shaar, M., & Tohme, C. (2025). Urban morphology and climate vulnerability assessment in Kuwait: A spatio-temporal predictive analysis utilizing deep neural network-enhanced Markov chain models for 2050 and 2100. *PLoS One*, *20*(8), e0318604. DOI: 10.1371/journal.pone.0318604 PMID: 40824951

Albahar, M. (2023). A survey on deep learning and its impact on agriculture: Challenges and opportunities. *Agriculture*, *13*(3), 540. DOI: 10.3390/agriculture13030540

Allu, A. R., & Mesapam, S. (2025). Impact of remote sensing data fusion on agriculture applications: A review. *European Journal of Agronomy*, *164*, 127478. DOI: 10.1016/j.eja.2024.127478

Ampatzidis, Y., Partel, V., & Costa, L. (2020). Agroview: Cloud-based application to process, analyze, and visualize UAV-collected data for precision agriculture applications utilizing artificial intelligence. *Computers and Electronics in Agriculture*, *174*, 105457. DOI: 10.1016/j.compag.2020.105457

Amritha, K. S., Sreedaya, G. S., & Shalini, J. R. (2025). Sustainable Intensification in Agriculture: A Review of Focus Areas, Approaches, and Constraints. *Journal of Experimental Agriculture International, 47*(2), 187–203. DOI: 10.9734/jeai/2025/v47i23280

Aslan, M. F., Sabanci, K., & Aslan, B. (2024). Artificial intelligence techniques in crop yield estimation based on Sentinel-2 data: A comprehensive survey. *Sustainability (Basel), 16*(18), 8277. DOI: 10.3390/su16188277

Atzberger, C. (2013). Advances in remote sensing of agriculture: Context description, existing operational monitoring systems and major information needs. *Remote Sensing (Basel), 5*(2), 949–981. DOI: 10.3390/rs5020949

Auernhammer, H. (2001). Precision farming—The environmental challenge. *Computers and Electronics in Agriculture, 30*(1–3), 31–43. DOI: 10.1016/S0168-1699(00)00153-8

Barbedo, J. G. A., da Silva, M. S., Cazzolato, M. T., Valem, L. P., Tinós, R., Romero, R. A. F., Murta, L. O.Junior, & Holanda, A. (2026). Explainability and privacy in AI-enabled crop monitoring: Trends and future directions in soybean research. *Computers and Electronics in Agriculture, 243*, 111392. DOI: 10.1016/j.compag.2025.111392

Barbosa Júnior, M. R., Moreira, B. R., Carreira, V. S., Brito Filho, A. L., Trentin, C., Souza, F. L. P., Tedesco, D., Setiyono, T., Flores, J. P., Ampatzidis, Y., Silva, R. P., & Shiratsuchi, L. S. (2024). Precision agriculture in the United States: A comprehensive meta-review inspiring further research, innovation, and adoption. *Computers and Electronics in Agriculture, 221*, 108993. DOI: 10.1016/j.compag.2024.108993

Barroso-Barroso, C., Vega-Muñoz, A., Maradiaga-López, J., Salazar-Sepúlveda, G., & Carabantes-Silva, R. (2026). Smart Farming and the SDGs: Emerging Research Patterns and Sustainability Implications. *Agriculture, 16*(1), 81. DOI: 10.3390/agriculture16010081

Bazrafkan, A., Igathinathane, C., Bandillo, N., & Flores, P. (2025). Optimizing integration techniques for UAS and satellite image data in precision agriculture — A review. *Frontiers in Remote Sensing, 6*, 1622884. Advance online publication. DOI: 10.3389/frsen.2025.1622884

Bhagwan, P. V., Anjaiah, T., Ravali, C., Chary, D. S., Zamani, A. T., Ullah, S., Rebouh, N. Y., & Tariq, A. (2025). Delineation and evaluation of management zones for site-specific nutrient management using a geostatistical and fuzzy C mean cluster approach. *Scientific Reports, 15*(1), 20991. DOI: 10.1038/s41598-025-07283-0 PMID: 40594516

Bordogna, G., Kliment, T., Frigerio, L., Brivio, P. A., Crema, A., Stroppiana, D., Boschetti, M., & Sterlacchini, S. (2016). A Spatial Data Infrastructure Integrating Multisource Heterogeneous Geospatial Data and Time Series: A Study Case in Agriculture. *ISPRS International Journal of Geo-Information, 5*(5), 73. DOI: 10.3390/ijgi5050073

Breunig, F. M., Galvão, L. S., Dalagnol, R., Dauve, C. E., Parraga, A., Santi, A. L., Della Flora, D. P., & Chen, S. (2020). Delineation of management zones in agricultural fields using cover–crop biomass estimates from PlanetScope data. *International Journal of Applied Earth Observation and Geoinformation, 85*, 102004. DOI: 10.1016/j.jag.2019.102004

Brinkhoff, J., Hornbuckle, J., & Dowling, T. (2018). Multisensor capacitance probes for simultaneously monitoring rice field soil-water-crop-ambient conditions. *Sensors (Basel), 18*(1), 53. DOI: 10.3390/s18010053 PMID: 29278397

Budah, A. G., Habeeb, T., & Faiza, S. (2025). Integrating GeoAI and deep learning for sustainable agriculture: Sentinel-2 based crop mapping and yield prediction for selected crops in Sokoto, Nigeria. Journal of Geography. *Environment and Earth Science International, 29*(12), 200–212. DOI: 10.9734/jgeesi/2025/v29i12992

Burabe, B. I., Dutse, F., Shuaibu, H., Ahmed, B., & Yusuf, H. O. (2024). Influence of Socio-Economics and Institutional Factors on Adoption of Agricultural Information Technologies of Irrigation Rice Farmers in Northeastern Nigeria. *Journal of Arid Agriculture, 25*(3), 14–21. https://jaaunimaid.ng/index.php/home/article/view/37

Cerjak, M., Medici, M., Faletar, I., Sundeep, J. V., & Canavari, M. (2025). Adoption of mobile-based agricultural extension services: Evidence from South India. *Journal of Rural Studies, 120*, 103851. DOI: 10.1016/j.jrurstud.2025.103851

Chen, X. (2025). The Role of Modern Agricultural Technologies in Improving Agricultural Productivity and Land Use Efficiency. *Frontiers in Plant Science, 16*, 1675657. DOI: 10.3389/fpls.2025.1675657 PMID: 41036393

Chen, X., Zhang, H., & Wong, C. U. I. (2025). Dynamic monitoring and precision fertilization decision system for agricultural soil nutrients using UAV remote sensing and GIS. *Agriculture, 15*(15), 1627. DOI: 10.3390/agriculture15151627

Chi, M., Plaza, A., Benediktsson, J. A., Sun, Z., Shen, J., & Zhu, Y. (2016). Big data for remote sensing: Challenges and opportunities. *Proceedings of the IEEE, 104*(11), 2207–2219. DOI: 10.1109/JPROC.2016.2598228

Choi, Y. (2023). GeoAI: Integration of Artificial Intelligence, Machine Learning, and Deep Learning with GIS. *Applied Sciences (Basel, Switzerland), 13*(6), 3895. DOI: 10.3390/app13063895

Dara, R., Hazrati Fard, S. M., & Kaur, J. (2022). Recommendations for Ethical and Responsible Use of Artificial Intelligence in Digital Agriculture. *Frontiers in Artificial Intelligence, 5*, 884192. DOI: 10.3389/frai.2022.884192 PMID: 35968036

Demissie, W. A., Sebastiani, L., & Rossetto, R. (2026). Integration of artificial intelligence and remote sensing for crop yield prediction and crop growth parameter estimation in Mediterranean agroecosystems: Methodologies, emerging technologies, research gaps, and future directions. *European Journal of Agronomy, 173*, 127894. DOI: 10.1016/j.eja.2025.127894

Dong, Y., Huang, W., Li, H., & Han, L. (2025). Editorial: Advancements in Agricultural Monitoring with AI Enhanced Remote Sensing Techniques. *Frontiers in Remote Sensing, 6*, 1664060. DOI: 10.3389/frsen.2025.1664060

Dritsas, E., & Trigka, M. (2025). Remote sensing and geospatial analysis in the big data era: A survey. *Remote Sensing (Basel), 17*(3), 550. DOI: 10.3390/rs17030550

Drusch, M., Del Bello, U., Carlier, S., Colin, O., Fernandez, V., Gascon, F., Hoersch, B., Isola, C., Laberinti, P., Martimort, P., Meygret, A., Spoto, F., Sy, O., Marchese, F., & Bargellini, P. (2012). Sentinel-2: ESA's optical high-resolution mission for GMES operational services. *Remote Sensing of Environment*, *120*, 25–36. DOI: 10.1016/j.rse.2011.11.026

Espinel, R., Herrera-Franco, G., Rivadeneira García, J. L., & Escandón-Panchana, P. (2024). Artificial intelligence in agricultural mapping. *Agriculture*, *14*(7), 1071. DOI: 10.3390/agriculture14071071

FAO, Food and Agriculture Organization of the United Nations. (2013). Climate-smart agriculture: Policies, practices and financing for food security, adaptation and mitigation. FAO. https://www.fao.org/3/i3325e/i3325e.pdf

FAO, Food and Agriculture Organization of the United Nations. (2018). Sustainable food and agriculture: Guidance for policymakers. FAO. https://www.fao.org/3/CA2096EN/ca2096en.pdf

Farhan, S. M., Yin, J., Chen, Z., & Memon, M. S. (2024). A comprehensive review of LiDAR applications in crop management for precision agriculture. *Sensors (Basel)*, *24*(16), 5409. DOI: 10.3390/s24165409 PMID: 39205103

Filintas, A., Gougoulias, N., Kourgialas, N., & Hatzichristou, E. (2023). Management zones delineation, correct and incorrect application analysis in a coriander field using precision agriculture, soil chemical, granular and hydraulic analyses, fuzzy k-means zoning, factor analysis and geostatistics. *Water (Basel)*, *15*(18), 3278. DOI: 10.3390/w15183278

Foley, J. A., Ramankutty, N., Brauman, K. A., Cassidy, E. S., Gerber, J. S., Johnston, M., Mueller, N. D., O'Connell, C., Ray, D. K., West, P. C., Balzer, C., Bennett, E. M., Carpenter, S. R., Hill, J., Monfreda, C., Polasky, S., Rockström, J., Sheehan, J., Siebert, S., & Zaks, D. P. M. (2011). Solutions for a Cultivated Planet. *Nature*, *478*(7369), 337–342. DOI: 10.1038/nature10452 PMID: 21993620

Gao, J., Gujarati, K., Hegde, M., Arra, P., Gupta, S., & Buch, N. (2025). Integration of UAV and remote sensing data for early diagnosis and severity mapping of diseases in maize crop through deep learning and reinforcement learning. *Remote Sensing (Basel)*, *17*(20), 3427. DOI: 10.3390/rs17203427

Getahun, S., Kefale, H., & Gelaye, Y. (2024). Application of precision agriculture technologies for sustainable crop production and environmental sustainability: A systematic review. *TheScientificWorldJournal*, *2126734*(1). Advance online publication. DOI: 10.1155/2024/2126734 PMID: 39421732

Godfray, H. C. J., Beddington, J. R., Crute, I. R., Haddad, L., Lawrence, D., Muir, J. F., Pretty, J., Robinson, S., Thomas, S. M., & Toulmin, C. (2010). Food Security: The Challenge of Feeding 9 Billion People. *Science*, *327*(5967), 812–818. DOI: 10.1126/science.1185383 PMID: 20110467

Gokool, S., Mahomed, M., Brewer, K., Naiken, V., Clulow, A., Sibanda, M., & Mabhaudhi, T. (2024). Crop mapping in smallholder farms using unmanned aerial vehicle imagery and geospatial cloud computing infrastructure. *Heliyon*, *10*(5), e26913. DOI: 10.1016/j.heliyon.2024.e26913 PMID: 38455562

Guebsi, R., Mami, S., & Chokmani, K. (2024). Drones in precision agriculture: A comprehensive review of applications, technologies, and challenges. *Drones (Basel)*, *8*(11), 686. DOI: 10.3390/drones8110686

Guimarães, N., Sousa, J. J., Couto, P., Bento, A., & Pádua, L. (2024). Combining UAV-based multispectral and thermal infrared data with regression modeling and SHAP analysis for predicting stomatal conductance in almond orchards. *Remote Sensing (Basel)*, *16*(13), 2467. DOI: 10.3390/rs16132467

Gupta, G., & Kumar Pal, S. (2025). Applications of AI in precision agriculture. *Discover Agriculture*, *3*(1), 61. DOI: 10.1007/s44279-025-00220-9

Hamidi, M., Homayouni, S., Safari, A., & Hasani, H. (2024). Deep learning-based crop-type mapping using SAR and optical data fusion. *International Journal of Applied Earth Observation and Geoinformation*, *129*, 103860. DOI: 10.1016/j.jag.2024.103860

Han, H., Liu, Z., Li, J., & Zeng, Z. (2024). Challenges in remote sensing–based climate and crop monitoring: Navigating the complexities using AI. *Journal of Cloud Computing (Heidelberg, Germany)*, *13*(1), 34. DOI: 10.1186/s13677-023-00583-8

Hardie, M. (2020). Review of novel and emerging proximal soil moisture sensors for use in agriculture. *Sensors (Basel)*, *20*(23), 6934. DOI: 10.3390/s20236934 PMID: 33291606

Hosseini, F. S., Seo, M. B., Razavi-Termeh, S. V., Sadeghi-Niaraki, A., Jamshidi, M., & Choi, S.-M. (2023). Geospatial Artificial Intelligence (GeoAI) and Satellite Imagery Fusion for Soil Physical Property Predicting. *Sustainability (Basel)*, *15*(19), 14125. DOI: 10.3390/su151914125

Hu, C., Ru, Y., Fang, S., Rong, Z., Zhou, H., Yan, X., & Liu, M. (2025). Orchard variable rate spraying method and experimental study based on multidimensional prescription maps. *Computers and Electronics in Agriculture*, *235*, 110379. DOI: 10.1016/j.compag.2025.110379

Hu, X., Li, L., Huang, J., Zeng, Y., Zhang, S., Su, Y., Hong, Y., & Hong, Z. (2024). Radar Vegetation Indices for Monitoring Surface Vegetation: Developments, Challenges, and Trends. *The Science of the Total Environment*, *945*, 173974. DOI: 10.1016/j.scitotenv.2024.173974 PMID: 38897467

Islam, M. R., Oliullah, K., Kabir, M. M., Alom, M., & Mridha, M. F. (2023). Machine learning enabled IoT system for soil nutrients monitoring and crop recommendation. *Journal of Agriculture and Food Research*, *14*, 100880. DOI: 10.1016/j.jafr.2023.100880

Jagan Mohan, R. N. V., Rayanoothala, P. S., & Sree, R. P. (2025). Next-gen agriculture: Integrating AI and XAI for precision crop yield predictions. *Frontiers in Plant Science*, *15*, 1451607. DOI: 10.3389/fpls.2024.1451607 PMID: 39845494

Javadi, S. H., Guerrero, A., & Mouazen, A. M. (2022). Clustering and smoothing pipeline for management zone delineation using proximal and remote sensing. *Sensors (Basel)*, *22*(2), 645. DOI: 10.3390/s22020645 PMID: 35062606

Jena, R. K., Bandyopadhyay, S., Pradhan, U. K., Moharana, P. C., Kumar, N., Sharma, G. K., Roy, P. D., Ghosh, D., Ray, P., Padua, S., Ramachandran, S., Das, B., Singh, S. K., Ray, S. K., Alsuhaibani, A. M., Gaber, A., & Hossain, A. (2022). Geospatial modelling for delineation of crop management zones using local terrain attributes and soil properties. *Remote Sensing (Basel)*, *14*(9), 2101. DOI: 10.3390/rs14092101

Jennewein, J. S., Davis, B. W., Seehaver-Eagen, S., Nicolette, J., Pittman, J., Hively, W. D., Goldsmith, A., Hidalgo, C., Reberg-Horton, C., & Mirsky, S. B. (2025). Multi-sensor proximal remote sensing for cover crop biomass estimation at high and moderate spatial resolutions. *Smart Agricultural Technology, 12*, 101201. DOI: 10.1016/j.atech.2025.101201

Jiang, C., Guo, X., Li, Y., Lai, N., Peng, L., & Geng, Q. (2025). Multimodal deep learning models in precision agriculture: Cotton yield prediction based on unmanned aerial vehicle imagery and meteorological data. *Agronomy (Basel), 15*(5), 1217. DOI: 10.3390/agronomy15051217

Jin, N., Tao, B., Ren, W., Feng, M., Sun, R., He, L., Zhuang, W., & Yu, Q. (2016). Mapping irrigated and rainfed wheat areas using multi-temporal satellite data. *Remote Sensing (Basel), 8*(3), 207. DOI: 10.3390/rs8030207

Joshi, A., Pradhan, B., Gite, S., & Chakraborty, S. (2023). Remote-sensing data and deep-learning techniques in crop mapping and yield prediction: A systematic review. *Remote Sensing (Basel), 15*(8), 2014. Advance online publication. DOI: 10.3390/rs15082014

Kamilaris, A., & Prenafeta-Boldú, F. X. (2018). Deep learning in agriculture: A survey. *Computers and Electronics in Agriculture, 147*, 70–90. DOI: 10.1016/j.compag.2018.02.016

Karmakar, P., Teng, S. W., Murshed, M., Pang, S., Li, Y., & Lin, H. (2024). Crop monitoring by multimodal remote sensing: A review. *Remote Sensing Applications: Society and Environment, 33*, 101093. DOI: 10.1016/j.rsase.2023.101093

Khanal, S., Kc, K., Fulton, J. P., Shearer, S., & Ozkan, E. (2020). Remote sensing in agriculture—Accomplishments, limitations, and opportunities. *Remote Sensing (Basel), 12*(22), 3783. DOI: 10.3390/rs12223783

Khaspuria, G., Khandelwal, A., Agarwal, M., Bafna, M., Yadav, R., & Yadav, A. (2024). Adoption of precision agriculture technologies among farmers: A comprehensive review. *Journal of Scientific Research and Reports, 30*(7), 671–686. DOI: 10.9734/jsrr/2024/v30i72180

Kroupová, Z. Ž., Aulová, R., Rumánková, L., Bajan, B., Čechura, L., Šimek, P., & Jarolímek, J. (2025). Drivers and Barriers to Precision Agriculture Technology and Digitalisation Adoption: Meta-Analysis of Decision Choice Models. *Precision Agriculture, 26*(1), 17. DOI: 10.1007/s11119-024-10213-1

Kumari, S., Ali, N., Dagati, M., & Dong, Y. (2025). IoT-enabled soil moisture and conductivity monitoring under controlled and field fertigation systems. *AgriEngineering, 7*(7), 207. DOI: 10.3390/agriengineering7070207

Kussul, N., Lavreniuk, M., Skakun, S., & Shelestov, A. (2017). Deep learning classification of land cover and crop types using remote sensing data. *IEEE Geoscience and Remote Sensing Letters, 14*(5), 778–782. DOI: 10.1109/LGRS.2017.2681128

Lacerda, L. N., Snider, J. L., Cohen, Y., Liakos, V., Gobbo, S., & Vellidis, G. (2022). Using UAV-based thermal imagery to detect crop water status variability in cotton. *Smart Agricultural Technology, 2*, 100029. DOI: 10.1016/j.atech.2021.100029

Lee, C.-H., Chen, K.-Y., & Liu, L.-Y. D. (2024). Effect of texture feature distribution on agriculture field type classification with multitemporal UAV RGB images. *Remote Sensing (Basel)*, *16*(7), 1221. DOI: 10.3390/rs16071221

Li, J., Hong, D., Gao, L., Yao, J., Zheng, K., Zhang, B., & Chanussot, J. (2022). Deep learning in multimodal remote sensing data fusion: A comprehensive review. *International Journal of Applied Earth Observation and Geoinformation*, *112*, 102926. DOI: 10.1016/j.jag.2022.102926

Li, K., Zhao, W., Chen, J., Zhang, L., Hu, D., & Wang, Q. (2023). Predicting crop growth patterns with spatial–temporal deep feature exploration for early mapping. *Remote Sensing (Basel)*, *15*(13), 3285. DOI: 10.3390/rs15133285

Li, Q., Tian, J., & Tian, Q. (2023). Deep learning application for crop classification via multi-temporal remote sensing images. *Agriculture*, *13*(4), 906. DOI: 10.3390/agriculture13040906

Li, S., Huang, J., Xiao, G., Huang, H., Sun, Z., & Li, X. (2024). Improved winter wheat yield estimation by combining remote sensing data, machine learning, and phenological metrics. *Remote Sensing (Basel)*, *16*(17), 3217. DOI: 10.3390/rs16173217

Li, X., Pan, Y.-C., Ge, Z.-Q., & Zhao, C.-J. (2007). Delineation and scale effect of precision agriculture management zones using yield monitor data over four years. *Agricultural Sciences in China*, *6*(2), 180–188. DOI: 10.1016/S1671-2927(07)60033-9

Liu, R., Sun, Y., Li, M., Zhang, M., Zhang, Z., Li, H., & Yang, W. (2022). Development and application experiments of a grain yield monitoring system. *Computers and Electronics in Agriculture*, *195*, 106851. DOI: 10.1016/j.compag.2022.106851

Liu, Y., Fan, Y., Yue, J., Ma, Y., Yang, F., Fan, J., Chen, R., Bian, M., & Feng, H. (2025). UAV-based remote sensing estimation of above-ground biomass in different crops: A review. *International Journal of Applied Earth Observation and Geoinformation*, *144*, 104938. DOI: 10.1016/j.jag.2025.104938

Maestrini, B., & Basso, B. (2021). Subfield crop yields and temporal stability in thousands of US Midwest fields. *Precision Agriculture*, *22*(6), 1749–1767. DOI: 10.1007/s11119-021-09810-1 PMID: 34744492

Mai, G., Xie, Y., Jia, X., Lao, N., Rao, J., Zhu, Q., Liu, Z., Chiang, Y.-Y., & Jiao, J. (2025). Towards the Next Generation of Geospatial Artificial Intelligence. *International Journal of Applied Earth Observation and Geoinformation*, *136*, 104368. DOI: 10.1016/j.jag.2025.104368

Maimaitijiang, M., Sagan, V., Sidike, P., Daloye, A. M., Erkbol, H., & Fritschi, F. B. (2020). Crop monitoring using satellite/UAV data fusion and machine learning. *Remote Sensing (Basel)*, *12*(9), 1357. DOI: 10.3390/rs12091357

McBratney, A., Whelan, B., Ancev, T., & Bouma, J. (2005). Future directions of precision agriculture. *Precision Agriculture*, *6*(1), 7–23. DOI: 10.1007/s11119-005-0681-8

Meghraoui, K., Sebari, I., Pilz, J., Ait El Kadi, K., & Bensiali, S. (2024). Applied deep learning-based crop yield prediction: A systematic analysis of current developments and potential challenges. *Technologies*, *12*(4), 43. DOI: 10.3390/technologies12040043

Mena, F., Pathak, D., Najjar, H., Sanchez, C., Helber, P., Bischke, B., Habelitz, P., Miranda, M., Siddamsetty, J., Nuske, M., Charfuelan, M., Arenas, D., Vollmer, M., & Dengel, A. (2025). Adaptive fusion of multi-modal remote sensing data for optimal sub-field crop yield prediction. *Remote Sensing of Environment, 318*, 114547. DOI: 10.1016/j.rse.2024.114547

Mendes, J., Pinho, T. M., dos Santos, F. N., Sousa, J. J., Peres, E., Boaventura-Cunha, J., Cunha, M., & Morais, R. (2020). Smartphone applications targeting precision agriculture practices—A systematic review. *Agronomy (Basel), 10*(6), 855. DOI: 10.3390/agronomy10060855

Messina, G., & Modica, G. (2020). Applications of UAV thermal imagery in precision agriculture: State of the art and future research outlook. *Remote Sensing (Basel), 12*(9), 1491. DOI: 10.3390/rs12091491

Miller, T., Mikiciuk, G., Durlik, I., Mikiciuk, M., Łobodzińska, A., & Śnieg, M. (2025). The IoT and AI in agriculture: The time is now—A systematic review of smart sensing technologies. *Sensors (Basel), 25*(12), 3583. DOI: 10.3390/s25123583 PMID: 40573470

Mirzaei, A., Bagheri, H., & Khosravi, I. (2023). Enhancing crop classification accuracy through synthetic SAR-optical data generation using deep learning. *ISPRS International Journal of Geo-Information, 12*(11), 450. DOI: 10.3390/ijgi12110450

Mohanty, S. P., Hughes, D. P., & Salathé, M. (2016). Using deep learning for image-based plant disease detection. *Frontiers in Plant Science, 7*, 1419. DOI: 10.3389/fpls.2016.01419 PMID: 27713752

Morales-Ona, A., Camberato, J., Nielsen, R., Paul, S., & Quinn, D. (2025). Integration of satellite, UAV, soil, and topographic data for assessing corn nitrogen uptake at early vegetative growth stages. *Precision Agriculture, 26*(6), 97. DOI: 10.1007/s11119-025-10293-7

Moravec, D., Komárek, J., López-Cuervo Medina, S., & Molina, I. (2021). Effect of atmospheric corrections on NDVI: Intercomparability of Landsat 8, Sentinel-2, and UAV sensors. *Remote Sensing (Basel), 13*(18), 3550. DOI: 10.3390/rs13183550

Mulla, D. J. (2013). Twenty-five years of remote sensing in precision agriculture: Key advances and remaining knowledge gaps. *Biosystems Engineering, 114*(4), 358–371. DOI: 10.1016/j.biosystemseng.2012.08.009

Muromba, P., Keeni, M., & Fuyuki, K. (2025). A systematic review of mobile agricultural service applications for smallholder farmers in sub-Saharan Africa: Perspectives from the technology acceptance model. *Agriculture & Food Security, 14*(1), 34. DOI: 10.1186/s40066-025-00563-y

Muruganantham, P., Wibowo, S., Grandhi, S., Samrat, N. H., & Islam, N. (2022). A systematic literature review on crop yield prediction with deep learning and remote sensing. *Remote Sensing (Basel), 14*(9), 1990. DOI: 10.3390/rs14091990

Nenkam, A. M., Wadoux, A. M. J.-C., Minasny, B., Silatsa, F. B. T., Yemefack, M., Ugbaje, S. U., Akpa, S., Van Zijl, G., Bouasria, A., Bouslihim, Y., Chabala, L. M., Ali, A., & McBratney, A. B. (2024). Applications and challenges of digital soil mapping in Africa. *Geoderma, 449*, 117007. DOI: 10.1016/j.geoderma.2024.117007

Ngugi, H. N., Akinyelu, A. A., & Ezugwu, A. E. (2024). Machine learning and deep learning for crop disease diagnosis: Performance analysis and review. *Agronomy (Basel)*, *14*(12), 3001. DOI: 10.3390/agronomy14123001

Nikolić, N., Mattivi, P., Pappalardo, S. E., De Marchi, M., & Masin, R. (2025). From detection to action: Creating operational prescription maps for weed management using low-cost UAVs. *Italian Journal of Agronomy*, *20*(3), 100047. DOI: 10.1016/j.ijagro.2025.100047

Nxumalo, G. S., & Chauke, H. (2025). Challenges and opportunities in smallholder agriculture digitization in South Africa. *Frontiers in Sustainable Food Systems*, *9*, 1583224. DOI: 10.3389/fsufs.2025.1583224

Omia, E., Bae, H., Park, E., Kim, M. S., Baek, I., Kabenge, I., & Cho, B.-K. (2023). Remote Sensing in Field Crop Monitoring: A Comprehensive Review of Sensor Systems, Data Analyses and Recent Advances. *Remote Sensing (Basel)*, *15*(2), 354. DOI: 10.3390/rs15020354

Orellana, E., Gonzalez, T., Álvarez, A., Fernández-Campusano, C., Muñoz, M., & Carrasco, R. (2025). Precision farming: Exploring the challenges and opportunities for smallholder farmers in Chile. *Agriculture*, *15*(24), 2555. DOI: 10.3390/agriculture15242555

Pallottino, F., Antonucci, F., Costa, C., Bisaglia, C., Figorilli, S., & Menesatti, P. (2019). Optoelectronic proximal sensing vehicle-mounted technologies in precision agriculture: A review. *Computers and Electronics in Agriculture*, *162*, 859–873. DOI: 10.1016/j.compag.2019.05.034

Pandit, V., Sravya, M., Ravali, C., Fiaz, S., Tariq, A., & Chatterjee, S. (2026). Precision agriculture in the era of technological advancement. In Fiaz, S., Nadeem, M. A., Baloch, F. S., & Chung, Y. S. (Eds.), *Artificial intelligence and data sciences for precision agriculture* (pp. 159–186). Springer. DOI: 10.1007/978-3-032-12770-9_8

Pantazi, X.-E., Moshou, D., & Bravo, C. (2016). Active learning system for weed species recognition based on hyperspectral sensing. *Biosystems Engineering*, *146*, 193–202. DOI: 10.1016/j.biosystemseng.2016.01.014

Pathak, D., Miranda, M., Mena, F., Sanchez, C., Helber, P., Bischke, B., Habelitz, P., Najjar, H., Siddamsetty, J., Arenas, D., Vollmer, M., Charfuelan, M., Nuske, M., & Dengel, A. (2023). Predicting crop yield with machine learning: An extensive analysis of input modalities and models on a field and sub-field level. In IGARSS 2023 - 2023 IEEE International Geoscience and Remote Sensing Symposium (pp. 2767–2770). Pasadena, CA, USA. DOI: 10.1109/IGARSS52108.2023.10282318

Pedersen, S. M., & Lind, K. M. (2017). Precision agriculture – From mapping to site-specific application. In Pedersen, S. M., & Lind, K. M. (Eds.), *Precision agriculture: Technology and economic perspectives. Progress in Precision Agriculture* (pp. 1–18). Springer. DOI: 10.1007/978-3-319-68715-5_1

Peeters, A., Cohen, Y., Bahat, I., Ohana-Levi, N., Goldshtein, E., Netzer, Y., Tenreiro, T. R., Alchanatis, V., & Ben-Gal, A. (2024). A spatial machine-learning model for predicting crop water stress index for precision irrigation of vineyards. *Computers and Electronics in Agriculture*, *227*(Part 2), 109578. DOI: 10.1016/j.compag.2024.109578

Pierce, F. J., & Nowak, P. (1999). *Aspects of precision agriculture, Advances in agronomy* (Vol. 67). Academic Press. DOI: 10.1016/S0065-2113(08)60513-1

Pierdicca, R., & Paolanti, M. (2022). GeoAI: A Review of Artificial Intelligence Approaches for the Interpretation of Complex Geomatics Data. *Geoscientific Instrumentation, Methods and Data Systems, 11*(1), 195–218. DOI: 10.5194/gi-11-195-2022

Romero, K. F., & Heenkenda, M. K. (2024). Developing site-specific prescription maps for sugarcane weed control using high-spatial-resolution images and light detection and ranging (LiDAR). *Land (Basel), 13*(11), 1751. DOI: 10.3390/land13111751

Ruder, S. L., Faxon, H. O., Orzel, E. C., Devkota, R., & Bronson, K. (2026). Reviewing the Evidence on Precision Agriculture and Environmental Sustainability. NPJ Sustainable Agriculture, 4, 9. DOI: 10.1038/s44264-026-00128-x

Ryo, M. (2022). Explainable Artificial Intelligence and Interpretable Machine Learning for Agricultural Data Analysis. *Artificial Intelligence in Agriculture, 6*, 257–265. DOI: 10.1016/j.aiia.2022.11.003

Saha, S., Kucher, O. D., Utkina, A. O., & Rebouh, N. Y. (2025). Precision agriculture for improving crop yield predictions: A literature review. *Frontiers in Agronomy, 7*, 1566201. DOI: 10.3389/fagro.2025.1566201

Sangeetha, C., Moond, V., Rajesh, G. M., Damor, J. S., Pandey, S. K., Kumar, P., & Singh, B. (2024). Remote sensing and geographic information systems for precision agriculture: A review. *International Journal of Environment and Climate Change, 14*(2), 287–309. DOI: 10.9734/ijecc/2024/v14i23945

Sankaran, S., Khot, L. R., Zúñiga Espinoza, C., Jarolmasjed, S., Sathuvalli, V. R., Vandemark, G. J., Miklas, P. N., Carter, A. H., Pumphrey, M. O., Knowles, N. R., & Pavek, M. J. (2015). Low-altitude, high-resolution aerial imaging systems for row and field crop phenotyping: A review. *European Journal of Agronomy, 70*, 112–123. DOI: 10.1016/j.eja.2015.07.004

Sapna, G., Gauttam, H., Chauhan, V., Pattanaik, K. K., Trivedi, A., & Ghosh, H. (2026). A comprehensive review of edge computing empowered smart agriculture: Trends, opportunities, and future directions. *Computers and Electronics in Agriculture, 241*, 111252. DOI: 10.1016/j.compag.2025.111252

Schauberger, B., Jägermeyr, J., & Gornott, C. (2020). A systematic review of local to regional yield forecasting approaches and frequently used data resources. *European Journal of Agronomy, 120*, 126153. DOI: 10.1016/j.eja.2020.126153

Schuster, J., Mittermayer, M., Maidl, F. X., Nätscher, L., & Hülsbergen, K.-J. (2023). Spatial variability of soil properties, nitrogen balance and nitrate leaching using digital methods on heterogeneous arable fields in southern Germany. *Precision Agriculture, 24*(2), 647–676. DOI: 10.1007/s11119-022-09967-3

Shawon, S. M., Ema, F. B., Mahi, A. K., Niha, F. L., & Zubair, H. T. (2025). Crop yield prediction using machine learning: An extensive and systematic literature review. *Smart Agricultural Technology, 10*, 100718. DOI: 10.1016/j.atech.2024.100718

Shen, J., Wang, H., & Tareque, H. (2025). Toward resilience in broadacre agriculture: A methodological review of remote sensing in crop productivity, phenology, and environmental stress detection. *Remote Sensing (Basel), 17*(23), 3886. DOI: 10.3390/rs17233886

Sishodia, R. P., Ray, R. L., & Singh, S. K. (2020). Applications of Remote Sensing in Precision Agriculture: A Review. *Remote Sensing (Basel), 12*(19), 3136. DOI: 10.3390/rs12193136

Sudha, S. P., & Loret, J. B. S. (2026). A review on machine learning-based precision agriculture techniques for crop farming monitoring with IoT. *Discover Environment, 4*(1), 10. DOI: 10.1007/s44274-025-00305-8

Sullivan, C. S., Gemtou, M., Anastasiou, E., & Fountas, S. (2024). Building Trust: A Systematic Review of the Drivers and Barriers of Agricultural Data Sharing. *Smart Agricultural Technology, 8*, 100477. DOI: 10.1016/j.atech.2024.100477

Tan, Y.-C., Duarte, L., & Teodoro, A. C. (2024). Comparative Study of Random Forest and Support Vector Machine for Land Cover Classification and Post-Wildfire Change Detection. *Land (Basel), 13*(11), 1878. DOI: 10.3390/land13111878

Tefera, G. W., Ray, R. L., Jackson, R., Deegala, B., & Akenzua, O. (2026). Integrating machine learning models with ground sensors to enhance soil moisture prediction in agroecosystems of Texas. *Computers and Electronics in Agriculture, 242*, 111358. DOI: 10.1016/j.compag.2025.111358

Thorp, K. R., Gore, M. A., Andrade-Sanchez, P., Carmo-Silva, A. E., Welch, S. M., White, J. W., & French, A. N. (2015). Proximal hyperspectral sensing and data analysis approaches for field-based plant phenomics. *Computers and Electronics in Agriculture, 118*, 225–236. DOI: 10.1016/j.compag.2015.09.005

Togneri, R., dos Santos, D. F., Camponogara, G., Nagano, H., Custódio, G., Prati, R., Fernandes, S., & Kamienski, C. (2022). Soil moisture forecast for smart irrigation: The primetime for machine learning. *Expert Systems with Applications, 207*, 117653. DOI: 10.1016/j.eswa.2022.117653

Tsouros, D. C., Bibi, S., & Sarigiannidis, P. G. (2019). A review on UAV-based applications for precision agriculture. *Information (Basel), 10*(11), 349. DOI: 10.3390/info10110349

Tufail, R., Ahmad, A., Javed, M. A., & Ahmad, S. R. (2022). A machine learning approach for accurate crop type mapping using combined SAR and optical time series data. *Advances in Space Research, 69*(1), 331–346. DOI: 10.1016/j.asr.2021.09.019

United Nations. (2015). Transforming our world: The 2030 Agenda for Sustainable Development. United Nations. https://sdgs.un.org/2030agenda

Vidican, R., Mălina , A., Ranta, O., Moldovan, C., Marian, O., Ghe e, A., Ghi e, C. R., Popovici, F., & Cătunescu, G. M. (2023). Using remote sensing vegetation indices for the discrimination and monitoring of agricultural crops: A critical review. *Agronomy (Basel), 13*(12), 3040. DOI: 10.3390/agronomy13123040

Vissarapu, S. (2025). AI in agriculture: Cloud-powered precision farming with real-time data analytics. *Journal of Computer Science and Technology Studies, 7*(5), 298–306. DOI: 10.32996/jcsts.2025.7.5.37

VoPham, T., Hart, J. E., Laden, F., & Chiang, Y.-Y.VoPham. (2018). Emerging Trends in Geospatial Artificial Intelligence (GeoAI): Potential Applications for Environmental Epidemiology. *Environmental Health : A Global Access Science Source, 17*(1), 40. DOI: 10.1186/s12940-018-0386-x PMID: 29665858

Wang, C., Xia, Y., Xia, L., Wang, Q., & Gu, L. (2025). Dual discriminator GAN-based synthetic crop disease image generation for precise crop disease identification. *Plant Methods, 21*(1), 46. DOI: 10.1186/s13007-025-01361-0 PMID: 40159478

Wang, X., Zeng, H., Yang, X., Shu, J., Wu, Q., Que, Y., Yang, X., Yi, X., Khalil, I., & Zomaya, A. Y. (2025). Remote sensing revolutionizing agriculture: Toward a new frontier. *Future Generation Computer Systems, 166*, 107691. DOI: 10.1016/j.future.2024.107691

Weiss, M., Jacob, F., & Duveiller, G. (2020). Remote sensing for agricultural applications: A meta-review. *Remote Sensing of Environment, 236*, 111402. DOI: 10.1016/j.rse.2019.111402

Wu, B., Zhang, M., Zeng, H., Tian, F., Potgieter, A. B., Qin, X., Yan, N., Chang, S., Zhao, Y., Dong, Q., Boken, V., Plotnikov, D., Guo, H., Wu, F., Zhao, H., Deronde, B., Tits, L., & Loupian, E. (2023). Challenges and Opportunities in Remote Sensing-Based Crop Monitoring: A Review. *National Science Review, 10*(4), nwac290. Advance online publication. DOI: 10.1093/nsr/nwac290 PMID: 36960224

Xiao, J., Aggarwal, A. K., Duc, N. H., Arya, A., Rage, U. K., & Avtar, R. (2023). A review of remote sensing image spatiotemporal fusion: Challenges, applications, and recent trends. *Remote Sensing Applications: Society and Environment, 32*, 101005. DOI: 10.1016/j.rsase.2023.101005

Xing, Y., Liu, X., & Wang, X. (2026). Integrating UAVs, satellite remote sensing, and machine learning in precision agriculture: Pathways to sustainable food production, resource efficiency, and scalable innovation. *Frontiers in Agronomy, 7*, 1670380. DOI: 10.3389/fagro.2025.1670380

Xu, J., Quackenbush, L. J., Volk, T. A., & Im, J. (2020). Forest and crop leaf area index estimation using remote sensing: Research trends and future directions. *Remote Sensing (Basel), 12*(18), 2934. DOI: 10.3390/rs12182934

Yang, X., Chen, J., Lu, X., Liu, H., Liu, Y., Bai, X., Qian, L., & Zhang, Z. (2025). Advances in UAV remote sensing for monitoring crop water and nutrient status: Modeling methods, influencing factors, and challenges. *Plants, 14*(16), 2544. DOI: 10.3390/plants14162544 PMID: 40872167

Yange, T. S., Egbunu, C. O., Rufai, M. A., Onyekwere, O., Abdulrahman, A. A., & Abdulkadri, I. (2020). Using prescriptive analytics for the determination of optimal crop yield. *International Journal of Data Science and Analytics, 6*(3), 72–82. DOI: 10.11648/j.ijdsa.20200603.11

Yenkikar, A., Mishra, V. P., Bali, M., & Ara, T. (2025). An Explainable AI-Based Hybrid Machine Learning Model for Interpretability and Enhanced Crop Yield Prediction. *MethodsX, 15*, 103442. DOI: 10.1016/j.mex.2025.103442 PMID: 40612261

Yu, H., Kong, B., Hou, Y., Xu, X., Chen, T., & Liu, X. (2022). A critical review on applications of hyperspectral remote sensing in crop monitoring. *Experimental Agriculture, 58*, e26. DOI: 10.1017/S0014479722000278

Yu, P., Teng, F., Zhu, W., Shen, C., Chen, Z., & Song, J. (2025). Cloud–edge–device collaborative computing in smart agriculture: Architectures, applications, and future perspectives. *Frontiers in Plant Science, 16*, 1668545. DOI: 10.3389/fpls.2025.1668545 PMID: 41164258

Zhang, C., & Kovacs, J. M. (2012). The application of small unmanned aerial systems for precision agriculture: A review. *Precision Agriculture, 13*(6), 693–712. DOI: 10.1007/s11119-012-9274-5

Zhang, J., Xu, S., Zhao, Y., Sun, J., Xu, S., & Zhang, X. (2022). Aerial orthoimage generation for UAV remote sensing [Review]. *Information Fusion, 89*, 91–120. DOI: 10.1016/j.inffus.2022.08.007

Zhang, R., Wu, X., Li, J., Zhao, P., Zhang, Q., Wuri, L., Zhang, D., Zhang, Z., & Yang, L. (2025). A bibliometric review of deep learning in crop monitoring: Trends, challenges, and future perspectives. *Frontiers in Artificial Intelligence*, *8*, 1636898. DOI: 10.3389/frai.2025.1636898 PMID: 41041084

Zhang, X., Feng, G., & Sun, X. (2024). Advanced technologies of soil moisture monitoring in precision agriculture: A review. *Journal of Agriculture and Food Research*, *18*, 101473. DOI: 10.1016/j.jafr.2024.101473

Zhang, X., Shi, L., Jia, X., Seielstad, G., & Helgason, C. (2010). Zone mapping application for precision-farming: A decision support tool for variable rate application. *Precision Agriculture*, *11*(2), 103–114. DOI: 10.1007/s11119-009-9130-4

Zhao, T., Wang, S., Ouyang, C., Chen, M., Liu, C., Zhang, J., Yu, L., Wang, F., Xie, Y., Li, J., Wang, F., Grunwald, S., Wong, B. M., Zhang, F., Qian, Z., Xu, Y., Yu, C., Han, W., Sun, T., & Wang, L. (2024). Artificial intelligence for geoscience: Progress, challenges, and perspectives. *Innovation (Cambridge (Mass.)*, *5*(5), 100691. DOI: 10.1016/j.xinn.2024.100691 PMID: 39285902

Zheng, W., Li, K., & Liu, X. (2025). GeoAI for land use observations, analysis, and forecasting. *Land (Basel)*, *14*(10), 2058. DOI: 10.3390/land14102058

Zhu, X., Guo, R., Liu, T., & Xu, K. (2021). Crop yield prediction based on agrometeorological indexes and remote sensing data. *Remote Sensing (Basel)*, *13*(10), 2016. DOI: 10.3390/rs13102016

Zhu, X. X., Tuia, D., Mou, L., Xia, G. S., Zhang, L., Xu, F., & Fraundorfer, F. (2017). Deep learning in remote sensing: A comprehensive review and list of resources. *IEEE Geoscience and Remote Sensing Magazine*, *5*(4), 8–36. DOI: 10.1109/MGRS.2017.2762307

Chapter 12
Sustainable Land Management With Geospatial AI

Atanu Nag
http://orcid.org/0000-0003-0561-1946
Department of Physics, School of Sciences, IFTM University, Moradabad, India

Debika Chaudhuri
http://orcid.org/0000-0002-9026-3191
Department of Electronics and Communication Engineering, School of Engineering and Technology, IFTM University, Moradabad, India

ABSTRACT

Sustainable land management (SLM) seeks to balance economic development, environmental protection, and social equity under growing pressure on limited land resources. It promotes efficient land use that maintains productivity, conserves ecosystems, and supports livelihoods for present and future generations. Advances in geospatial data and artificial intelligence have strengthened SLM through Geospatial Artificial Intelligence (GeoAI), which integrates remote sensing, GIS, and machine learning to analyze complex spatial information. GeoAI enables accurate monitoring of land-use change, supports precision agriculture, and helps predict land degradation risks, improving decision-making and resource efficiency. Despite its potential, challenges such as data quality, algorithm transparency, limited institutional capacity, and ethical concerns persist. Addressing these through inclusive governance, capacity building, and transparent frameworks can ensure GeoAI complements human judgment and contributes to resilient, equitable, and sustainable land governance.

1. INTRODUCTION

Land has always remained a finite and irreplaceable resource that provides the foundation for the survival of the human race, economic growth, and the functioning of the natural environment (Foley et al., 2005; Cabernard et al., 2024; Verburg et al., 2015). However, the challenge of land management has become increasingly complicated in the context of growing concerns related to the finite nature of the earth's resources, the increasing rate of population growth, and the consequences of urbanization and changing consumption patterns on the environment. In this context, sustainable land management

DOI: 10.4018/979-8-3373-8037-7.ch012

(SLM) has been recognized as an important approach that emphasizes the sustainable management of land resources to meet the needs of the present without compromising the ability of future generations to meet their own needs (Lambin & Meyfroidt, 2011; Salvati et al., 2009; Ziadat et al., 2022; von Keyserlingk et al., 2023).

Over the last few decades, the occurrence of land degradation has highlighted the need to resort to SLM practices. Soil erosion, salinization, nutrient depletion, deforestation, desertification, and the fragmentation of natural habitats have resulted in the degradation of land productivity in different parts of the world (Arneth et al., 2021; Haregeweyn et al., 2023). According to international organizations such as the Food and Agriculture Organization (FAO, 2023; FAO, 2025), it has been estimated that a considerable portion of the agricultural land around the globe has already been subjected to degradation. However, these issues are further complicated by climate change, which affects precipitation patterns, frequency, and severity of extreme weather events, and also impacts existing vulnerabilities in land resources. SLM, therefore, is not only an environmental issue; rather, it has emerged as an important aspect of sustainable development, climate change mitigation/adaption, and socio-economic security.

Traditionally, land management practices have largely been based on field surveys, local observations, and expert judgments. Although conventional land management practices are important in this context, they are inevitably handicapped by the availability of data in terms of spatial extent, frequency of data collection, and the ability to analyze the data. Land management in the twenty-first century is inevitably handicapped by the need for continuous observation and the ability to forecast the changes that might occur in the future. This demand has coincided with a significant shift in the geospatial landscape (Dong et al., 2019). Advancements in remote sensing (RS) technologies, the availability of numerous earth observation satellites, the increased use of unmanned aerial vehicles (UAVs), the establishment of in situ sensor networks, and the development of geographic information systems (GIS) have collectively provided the impetus for an unprecedented rise in the quantity, variety, and velocity of spatial data (Colpaert, 2022; Dritsas & Trigka, 2025; See et al., 2024). Such developments have provided the basis for a more data-driven and spatially explicit approach to land management.

Parallelly, there has been rapid evolution of artificial intelligence (AI), particularly in the domains of machine learning (ML) and deep learning (DL). AI has increasingly demonstrated the potential to detect intricate non-linear patterns in large and heterogeneous datasets, automate analytical processes, and develop predictive capabilities that are not possible with traditional statistical tools. However, the fusion of geospatial big data and AI has given rise to the concept of Geospatial Artificial Intelligence (GeoAI), which is now recognized as a separate scientific discipline (Choi, 2023; Pierdicca & Paolanti, 2022). The concept of GeoAI is based on the integration of spatial data infrastructures and AI to further improve the extraction of significant information from geospatial data, which in turn improves the accuracy of the interpretation, modeling, and forecasting of land system dynamics (Song et al., 2023; Mai, G et al., 2025).

GeoAI is a new paradigm for the processing and utilization of land-related data. This is in contrast to the conventional approaches of analyzing land-related data through the application of GIS-based approaches. In the case of GeoAI, the systems have the potential to learn directly from the data and can operate across scales from individual fields to the entire continents. By relying on satellite images, hyperspectral images, climate datasets, soil datasets, socio-economic datasets, and crowdsourced datasets, GeoAI has the potential to analyze the entire land systems as coupled human and environmental systems (Zheng et al., 2025; Wang, S et al., 2024). Such an integration is also pertinent in the context of SLM, which inherently demands the consideration of biophysical, socio-economic, and institutional aspects of land resources. The importance of GeoAI in the context of SLM can be illustrated by its relevance to the entire gamut of land management issues (Wang, S et al., 2024; Hosseini et al., 2023). For example, in the context of agricultural resources, the application of GeoAI-driven approaches can be made in precision farming, which enables the accurate identification of spatial variability in soil moisture, nutrient status, and crop status (Sishodia et al., 2020). In the field of forest and rangeland management, DL models, when applied to multi-temporal satellite images, can effectively identify forest loss, forest degradation, biomass, and fire risk (Tian et al., 2024). In urban and peri-urban regions, GeoAI helps in the evaluation of land use/cover change, urban growth, and infrastructure development, which are important for supporting sustainable urbanization (Liu & Biljecki, 2022). Likewise, in ecosystem restoration and conservation, GeoAI helps in identifying areas of intervention, continuous monitoring of restoration activities, and evaluating ecosystem service provision (Mai, G et al., 2025).

From a policy perspective, the integration of GeoAI in SLM practices is also in line with the United Nations' Sustainable Development Goals (SDGs), which include those related to Zero Hunger, Climate Action, Life on Land, and Sustainable Cities (Somvanshi et al., 2025). The spatial and data-driven information that is made available through the use of GeoAI is also deemed to have strong potential in supporting evidence-based policymaking and ensuring transparency in land governance, which could be done in ways that are tailored to the unique characteristics of each place and its problems (Zheng et al., 2025; Eshetie et al., 2024). As such, the use of GeoAI in land management is not only an innovation in technology but also in supporting policymaking in ways that are more adaptive and inclusive in their approach.

However, the use of GeoAI in SLM practices is also faced with some challenges that need to be considered in order to ensure its effective and equitable use in land management practices. One of the biggest problems facing the application of GeoAI in SLM is related to the availability of quality and accessible data. In many developing countries of the world, lack of access to high-resolution data and lack of effective mechanisms for sharing such information are considered to be some of the biggest problems facing the application of GeoAI (Wang, S et al., 2024; Eshetie et al., 2024). Additionally, the "black box" problem of many advanced ML and DL approaches poses a problem of transparency and interpretability of results, especially when GeoAI outputs are used to inform policy decisions that have important implications for livelihoods and land rights (Marasinghe et al., 2024). Institutional and capacity-related factors also appear to be important. The successful integration of GeoAI in SLM practices is also subject to the availability of adequate computational power, financial investment, and supportive institutional frameworks (Zheng et al., 2025; Marasinghe et al., 2024).

In the absence of these factors, the application of GeoAI technologies might be underutilized or even become another tool for the perpetuation of inequality. In addition, there are several critical issues that affect the application of GeoAI-based approaches. For instance, there are ethical issues that relate to data privacy and surveillance, and the marginalization of indigenous knowledge systems (Marasinghe et al., 2024; Oluoch, 2024). These challenges underscore the need to see GeoAI not as a substitute but rather as a complementary decision support system (Marasinghe et al., 2024). The SLM approach requires the integration of scientific knowledge with the perceptions and values of the stakeholders and the socio-cultural values of the people (Löbmann et al., 2022). In this context, the GeoAI approach can be seen as an enabler that facilitates the making of decisions in a timely and objective manner. However, it is important that such an approach be supported by appropriate frameworks. The important principles that can be adopted in this context are the principles of transparency in the development of the models and the decision-making process, and the principle of inclusiveness in the decision-making process itself (Marasinghe et al., 2024; Oluoch, 2024).

Thus, the purpose of this chapter is to critically assess the relationship that exists between SLM and GeoAI. It attempts to conceptualize the role of GeoAI in enhancing SLM approaches through the effective monitoring and analysis of land systems. It also aims to explore the existing uses of GeoAI in various land management fields, with emphasis on methodological improvements in image classification using convolutional neural networks (CNNs) and planning using reinforcement learning. It also attempts to present various case studies on the application of GeoAI in real-life scenarios to address various land management issues such as agriculture, forestry, and urban development. Moreover, it critically assesses the constraints associated with GeoAI while identifying the key enablers for its effective implementation. Finally, it provides insights on future directions in terms of research, policy, and practice, with emphasis on the need for inter-disciplinary work, capacity building, and development of transparent GeoAI frameworks. In this way, by locating GeoAI in the overall goals of sustainable development and land management, the chapter suggests that the value of GeoAI ultimately rests in its capacity to assist human decision-making and promote more resilient and just outcomes in land management. In this way, the chapter locates GeoAI as a key ally in the global quest to manage land in a sustainable and responsible fashion in the face of unprecedented environmental and social change.

2. CONCEPTUAL FOUNDATIONS: LINKING GEOAI AND SLM

SLM is guided by a variety of well-established conceptual frameworks that highlight the intricate interactions between human activities and natural systems. The most important of these frameworks is perhaps the Driver-Pressure-State-Impact-Response (DPSIR) framework (Chowdhury et al., 2025; Obubu et al., 2022), which has been widely promoted by many international institutions such as the European Environment Agency. The DPSIR model describes land systems as socio-ecological systems. In this context, socio-economic pressures such as increases in human populations, market demands, and policy incentives drive pressures on land systems, which include land cover change, over-exploitation of water resources, and intensified land use (Obubu et al., 2022). These pressures then drive changes in land systems' states, which in turn influence key components such as soil properties, vegetation cover, hydrological processes, and biodiversity. The change of state then leads to impacts on ecosystem services and human well-being, such as food security, climate regulation, and livelihoods, which then leads to responses through management practices and interventions (Polasky et al., 2011; Liu et al., 2022). In addition to the above-mentioned DPSIR approach, land system science and ecosystem services approaches also reinforce the above-stated facts that land is not simply a biophysical substrate; rather, it is a product of natural processes and human decision-making processes at different scales of operation (Verburg et al., 2015; Aspinall & Staiano, 2017). Ecosystem services approach also link land conditions with the services people receive from land; therefore, it provides a normative foundation for decision-making for sustainability. All of these above-stated approaches for understanding land reinforce the fact that SLM practices require constant observation, analysis, and decision-making that are capable of dealing with non-linearities and uncertainties in land systems (Martinez-Harms et al., 2015).

Traditionally applied methods for SLM monitoring and assessment, such as periodic field surveys and static land use maps, as well as individual indicators, have been reported to be challenged in effectively capturing such complex land system dynamics. Such methods are often inadequate for early warning of land degradation and impacts, as well as for exploring alternative futures under changing climatic and socio-economic conditions (Schwilch et al., 2011). It is within this context that the need for advanced monitoring and modeling techniques and tools for SLM is becoming increasingly obvious. It is at this point that GeoAI comes into play as a bridge between land system theory and SLM practice (Zheng et al., 2025). GeoAI also contributes to SLM by applying the fundamental components of socio-ecological models in a data-informed way. By combining large amounts of geospatial data obtained from satellites, UAVs, in situ sensors, and socio-economic datasets with ML and DL techniques, GeoAI is able to identify patterns, trends, and anomalies in land systems (Mai, G et al., 2025; Zheng et al., 2025).

For example, classification and change detection techniques may be employed to track the spatial and temporal patterns of land use and land cover dynamics, which would directly assist in the 'state' part of the DPSIR framework (Usman et al., 2015). Moreover, techniques such as regression analysis, clustering algorithms, and prediction models may be utilized to establish potential relationships between the state of the land system and the underlying pressure/drivers. This would improve the understanding of the land system and enable more proactive land management practices (Estacio et al., 2023; Pourmohammadi et al., 2021). One of the important conceptual contributions of GeoAI is the move from descriptive and static representations of land to dynamic and predictive analytics. Instead of merely describing what has happened to land systems in the past, GeoAI approaches have the potential to describe what will happen in the future under various scenarios of land management, climate change, and other land-related policy interventions (Zheng et al., 2025). Such predictive capacity is of particular significance in the case of SLM, as timely intervention is often critical to the effectiveness of the practice. To exemplify, early warning of the onset of soil degradation, drought stress, and deforestation risks could allow stakeholders to take early intervention measures to prevent the damage from becoming irreversible, thereby increasing the efficiency and effectiveness of SLM practices (Satapathy et al., 2024).

GeoAI also enhances the 'response' part of SLM approaches by converting complex analytical results into useful decision-support tools like risk maps, suitability evaluations, predictions, and interactive tools that address specific information requirements for land managers and decision-makers (Zheng et al., 2025). By integrating these tools in governance processes like participatory governance, it is possible to achieve greater transparency and develop more flexible and responsive approaches to land management. In this regard, GeoAI may be considered as a bridge that connects scientific analysis and decision-making, where insights from the sciences are effectively linked to land management needs. The difference between GeoAI and conventional GIS and RS techniques, however, does not lie in the utilization of spatial data, but in how this information is processed, analyzed, and interpreted (Choi, 2023; Pierdicca & Paolanti, 2022; Zheng et al., 2025; Wang, S et al., 2024). Conventional GIS/RS techniques, for instance, often involve a set of predetermined rules and sometimes require manual feature extraction. Although successful, this technique may not be efficient, especially considering the extensive spatial scope and/or rapid temporal dynamics of spatial information (Mehmood et al., 2022). In the GeoAI approach, AI/ML algorithms have the ability to automatically learn features of interest. (Choi, 2023; Pierdicca & Paolanti, 2022).

This ability is especially critical for SLM since land processes are controlled by interacting biophysical and socio-economic factors (Zheng et al., 2025; Herath & Nayanajith, 2025). Another important characteristic of GeoAI is the ability to integrate diverse datasets from different sensors and over different timescales. Satellite images from optical and radar sensors, LiDAR, climate records, soil information, and socio-economic information can be integrated within a single analytical framework (Pierdicca & Paolanti, 2022; Mai, G et al., 2025). When integrated with cloud-based geospatial platforms such as

Google Earth Engine, GeoAI can offer near real-time, large-scale data processing, which was beyond human capabilities in the past (Tamiminia et al., 2020; Yang et al., 2022; Mashala et al., 2023a). From the SLM perspective, such an approach can allow for faster detection of land degradation process, better modeling of soil-water-vegetation interactions, and improved capacities for scenario analysis under uncertainty (Zheng et al., 2025; Jahangeer et al., 2025). From a conceptual point of view, GeoAI can be said to be consistent with the principles of adaptive management, which is an integral component of SLM. Land systems are naturally dynamic, uncertain, and complex, making it imperative to have management approaches that can be adapted to new circumstances. Adaptive management is an approach to management that seeks to learn from experience, refine decisions, and improve them through new information. GeoAI can support this approach well, considering that models can be updated, improved, and refined as new information becomes available (Zheng et al., 2025; Song et al., 2025).

In summary, the conceptual connection between GeoAI and SLM can be established on the premise of the emphasis on systems thinking, integration, and flexibility (Löbmann et al., 2022; Song et al., 2025). The SLM framework offers the theoretical foundation for understanding land as a socio-ecological system, driven by the interplay of drivers, pressures, states, impacts, and responses (Kgaphola et al., 2023; Eckebil et al., 2025). GeoAI, on the other hand, operationalizes these theoretical constructs through the development of cutting-edge tools for monitoring, modeling, and decision-support, converting static analysis into dynamic, predictive, and actionable intelligence (Choi, 2023; Zheng et al., 2025).

The workflow of GeoAI-enabled SLM, as depicted in Figure 1, may be viewed as an ongoing process in which various data inputs are fed into the GeoAI system with outputs that have a direct bearing on land management decisions and policy interventions. This provides the conceptual foundation for the applications and case studies discussed in the following sections of the chapter.

Figure 1. GeoAI-enabled conceptual SLM workflow

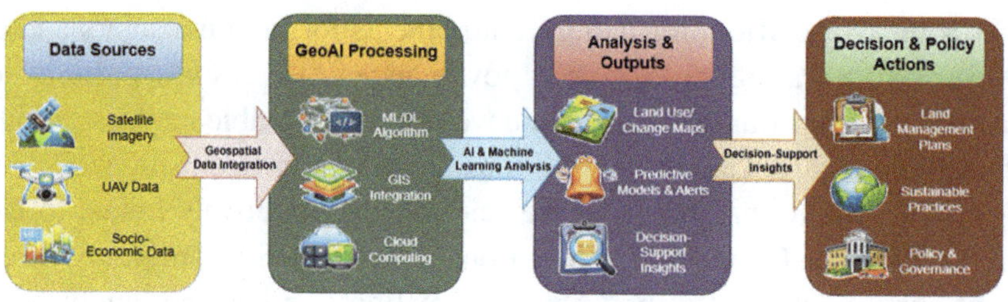

3. STATE OF THE ART: APPLICATIONS OF GEOAI FOR SLM

The fast development of GeoAI has greatly enhanced the analytical and operational potential of the toolset for SLM. GeoAI has enabled the development of innovative applications in SLM, which have enabled the transition from descriptive analysis in land management to predictive, prescriptive, and adaptive approaches (Song et al., 2023; Zheng et al., 2025). The following section presents the major application domains of GeoAI in relation to SLM.

3.1 Land Use and Land Cover (LULC) and Land-Cover Change

One of the most mature and influential application areas of GeoAI in LULC mapping, as well as LULC changes, is to support the process of SLM by providing accurate and timely information on LULC changes, which are an integrated measure of human activities and environmental processes on the land surface (Deressu et al., 2025; Sertel & Huang, 2025). Recently, the use of DL methods, such as CNNs, has shown better performance in comparison to conventional approaches such as pixel- or object-based image classification methods in terms of their accuracy, as well as generalization potential (Deressu et al., 2025; Vali et al., 2020). Moreover, by automatically learning hierarchical spatial features from high-resolution satellite and aerial images, CNNs can greatly enhance the accuracy of classification in complex environments characterized by heterogeneous land use patterns (Zhou et al., 2019). GeoAI-based LULC analysis also allows for near-real-time monitoring of activities such as deforestation, urbanization, agricultural intensification, and land abandonment over different scales, from local to global (Zheng et al., 2025; Deressu et al., 2025). Moreover, multi-temporal DL models can also identify changes that may not be easily noticeable but could indicate future land degradation risks, thus allowing for more proactive land management (Cheng et al., 2023; Masolele et al., 2021). Most significantly, recent studies on GeoAI-based land use observation, analysis, and forecasting indicate a paradigm shift from traditional land cover-based studies to land use-based studies (Zheng et al., 2025). This transition enables practitioners not only to identify areas where changes have occurred but also areas where changes are most likely to occur under various socio-economic or policy situations, hence improving proactive land management and spatial planning (e.g., scenario-based and regression-driven spatio-temporal modeling frameworks applied to urban dynamics and land-use prediction) (Al-Shaar et al., 2022a, 2022b). For SLM, this capability is useful for identifying areas of unsustainable land conversion, measuring fragmentation, and supporting compliance assessment for land use regulations. When integrated with governance systems, GeoAI-derived LULC products have the potential to improve transparency and accountability in land management decisions (Zheng et al., 2025; Ribeiro et al., 2024).

3.2 Soil Health, Nutrient Status, and Land Degradation

Soil health is the foundation of the biophysical component of SLM practices and affects land productivity, water regulation, carbon sequestration, and land resilience (Kibblewhite et al., 2008; Kabato et al., 2025). Nevertheless, the spatial and temporal variability of soils is large, and traditional surveying techniques are expensive, labor-intensive, and often sparse (De Caires et al., 2025; Maynard et al., 2023). GeoAI provides new opportunities for digital soil mapping by combining RS information, digital elevation models (DEMs), climate variables, and terrain attributes with ML techniques (Hosseini et al., 2023). Recent research has shown the potential of GeoAI models in predicting various soil physical and chemical properties, including texture, organic carbon, and nutrient status, on a large scale with improved accuracy. Using ML models, relationships can be established between surface characteristics and subsurface soil properties, which can be used to produce continuous soil maps for specific land management decisions (Hosseini et al., 2023; Sakhaee et al., 2022). Similarly, GeoAI-based erosion and land degradation assessment tools use vegetation indices, rainfall erosivity, slopes, and land use data to identify areas of land degradation and prioritize intervention areas (Priyadharshini et al., 2025; Uyar, 2025). ML models can be used in the framework of SLM to develop specific strategies for soil conservation, fertilizer use, and land restoration. These models can also be used in climate-smart land management to improve the estimation of soil carbon stocks in changing climatic conditions (Qian et al., 2025; Ma et al., 2022).

3.3 Water-Land Interaction, Watershed, and Ecosystem Monitoring

The SLM is closely and intrinsically linked with the availability of water and the functioning of ecosystem services. Land use and management practices significantly impact important hydrological processes such as infiltration, runoff, evaporation, and recharge. Similarly, water-related extremes such as floods and droughts impact land productivity (Shiferaw et al., 2025; Kalogiannidis et al., 2023). GeoAI has been increasingly used to better address the complex interactions between land and water by integrating satellite imagery and hydrological models with GIS-based data and ML. Some examples include the monitoring of land cover dynamics in wetlands, riparian zones, reservoirs, and floodplains, the mapping of drought or flood susceptibility, and the evaluation of ecosystem health metrics in watersheds (Cheng et al., 2025). Thus, the incorporation of AI-based analysis in time series data improves the recognition of seasonal and interannual patterns, thereby improving the understanding of land-water interactions (Fu et al., 2024). For SLM practices, integrated monitoring of land and water helps in watershed planning and restoration activities, along with evaluating trade-offs between land management practices and water-related ecosystem services (Zerga, 2025; Das & Edgaonkar, 2025). This also improves adaptive management by providing timely information on hydrological extremes and associated responses in ecosystems.

3.4 Precision Agriculture and Sustainable Land Use Practices

In production landscapes, GeoAI plays an important role in bridging the gap between SLM and precision agriculture (PA). GeoAI helps in making accurate evaluations of land use efficiency and sustainability, thus improving responsible land use practices (Wang, J et al., 2024). Some of the applications include the demarcation of boundaries, identification of crops, yield estimation, identification of pests and diseases, and optimization of water and fertilizer application. The application of PA, which is based on the use of GeoAI, is also likely to reduce the level of waste, minimize the leaching of fertilizers, and make the process more efficient, thus making it more compatible with the objective of SLM. Although the application of GeoAI in PA has been discussed, the philosophy of site-specific management, optimization of land use, and minimization of environmental externalities can be applied in other land use practices such as forestry, rangeland, and mixed land use (Bayar et al., 2025; Moomen et al., 2024; Ijaware, 2024).

3.5 Decision Support Systems and Scenario Modelling

Apart from mapping and monitoring, arguably the most significant contribution that GeoAI makes to SLM is in the development of decision support systems or scenario modeling approaches. Decision support systems developed using GeoAI integrate spatial data with predictive models and visualization tools, which allow stakeholders to assess different land use options. The systems have the ability to produce 'what-if' scenarios that evaluate the effects of policy interventions, climate change, and land management options on land productivity and socio-economic outcomes (Song et al., 2023; Zheng et al., 2025; Herath & Nayanajith, 2025; Wang et al., 2022) through integrative approaches combining variance analysis, frequency-based methods, and geographically weighted regression for spatially explicit scenario construction (Al-Shaar et al., 2022a, 2022b). Cloud-based geospatial platforms and data cube structures facilitate the execution of reproducible GeoAI applications in various areas, including anomaly detection, risk assessment, and supervised validation. The results are communicated to end users through easy-to-understand interfaces, dashboards, alerts, and planning tools (Marshall et al., 2025). This move from analytical results to decision-support information is an important milestone in the operationalization of GeoAI for SLM. In this context, GeoAI is no longer simply an innovative technique for experimentation; it is now an integral part of SLM practices. In fact, it improves the ability to understand, anticipate, and manage land-system dynamics more holistically, efficiently, and adaptively across LULC change monitoring, soil and water assessment, production systems, and decision support (Mai, G et al., 2025; Zheng et al., 2025).

Figure 2. GeoAI application landscape for SLM. Geospatial data sources, such as satellite Earth observation, UAV, in situ sensor, climate, and socio-economic data, are integrated using various DL, ML, and data fusion technologies within a GeoAI framework. The integrated process produces analytical outputs in various SLM domains, such as land use change detection, soil health analysis, ecosystem service assessment, PA, and decision support systems. The analytical outputs, such as risk, forecasts, warnings, and indicators, enable the development of effective SLM through informed decision-making

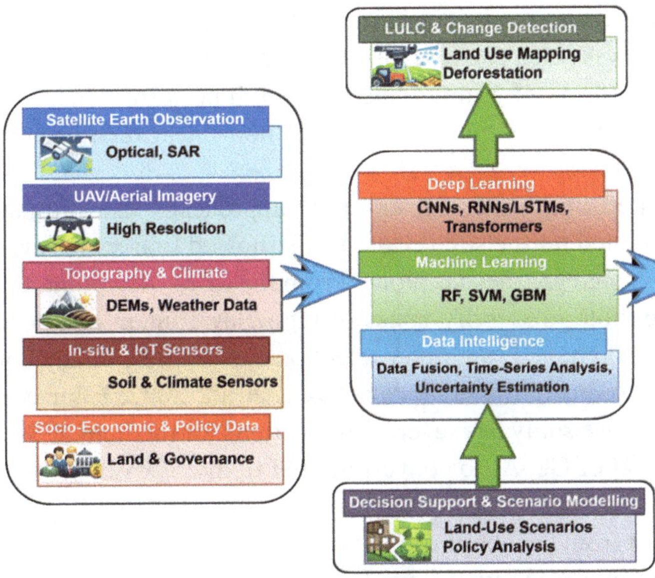

In Figure 2, GeoAI is portrayed as a core integrative intelligence layer that leverages diverse geospatial information with sophisticated ML and DL techniques to generate actionable intelligence for sustainable land and ecosystem management.

4. CASE STUDY ILLUSTRATIONS

In order to illustrate the relevance and applicability of the conceptual and methodological discussions presented so far, this section presents two case study illustrations of how GeoAI is being used within SLM-relevant contexts, albeit with different thematic emphases, one on soil health and the other on land use change under urbanization processes, but both providing insights on the potential, needs, and limitations of GeoAI, as well as its implications for SLM.

4.1 Soil Physical Property Prediction via GeoAI

The soil physical properties, such as sand content, silt content, and clay content, are considered essential soil characteristics that play a vital role in land productivity, water retention, erosion potential, and soil health (Mandal et al., 2023). Nevertheless, the process of measuring soil physical properties directly through soil sampling is considered expensive and limited (De Caires et al., 2025; Dai et al., 2019). Recently, in a study (Hosseini et al., 2023), it has been demonstrated that an effective GeoAI approach for predicting soil texture components in a specific region is possible through the integration of satellite imagery and environmental data. The study was conducted using a dataset consisting of 317 georeferenced soil samples and various predictor variables from different sources, including optical satellite imagery, vegetation indices, DEM-derived terrain attributes, and meteorological data. Moreover, in this study, a new hybrid modeling approach was also proposed, in which CNNs were used for automatically extracting spatial features from satellite imagery for improving the overall performance in soil property mapping. The model also utilized a random forest model to identify non-linear relationships between predictors and soil properties.

The introduced model was compared with other geostatistical approaches like ordinary kriging (Suneetha et al., 2025). The results obtained for the soil texture maps showed fine-scale variability that would have gone undetected with traditional approaches alone. From the SLM viewpoint, these results are of particular interest since it is possible to identify zones of high clay content and poor drainage characteristics, or zones of high sand content and low water retention capabilities, or any combination of these characteristics that are prone to erosion (Mirzaeitalarposhti et al., 2022; Ellur et al., 2025). Apart from the performance aspect, the case study has several implications for the field of SLM. Firstly, the case study demonstrates the potential of GeoAI to convert limited field observations to exhaustive information products on soils, thereby minimizing the monitoring costs while maximizing the decision-support potential. Secondly, it is evident from the above case study that there is significant emphasis on the integration of ground truth data in the GeoAI workflow. If not done, there would be considerable limitations in the outputs produced. Lastly, it is evident from the above case study that there is significant potential in using GeoAI for creating detailed soil maps, which could be used in addressing other issues such as land degradation, climate change, etc.

4.2 Land Use Prediction for Urban Expansion and SLM

The second case study concerns LULC change prediction under conditions of rapid urbanization, taking Indonesia's new capital city (Abdillah et al., 2025; Syaban & Appiah-Opoku, 2024) as a specific case. However, it is worth noting that urbanization is considered independent of rural land management. Despite this, it plays an important role in influencing SLM through processes such as forest and agricultural land conversion, water demand, and habitat fragmentation. Therefore, it is imperative to predict LULC changes in order to ensure effective SLM (Gaur & Singh, 2023). For example, in the recent study (Abdillah et al., 2025), LULC changes were examined in an area designated for the establishment of Indonesia's future capital, Nusantara. In addition, the research concentrated on forest and shrubland conversion to built-up areas. In this case, an analysis of LULC data with updated datasets in a regional climate modeling framework was carried out. By considering pre-urbanization and post-urbanization scenarios, in addition to considering the effects of global warming, spatially explicit outputs were generated. These outputs offered valuable insights concerning land transformation, differences in air temperatures, and the effects of urban heat island.

For example, regions that have been classified at high risk of deforestation-driven urbanization may be targeted for conservation or regulatory efforts, whereas regions that have been classified as appropriate for development may be aligned with infrastructure development strategies in order to mitigate environmental impacts (Wen et al., 2024; Veach et al., 2017). While such research is necessarily positioned in an urbanization paradigm rather than an agricultural or rural setting, the underlying methodological approach—encompassing spatio-temporal prediction, scenario-based analysis, and risk mapping—has significant cross-over potential for application in SLM more broadly. In this regard, there is significant potential for GeoAI in supporting an anticipatory, rather than reactive, approach to land management. It also demonstrates how the GeoAI approach might be made more relevant by being integrated into a broader institutional framework of planning rather than being considered a standalone technical product.

From the two case studies presented above, there are a number of lessons that might be drawn for the effective application of GeoAI in SLM studies.

- Firstly, GeoAI facilitates higher resolution and dynamism in mapping and forecasting land system characteristics than many traditional approaches. In forecasting soil characteristics or future land use changes, AI-based models take into account the spatial and temporal dynamics that are critical for sustainability-oriented decisions (Song et al., 2023; Zheng et al., 2025; Hosseini et al., 2023).
- Second, the integration of multi-sensor and multi-source data, e.g., satellite images, terrain data, climate data, and field observations, is another important factor affecting model performance. GeoAI is data-intensive, and its value increases when data from multiple sources are integrated within a consistent analytical paradigm (Song et al., 2023; Mai, G et al., 2025; Hosseini et al., 2023).
- Third, GeoAI models have to be properly customized according to local environment and socio-economic contexts. While it is possible for models to be transferable across different geographic contexts, calibration and validation are crucial in order for the accuracy and relevance of results to be achieved (Li et al., 2024).
- Fourth, if GeoAI is to contribute to SLM in an important way, it must be linked to decision processes. Sophisticated maps and models may be useless if integration with decision-making is not properly executed (Shalwee et al., 2025).
- Finally, institutional capacity, data governance structures, and stakeholder engagement are critical success factors for turning the insights of GeoAI into action. Investments in technical capacity building and data sharing are as important as advances in algorithms for GeoAI (Marasinghe et al., 2024; Mkhitaryan et al., 2025).

Figure 3 presents a conceptual model of an end-to-end process through which various sources of geospatial data are subjected to GeoAI analytics, which are locally contextualized and validated, and ultimately linked to decision-making processes with the goal of improved SLM outcomes. Overall, these case studies demonstrate that the value of GeoAI in SLM is not only related to its capacity to perform analytics but also to its ability to inform and improve land management decisions in an appropriate context of governance and society.

Figure 3. Conceptual framework linking GeoAI analytics to SLM outcomes

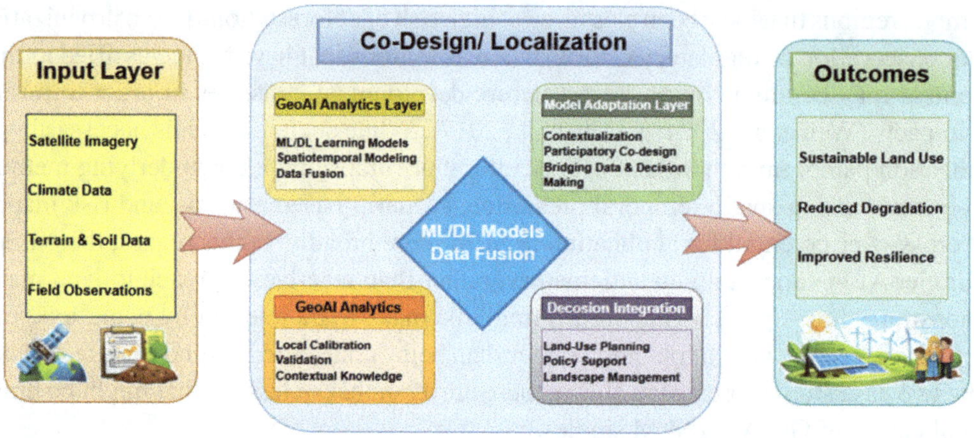

5. CHALLENGES AND ENABLERS

The successful uptake of GeoAI in SLM is subject to a complex array of challenges and enablers. These challenges are no longer just technical in scope but also extend to the institutional and governance aspects. However, the recent advancements in open data policy, cloud computing technologies, and collaborative governance models are aiding in the mitigation of these issues and making the application of GeoAI in land management more equitable and accessible (Zheng et al., 2025; Eshetie et al., 2024; Marasinghe et al., 2024).

5.1 Challenges

Availability and quality of data: The efficacy and accuracy of GeoAI models are largely data-dependent. The quality and availability of data are issues that affect the reliability and accuracy of the output. In most parts of the world, particularly in the Global South, the availability of cloud-free high-resolution satellite imagery is low. Similarly, the availability of long-term and standardized ground-truth data is also low (Marasinghe et al., 2024; Christofi et al., 2025). The reliability of GeoAI-based SLM decisions is at stake when these models are trained on incomplete or biased data sets and may propagate such uncertainties or inaccuracies in their predictions regarding certain land conditions, thereby calling into question their credibility for SLM decisions. This problem is most evident in the application of such models in soil property prediction and land degradation assessment. For such applications, it is important that the data be of high quality and consistent (Li et al., 2024; Koldasbayeva et al., 2024).

Model-related challenges: Despite the rapid developments in ML and DL techniques, there are a number of limitations that are inherent in the application of GeoAI models. Another limitation of GeoAI models is that they are not easily transferable. For instance, when such models are developed under specific environmental conditions, they might perform poorly in different conditions. This is likely to limit the scalability of such models in supporting SLM activities at the national and global levels (Mai, G et al., 2025; Wang, S et al., 2024; Ma et al., 2024). Another limitation of AI models in general is the problem of transparency and explainability. For instance, most AI models are referred to as 'black box' models. These models are often considered to be 'black boxes,' and it becomes difficult for land management agencies or policymakers to trust these models when they are applied for decision-making scenarios that have legal, economic, or social implications. In addition to these limitations, it is also important to note that these models require a large amount of computational resources for effective implementation; therefore, it becomes difficult to maintain these models outside of research environments (Pierdicca & Paolanti, 2022; Marasinghe et al., 2024; Ayadi et al., 2025). Uncertainty quantification is another important limitation of these models; it is rarely considered in the application of these models.

Institutional and governance challenges: The capacity of institutions may also limit the adoption of advanced AI and geospatial technologies in land management, with many public sector agencies being unable to provide the infrastructure and institutional environment to adopt GeoAI approaches in land management (Eshetie et al., 2024). In addition, fragmented data holdings, inadequate data sharing protocols, and lack of interoperability among agencies have also been identified as limiting the adoption of geospatial data and analysis results (Gelagay, 2017). The issue of the digital divide also arises, which may affect the adoption of geospatial technologies in low and middle-income countries due to inequalities in access to digital literacy and infrastructure (Huang et al., 2025; Manzoor et al., 2025). Without institutional support and alignment with SLM systems, GeoAI may continue to be used as an external approach rather than being fully integrated with SLM systems.

Ethical and societal concerns: The application of GeoAI in land systems is also a critical issue in the context of ethics. The biases in the datasets, such as the absence of representation of marginal landscapes or communities, might lead to biased analysis and decision-making. The application of high-resolution spatial datasets also creates privacy concerns, especially when land use or location is linked to individuals or communities (Mai, G et al., 2025; Marasinghe et al., 2024; Oluoch, 2024). Moreover, disparities in access to GeoAI technologies and the derived insights might exacerbate power disparities, allowing powerful groups to reap more benefits from data-driven land governance. These issues underscore the importance of developing effective ethical standards that emphasize transparency, accountability, and inclusiveness in GeoAI-driven SLM (Oluoch, 2024; Micheli et al., 2022).

Gaps in the operationalization of GeoAI: One of the challenges in the operationalization of GeoAI is the gap between the proof of concept and the actual operationalization of the technology in real-world scenarios. Although there is significant promise in the performance of GeoAI models in controlled and experimental scenarios, their actual operationalization in real-world scenarios is still in its infancy. While many GeoAI studies have shown promising results in terms of analysis, they still remain at the stage of publications or pilot tests. The deployment of these GeoAI models into useful tools, decision support systems, or policy instruments requires extra effort in software development, user training, and integration. Without this, GeoAI applications might remain one-time tests instead of becoming integral parts of land management practice (Pierdicca & Paolanti, 2022; Wang, S et al., 2024; Marasinghe et al., 2024).

Table 1 summarizes the multidimensional challenges of GeoAI adoption in SLM.

Table 1. Key challenges limiting the effective use of GeoAI in SLM

Challenge Category	Core Issues	Implications for SLM	Typical Examples	References
Data availability & quality	Sparse ground truth, inconsistent temporal coverage, sensor heterogeneity, missing metadata	Reduced model reliability; biased or uncertain land-condition estimates	Soil property mapping without field calibration; cloud-contaminated optical imagery	(Zhang et al., 2024; Mashala et al., 2023b)
Model-related limitations	Limited transferability, black-box behavior, high computational demands, lack of uncertainty quantification	Poor scalability; low trust among decision-makers; difficulty in risk-informed planning	Models trained in temperate regions failing in tropical agroecosystems	(Koldasbayeva et al., 2024)
Institutional & governance constraints	Limited technical capacity, fragmented data ownership, weak interoperability, digital divide	Underutilization of GeoAI outputs; weak integration into planning workflows	Agencies unable to operationalize AI-based land-degradation maps	(Eshetie et al., 2024; Mkhitaryan et al., 2025)
Ethical & societal concerns	Dataset bias, privacy risks, unequal access to technology	Risk of inequitable outcomes; erosion of social legitimacy	Underrepresentation of marginal lands or smallholder systems	(Mai, G et al., 2025; Marasinghe et al., 2024; Oluoch, 2024; Omotayo et al., 2025)
Deployment gaps	Research–practice disconnect, lack of user-friendly tools, limited long-term support	GeoAI remains experimental rather than operational	Pilot projects not maintained beyond funding cycles	(Wang, S et al., 2024; Marasinghe et al., 2024; Richter & Scheider, 2023)

5.2 Enablers

Open-source data and tools: One of the important factors that have facilitated the use of GeoAI in SLM is the emergence of open source tools and data. For instance, satellite imagery can be accessed for free through various significant initiatives, such as the Landsat program led by the NASA (Zhu et al., 2019), and the Sentinel initiative led by the European Space Agency (Sawyer et al., 2022). In addition, tools such as Google Earth Engine play a significant role in supporting the use of GeoAI in SLM, as this tool allows access to a wide range of geospatial data repositories along with significant analytical capabilities, as observed in various studies (Yang et al., 2022; Christofi et al., 2025; Chen et al., 2023).

Cloud computing and data-cube architectures: Recent breakthroughs in cloud computing and geospatial data cube frameworks have significantly improved the scalability of GeoAI. Such computing models allow for processing and analysis of vast multi-temporal data sets without the need for local high-performance computing capabilities (Kopp et al., 2019; Xu et al., 2022a, 2022b; Munteanu, 2024). In the context of SLM, such capabilities allow for continuous monitoring, analysis in near real-time, and large-scale analysis, which were previously difficult to achieve due to computational limitations (Xu et al., 2022a; Munteanu, 2024). The data cube paradigm is also beneficial for the reproducibility and standardization of analyses over space and time.

Capacity building and partnerships: Human and institutional capacity development is another enabler of the adoption of GeoAI. Capacity development through training programs aimed at improving the geospatial and AI literacy of land management professionals is essential in bridging the knowledge and innovation-application divide (Kausika & van Altena, 2025). However, it is also important to build solid interdisciplinary synergies between geospatial scientists, AI specialists, ecologists, and social scientists. Such collaboration is not only important for ensuring the technical soundness of GeoAI tools but also for making them more context-relevant and congruent with the larger SLM agenda (Marasinghe et al., 2024; Herath & Nayanajith, 2025).

Governance frameworks and standardization: Establishing sound governance framework also helps in the enhancement of trust, reliability, and usability of GeoAI results. The use of standard formats for data, as well as protocols for interoperability and metadata, helps in the sharing and integration of data. Further, appropriate guidance on ethical aspects, such as data privacy, algorithmic transparency, and accountability, helps in alleviating societal concerns. When incorporated within appropriate governance frameworks, GeoAI can be a legitimate part of the decision-making processes for land management, as discussed in various studies (Micheli et al., 2022; Epasto, 2025).

Stakeholder engagement and co-design: The active participation of various stakeholders, including government bodies, land management agencies, and local communities, can be recognized as an important facilitator for the successful implementation of GeoAI in SLM. Co-design methods, which emphasize the participation of end users from the early stages of the process, can guarantee that GeoAI technologies are aligned with real user needs, compatible with existing workflows, and able to produce outputs that are both intelligible and useful to users. Co-design methods can promote a sense of ownership of GeoAI-based SLM projects (Marasinghe et al., 2024; Mattijssen et al., 2024).

To summarize, while GeoAI has many enabling challenges to do with data, models, institutions, ethics, and operationalization, there is an increasingly large number of enabling factors being developed. It is therefore imperative that these enabling factors be maximized in an integrated and inclusive manner in order to tap into the tremendous transformative potential that GeoAI holds for SLM. Table 2 is a thematic synthesis of the key enablers that facilitate the effective adoption and scaling up of GeoAI for SLM.

Table 2. Key enablers facilitating the adoption and scaling up of GeoAI for SLM

Enabler Category	Description	Contribution to SLM	Illustrative Examples	References
Open-source data & tools	Free access to satellite data and open-source software ecosystems	Lowers entry barriers; enables reproducible and scalable analyses	NASA Landsat, European Space Agency Sentinel	(Christofi et al., 2025; Zhu et al., 2019; Sawyer et al., 2022; Radočaj et al., 2020)
Cloud platforms & data cubes	Cloud-based processing and standardized spatio-temporal data structures	Supports large-area, multi-temporal monitoring and near-real-time analysis	Google Earth Engine	(Tamiminia et al., 2020; Pham-Duc et al., 2023; Gorelick et al., 2017)
Capacity building & partnerships	Training, skill development, and interdisciplinary collaboration	Bridges research–practice gap; improves contextual relevance	Joint programs linking AI scientists and land managers	(Thapa et al., 2019)

Enabler Category	Description	Contribution to SLM	Illustrative Examples	References
Governance & standardization	Interoperability standards, metadata conventions, ethical guidelines	Enhances trust, transparency, and institutional uptake	Open data policies; standardized land-cover classifications	(Yang et al., 2017)
Stakeholder engagement & co-design	Participatory design involving end users and communities	Improves usability, legitimacy, and long-term sustainability	Co-designed decision-support tools for land planners	(Arulnathan et al., 2026)

Figure 4. Challenge–Enabler–Impact framework for the adoption of GeoAI in SLM

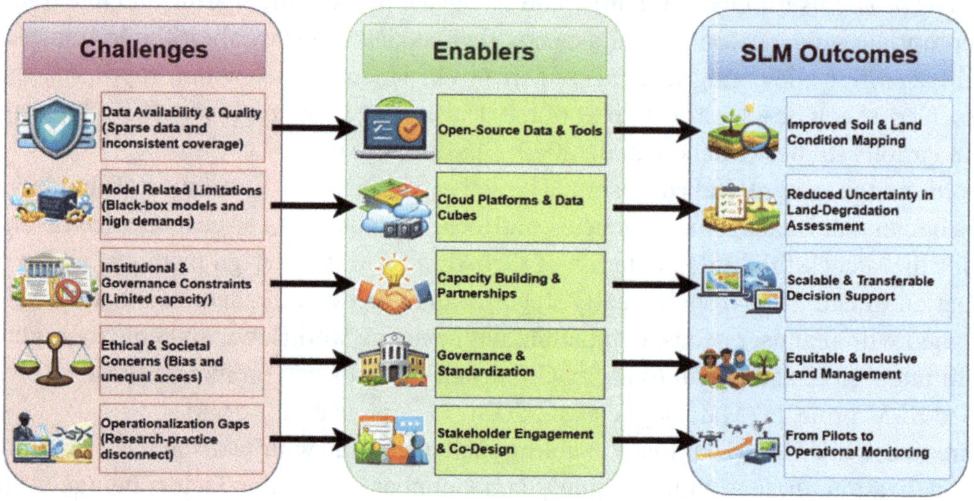

The conceptual framework as shown in Figure 4 demonstrates the role of enabling technological, institutional, and participatory mechanisms in addressing the major challenges to the adoption of GeoAI in SLM.

6. FUTURE DIRECTIONS AND RECOMMENDATIONS

The accelerated development of GeoAI creates a promising platform to reassess the way in which SLM is defined, delivered, and assessed. In the future, the challenge is not only to improve the science of the analysis, but also to integrate GeoAI in the institutional, policy, and social context to prioritize sustainability and long-term resilience (Zheng et al., 2025; Marasinghe et al., 2024; Kausika & van Altena, 2025). This section provides an overview of the major future directions and recommendations, which include both research and practice- and policy-related recommendations.

6.1 Research Priorities

Multimodal data fusion: One of the promising areas of future research is the integration of varied data streams to effectively address the complexity of land systems. Therefore, future GeoAI studies should focus on integrating satellite imagery with UAV/drone data, in situ sensors (soil moisture, nutrient, and microclimate sensors), socio-economic spatial data, and Internet of Things (IoT) data streams (Dritsas & Trigka, 2025; Ogungbuyi et al., 2024). Such integrated approaches are likely to enhance the robustness of models, minimize uncertainties associated with land systems, and provide a more holistic view of land system changes (Mansoor et al., 2025). In the context of SLM, such integration is vital to effectively relate land condition indicators with management practices and market and livelihoods.

Explainable and trustworthy GeoAI: As GeoAI outputs become more integrated into land use planning and policy, the need for transparency and interpretability will increase. Future research should emphasize explainable GeoAI models, where users are able to understand why a prediction or classification was made by the model. Hybrid models, where ML is constrained by physical, ecological, or process-based constraints, may be one avenue toward making GeoAI models more credible and useful for land use policy and management. By incorporating domain knowledge into AI models, researchers may be able to make them more useful and increase their acceptance by land use practitioners and policymakers (Roussel & Böhm, 2023; Papadakis et al., 2022).

Dynamic land-system modelling and digital twins: The development of dynamic land system models, including digital twins of landscapes, is considered one of the most promising frontiers for SLM research and practice. Digital twins are essentially virtual models of actual land systems, with data being constantly updated in near-real time. Such models have great potential for continuous monitoring, testing, and adaptive management of land systems. By simulating the outcomes of various land use options, climate change, or policy options, digital twins could be used by stakeholders to better plan and prepare for the future. Recent studies have successfully tested the feasibility of spatio-temporal models enhanced by deep neural networks for conducting scenario analysis for distant futures, using Markov Chain frameworks and DL models to project urban morphology and climate vulnerability for 2050 and 2100, respectively (Al-Shaar et al., 2025). The future research in this area should be focused on scaling up the GeoAI applications, conducting better analyses of uncertainties, and improving the integration with participatory decision support systems (Wang, Y et al., 2024).

Transferability and generalisation: The limitation of the current GeoAI applications is that they are not able to generalize across various regions, seasons, sensor modalities, and socio-ecological contexts. The solution to this problem requires conducting research in the area of domain adaptation and transfer learning. Generalization of GeoAI applications is also a major challenge that has to be addressed to ensure the expansion of GeoAI benefits to under-monitored areas and to support large-scale SLM practices that operate across heterogeneous landscapes (Sharma & Kaur, 2025; Du et al., 2024; Mai, J et al., 2025).

Integration of social and economic dimensions: Therefore, it is important that research in the field of GeoAI in the future does not only concentrate on biophysical factors but also includes social, economic, and governance aspects of land systems for a more holistic approach to assessing sustainability. This socio-ecological integration of GeoAI makes it more consistent with the fundamental principles of SLM that include social equity and economic viability in addition to environmental soundness (Eshetie et al., 2024; Rosier et al., 2022).

6.2 Practice and Policy Recommendations

National and regional GeoAI platforms for land management: Strategic investments in national or regional GeoAI platforms that integrate geospatial data repositories, cloud computing infrastructures, and AI-driven analytics that are specific to land management practices have the potential to greatly improve SLM practices. Such platforms have the potential to deliver standardized monitoring tools for land use, soil, water, and ecosystem services. Such platforms also have the potential to improve coordination at different levels if they are integrated with national land policies and planning. The development and operationalization of such platforms can be driven by governments, regional bodies, or development agencies (Tamiminia et al., 2020; Gorelick et al., 2017).

Capacity enhancement and human capital development: Innovation in technology is not sufficient without parallel investments in human capacity. Training programs for land use planners, GIS and remote sensing experts, and policymakers are also significant in utilizing GeoAI results effectively. Such training programs should not only include training on using GeoAI tools but also include training on assessing the underlying assumptions, constraints, and ethics of using GeoAI. Enhancing partnerships among academic institutions, government agencies, and international organizations could also be instrumental in supporting SLM efforts in utilizing GeoAI (Thapa et al., 2019; Kumar, 2022).

Data governance and ethical frameworks: A strong data governance structure is also necessary for building trust in GeoAI-based SLM. It is recommended that appropriate policies be developed for data sharing, access, and interoperability, while also addressing privacy concerns and respecting local and indigenous data sovereignty. The development of appropriate ethical standards for addressing bias, transparency, and accountability in GeoAI development and deployment processes is also important for building trust in GeoAI-based SLM (Marasinghe et al., 2024; Micheli et al., 2022).

Focus on the Global South and under-resourced regions: Some of the most critical land degradation issues occur in developing countries, where there is a scarcity of technological and financial expertise. There is a need for targeted investments in ensuring that there is equity of access to GeoAI tools, data, and training in these countries. Global cooperation, the use of open data, and south-south cooperation can help address the equity gap and allow GeoAI to contribute significantly to SLM in these areas (Haregeweyn et al., 2023; Urdánigo-Zambrano et al., 2026; Giuliani et al., 2020).

Outcome-oriented evaluation and impact metrics: Lastly, there is an urgent need to look beyond tool-based evaluations and create outcome-based measures for assessing the effectiveness of GeoAI in land management. The new initiatives should aim at establishing a link between the application and implementation of GeoAI and specific results in terms of improved land condition, reduced land degradation, and increased food security and livelihoods. By establishing a link between analytical results and land management outcomes, it will be easy to demonstrate the value of investing in GeoAI and consequently increase the overall effectiveness in land management strategies (Khattri et al., 2025; Li et al., 2025).

The multi-level conceptual framework in Figure 5 is intended to illustrate the flow of the benefits of GeoAI research and innovation to the development of platforms and ultimately to policy and societal outcomes. The cross-cutting enablers of data governance and ethics, human capacity development, and Global South inclusion play a crucial role in ensuring the responsible development of GeoAI for the management of agricultural and land resources.

Figure 5. Bridging the research innovation in GeoAI, operational platforms, and policy outcomes to support the achievement of sustainable, equitable, and resilient land governance

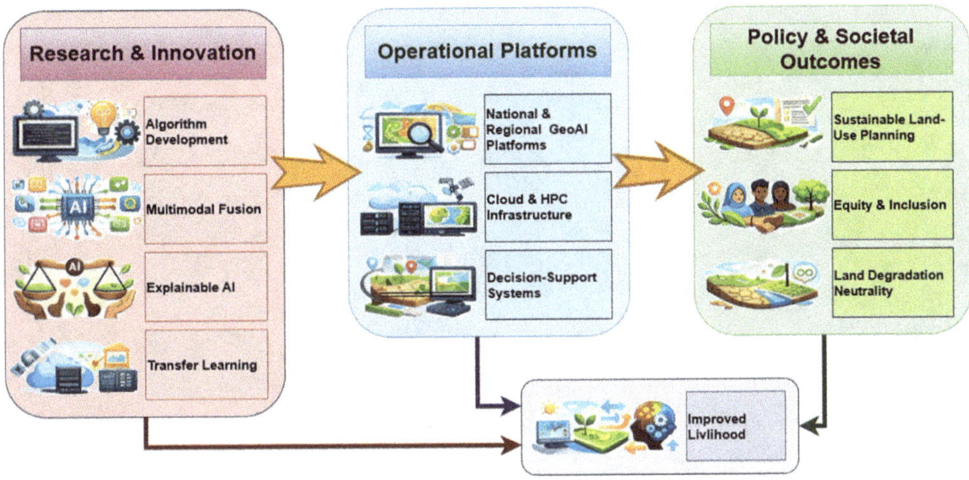

7. CONCLUSION

GeoAI marks a paradigm shift in the observation, analysis, and governance of land systems, presenting transformative opportunities for SLM. The capacity of GeoAI to carry out high-frequency, high-resolution, and predictive analysis on land use, soil, water, and ecosystem processes improves the ability of land managers and policymakers to look beyond the current paradigm of reacting to land system changes, trends, and challenges. The capacity of GeoAI to incorporate various geospatial data sources using advanced machine learning algorithms provides opportunities for the development of a more nuanced understanding of spatial, temporal, and risk dynamics in various land systems.

At the same time, this chapter has underscored the fact that GeoAI is not a panacea for the complex challenges facing land resources on its own. SLM is, by its very nature, a socio-ecological and governance challenge, and its success will always be dependent on the enabling environment within which it is being implemented. This means that without data, human resources, good governance, and effective engagement, even the most advanced GeoAI has little chance of being used to its maximum potential.

One of the key takeaways is that the GeoAI should be thought of as a tool for decision support, rather than replacing human judgment and land stewardship. GeoAI has significant potential to support better evidence-based planning, monitoring, accountability, and scenario exploration of land use futures. The results of GeoAI, however, should be contextualized to local settings and aligned with societal values to make them relevant and socially acceptable. Thus, the integration of GeoAI and participatory land governance ensures that innovation in land technology is complementary rather than undermining the principles of inclusiveness, legitimacy, and trust in land management. As land management continues to be challenged by issues of agricultural expansion, climate change, rapid urbanization, and ecosystem degradation, the timely application of GeoAI in land management in combination with SLM represents a unique opportunity to bridge the knowledge-action gap in land management. The advancement of land management in an uncertain world hinge on the judicious integration of GeoAI with good land governance, ethics, and sustainability principles.

REFERENCES

Abdillah, M. R., Adysti, R. T., Wijaya, W., Junnaedhi, I. D. G. A., Trilaksono, N. J., Suwarman, R., Marzuki, M., Hidayat, R., Miftahuddin, Y. I., Kombara, P. Y., & Mukhsinin, H. A. (2025). Urbanization and global warming impacts of Indonesia's future capital of Nusantara on air temperature and urban heat island. *Scientific Reports*, *15*(1), 41543. DOI: 10.1038/s41598-025-25500-8 PMID: 41285888

Al-Shaar, W., Bonin, O., & de Gouvello, B. (2022a). Scenario-based predictions of urban dynamics in Île-de-France region: A new combinatory methodologic approach of variance analysis and frequency ratio. *Sustainability (Basel)*, *14*(11), 6806. DOI: 10.3390/su14116806

Al-Shaar, W., Bonin, O., de Gouvello, B., Chatellier, P., & Hendel, M. (2022b). Geographically weighted regression-based predictions of water–soil–energy nexus solutions in Île-de-France. *Urban Science (Basel, Switzerland)*, *6*(4), 81. DOI: 10.3390/urbansci6040081

Al-Shaar, W., Lehmann, X., Saad, N., Elmazi, G., Al-Shaar, M., & Tohme, C. (2025). Urban morphology and climate vulnerability assessment in Kuwait: A spatio-temporal predictive analysis utilizing deep neural network-enhanced Markov chain models for 2050 and 2100. *PLoS One*, *20*(8), e0318604. DOI: 10.1371/journal.pone.0318604 PMID: 40824951

Arneth, A., Olsson, L., Cowie, A., Erb, K.-H., Hurlbert, M., Kurz, W. A., Mirzabaev, A., & Rounsevell, M. D. A. (2021). Restoring degraded lands. *Annual Review of Environment and Resources*, *46*(1), 569–599. DOI: 10.1146/annurev-environ-012320-054809

Arulnathan, V., Li, E., & Pelletier, N. (2026). Implementation of a Participatory Design Approach to the Development of a Sustainability Decision Support Tool for Canadian Egg Farmers. *Sustainability (Basel)*, *18*(1), 69. DOI: 10.3390/su18010069

Aspinall, R., & Staiano, M. (2017). A conceptual model for land system dynamics as a coupled human–environment system. *Land (Basel)*, *6*(4), 81. DOI: 10.3390/land6040081

Ayadi, R., Forouheshfar, Y., & Moghadas, O. (2025). Enhancing System Resilience to Climate Change through Artificial Intelligence: A Systematic Literature Review. *Frontiers in Climate*, *7*, 1585331. DOI: 10.3389/fclim.2025.1585331

Bayar, J., Ali, N., Cao, Z., Ren, Y., & Dong, Y. (2025). Artificial intelligence of things (AIoT) for precision agriculture: Applications in smart irrigation, nutrient and disease management. *Smart Agricultural Technology*, *12*, 101629. DOI: 10.1016/j.atech.2025.101629

Cabernard, L., Pfister, S., & Hellweg, S. (2024). Biodiversity impacts of recent land-use change driven by increases in agri-food imports. *Nature Sustainability*, *7*(11), 1512–1524. DOI: 10.1038/s41893-024-01433-4

Chen, H., Yang, L., & Wu, Q. (2023). Enhancing Land Cover Mapping and Monitoring: An Interactive and Explainable Machine Learning Approach Using Google Earth Engine. *Remote Sensing (Basel)*, *15*(18), 4585. DOI: 10.3390/rs15184585

Cheng, K., Zhang, C., Fan, Y., Diao, H., & Xia, S. (2025). AI-based approaches for wetland mapping and classification: A review of current practices and future perspectives. *Artificial Intelligence in Geosciences*, *6*(2), 100165. DOI: 10.1016/j.aiig.2025.100165

Cheng, X., Sun, Y., Zhang, W., Wang, Y., Cao, X., & Wang, Y. (2023). Application of deep learning in multitemporal remote sensing image classification. *Remote Sensing (Basel)*, *15*(15), 3859. DOI: 10.3390/rs15153859

Choi, Y. (2023). GeoAI: Integration of artificial intelligence, machine learning, and deep learning with GIS. *Applied Sciences (Basel, Switzerland)*, *13*(6), 3895. DOI: 10.3390/app13063895

Chowdhury, S., Paul, C., Hamidov, A., Bayer, L., Arndt, M., McPherson, J., & Helming, K. (2025). Developing a generic DPSIR framework for land use and soil management: A systems approach to maximise soil health. *European Journal of Soil Science*, *76*(6), e70230. DOI: 10.1111/ejss.70230

Christofi, D., Mettas, C., Evagorou, E., Stylianou, N., Eliades, M., Theocharidis, C., Chatzipavlis, A., Hasiotis, T., & Hadjimitsis, D. (2025). A Review of Open Remote Sensing Data with GIS, AI, and UAV Support for Shoreline Detection and Coastal Erosion Monitoring. *Applied Sciences (Basel, Switzerland)*, *15*(9), 4771. DOI: 10.3390/app15094771

Colpaert, A. (2022). Satellite and UAV platforms, remote sensing for geographic information systems. *Sensors (Basel)*, *22*(12), 4564. DOI: 10.3390/s22124564 PMID: 35746346

Dai, Y., Shangguan, W., Wei, N., Xin, Q., Yuan, H., Zhang, S., Liu, S., Lu, X., Wang, D., & Yan, F. (2019). A review of the global soil property maps for Earth system models. *Soil (Göttingen)*, *5*(2), 137–158. DOI: 10.5194/soil-5-137-2019

Das, N., & Edgaonkar, A. (2025). Economic valuation of ecosystem services at the watershed scale for strategic land-use planning. *Environmental Conservation*, *53*(1), 30–39. DOI: 10.1017/S0376892925100271

De Caires, S. A., St. Martin, C., Atwell, M. A., Kaya, F., Wuddivira, G. A., & Wuddivira, M. N. (2025). Advancing soil mapping and management using geostatistics and integrated machine learning and remote sensing techniques: A synoptic review. *Discover Soil*, *2*(1), 53. DOI: 10.1007/s44378-025-00082-z

Deressu, T. F., Bojer, A. K., Debelee, T. G., Negera, W. G., Nadarajah, S., & Gebissa, K. W. (2025). Enhancing land use and land cover classification with deep learning-based satellite imagery segmentation. *International Journal of Applied Earth Observation and Geoinformation*, *144*, 104839. DOI: 10.1016/j.jag.2025.104839

Dong, J., Metternicht, G., Hostert, P., Fensholt, R., & Roy Chowdhury, R. (2019). Remote sensing and geospatial technologies in support of a normative land system science: Status and prospects. *Current Opinion in Environmental Sustainability*, *38*, 44–52. DOI: 10.1016/j.cosust.2019.05.003

Dritsas, E., & Trigka, M. (2025). Remote sensing and geospatial analysis in the big data era: A survey. *Remote Sensing (Basel)*, *17*(3), 550. DOI: 10.3390/rs17030550

Du, A., Doan, A.-D., Law, Y. W., & Chin, T.-J. (2024). Domain adaptation for satellite-borne multispectral cloud detection. *Remote Sensing (Basel)*, *16*(18), 3469. DOI: 10.3390/rs16183469

Eckebil, P. P. T., Mintah, F., Bürgi, M., Akinyemi, F. O., Sonwa, D. J., & Speranza, C. I. (2025). Tropical ecosystem multifunctionality assessment and insights for sustainable land management: A systematic literature review using the Drivers-Pressures-State-Impacts-Responses framework. *Frontiers in Forests and Global Change*, *8*, 1623266. DOI: 10.3389/ffgc.2025.1623266

Ellur, R., Ananthakumar, M. A., & Desai, K. (2025). Prediction and mapping of soil texture at high spatial resolution in a canal irrigated region using machine learning. Journal of Geography. *Environment and Earth Science International*, *29*(6), 114–125. DOI: 10.9734/jgeesi/2025/v29i6911

Epasto, S. (2025). Spatializing Trust: A GeoAI-Based Model for Mapping Digital Trust Ecosystems in Mediterranean Smart Regions. *ISPRS International Journal of Geo-Information*, *14*(12), 491. DOI: 10.3390/ijgi14120491

Eshetie, G. G., Alemie, B. K., & Wubie, A. M. (2024). Geospatial technologies in support of responsible land tenure governance: A systematic review. *Geomatica*, *76*(2), 100014. DOI: 10.1016/j.geomat.2024.100014

Estacio, I., Sianipar, C. P. M., Onitsuka, K., Basu, M., & Hoshino, S. (2023). A statistical model of land use/cover change integrating logistic and linear models: An application to agricultural abandonment. *International Journal of Applied Earth Observation and Geoinformation*, *120*, 103339. DOI: 10.1016/j.jag.2023.103339

FAO (Food and Agriculture Organization of the United Nations). (2023). Land degradation assessment and its impact on agricultural productivity. https://www.fao.org/land-water/land/land-assessment/en/

FAO (Food and Agriculture Organization of the United Nations). (2025). The state of the world's land and water resources for food and agriculture 2025: Managing systems at risk. Rome, Italy: https://www.fao.org/publications/fao-flagship-publications/the-state-of-the-worlds-land-and-water-resources-for-food-and-agriculture/en

Foley, J. A., DeFries, R., Asner, G. P., Barford, C., Bonan, G., Carpenter, S. R., Chapin, F. S., Coe, M. T., Daily, G. C., Gibbs, H. K., Helkowski, J. H., Holloway, T., Howard, E. A., Kucharik, C. J., Monfreda, C., Patz, J. A., Prentice, I. C., Ramankutty, N., & Snyder, P. K. (2005). Global consequences of land use. *Science*, *309*(5734), 570–574. DOI: 10.1126/science.1111772 PMID: 16040698

. Fu, Y., Zhu, Z., Liu, L., Zhan, W., He, T., Shen, H., Zhao, J., Liu, Y., Zhang, H., Liu, Z., Xue, Y., & Ao, Z. (2024). Remote sensing time series analysis: A review of data and applications. Journal of Remote Sensing, 4, 0285. https://doi.org/DOI: 10.34133/remotesensing.0285

Gaur, S., & Singh, R. (2023). A comprehensive review on land use/land cover (LULC) change modeling for urban development: Current status and future prospects. *Sustainability (Basel)*, *15*(2), 903. DOI: 10.3390/su15020903

Gelagay, H. S. (2017). Geospatial Data Sharing Barriers across Organisations and the Possible Solution for Ethiopia. *International Journal of Spatial Data Infrastructures Research*, *12*, 62–84. DOI: 10.2902/1725-0463.2017.12.art4

Giuliani, G., Mazzetti, P., Santoro, M., Nativi, S., Van Bemmelen, J., Colangeli, G., & Lehmann, A. (2020). Knowledge generation using satellite earth observations to support sustainable development goals (SDG): A use case on land degradation. *International Journal of Applied Earth Observation and Geoinformation, 88*, 102068. DOI: 10.1016/j.jag.2020.102068

Gorelick, N., Hancher, M., Dixon, M., Ilyushchenko, S., Thau, D., & Moore, R. (2017). Google Earth Engine: Planetary-Scale Geospatial Analysis for Everyone. *Remote Sensing of Environment, 202*, 18–27. DOI: 10.1016/j.rse.2017.06.031

Haregeweyn, N., Tsunekawa, A., Tsubo, M., Almaw Fenta, A., Ebabu, K., Vanmaercke, M., Borrelli, P., Panagos, P., Berihun, M. L., Langendoen, E. J., Nigussie, Z., Setargie, T. A., Maurice, B. N., Minichil, T., Elias, A., Sun, J., & Poesen, J. (2023). Progress and challenges in sustainable land management initiatives: A global review. *The Science of the Total Environment, 858*(Part 3), 160027. DOI: 10.1016/j.scitotenv.2022.160027 PMID: 36356757

Herath, H. M. B. S., & Nayanajith, J. (2025). Land use changes and sustainability: A review of geospatial, AI and socioeconomic approaches. Journal of Geography. *Environment and Earth Science International, 29*(11), 144–158. DOI: 10.9734/jgeesi/2025/v29i11973

Hosseini, F. S., Seo, M. B., Razavi-Termeh, S. V., Sadeghi-Niaraki, A., Jamshidi, M., & Choi, S.-M. (2023). Geospatial artificial intelligence (GeoAI) and satellite imagery fusion for soil physical property predicting. *Sustainability (Basel), 15*(19), 14125. DOI: 10.3390/su151914125

Huang, X., Wang, S., & Song, Y. (2025). Digital Divide and GeoAI. In X. Huang, S. Wang, J. Wilson, & P. Kedron (Eds.), GeoAI and Human Geography (pp. 361-371). Springer Geography, Springer, Cham. DOI: 10.1007/978-3-031-87421-5_25

Ijaware, V. A. (2024). GIS and artificial intelligence application in smart forest ecosystem sustainability evaluation of Olokemeji Forest Reserve, Ogun State, Nigeria. *American Journal of Geospatial Technology, 3*(1), 9–16. DOI: 10.54536/ajgt.v3i1.2621

Jahangeer, J., Joshi, P., Kapoor, A., & Tang, Z. (2025). A review of AI-driven Google Earth Engine applications in surface water monitoring, assessment, and management. *Discover Geoscience, 3*(1), 140. DOI: 10.1007/s44288-025-00255-x PMID: 41018550

Kabato, W., Hailegnaw, N., Mutum, L., & Molnar, Z. (2025). Managing soil health for climate resilience and crop productivity in a changing environment. *The Science of the Total Environment, 1000*, 180460. DOI: 10.1016/j.scitotenv.2025.180460 PMID: 40946612

Kalogiannidis, S., Kalfas, D., Giannarakis, G., & Paschalidou, M. (2023). Integration of water resources management strategies in land use planning towards environmental conservation. *Sustainability (Basel), 15*(21), 15242. DOI: 10.3390/su152115242

Kausika, B. B., & van Altena, V. (2025). GeoAI in Topographic Mapping: Navigating the Future of Opportunities and Risks. *ISPRS International Journal of Geo-Information, 14*(8), 313. DOI: 10.3390/ijgi14080313

Kgaphola, M. J., Ramoelo, A., Odindi, J., Mwenge Kahinda, J.-M., Seetal, A., & Musvoto, C. (2023). Social–ecological system understanding of land degradation in response to land use and cover changes in the Greater Sekhukhune District Municipality. *Sustainability (Basel), 15*(4), 3850. DOI: 10.3390/su15043850

Khattri, P., Sayers, R., Singh, K. K., Slapikas, R., Tamang, C. B., Tamang, D., Sagara, B., & BenYishay, A. (2025). Geospatial impact evaluation of a low-cost agricultural intervention for enhancing environmental resilience. *International Journal of Applied Earth Observation and Geoinformation, 141*, 104657. DOI: 10.1016/j.jag.2025.104657 PMID: 40599674

Kibblewhite, M. G., Ritz, K., & Swift, M. J. (2008). Soil health in agricultural systems. *Philosophical Transactions of the Royal Society of London. Series B, Biological Sciences, 363*(1492), 685–701. DOI: 10.1098/rstb.2007.2178 PMID: 17785275

Koldasbayeva, D., Tregubova, P., Gasanov, M., Zaytsev, A., Petrovskaia, A., & Burnaev, E. (2024). Challenges in Data-Driven Geospatial Modeling for Environmental Research and Practice. *Nature Communications, 15*(1), 10700. DOI: 10.1038/s41467-024-55240-8 PMID: 39702456

Kopp, S., Becker, P., Doshi, A., Wright, D. J., Zhang, K., & Xu, H. (2019). Achieving the Full Vision of Earth Observation Data Cubes. *Data, 4*(3), 94. DOI: 10.3390/data4030094

Kumar, A. S. (2022). Capacity development in geospatial technology – An overview. ISPRS Annals of the Photogrammetry, Remote Sensing and Spatial Information Sciences, 5, 9–16. DOI: 10.5194/isprs-annals-V-5-2022-9-2022

Lambin, E. F., & Meyfroidt, P. (2011). Global land use change, economic globalization, and the looming land scarcity. *Proceedings of the National Academy of Sciences of the United States of America, 108*(9), 3465–3472. DOI: 10.1073/pnas.1100480108 PMID: 21321211

Li, M., Zhu, Z., Deng, J., Zhang, J., & Li, Y. (2025). Geospatial explainable AI uncovers eco-environmental effects and its driving mechanisms—Evidence from the Poyang Lake Region, China. *Land (Basel), 14*(7), 1361. DOI: 10.3390/land14071361

Li, W., Hsu, C. Y., Wang, S., & Kedron, P. (2024). GeoAI Reproducibility and Replicability: A Computational and Spatial Perspective. *Annals of the American Association of Geographers, 114*(9), 2085–2103. DOI: 10.1080/24694452.2024.2373787

Liu, M., Wei, H., Dong, X., Wang, X.-C., Zhao, B., & Zhang, Y. (2022). Integrating land use, ecosystem service, and human well-being: A systematic review. *Sustainability (Basel), 14*(11), 6926. DOI: 10.3390/su14116926

Liu, P., & Biljecki, F. (2022). A review of spatially-explicit GeoAI applications in urban geography. *International Journal of Applied Earth Observation and Geoinformation, 112*, 102936. DOI: 10.1016/j.jag.2022.102936

Löbmann, M. T., Maring, L., Prokop, G., Brils, J., Bender, J., Bispo, A., & Helming, K. (2022). Systems knowledge for sustainable soil and land management. *The Science of the Total Environment, 822*, 153389. DOI: 10.1016/j.scitotenv.2022.153389 PMID: 35104520

Ma, J., Rabin, S. S., Anthoni, P., Bayer, A. D., Nyawira, S. S., Olin, S., Xia, L., & Arneth, A. (2022). Assessing the impacts of agricultural managements on soil carbon stocks, nitrogen loss, and crop production – a modelling study in eastern Africa. *Biogeosciences*, *19*(8), 2145–2169. DOI: 10.5194/bg-19-2145-2022

Ma, Y., Chen, S., Ermon, S., & Lobell, D. B. (2024). Transfer Learning in Environmental Remote Sensing. *Remote Sensing of Environment*, *301*, 113924. DOI: 10.1016/j.rse.2023.113924

Mai, G., Xie, Y., Jia, X., Lao, N., Rao, J., Zhu, Q., Liu, Z., Chiang, Y.-Y., & Jiao, J. (2025). Towards the next generation of geospatial artificial intelligence. *International Journal of Applied Earth Observation and Geoinformation*, *136*, 104368. DOI: 10.1016/j.jag.2025.104368

Mandal, D., Patra, S., Sharma, N. K., Alam, N. M., Jana, C., & Lal, R. (2023). Impacts of soil erosion on soil quality and agricultural sustainability in the north-western Himalayan region of India. *Sustainability (Basel)*, *15*(6), 5430. DOI: 10.3390/su15065430

Mansoor, S., Iqbal, S., Popescu, S. M., Kim, S. L., Chung, Y. S., & Baek, J. H. (2025). Integration of Smart Sensors and IoT in Precision Agriculture: Trends, Challenges and Future Prospectives. *Frontiers in Plant Science*, *16*, 1587869. DOI: 10.3389/fpls.2025.1587869 PMID: 40438737

Manzoor, F., Wei, L., Siraj, M., Lu, X., & Qiyang, G. (2025). Digital Agriculture Technology Adoption in Low- and Middle-Income Countries—A Review of Contemporary Literature. *Frontiers in Sustainable Food Systems*, *9*, 1621851. DOI: 10.3389/fsufs.2025.1621851

Marasinghe, R., Yigitcanlar, T., Mayere, S., Washington, T., & Limb, M. (2024). Towards responsible urban geospatial AI: Insights from the white and grey literatures. *Journal of Geovisualization and Spatial Analysis*, *8*(2), 24. DOI: 10.1007/s41651-024-00184-2

Marshall, E., Cherian, D., Henderson, S., Scheick, J., & Forster, R. (2025). Cloud-native geospatial data cube workflows with open-source tools. *The Journal of Open Source Education*, *8*(89), 267. DOI: 10.21105/jose.00267

Martinez-Harms, M. J., Bryan, B. A., Balvanera, P., Law, E. A., Rhodes, J. R., Possingham, H. P., & Wilson, K. A. (2015). Making decisions for managing ecosystem services. *Biological Conservation*, *184*, 229–238. DOI: 10.1016/j.biocon.2015.01.024

Mashala, M. J., Dube, T., Ayisi, K. K., & Ramudzuli, M. R. (2023a). Using the Google Earth Engine cloud-computing platform to assess the long-term spatial temporal dynamics of land use and land cover within the Letaba watershed, South Africa. *Geocarto International*, *38*(1), 2252781. DOI: 10.1080/10106049.2023.2252781

Mashala, M. J., Dube, T., Mudereri, B. T., Ayisi, K. K., & Ramudzuli, M. R. (2023b). A Systematic Review on Advancements in Remote Sensing for Assessing and Monitoring Land Use and Land Cover Changes Impacts on Surface Water Resources in Semi-Arid Tropical Environments. *Remote Sensing (Basel)*, *15*(16), 3926. DOI: 10.3390/rs15163926

Masolele, R. N., De Sy, V., Herold, M., Marcos, D., Verbesselt, J., Gieseke, F., Mullissa, A. G., & Martius, C. (2021). Spatial and temporal deep learning methods for deriving land-use following deforestation: A pan-tropical case study using Landsat time series. *Remote Sensing of Environment, 264*, 112600. DOI: 10.1016/j.rse.2021.112600

Mattijssen, T. J. M., Hennen, W., Buijs, A. E., De Dooij, P., Van Lammeren, R., & Walet, L. (2024). Urban greening co-creation: Participatory spatial modelling to bridge data-driven and citizen-centred approaches. *Urban Forestry & Urban Greening, 94*, 128257. DOI: 10.1016/j.ufug.2024.128257

Maynard, J. J., Yeboah, E., Owusu, S., Buenemann, M., Neff, J. C., & Herrick, J. E. (2023). Accuracy of regional-to-global soil maps for on-farm decision-making: Are soil maps "good enough"? *Soil (Göttingen), 9*(1), 277–300. DOI: 10.5194/soil-9-277-2023

Mehmood, M., Shahzad, A., Zafar, B., Shabbir, A., & Ali, N. (2022). Remote sensing image classification: A comprehensive review and applications. *Mathematical Problems in Engineering, 5880959*, 1–24. Advance online publication. DOI: 10.1155/2022/5880959

Micheli, M., Gevaert, C. M., Carman, M., Craglia, M., Daemen, E., Ibrahim, R. E., Kotsev, A., Mohamed-Ghouse, Z., Schade, S., Schneider, I., Shanley, L. A., Tartaro, A., & Vespe, M. (2022). AI Ethics and Data Governance in the Geospatial Domain of Digital Earth. *Big Data & Society, 9*(2). Advance online publication. DOI: 10.1177/20539517221138767

Mirzaeitalarposhti, R., Shafizadeh-Moghadam, H., Taghizadeh-Mehrjardi, R., & Demyan, M. S. (2022). Digital soil texture mapping and spatial transferability of machine learning models using Sentinel-1, Sentinel-2, and terrain-derived covariates. *Remote Sensing (Basel), 14*(23), 5909. DOI: 10.3390/rs14235909

Mkhitaryan, K., Sanamyan, A., Mnatsakanyan, M., Kirakosyan, E., & Ratner, S. (2025). Integrating AI and Geospatial Technologies for Sustainable Smart City Development: A Case Study of Yerevan. *Urban Science (Basel, Switzerland), 9*(10), 389. DOI: 10.3390/urbansci9100389

Moomen, A.-W., Yevugah, L. L., Boakye, L., Osei, J. D., & Muthoni, F. (2024). Review of applications of remote sensing towards sustainable agriculture in the northern savannah regions of Ghana. *Agriculture, 14*(4), 546. DOI: 10.3390/agriculture14040546

Munteanu, A. (2024). Data Cubes and Cloud-Native Environments for Earth Observation: An Overview. Scalable Computing. *Practice and Experience, 25*(6), 5745–5759. DOI: 10.12694/scpe.v25i6.4999

Obubu, J. P., Odong, R., Alamerew, T., Fetahi, T., & Mengistou, S. (2022). Application of DPSIR model to identify the drivers and impacts of land use and land cover changes and climate change on land, water, and livelihoods in the L. Kyoga basin: Implications for sustainable management. *Environmental Systems Research, 11*(1), 11. DOI: 10.1186/s40068-022-00254-8 PMID: 35607482

Ogungbuyi, M. G., Mohammed, C., Fischer, A. M., Turner, D., Whitehead, J., & Harrison, M. T. (2024). Integration of Drone and Satellite Imagery Improves Agricultural Management Agility. *Remote Sensing (Basel), 16*(24), 4688. DOI: 10.3390/rs16244688

Oluoch, I. (2024). Crossing boundaries: The ethics of AI and geographic information technologies. *ISPRS International Journal of Geo-Information, 13*(3), 87. DOI: 10.3390/ijgi13030087

Omotayo, A. O., Adediran, S. A., Omotoso, A. B., Olagunju, K. O., & Omotayo, O. P. (2025). Artificial Intelligence in Agriculture: Ethics, Impact Possibilities, and Pathways for Policy. Computers and Electronics in Agriculture, 239(Part A), 110927. DOI: 10.1016/j.compag.2025.110927

Papadakis, E., Adams, B., Gao, S., Martins, B., Baryannis, G., & Ristea, A. (2022). Explainable artificial intelligence in the spatial domain (X-GeoAI). *Transactions in GIS*, *26*(6), 2413–2414. DOI: 10.1111/tgis.12996

Pham-Duc, B., Nguyen, H., Phan, H., & Tran-Anh, Q. (2023). Trends and Applications of Google Earth Engine in Remote Sensing and Earth Science Research: A Bibliometric Analysis Using Scopus Database. *Earth Science Informatics*, *16*(3), 2355–2371. DOI: 10.1007/s12145-023-01035-2

Pierdicca, R., & Paolanti, M. (2022). GeoAI: A review of artificial intelligence approaches for the interpretation of complex geomatics data. *Geoscientific Instrumentation, Methods and Data Systems*, *11*(1), 195–218. DOI: 10.5194/gi-11-195-2022

Polasky, S., Nelson, E., Pennington, D., & Johnson, K. A. (2011). The impact of land-use change on ecosystem services, biodiversity and returns to landowners: A case study in the State of Minnesota. *Environmental and Resource Economics*, *48*(2), 219–242. DOI: 10.1007/s10640-010-9407-0

Pourmohammadi, P., Strager, M. P., Dougherty, M. J., & Adjeroh, D. A. (2021). Analysis of land development drivers using geographically weighted ridge regression. *Remote Sensing (Basel)*, *13*(7), 1307. DOI: 10.3390/rs13071307

Priyadharshini, V. M., Aldehim, G., Negm, N., & Subathradevi, S. (2025). Integrating geospatial techniques and machine learning for assessing soil erosion and associated geomorphic risks. *Journal of South American Earth Sciences*, *156*, 105463. DOI: 10.1016/j.jsames.2025.105463

Qian, J., Yang, J., Sun, W., Zhao, L., Shi, L., Shi, H., Liao, L., Dang, C., & Dou, Q. (2025). Soil organic carbon estimation and transfer framework in agricultural areas based on spatiotemporal constraint strategy combined with active and passive remote sensing. *Remote Sensing (Basel)*, *17*(2), 333. DOI: 10.3390/rs17020333

Radočaj, D., Obhođaš, J., Jurišić, M., & Gašparović, M. (2020). Global Open Data Remote Sensing Satellite Missions for Land Monitoring and Conservation: A Review. *Land (Basel)*, *9*(11), 402. DOI: 10.3390/land9110402

Ribeiro, V., De Souza Mendes, F., Singh, C., & Heilmayr, R. (2024). Leveraging remote sensing for transparency and accountability in Amazonian commodity supply chains. *One Earth*, *7*(11), 1935–1940. DOI: 10.1016/j.oneear.2024.10.011

Richter, K.-F., & Scheider, S. (2023). Current Topics and Challenges in GeoAI. *Künstliche Intelligenz*, *37*(1), 11–16. DOI: 10.1007/s13218-022-00796-0

Rosier, J. F., Taubenböck, H., Verburg, P. H., & van Vliet, J. (2022). Fusing Earth observation and socioeconomic data to increase the transferability of large-scale urban land use classification. *Remote Sensing of Environment*, *278*, 113076. DOI: 10.1016/j.rse.2022.113076

Roussel, C., Böhm, K., & Reiterer, A. (2025). Q-GGXAI: A framework for quality-aware explainable AI in geospatial analysis. *Journal of Photogrammetry, Remote Sensing and Geoinformation Science*. Advance online publication. DOI: 10.1007/s41064-025-00360-z

Sakhaee, A., Gebauer, A., Ließ, M., & Don, A. (2022). Spatial prediction of organic carbon in German agricultural topsoil using machine learning algorithms. *Soil (Göttingen)*, *8*(2), 587–604. DOI: 10.5194/soil-8-587-2022

Salvati, L., Venezian Scarascia, M. E., Zitti, M., Ferrara, A., Urbano, V., Sciortino, M., & Giupponi, C. (2009). The integrated assessment of land degradation. *Italian Journal of Agronomy*, *4*(3), 77–90. DOI: 10.4081/ija.2009.3.77

Satapathy, T., Dietrich, J., & Ramadas, M. (2024). Agricultural drought monitoring and early warning at the regional scale using a remote sensing-based combined index. *Environmental Monitoring and Assessment*, *196*(11), 1132. DOI: 10.1007/s10661-024-13265-y PMID: 39476296

Sawyer, G., Mamais, E., & Papadakis, D. (2022). The Six Dimensions of Value Associated to the Use of Copernicus Sentinel Data: Key Findings from the Sentinel Benefits Study. *Frontiers in Environmental Science*, *10*, 804862. DOI: 10.3389/fenvs.2022.804862

Schwilch, G., Bestelmeyer, B., Bunning, S., Critchley, W., Herrick, J., Kellner, K., Liniger, H. P., Nachtergaele, F., Ritsema, C. J., Schuster, B., Tabo, R., van Lynden, G., & Winslow, M. (2011). Experiences in monitoring and assessment of sustainable land management. *Land Degradation & Development*, *22*(2), 214–225. DOI: 10.1002/ldr.1040

See, L., Lesiv, M., & Schepaschenko, D. (2024). Integrating remote sensing and geospatial big data for land cover and land use mapping and monitoring. *Land (Basel)*, *13*(6), 769. DOI: 10.3390/land13060769

Sertel, E., & Huang, X. (2025). Advancing human geography with GeoAI: Transforming land use and land cover analysis. In X. Huang, S. Wang, J. Wilson, & P. Kedron (Eds.), GeoAI and human geography (pp. 373–386). Springer Geography, Springer, Cham. DOI: 10.1007/978-3-031-87421-5_26

Shalwee, D., Dhupper, R., Kumari, M., Gupta, A. K., & Kumar, D. (2025). Implementing Multi-Criteria Decision-Making Approaches to Evaluate and Map Drought Vulnerability with Geospatial Artificial Intelligence (Geo-AI) Based Investigations. *Discover Applied Sciences*, *7*(10), 1185. DOI: 10.1007/s42452-025-07189-6

Sharma, H., & Kaur, S. (2025). Deep learning for sustainable development across climate, energy, agriculture and urban systems. *Discover Sustainability*, *6*(1), 1408. DOI: 10.1007/s43621-025-02186-6

Shiferaw, N., Habte, L., & Waleed, M. (2025). Land use dynamics and their impact on hydrology and water quality of a river catchment: A comprehensive analysis and future scenario. *Environmental Science and Pollution Research International*, *32*(7), 4124–4136. DOI: 10.1007/s11356-025-35946-y PMID: 39862375

Sishodia, R. P., Ray, R. L., & Singh, S. K. (2020). Applications of remote sensing in precision agriculture: A review. *Remote Sensing (Basel)*, *12*(19), 3136. DOI: 10.3390/rs12193136

Somvanshi, S., Kumar, D., & Kumari, M. (2025). Trends of recent data, AI/ML approaches for geospatial AI in Earth observation towards sustainable development goals. *ITU Journal : ICT Discoveries*, *6*(1), 11–28. DOI: 10.52953/CDIE7940

Song, Y., Biljecki, F., Camps-Valls, G., & Atkinson, P. M. (2025). GeoAI: Beyond mapping earth and cities through explainability, adaptability, and sustainability. *iScience*, *29*(1), 114407. DOI: 10.1016/j.isci.2025.114407 PMID: 41536980

Song, Y., Kalacska, M., Gašparović, M., Yao, J., & Najibi, N. (2023). Advances in geocomputation and geospatial artificial intelligence (GeoAI) for mapping. *International Journal of Applied Earth Observation and Geoinformation*, *120*, 103300. DOI: 10.1016/j.jag.2023.103300

Suneetha, C., Kumar, L. S., Sreenivas, K., & Mitran, T. (2025). Deep learning-driven soil texture classifier using Landsat 8 images. *Remote Sensing Applications: Society and Environment*, *38*, 101568. DOI: 10.1016/j.rsase.2025.101568

Syaban, A. S. N., & Appiah-Opoku, S. (2024). Unveiling the complexities of land use transition in Indonesia's new capital city IKN Nusantara: A multidimensional conflict analysis. *Land (Basel)*, *13*(5), 606. DOI: 10.3390/land13050606

Tamiminia, H., Salehi, B., Mahdianpari, M., Quackenbush, L., Adeli, S., & Brisco, B. (2020). Google Earth Engine for geo-big data applications: A meta-analysis and systematic review. *ISPRS Journal of Photogrammetry and Remote Sensing*, *164*, 152–170. DOI: 10.1016/j.isprsjprs.2020.04.001

Thapa, R. B., Matin, M. A., & Bajracharya, B. (2019). Capacity Building Approach and Application: Utilization of Earth Observation Data and Geospatial Information Technology in the Hindu Kush Himalaya. *Frontiers in Environmental Science*, *7*, 165. DOI: 10.3389/fenvs.2019.00165

Tian, X., Li, J., Zhang, F., Zhang, H., & Jiang, M. (2024). Forest aboveground biomass estimation using multisource remote sensing data and deep learning algorithms: A case study over Hangzhou area in China. *Remote Sensing (Basel)*, *16*(6), 1074. DOI: 10.3390/rs16061074

Urdánigo-Zambrano, J., Torres, B., De-Pablos-Heredero, C., Herrera-Feijoo, R., & García, A. (2026). Open data reuse in agricultural, livestock, and environmental systems: A global scoping review with a case analysis of Ecuador. *Land (Basel)*, *15*(1), 13. DOI: 10.3390/land15010013

Usman, M., Liedl, R., Shahid, M. A., & Abbas, A. (2015). Land use/land cover classification and its change detection using multi-temporal MODIS NDVI data. *Journal of Geographical Sciences*, *25*(12), 1479–1506. DOI: 10.1007/s11442-015-1247-y

Uyar, N. (2025). Index-driven soil loss mapping across environmental scenarios: Insights from a remote sensing approach. *Sustainability (Basel)*, *17*(17), 7913. DOI: 10.3390/su17177913

Vali, A., Comai, S., & Matteucci, M. (2020). Deep learning for land use and land cover classification based on hyperspectral and multispectral Earth observation data: A review. *Remote Sensing (Basel)*, *12*(15), 2495. DOI: 10.3390/rs12152495

Veach, V., Moilanen, A., & Di Minin, E. (2017). Threats from Urban Expansion, Agricultural Transformation and Forest Loss on Global Conservation Priority Areas. *PLoS One*, *12*(11), e0188397. DOI: 10.1371/journal.pone.0188397 PMID: 29182662

Verburg, P. H., Crossman, N., Ellis, E. C., Heinimann, A., Hostert, P., Mertz, O., Nagendra, H., Sikor, T., Erb, K.-H., Golubiewski, N., Grau, R., Grove, M., Konaté, S., Meyfroidt, P., Parker, D. C., Roy Chowdhury, R., Shibata, H., Thomson, A., & Zhen, L. (2015). Land system science and sustainable development of the earth system: A global land project perspective. *Anthropocene*, *12*, 29–41. DOI: 10.1016/j.ancene.2015.09.004

von Keyserlingk, J., Thieken, A. H., & Paton, E. N. (2023). Approaches to assess land degradation risk: A synthesis. *Ecology and Society*, *28*(1), 53. DOI: 10.5751/ES-13951-280153

Wang, J., Bretz, M., Dewan, M. A. A., & Delavar, M. A. (2022). Machine learning in modelling land-use and land cover-change (LULCC): Current status, challenges and prospects. *The Science of the Total Environment*, *822*, 153559. DOI: 10.1016/j.scitotenv.2022.153559 PMID: 35114222

Wang, J., Wang, Y., Li, G., & Qi, Z. (2024). Integration of remote sensing and machine learning for precision agriculture: A comprehensive perspective on applications. *Agronomy (Basel)*, *14*(9), 1975. DOI: 10.3390/agronomy14091975

Wang, S., Huang, X., Liu, P., Zhang, M., Biljecki, F., Hu, T., Fu, X., Liu, L., Liu, X., Wang, R., Huang, Y., Yan, J., Jiang, J., Chukwu, M., Naghedi, S. R., Hemmati, M., Shao, Y., Jia, N., Xiao, Z., & Bao, S. (2024). Mapping the landscape and roadmap of geospatial artificial intelligence (GeoAI) in quantitative human geography: An extensive systematic review. *International Journal of Applied Earth Observation and Geoinformation*, *128*, 103734. DOI: 10.1016/j.jag.2024.103734

Wang, Y., Wang, L., Han, W., & Yan, J. (2024). Digital twin of Earth: A novel information framework for managing a sustainable Earth. *The Innovation Geoscience*, *2*(4), 100092. DOI: 10.59717/j.xinn-geo.2024.100092

Wen, J., Xi, J., Pan, Y., Wang, S., Fan, Z., & Fu, W. (2024). Spatial Conservation Prioritization for Land in Megacity Facing Climate Change and Biodiversity Loss. *Sustainability (Basel)*, *16*(4), 1392. DOI: 10.3390/su16041392

Xu, C., Du, X., Fan, X., Giuliani, G., Hu, Z., Wang, W., Liu, J., Wang, T., Yan, Z., Zhu, J., Jiang, T., & Guo, H. (2022a). Cloud-Based Storage and Computing for Remote Sensing Big Data: A Technical Review. *International Journal of Digital Earth*, *15*(1), 1417–1445. DOI: 10.1080/17538947.2022.2115567

Xu, C., Du, X., Jian, H., Dong, Y., Qin, W., Mu, H., Yan, Z., Zhu, J., & Fan, X. (2022b). Analyzing Large-Scale Data Cubes with User-Defined Algorithms: A Cloud-Native Approach. *International Journal of Applied Earth Observation and Geoinformation*, *109*, 102784. DOI: 10.1016/j.jag.2022.102784

Yang, H., Li, S., Chen, J., Zhang, X., & Xu, S. (2017). The Standardization and Harmonization of Land Cover Classification Systems Towards Harmonized Datasets: A Review. *ISPRS International Journal of Geo-Information*, *6*(5), 154. DOI: 10.3390/ijgi6050154

Yang, L., Driscol, J., Sarigai, S., Wu, Q., Chen, H., & Lippitt, C. D. (2022). Google Earth Engine and artificial intelligence (AI): A comprehensive review. *Remote Sensing (Basel), 14*(14), 3253. DOI: 10.3390/rs14143253

Zerga, B. (2025). Integrated watershed management: A review. *Discover Sustainability, 6*(1), 657. DOI: 10.1007/s43621-025-01352-0

Zhang, B., Liu, L., Zhang, Y., Wei, B., Gong, D., & Li, L. (2024). Spatial Consistency and Accuracy Analysis of Multi-Source Land Cover Products on the Southeastern Tibetan Plateau, China. *Remote Sensing (Basel), 16*(17), 3219. DOI: 10.3390/rs16173219

Zheng, W., Li, K., & Liu, X. (2025). GeoAI for land use observations, analysis, and forecasting. *Land (Basel), 14*(10), 2058. DOI: 10.3390/land14102058

Zhou, K., Ming, D., Lv, X., Fang, J., & Wang, M. (2019). CNN-based land cover classification combining stratified segmentation and fusion of point cloud and very high-spatial resolution remote sensing image data. *Remote Sensing (Basel), 11*(17), 2065. DOI: 10.3390/rs11172065

Zhu, Z., Wulder, M. A., Roy, D. P., Woodcock, C. E., Hansen, M. C., Radeloff, V. C., Healey, S. P., Schaaf, C., Hostert, P., Strobl, P., Pekel, J.-F., Lymburner, L., Pahlevan, N., & Scambos, T. A. (2019). Benefits of the Free and Open Landsat Data Policy. *Remote Sensing of Environment, 224*, 382–385. DOI: 10.1016/j.rse.2019.02.016

Ziadat, F. M., Zdruli, P., Christiansen, S., Caon, L., Abdel Monem, M., & Fetsi, T. (2022). An overview of land degradation and sustainable land management in the Near East and North Africa. *Sustainable Agriculture Research, 11*(1), 11–24. DOI: 10.5539/sar.v11n1p11

Chapter 13

GeoAI-Powered Soil Health Mapping for Enhancing Nutrient Management and Reducing Environmental Impact

Supaprawat Siripipatthanakul
http://orcid.org/0000-0001-6671-2682
Bangkokthonburi University, Thailand

Tamonwan Sitthipon
http://orcid.org/0009-0008-9939-5546
GlobalNxt University, Malaysia

Sutithep Siripipattanakul
http://orcid.org/0000-0002-5477-6723
Kasetsart University, Thailand

Pichakoon Auttawechasakoon
http://orcid.org/0000-0002-2247-6750
Sukhothai Thammathirat Open University, Thailand

Parichat Jaipong
http://orcid.org/0000-0002-9249-3169
GlobalNxt University, Malaysia

ABSTRACT

To provide high-resolution, useful representations of soil characteristics such as pH, moisture content, and nutrient content, geoAI-powered soil health mapping combines cutting-edge machine learning algorithms with geospatial data from satellites, drones, and Internet of Things sensors. This chapter employs a systematic review of papers published from 2022 to 2026 in Scopus, WOS, and Google Scholar. The finding shows that farmers can switch from uniform field treatments to variable-rate fertilization—applying precise nutrient quantities suited to particular soil conditions—by examining these digital

DOI: 10.4018/979-8-3373-8037-7.ch013

maps. By eliminating excessive nutrient runoff and minimizing chemical inputs, this focused approach reduces the environmental impact of agriculture while optimizing nutrient management, boosting crop yields, and promoting long-term soil sustainability.

1. INTRODUCTION

A GeoAI-powered soil health mapping framework uses a combination of AI, machine learning, and geospatial big data to monitor, analyze, and predict soil quality metrics, including texture (clay, sand, silt), moisture, and organic carbon. This is often done with satellite images or UAV sensors. This method changes the way we traditionally monitor agriculture by providing high-resolution, dynamic, and geographically precise information on soil parameters.

GeoAI-powered soil health mapping uses geospatial AI, satellite imagery, and IoT sensors to provide real-time, high-resolution information on soil conditions. This is better than standard static sample approaches. By applying deep learning methods, like CNNs for spatial patterns, these models can effectively predict soil moisture using data from satellites, soil, and weather.

This method outperforms traditional hydrological and statistical models because it is more accurate and reliable across a wider range of agro-climatic conditions. Because of this, GeoAI-based systems help with precision irrigation, flexible crop planning, and more resilient resource allocation. This all leads to more sustainable farming and food security in the face of climate change and water shortages. GeoAI-powered soil health mapping is a game-changing way to farm. It uses geospatial artificial intelligence (GeoAI), satellite imagery, and Internet of Things (IoT) sensors to provide you with real-time, high-resolution data that outperforms traditional, static sampling approaches. This technique, which frequently uses Convolutional Neural Networks (CNNs) to recognize spatial patterns and integrate data from multiple sources, enables accurate prediction of soil nutrient and moisture levels (Hosseini et al., 2023).

The revolutionary potential of precision agriculture and proximate sensing in transforming crop management approaches. It examines how these cutting-edge technologies can help mitigate the negative effects of agrochemical use while improving the accuracy of crop health monitoring. Precision agriculture can help solve both environmental and economic problems caused by the overuse of agrochemicals. It does this by improving production while protecting the environment. Several proximal sensing methods, such as spectral imaging, thermal imaging, and fluorescence sensors, to demonstrate their effectiveness in detecting and identifying signs of crop health problems, including stress, nutritional deficits, illnesses, and pests. It is important to connect monitoring sensors to real-time decision-making and to improve image-processing and data-management systems to fully realize their potential for sustainable crop management practices (Laveglia et al., 2024).

GeoAI-Powered Soil Health Mapping is connected to (1) Technological Integration, since these systems commonly use Convolutional Neural Networks (CNNs) and other machine-learning methods to look at spatial patterns and combine data from many sources. (2) GeoAI's Real-Time Data system continuously and at high resolution checks soil moisture, nutrient levels, and overall health. This is different from older systems that only tested the soil occasionally. (3) Precision Agriculture makes decisions "plant-by-plant," which lets users provide each plant the best amount of water, fertilizer, and pesticides. This cuts down on waste and costs. This technology helps the environment by providing early warnings of droughts and improving irrigation management, saving a lot of water. GeoAI models are revolutionizing how soil is mapped by using up-to-date, highly detailed data from IoT devices and remote sensing technology. This enables more active and flexible soil management than prior approaches, which only support static capabilities. Convolutional Neural Networks (CNNs) are important to this breakthrough because they can quickly capture complex spatial patterns in soil texture and other properties. Combining data from several sources, such as satellite photos, weather data, and ground-based IoT sensor readings, also makes soil property forecasts far more accurate and reliable.

1.1 Research Question

How can a GeoAI model that combines satellite images with topographic and climatic data make soil health mapping (like pH, nitrogen, and soil texture) more accurate than traditional spatial interpolation methods?

And how much can this help with site-specific nutrient management to reduce agricultural runoff?

1.2 Research Objective

To discuss a GeoAI-based soil health mapping system that uses satellite images, topographical data, and localized soil samples to make high-resolution prediction maps that help with nutrient management and have less of an effect on the environment.

2. LITERATURE REVIEW

The literature review covers (1) soil health monitoring and its limitations; (2) digital soil mapping and remote sensing approaches; (3) IoT sensor networks for soil nutrient assessment; (4) machine learning and deep learning methods applied to soil data; and (5) GeoAI frameworks for precision agriculture. Each subsection must engage critically with the cited literature, not merely summarize individual papers.

2.1 AI in Soil Health Monitoring and Its Limitations

Most soil quality measures have been limited to laboratory-based procedures that are slow, expensive, labor-intensive, collect limited data, and require extensive sample preparation. Real-time continuous soil monitoring (RTCSM) has the potential to change how we measure in the field by providing first-hand information on changes in a wide range of soil parameters and pollutants. This would allow for constant updates that are necessary for system control and decision-making. This critical review fills gaps in our knowledge of spatiotemporal uninterrupted soil monitoring and the execution of soil management by conducting a systematic literature search, a thorough analysis of the latest RTCSM technologies, an extensive discussion of their most important problems, and a look at our future plans. First, the difficulties in obtaining valid RTCSM data are made clear by examining common soil-monitoring methods, such as electrochemical and spectroscopic sensors. Next, we talk in depth about the main problems with the RTCSM sensor network, data transmission, data processing, and tailored data management. This paper also examines how RTCSM data can be used to improve a range of applications, including high-fidelity soil process models, control methods, digital soil mapping, soil deterioration, food security, and climate change mitigation. Finally, the importance of implementing RTCSM in farming (Fan et al., 2022).

Despite soil and crop diversity, uniform fertilizer and pesticide application dominate worldwide agriculture. This mismatch causes yield gaps, chemical waste, nutrient leaching, and greenhouse gas emissions. AI and Remote Sensing (RS) have transformed variability diagnosis and site-specific, climate-responsive management. This comprehensive synthesis examines how AI–RS technologies improve agrochemical efficiency. Random Forests, CNNs, Physics-Informed Neural Networks, and operational outcomes were the focus of data extraction. Depending on spectral resolution and algorithm choice, AI–RS systems may accurately forecast salinity, moisture, and nutrient levels. These predictive maps led to Variable-Rate Application (VRA) deployments that reduced fertilizer consumption. AI–RS technologies have evolved from experimental methods to predictive, precise tools that can change agrochemical science from reactive, consistent processes. Data uniformity, model transferability, and regulatory alignment hinder adoption. Interoperable data infrastructures, digital soil twins, and multi-sensor fusion pipelines are needed to make these technologies the foundation of sustainable agricultural intensification (Howari, 2025).

Due to growing environmental and systemic problems, such as soil degradation, climate variability, and falling resource efficiency, global agriculture is changing quickly thanks to new technologies. The coming together of Geographic Geo-intelligent agriculture is a data-driven framework that combines spatial analytics, real-time sensing, and predictive modeling for sustainable farm management. It has arisen from Information Systems (GIS), Remote Sensing (RS), and the Internet of Things (IoT). This work integrates recent developments in spatially explicit decision-making, emphasizing how multisource data fusion facilitates real-time monitoring of soil and crop health, early anomaly detection, and predictive diagnoses. Geostatistical modeling using GIS defines management zones for individual sites, while remote sensing from satellites provides ongoing spectral and thermal information on crop health and stress levels. IoT-enabled sensor networks also provide real-time data streams on soil and microclimate that support adaptive irrigation, nutrient optimization, and insect control. By connecting these systems with cloud and edge computing, we get a hyperconnected, self-learning agroecosystem that can provide real-time diagnostics and localized decision support. New frameworks like Geo-Conscious Agriculture build on this idea by adding ethical data governance, explainable AI, and digital twin simulations to make the climate more resilient and open. The combination of geospatial information, the Internet of Things (IoT), and AI is a game-changer for agriculture, delivering predictive, site-specific, and sustainable outcomes. It changes the way technology, ecology, and food security are related. GeoAI uses modern AI techniques such as machine learning and deep learning, along with geospatial technologies like GIS and remote sensing, to analyze and anticipate spatial patterns. GeoAI uses satellite images, sensor data, and historical soil records to produce high-resolution, accurate maps of soil attributes such as nutrient levels, pH, and moisture. GeoAI-powered soil health mapping combines and analyzes data from multiple sources to enable accurate, scalable, and efficient assessment of soil conditions. This helps with sustainable agriculture, environmental management, and making smart decisions.

2.2 Digital Soil Mapping and Remote Sensing Approaches

DSM is gaining traction and applications for assessing and mapping the world's soil resources. The DSM describes quantitative tools for mapping soil attributes and types. It combines pedology, statistical modeling, geostatistics, remote sensing, and GIS as a pedometric method. Recent advances in computational power, readily available GIS software, improved GPS accuracy, a variety of remote-sensing data-capture platforms with extensive time-series data availability, and rapid advances in machine- and deep-learning algorithms have made it the most popular soil-mapping technique. The DSM creates quantitative, gridded soil property maps in 2.5D or 3D and estimates forecast uncertainty at high resolution. Traditional mapping methods only produce categorical soil property values. The Indian Agricultural Research Institute, New Delhi, developed digital maps of surface soil moisture (SSM) using optical, thermal, and microwave remote sensing, along with four machine learning (ML) models. In SSM predictive mapping, the random forest model was superior, and the backscatter coefficient (Sigma0_VV) was significant. Due to its dynamic nature and great spatiotemporal variability, SSM prediction uncertainty was moderate to high. DSM struggles because ML cannot define complicated soil-based systems. The urging soil scientists to go beyond model predictions and explain soil processes (Mondal et al., 2024).

Food, nutrition, fodder security, environmental sustainability, and farm revenue depend on soil resource studies. Digital soil mapping (DSM) uses limited soil observations and secondary data to spatially manage soil parameters in a timely manner. DSM is needed for soil and crop monitoring due to land degradation and rising food demand. Remote sensing (RS) is a reliable technology that helps optimize soil datasets and quickly acquire fresh soil resource databases. Over the past few decades, RS has greatly enhanced the availability of space resources and will continue to do so. RS can quickly and cheaply collect data over large areas. Soil mappers can create high-resolution digital soil maps using fine-resolution remotely sensed products. ASTER, MODIS, Landsat 8 and 9, ResourceSat, Sentinel-1, Sentinel-2, WorldView-2, SPOT-6 and 7, Cartosat, KOMPSAT-3, RADARSAT-2, RISAT, and ALOS-2 have been extensively employed as covariates in DSM activities for two decades. Digital terrain features for the DSM are derived from the Cartosat, ALOS, and SRTM digital elevation models. RS product data may serve as primary or secondary DSM covariates. Although promising, RS applications for soil sensing are generally hampered by vegetation, clouds, weather, coarse resolutions, and diverse terrains. With advances in multispectral and hyperspectral sensors, these concerns are somewhat resolved. RS products' spatial, spectral, radiometric, and temporal resolutions must be improved to fully exploit their promise for fine-resolution soil and crop investigations. Thus, there is much potential to improve sensing techniques to make RS more useful for site-specific soil management utilizing digital soil maps (Dash, 2024).

A dynamic monitoring and precision fertilization decision system for agricultural soil nutrients using UAV remote sensing and GIS to overcome traditional soil nutrient assessment constraints. uses hyperspectral and multispectral UAV imagery and ground-sensor data to analyze soil nutrients at high resolution. A time-series dynamic model captures seasonal variations and influences on crop growth stages, improving prediction accuracy (with RMSE reductions for nitrogen, phosphorus, and potassium compared with conventional laboratory-based methods and satellite NDVI approaches). Real-time data-processing algorithms rapidly update soil nutrient status. The experimental validation compared the proposed technique with laboratory soil testing, uniform fertilizer recommendations, and satellite NDVI-based fertilizing. Field testing across three agroecological zones showed that the suggested technique reduced fertilizer inputs and increased crop yields, outperforming both conventional methods. An intelligent fertilization decision algorithm analyzes real-time soil conditions, crop demands, and climate factors to create customized fertilization plans, with continuous learning improving its accuracy. GIS-based visualization tools present nutrient distributions in intuitive geographic representations and offer interactive features for detailed insights. Precision agriculture is improved by automating data collection and decision-making, decreasing resource waste, and optimizing crop yields. The combination of UAV remote sensing, dynamic modeling, and machine learning creates a scalable and adaptive framework for sustainable farming (Chen et al., 2025).

Traditional geographic information systems (GIS) manage geographic and non-spatial data, such as urban/rural feature coordinates, border information, weather statistics, and ownership details. Collecting and maintaining this much data is difficult and expensive. Next-generation GIS (NexGen GIS) uses AI and RS to automate and scale GIS. This convergence enables real-time data processing, automated feature extraction, and predictive spatial analysis for urban planning, environmental monitoring, infrastructure construction, and precision agriculture. The fundamentals of GIS, RS, and AI, and how they creatively solve difficult spatial problems. RS provides accurate data via multispectral and hyperspectral photography on satellites and drones, while AI automates land classification, change detection, and 3D terrain modeling using machine learning and deep learning. Cloud-based GIS platforms enable large-scale analysis, and Digital Twin technology creates dynamic, real-time digital replicas of the real world. New technologies are incorporated into GIS systems, altering the field and enabling efficient, precise, and sustainable decision-making (Pranjal & Soni, 2025).

Monitoring soil nutrients is an important part of promoting sustainable agriculture, as it helps keep the soil healthy, increases crop yields, and reduces environmental harm. The five principal monitoring methodologies: classical techniques, Remote Sensing (RS), the Internet of Things (IoT) and smart systems, in situ sensors, and Artificial Intelligence (AI)-driven models, each offering distinct contributions to nutrient evaluation. A clear trend toward combining machine learning and deep learning with sensor technologies indicates we are moving toward precision agriculture based on real-time data. Soil nutrient monitoring is suggested to facilitate future research and standardization. The vital importance of soil nutrient-monitoring technology for sustainable agriculture, crop optimization, and environmental stewardship. It provides a complete picture of the methods used to monitor soil nutrients for precise soil management (Sobhy & Anandhi, 2025).

A big part of repairing and sustaining soil health stems from the growing focus on sustainable agriculture. Several research investigations and field trials have emphasized soil health through its physico-chemical and biological indicators, while also pinpointing management strategies that can improve it. Soil health is mostly about how well soil functions as a dynamic, life-supporting ecosystem that meets human needs while also supporting the development of plants and animals. Assessing soil health the old-fashioned way means considering many factors, which can be costly in terms of time, money, and effort. Modern technologies and machine learning-based soil health monitoring, on the other hand, offer a better, faster, cheaper, greener, and more real-time option. Digital soil health monitoring uses sensor-based technology and advanced artificial intelligence to comprehensively and accurately track soil health. Precision agriculture tools, used in soil management, crop selection, and farming methods, are very useful for improving soil health, reducing input costs, optimizing resource use, protecting the environment, and increasing crop yields (Roy et al., 2025).

Geographic Information Systems (GIS) and GeoAI (Geospatial Artificial Intelligence) are changing the way farmers manage soil moisture and nutrients by enabling real-time monitoring and decision-making. Farmers can see how soil conditions change over time and space, receive automated alerts, and get precise advice on when to water and fertilize their crops by combining data from in-field sensors, satellite images, and IoT devices into GIS platforms and using AI algorithms to analyze it. Cloud-based systems and mobile apps give customers practical, data-driven insights that help them use resources more efficiently, get the most out of their crops, and support sustainable agriculture.

2.3 IoT Sensor Networks for Soil Nutrient Assessment

Evaluating soil nutrients is very important in horticulture. There are many problems with setting up an information system for horticulture. For example, farms can have very different soil properties (such as water-holding capacity, amounts of sand, silt, clay, and soil organic matter, pH, and permeability) for different plants. Monoculture farms are usually very small, and many different components, agroecological zones, and socio-economic factors affect them. Improvements in information and communication technologies have enabled low-cost, effective information systems that help horticultural farms manage their resources more effectively and become more productive and sustainable. The information system for horticultural farms that uses multiple sensor types, the Internet of Things, and a mobile app. There is also an explanation of numerous methods and technologies for figuring out how fertile the soil is. A botanical garden that replicates the environmental and plant diversity of a horticultural farm is examined in light of the challenges identified in the existing literature and field studies. The theoretical framework and technical assistance advance technologies that empower horticultural farmers to enhance resource management (Postolache et al., 2022).

As the world's population grows, it's more important than ever for everyone to have enough healthy crops and food. A country's economy is greatly affected by crop production. In farming, it's vital to consider the soil, weather, and water supply when choosing the right crop. It assists in growing crops, finding seeds, and assessing market demand for crops. A lot of progress has been achieved recently, from choosing the right crops to cutting them down. The Internet of Things, cloud computing, and machine learning tools enable farmers to analyze data and make smarter decisions at every stage of crop production. After the farmer chooses the right seeds for the crop, they will plant them, monitor plant growth, look for signs of illness, check the produce's ripening stage, and finally harvest it. The primary goal is to provide farmers with a framework for ongoing support, enabling them to receive regular information about their fields and crops. Make the right choices at every step of farming. The decision support system will combine artificial intelligence, machine learning, the cloud, sensors, and other automated devices to provide you with the right information quickly. With the help of a support system, a farmer can make important decisions without relying entirely on local agricultural offices. We have put up an IoT-enabled soil nutrient categorization and crop recommendation (IoTSNA-CR) model to suggest crops. The model helps to reduce the amount of fertilizer used in soil to get the most out of it. The model comprises several phases: data collection via IoT sensors from agricultural lands; storage of this real-time data in cloud memory services; access to this cloud data through an Android application; and pre-processing and regular analysis using various learning techniques. An Arduino UNO board and a variety of sensors, including a soil temperature sensor, a soil moisture sensor, a water level indicator, a pH sensor, a GPS sensor, and a color sensor, were used to build a cost-effective sensing system. We could get the date, time, longitude, and latitude, as well as the moisture, temperature, water level, soil NPK color values, and more from this sensing system (Senapaty et al., 2023).

The integration of AI-Geospatial Intelligence (GEOINT) is changing how we think about modern approaches to environmental resilience and sustainable development. AI-based geospatial intelligence is a key tool for monitoring, predicting, and mitigating environmental hazards worsening due to climate change, urban growth, and environmental degradation. GEOINT uses satellite images, remote sensing technology, and Geographic Information Systems (GIS) to create important spatial data that helps with environmental protection, disaster response, and choices about how to distribute resources. Machine learning and deep learning techniques that use real-time analytics make geospatial analysis far more accurate and efficient. AI and GEOINT can be used together in important ways, such as protecting people from climate change and disasters, protecting micro-biodiversity, planning cities more effectively, improving precision farming, and optimizing renewable energy systems. New technologies, such as AI systems that anticipate floods, detect wildfires, track species, and monitor carbon emissions (Husain & Bhowmick, 2026).

2.4 Machine Learning and Deep Learning Methods Applied to Soil Data

Digital soil mapping (DSM) is a popular, cost-effective soil map technology. However, existing DSM data rarely include landscape context. Calibrated DSM models use point observations intersected with spatially equivalent point covariates. The convolutional neural network (CNN) model, which incorporates contextual information about an observation, improves prediction accuracy compared with standard DSM models. The CNN model uses covariate images to detect non-linear local spatial correlations among neighboring pixels and investigate spatial contextual information. The model's 3-D stack of input images, data augmentation to minimize overfitting, and simultaneous prediction of multiple outputs are unique. The CNN model was trained to predict soil organic carbon across multiple depths in Chile using soil-mapping data. The CNN model reduced error compared to conventional strategies that used only covariate point information. For country-wide mapping at 100 m resolution, a neighborhood size of 3 to 9 pixels is more useful than a point location or larger neighborhood sizes. The CNN model predicts soil carbon at deeper soil layers more precisely and with reduced prediction uncertainty. The CNN model provides a simple and effective framework for future DSM models by using images as covariates (Padarian et al., 2019).

Life depends on soil, therefore soil parameters are critical to soil health analysis. New proximal sensor technologies, such as portable X-ray fluorescence (pXRF) and Vis–NIR spectroscopy, can quickly and non-destructively measure soil profile data. Although these technologies reduce data collection time, analysis remains time-consuming, and existing analysis tools offer only basic representations. Due to its simplicity, the use of proximal sensor data to forecast high-level soil qualities has attracted widespread attention over the past decade. Following these standards, data visualizers collaborate with soil scientists to create interactive, pXRF-based soil profile visualizations. Domain specialists like these interactive visuals. It tests machine-learning and deep-learning methods for predicting soil parameters from spectral data. The proposed deep learning model, RDNet, delivers top-notch results in predicting from Vis-NIR spectra from globally scattered soil samples (Pham et al., 2021).

A structured six analytical areas that include crops, soil, weather and water, land use, animal systems, and the choices farmers make benefit from AI. ML-driven agricultural analytics have grown a lot. Random Forest and Convolutional Neural Networks are still widely used, but recent research shows a clear shift toward more advanced Deep Learning architectures and integrated cloud-edge-device infrastructures. Federated Learning frameworks for collaborative work that protect privacy, use explainable AI methods to make things clearer, and use governance-oriented methods to make sure that different systems can work together. Despite these improvements, there are still a number of problems that won't go away. It does not always generalize across different agroclimatic settings, the high costs of annotating high-quality data, the difficulty of combining different types of data, and the fact that infrastructure isn't always capable of handling high-speed connections. These advancements are integrated within the IAE conceptual framework, highlighting governance- and lifecycle-aware orchestration MLOps as a pivotal (Cravero et al., 2026).

Dynamic and causal impacts of climate stress and Artificial Intelligence-enabled agricultural monitoring on the quality, productivity, and sustainability of cultivated land in Saudi Arabia. The high-resolution climate variables, along with remote sensing indicators such as the Normalized Difference Vegetation Index, Enhanced Vegetation Index, Net Primary Productivity, Water-Use Efficiency, and crop water productivity. A comprehensive econometric framework that integrates the System Generalized Method of Moments, Difference-in-Differences, and event-study methodologies is utilized to tackle issues of persistence, endogeneity, and causal identification. Water availability, as measured by soil moisture and precipitation, substantially improves cultivated land outcomes, whereas heat stress and wind speed have pronounced negative impacts, underscoring the susceptibility of arid agricultural systems. Monitoring with artificial intelligence and smart irrigation consistently improves land quality and productivity. The biggest improvements are in how efficiently water is used and how much water crops can use. Adopting artificial intelligence makes crops use water more efficiently and increases crop water productivity. On the other hand, heat stress lowers vegetation indicators. Evidence from event studies shows that these effects occur after adoption and persist over time, supporting a causal interpretation. In general, the results show that AI technologies help reduce climate stress by improving water management and making decisions more flexible. The policy-relevant insights consistent with Vision 2030, highlighting digital agriculture as a pivotal tool for sustainable land governance, climate adaptation, and food security in arid regions (Hamdouni, 2026).

The big problem for people in the 21st century is how to grow more food to feed a population expected to reach 10 billion. It must be accomplished while preserving sustainable agricultural systems and addressing challenges such as climate change, water resource depletion, and the heightened risks of erosion and productivity loss from extreme weather events. Advances included the creation of key technologies such as GPS and satellite imagery. The effects of climate change will catalyze the forthcoming revolution in precision agriculture and agriculture at large, propelled by Sustainable Precision Agriculture and Environment (SPAE, akin to the 7 Rs), which may utilize historical technologies in conjunction with Big Data analysis. This new SPAE, which focuses on technology, shifts from site-specific management toward global sustainability. SPAE management can be done, and field, watershed, national, and global sustainability can be connected. Automation, AI, IoT, drones, robots, and Big Data are all parts of a global "Digital Twin." (Delgado et al., 2019).

AI, machine learning, computer vision, autonomous sensing, and data-driven decision support are transforming agriculture quicker than ever. Modern technologies are transforming how we monitor crops, test soils, predict weather, detect disease symptoms, anticipate yields, and evaluate animal welfare. Intelligent systems that deliver real-time insights and precision-guided solutions are replacing human labor, specialized knowledge, and frequent field visits. AI's advancement in agriculture and its potential. In agriculture, where illumination, impediments, weather, biological diversity, and limited resources make old algorithms unreliable, modern learning paradigms are applied. These books emphasize both fresh scientific ideas and their practical applications. They include agricultural-friendly robotic systems, mobile apps, edge devices, spectral analysis pipelines, and federated learning networks (Munir, 2026).

Spatially explicit regression approaches, such as Geographically Weighted Regression (GWR), have demonstrated the ability to capture local variability in soil-water-energy interactions, providing a methodological foundation for site-specific resource management in precision agriculture (Al-Shaar et al., 2022).

Geospatial Artificial Intelligence (GeoAI) is revolutionizing agriculture by integrating Geographic Information Systems (GIS), remote sensing technologies, and advanced machine learning algorithms to enable sustainable, data-driven farming practices. Through the analysis of satellite imagery, drone data, and IoT sensor inputs, GeoAI helps farmers make precise decisions such as optimizing irrigation, minimizing the use of fertilizers and pesticides, and closely monitoring crop health. This approach not only supports environmental sustainability but also has the potential to increase crop yields.

3. METHODOLOGY

3.1 Data Collection and Analysis

A systematic meta-analysis is typically used as the methodology for a review chapter. The PRISMA method is typically used for this research strategy. Scholarly papers published in English with peer reviews between 2020 and 2026 were selected using purposive sampling based on topics and keywords related to soil health mapping, such as GeoAI-Powered, Soil Health Mapping, Nutrient Management, and Environmental Impact.

The modified PRISMA method, adapted based on Page et al. (2021), comprises 4 steps that have been applied to this chapter as follows;

3.1.1 Identification

Purposeful sampling is used to choose scholarly journals. This is where the identifying process starts. The conclusion of the selection process is indicated by the identification of keywords and subjects in the examined articles. Information was gathered and analyzed from a collection of initially published English-language articles. These papers came from a wide range of sources; at this stage of the identification process, 548 academic articles on the subject and related topics were identified. Five specific databases (Scopus, Web of Science, Google Scholar, IEEE Xplore, MDPI) are identified, as PRISMA 2020 recommends authors present full search strategies for all databases, registers, and websites searched. Databases selected from 2019 to 2026.

3.1.2 Screening

During the screening process, articles are chosen for review. For this systematic review, secondary data were collected and evaluated using purposeful sampling. After this filtering process, 30 scholarly papers were picked from the first list. The keywords are "GeoAI-Powered," "Soil Health Mapping," "Nutrient Management," and "Environmental Impact."

3.1.3 Eligibility

The researchers excluded papers and copies that didn't have full versions of the papers in order to check if they were eligible. The last 30 scholarly papers chosen during the qualifying stage are now available to the public.

3.1.4 Inclusion

This step included putting together thirty scholarly articles. In the fourth phase, the list of studies to be included in the systematic review is finalized. Make sure all academic work follows the correct chapter order. Before the results were made public, four independent reviewers looked at the study's quality. Grammarly was used to identify instances of plagiarism and grammatical errors. Grammarly also found instances of computer-generated content and plagiarism. Before the final version is sent in, two English experts read it to ensure it is correct.

Each phase of reduction is documented with numbers and reasons: 548 identified → 412 after de-duplication → 97 after title/abstract screening → 47 after full-text review → 30 after quality appraisal. The PRISMA flow diagram shows the number of records identified, included, and excluded, along with the reasons for exclusion.

4. RESULTS

Figure 1. Conceptual framework

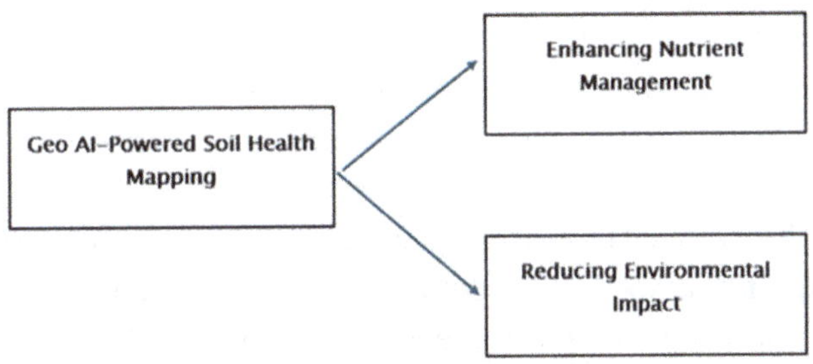

Figure 1 shows the conceptual framework proposed by the authors' own elaboration.

Geo AI-powered soil health mapping enhances nutrient management and reduces environmental impact.

The outcomes of the literature analysis, based on 30 valid scholarly papers, emerge from applying inclusion/exclusion criteria to the literature retrieved from databases, following the PRISMA 2020 guidelines (Page et al., 2021). It establishes the review's credibility and scope based on its geographic distribution, including China, India, the United States, Iran, and other countries. In this systematic review, the results section presents how the various researchers quantified and validated the performance of their models, tools, and systems. In the realm of GeoAI, remote sensing, soil health mapping, nutrient management, and environmental impact assessment, performance metrics function as the established criteria for comparing the accuracy, reliability, and generalizability of various methodologies across studies.

It could be summarized that Geo AI-powered soil health mapping employs artificial intelligence and advanced geospatial analytics to monitor soil conditions across large areas. This device helps farmers apply the proper quantities of fertilizer in the right places by determining exactly which nutrients the soil needs. Because of this, it reduces excess nutrient runoff, has less environmental impact, and encourages farming that is good for the long term.

5. PREVIOUS STUDIES

GeoAI in agriculture reduces fertilizer and pesticide use by precisely focusing application, minimizing chemical runoff, and safeguarding water sources. AI-driven smart irrigation systems analyze soil moisture and weather data to optimize water usage, solve water scarcity, and improve resource efficiency. GeoAI also helps identify soil degradation hotspots, enabling sustainable land management to reduce erosion, salinization, and waterlogging. GeoAI optimizes water and chemical use to reduce farming's environmental impact and boost agroecosystems. It also aids carbon sequestration and climate change mitigation by monitoring soil health and carbon levels. However, GeoAI's environmental benefits, especially in resource-constrained settings, may be limited by technology, farmer training, and data administration. GeoAI has enormous environmental benefits, but addressing practical and ethical issues is necessary to maximize its potential.

Geospatial Artificial Intelligence, or GeoAI, uses artificial intelligence, geographic data, scientific methodologies, and technological tools to analyze and understand complex location-based information. GeoAI automates the extraction, categorization, and identification of information from geographic data, such as photos, videos, point clouds, and text, using machine learning, deep learning, and computer vision, as well as Geographic Information Systems (GIS). GeoAI extends beyond simple mapping of item locations to include predictive spatial analytics. Generative Artificial Intelligence (GenAI) denotes a class of AI systems proficient in generating original content derived from patterns acquired from huge datasets. Recently, information generated by GenAI, including text, images, videos, and mathematical models, has become extensively available due to the widespread adoption of GenAI technology (Adduri & Ali, 2025).

Air pollution is one of the most serious problems for the environment and public health right now, especially in rapidly developing nations. Conventional air quality monitoring systems do not provide prediction capabilities essential for preventative measures and sustainable urban development. The study introduces a GeoAI-driven system that amalgamates real-time monitoring with machine-learning-based forecasting to mitigate urban air pollution. The project employs deep learning algorithms and geographic data to enhance air quality management; utilizes predictive models to forecast the Air Quality Index (AQI) based on climatic and historical data; features interactive dashboards with mapping visuals and alert capabilities; and incorporates an AI-driven chatbot designed for accessibility by non-experts. Python-based automation executes data processing and model deployment without stopping. This study offers a replicable paradigm for extensive urban air quality control through the integration of monitoring and predictive analytics (Adduri & Ali, 2025).

Rapid urbanization poses unparalleled challenges to ecosystem services, environmental quality, and sustainable development worldwide. It is crucial to understand how several Geospatial Artificial Intelligence (GeoAI) frameworks could work together to capture the spatial and temporal dynamics of urbanization. environmental impacts. It brings together new developments in deep learning, remote sensing, and methods for measuring ecosystem services. Current GeoAI methods assess how environmental impact assessments are conducted and identify areas that require a uniform framework. GeoAI-based methods, especially Random Forest and deep learning models, can classify urban land use with 89–96% accuracy. The InVEST model and machine learning work together to improve the accuracy of ecosystem service measurements, with spatial resolutions going up to 30 meters. GeoAI-based environmental impact assessments can forecast the effects of urbanization with 94% accuracy. There are six full tables that show information about classification accuracies, ecosystem service models, spatio-temporal frameworks, environmental indicators, urban growth scenarios, and ways to integrate these frameworks. The conversation highlights the revolutionary power of unified GeoAI frameworks, but also the challenges of integrating data, scaling models, and implementing them in real time (Tikariha & Tripathi, 2025).

Because natural and human-made disasters are occurring more frequently and with greater intensity, disaster risk reduction and emergency response are very important topics in Nigeria. Concentrating on the utilization of geospatial technologies, such as remote sensing, Geographic Information Systems (GIS), and GeoAI, for catastrophic risk mitigation, the restricted use of sophisticated geospatial instruments, and the insufficient representation of specific geopolitical regions. While several studies demonstrate the efficacy of geospatial techniques in hazard mapping, risk assessment, and early warning systems, a lack of cohesive geospatial datasets and inconsistent technical proficiency among practitioners persists. The increasing dependence on passive remote sensing data, with minimal use of active sensors such as SAR. The evaluation says that geospatial technology can change the way we plan for and respond to disasters, but we need to invest more in spatial data infrastructure, capacity building, and nationwide integration to improve our response and resilience. It emphasizes the essential need for inclusive, data-informed policies to reduce catastrophe risk, in line with global frameworks such as the SDGs (Adebayo, 2025).

In the past, assessing damage to buildings after natural disasters was mostly done by hand, a time-consuming and costly process. But recent advances in geospatial artificial intelligence (GeoAI) have made it possible to automate and scale up this important process. Using technology like GeoAI, which uses computer vision, remote sensing, and machine learning on geospatial data from satellites, drones, and other sensors. It could change the way communities assess building damage in disaster-stricken areas and help them access recovery resources more quickly and effectively. Nonetheless, efforts to use GeoAI to assess building damage face significant hurdles in data and model quality, necessitating further investigation. The opportunities and challenges of using GeoAI for building damage assessment. The novel applications of technologies, including deep learning, to high-resolution aerial images for the automated identification and categorization of structural damage. Identifying critical needs for developing strong GeoAI systems involves obtaining comprehensive training data that encompasses the full spectrum of potential damage patterns while accounting for environmental constraints. How humanitarian groups and businesses have sought to deploy GeoAI-powered damage assessment systems during real-world disasters, and what they have learned is crucial (Agbaje et al., 2024).

Artificial intelligence has completely changed how many jobs are done, reshaping beliefs about limits, outcomes, scale, and cost. GeoAI is a multidisciplinary field that includes computer science, engineering, statistics, and spatial science. GeoAI has a significant impact on society and the economy by addressing real-world issues. A broad framework that includes basic questions about theory, epistemology, and the scientific method brings together artificial intelligence (AI) and geography. This link could have big effects on geographic studies. GeoAI, the combination of artificial intelligence and geography, offers unique solutions to many of the challenges smart cities face. GeoAI technology, including a definition and a comparison of GeoAI and classical AI. To do a good job of analyzing geographic data, you need to use GeoAI tools and technologies and combine AI with GIS. The technology's main uses and GeoAI models are also presented, along with the pros and cons of its use in cities (Pritee & Garg, 2024).

Spatial prediction methods are a group of tools that let you get reliable information about spatial variables from a small number of observations. GeoAI is a new field within GIScience that combines AI and machine learning to extract geographic information. It offers a new and ambitious way to think about improving current approaches to spatial prediction and interpolation. This chapter discusses the GeoAI motivations of spatial data representation, geographical structure measurement, and spatial relationship modeling within the workflow of spatial prediction, emphasizing the application of AI techniques. GeoAI in the context of geographical prediction and interpolation techniques, emphasizing two principal domains: geostatistics and spatial regression (Zhu & Cao, 2023).

Mamatha & Shalini (2025); modern farming is a major problem. It affects food quality, environmental conditions, and the agricultural industry's ability to adapt to climate change. There are a lot of problems with traditional ways of managing fertility. They sometimes use too many chemicals, lack real-time, site-specific data, and waste resources. These problems weaken the soil, reduce crop yields, and harm the environment. It is hard to make data-driven judgments about soil management due to high implementation costs, limited knowledge of Machine Learning models, low global standards, and the fact that soil data is not always available. The growing importance of technologies such as Artificial Intelligence (AI) and Machine Learning (ML) in sustainable agricultural systems, with an emphasis on remote sensing, sensor networks integrated with the Internet of Things (IoT), robotics, and data-driven decision support systems. The current state of these technologies and their growing use to enhance diagnostic precision and the efficacy of soil management are also identified.

Obasi et al. (2024). Artificial Intelligence (AI) has become a game-changing tool in farming, especially for precision farming and geographic analysis. This paper examines the impact of AI on precision agriculture and spatial analysis, focusing on the benefits and challenges within the African agricultural context. landscape shows. AI is currently used, and the article discusses how technology could change farming practices in agro-ecological zones. The adaptation of precision farming to improve crop yields, soil health, and address climate change issues with AI technologies, including sensor-based monitoring and satellite imaging analysis. By examining the social and economic impacts of AI adoption in agriculture in Africa, we can see how this technology could support both economic growth and long-term development. This study deepens understanding of AI's transformative impact on agricultural practices in Africa by examining key elements of precision agriculture and spatial analysis, thereby facilitating innovative, efficient, and sustainable farming techniques. Some of the key areas of concentration include using AI to keep a close eye on crops, check soil conditions, and make better decisions by predicting the weather. The research examines the potential, challenges, and socioeconomic impacts of implementing AI in agriculture across several agro-ecological zones. It also provides policymakers with advice, outlines best practices, and suggests areas for further research to fully harness AI to improve agriculture in Africa in ways that are both resilient and sustainable.

Karmaoui et al. (2023); the future of farming may be fully realized by leveraging people's knowledge of artificial intelligence (AI). As climate change and population pressures grow, we need new technologies like AI and RS in key areas like agriculture to quickly identify which crops are best for the environment, boost productivity, and support long-term growth. This research includes bibliometric and review assessments of the application of AI and GIS analysis in agriculture. The number of publications has been going up during this time. Earth and Planetary Sciences is the most important field, remote sensing (RS) is the most important source, and the United States and India are the most important countries. The most common subjects in this search are RS, machine learning (ML), GIS, land use, decision-making, decision trees, and satellite images. Otherwise, research in this area is moving toward machine learning and deep learning methods, while agricultural robots and antennas are the most popular equipment. The productive time when AI and GIS work well together.

Akter et al. (2024). Artificial Intelligence (AI) is transforming the farming industry by improving precision farming and spatial analysis, particularly across the United States' agro-ecological zones. AI technologies, such as sensor-based monitoring and satellite imaging analysis, have the power to change things by making crops grow better, soil healthier, and the environment more resilient. AI can help agriculture in Africa in many ways, such as making it more efficient and sustainable. However, the agricultural sector in Africa also faces many challenges, including limited infrastructure and disparities in wealth and income. explore how AI may help solve these problems by encouraging long-term growth and development. It also shows how important AI is for accurately monitoring crops, assessing soil conditions, and making decisions based on improved weather forecasts. The need for supportive policies and best practices to fully leverage AI's potential in promoting resilient and sustainable farming methods in the USA by analyzing the socio-economic impacts of AI adoption. AI's impact on American agriculture through thorough analysis and strategic recommendations, facilitating the development of innovative, effective agricultural techniques.

According to previous studies, cutting-edge technologies, such as the Internet of Things (IoT), Artificial Intelligence (AI), and Machine Learning (ML), can boost productivity and help manage farms more effectively, meeting food needs. In recent years, researchers have been more interested in Smart Agriculture. Prior literature evaluations have conducted analogous bibliometric analyses; yet, there remains a lack of research in Operations Research (OR) on Smart Agriculture. It shows a bibliometric analysis of operations research knowledge in Agriculture 4.0 over the last two decades, aiming to identify trends and gaps. Governments and policymakers will benefit from understanding how robotic units are utilized in agriculture to enhance resource allocation. Countries with dry climates would learn how satellite imagery and maps could help them identify new land for irrigation to support their limited agricultural resources. Geospatial Artificial Intelligence (GeoAI) integrates artificial intelligence technologies such as machine learning and deep learning with geospatial science and big data analytics. This combination automates spatial analysis, models complex geographic relationships, and predicts future outcomes. By leveraging data from sources such as satellite imagery and textual information, GeoAI transforms processes that traditionally required weeks of manual work into fast, automated workflows, enabling the rapid extraction of valuable insights and actionable intelligence from vast geospatial datasets. AI has changed how smart cities are built, making them more efficient and improving the quality of life. But the widespread use of AI and the rapid growth of Internet of Things (IoT) technologies have made it much harder to protect data security and privacy in AI research for smart cities. Urban transportation has seen significant AI-driven advances, particularly in electric and driverless vehicles, as well as unmanned aerial systems. Even with these new ideas, unresolved security and privacy problems show that more focused research is needed. More research into AI applications in healthcare, environmental management, and industry to address important issues related to data governance and their societal impacts. The issue of the effects, limits, and future research approaches needed to ensure that smart cities continue to grow sustainably.

6. DISCUSSIONS

This chapter highlights a rapid transition towards AI-driven decision support systems in precision agriculture, emphasizing cross-cutting themes of real-time monitoring (IoT/drones), automated resource optimization (water, nutrients), and predictive yield modeling to boost sustainability. However, a significant gap exists between high-tech advancements and practical adoption by Agri-SMEs due to infrastructural deficiencies, high equipment costs, and data interoperability challenges, indicating a trend of fragmentation. Addressing the GeoAI mandate, current research leverages geospatial data to analyze local ecosystems, yet there is a tension between the accuracy of complex AI models and the need for low-cost, simplified applications suitable for variable field environments. These findings imply that future practice must focus on developing edge computing solutions to bypass connectivity issues, while policy must prioritize improving data access, incentivizing sustainable technology adoption, and providing training to bridge capacity gaps, ultimately transforming fragmented digital efforts into cohesive, sustainable farming strategies. An analysis of the reviewed literature demonstrates that GeoAI, at the intersection of artificial intelligence, geospatial science, and big data, has emerged as a transformative force across multiple domains, including precision agriculture, public health, urban sustainability, and disaster management. Several cross-cutting themes emerge from this body of the chapter as follows;

6.1 Integration and Automation of Geospatial Intelligence

Across studies, a unifying theme is the automation of spatial analysis through AI-driven tools that can rapidly process vast and diverse geospatial datasets (e.g., satellite imagery, sensor networks, LiDAR, street-level data). This has enabled substantial improvements in efficiency, scalability, and accuracy for applications ranging from monitoring air quality (Adduri & Ali, 2025; BREATHE system) to rapid disaster damage assessment (Agbaje et al., 2024). For instance, Deep Convolutional Neural Networks (DCNNs) and Bayesian Graph Attention Networks (BGATs) have been shown to outperform traditional rule-based or manual approaches in extracting actionable insights from complex spatial data.

6.2 Predictive Analytics for Proactive Decision-Making

A recurring advancement is the shift from retrospective analyses to predictive, real-time analytics—most notably in urban air quality forecasting and precision agriculture. Models such as Random Forest, deep learning, and spatial regression enable the anticipation of environmental changes, disease prevalence, or crop yield variations, facilitating preventive interventions and adaptive management. GeoAI-enabled precision farming exemplifies how GeoAI supports data-driven, dynamic decision-making.

6.3 Multidisciplinary Integration and Policy Implications

GeoAI inherently draws upon computer science, statistics, engineering, and geography, demanding collaboration across disciplines. Its application in urban planning, public health policy, and precision agriculture signals a move toward more holistic, systems-based approaches. Notably, GeoAI supports the development of inclusive, data-informed policies aligned with global frameworks like the SDGs by providing granular, actionable intelligence for practitioners, policymakers, and the general public.

6.4 Challenges: Data Quality, Integration, and Equity

Despite these advances, the literature identifies persistent challenges. There are significant data-related barriers, such as insufficient access to high-quality, comprehensive training datasets (critical for disaster assessment and environmental modeling), inconsistent technical capacity among practitioners, and limited adoption of active sensors (e.g., SAR) in certain regions. Model robustness, real-time scalability, and integration across diverse data sources remain ongoing issues. Additionally, the rapid expansion of GeoAI and IoT technologies raises unresolved concerns regarding data security, privacy, and equitable access—particularly in the context of smart cities and healthcare.

6.5 Contradictions and Gaps

Contradictions emerge in the relative maturity of GeoAI applications: while high-accuracy predictive models exist for tasks such as urban land-use classification, there is an evident gap in unified frameworks for integrating and operationalizing GeoAI at scale, especially for disaster risk reduction in underrepresented regions. Furthermore, while GeoAI offers unparalleled efficiency, reliance on complex AI models can limit interpretability, posing a challenge to transparent policymaking and community trust.

6.6 Implications for Precision Agriculture and Policy

In precision agriculture, integrating genetic, climatic, and topographic data using GeoAI enables adaptive, site-specific crop management that improves yield and sustainability. However, barriers such as high implementation costs, knowledge gaps in ML, and a lack of standardized soil data constrain broader adoption. Policy efforts must therefore focus on capacity building, investment in spatial data infrastructures, and the development of open, interoperable frameworks to democratize access to GeoAI tools.

GeoAI is reshaping the landscape of spatial analytics and decision-making across sectors, but realizing its full potential will require addressing data, governance, and equity challenges. The reviewed literature highlights the need for continued interdisciplinary research, investment in infrastructure and skills, and the development of transparent, inclusive frameworks to guide the responsible deployment of GeoAI in practice and policy.

7. CONCLUSION

GeoAI, or Geospatial Artificial Intelligence, is a game-changing combination of AI, geographic data, and cutting-edge technologies that is reshaping how we approach and solve complex location-based problems, especially in agriculture and environmental sustainability.

GeoAI uses machine learning, deep learning, and computer vision, along with Geographic Information Systems (GIS), to automate the collection, organization, and analysis of large volumes of geospatial data, including satellite and drone imagery, sensor data, and text. In farming, this potential enables digital soil mapping, remote sensing, and AI-driven analytics to assess factors such as organic carbon content, soil texture, and pH with greater precision. This not only shows where action is needed but also makes it possible to manage land more focused, use fertilizer more effectively, and produce higher agricultural yields, all while having less impact on the environment.

GeoAI can do more than just make static maps. It can also do dynamic, real-time, and predictive spatial analytics. This lets people respond right away to changes in crop health, land usage, and environmental conditions using IoT devices, drones, and real-time satellite data. The technology helps farmers use resources more effectively and sustainably, making them more productive and better equipped to address challenges such as climate change, food security, and resource scarcity.

GeoAI is also very important for monitoring climate indicators, mapping carbon sinks, reducing transportation emissions, and identifying the best locations for renewable energy. All of these things help us reach our net-zero goals and encourage people to take care of the environment.

GeoAI is not only making farming easier by providing useful information and automating decision-making, but also giving policymakers and stakeholders the tools they need to promote sustainability, protect natural ecosystems, and create smarter, more resilient systems for managing farms and monitoring the environment as a whole.

Thus, this chapter has demonstrated that GeoAI is a major advancement in spatial analytics, offering significant improvements in monitoring soil health and nutrient management in precision agriculture. The main conclusions imply that extremely accurate, automated detection of soil deficiencies and land-use changes over time can be achieved by combining deep learning with high-resolution geospatial imagery (e.g., drones and satellites). This chapter advances the field by showing how GeoAI bridges the gap between raw data collection and practical land management techniques, moving beyond conventional static GIS mapping to real-time predictive analytics. In contrast to earlier evaluations, this chapter focuses on combining data from multiple sensors to improve the efficiency of farming inputs.

GeoAI is a powerful tool for monitoring and managing the environment today. GeoAI helps improve agricultural and food security by making agriculture more sustainable through smart irrigation systems like Variable Rate Irrigation (VRI). These technologies quickly assess soil conditions and reduce water waste. Smart Crop Planning uses data-driven insights to help farmers choose the best crops and make more accurate estimates of their yields. This helps people better adapt to climate change and water scarcity. Using AI for efficient resource management ensures that water and fertilizers are used correctly, at the right times, and in the right places. This is very helpful for maintaining food production in dry, resource-limited areas.

GeoAI improves soil health mapping by simultaneously using spatial data, spatial analytics, and location-aware modeling. This helps us learn much more about how ecosystems evolve and how soil changes than standard remote sensing or non-spatial machine learning methods do.

GeoAI finds complex spatial patterns and contextual links that other methods often miss, using high-resolution geospatial datasets such as satellite imagery, terrain models, and field sensor data. Spatial analytics help uncover trends that occur only in certain places, spatial relationships, and how soil characteristics and environmental factors affect one another. AI models that know their location can adjust their forecasts accordingly, making soil health assessments more accurate, thorough, and helpful. This integration ultimately enables focused land management, precision agriculture, and sustainable soil conservation techniques that are unattainable with non-spatial or exclusively remote sensing-based approaches.

8. IMPLICATIONS

A GeoAI model that integrates satellite imagery with topographic and climatic data can make soil health mapping (such as pH, nitrogen, and soil texture) far more accurate than previous methods that rely on spatial interpolation.

8.1 More Detailed Data Inputs

Traditional spatial interpolation methods, such as kriging or inverse distance weighting, primarily use point-based soil measurements and their spatial locations. A hybrid GeoAI model, on the other hand, combines data from different sources, such as satellite images (which show land cover, vegetation health, and surface reflectance), topography (elevation, slope, aspect), and climate (rainfall, temperature, and moisture). This gives a more complete picture of the factors that affect soil properties.

8.2 Capturing Non-Linear Relationships

Soil health indices are affected by complicated, non-linear interactions between environmental factors. GeoAI models, particularly those employing machine learning or deep learning, can more effectively simulate these complex interactions than linear or geostatistical interpolation methods.

8.3 Spatial and Temporal Resolution

Satellite photos have high spatial and temporal resolution, which means that GeoAI models may detect subtle changes in the landscape that other methods would overlook. This makes soil health maps that are more detailed and changeable. Generalization and Prediction: GeoAI models can make more accurate predictions in areas without direct soil measurements by learning patterns from diverse data. This is because they can generalize better to areas that haven't been sampled.

8.4 Feature Extraction

Advanced AI and image-processing methods can extract useful information from satellite images, such as vegetation indices and texture metrics. These are often strongly linked to soil parameters.

8.5 Incorporating Environmental Context

Information on the land's shape and climate is important for understanding how soil forms and changes. By including these variables, the model can adjust its predictions in response to environmental gradients that standard interpolation wouldn't capture. In short, a hybrid GeoAI model leverages the best aspects of multiple data sources and advanced machine learning to produce soil health maps that are more accurate, context-aware, and high-resolution. This is better than traditional spatial interpolation, which mostly relies on spatial proximity and linear assumptions.

9. LIMITATIONS AND RECOMMENDATIONS

This chapter employs a systematic review and may not include respondents. Thus, biases in data collection and analysis could be incurred. Further studies suggest that respondents should be asked about surveys, questionnaires, and interviews.

The value of GenAI comes from how well it works together: spatial data that show the multi-dimensional geographic context of soil formation; spatial analytics that encode the structure of soil variability; and location-aware modeling that makes predictions sensitive to where they are made, not just to the inputs given. This integration creates soil health maps that are more accurate, cover a larger area, change over time, and account for uncertainty than standard remote sensing or non-spatial ML can on their own. It also enables the outputs to be directly actionable by linking soil health projections to spatially explicit management prescriptions, variable-rate fertilization zones, and environmental risk assessments. Non-spatial predictions can't do this. In this way, GeoAI does more than just improve soil health maps. It changes what soil health mapping can mean by making it a dynamic, spatially intelligent decision-support system, recognizing that soil is not a uniform medium but a spatially structured, temporally evolving, and geographically contextualized part of the agricultural landscape. Thus, this content is suggested for further study, drawing on respondents' perceptions.

REFERENCES

Adduri, S. D., & Ali, T. (2025). BREATHE: A GeoAI-Powered Air Quality Monitoring and Forecasting System for Urban Sustainability. *The International Archives of the Photogrammetry, Remote Sensing and Spatial Information Sciences*, *48*, 39–46. DOI: 10.5194/isprs-archives-XLVIII-G-2025-39-2025

Adebayo, I. A. (2025). Risk Reduction and Emergency Response: The Geospatial Assessment of Disaster Management in Nigeria. *SSRN*, *5815542*. https://papers.ssrn.com/sol3/papers.cfm?abstract_id=5815542

Agbaje, T. H., Abomaye-Nimenibo, N., Ezeh, C. J., Bello, A., & Olorunnishola, A. (2024). Building damage assessment in aftermath of disaster events by leveraging GeoAI (Geospatial Artificial Intelligence). *World Journal of Advanced Research and Reviews*, *23*(1), 667–687. DOI: 10.30574/wjarr.2024.23.1.2000

Akter, J., Kamruzzaman, M., Hasan, R., Khatoon, R., Farabi, S. F., & Ullah, M. W. (2024, September). Artificial intelligence in American agriculture: a comprehensive review of spatial analysis and precision farming for sustainability. In *2024 IEEE International Conference on Computing, Applications and Systems (COMPAS)* (pp. 1–7). IEEE. https://ieeexplore.ieee.org/document/10796075

Al-Shaar, W., Bonin, O., de Gouvello, B., Chatellier, P., & Hendel, M. (2022). Geographically weighted regression-based predictions of water–soil–energy nexus solutions in Île-de-France. *Urban Science (Basel, Switzerland)*, *6*(4), 81. DOI: 10.3390/urbansci6040081

Chen, X., Zhang, H., & Wong, C. U. I. (2025). Dynamic monitoring and precision fertilization decision system for agricultural soil nutrients using UAV remote sensing and GIS. *Agriculture*, *15*(15), 1627. DOI: 10.3390/agriculture15151627

Costa, D. G. (2026). The risks of GenAI in shaping our cities. *Cities (London, England)*, *173*, 107025. DOI: 10.1016/j.cities.2026.107025

Cravero, A., Sepúlveda, S., Gutiérrez, F., & Muñoz, L. (2026). From Precision Agriculture to Intelligent Agricultural Ecosystems: A Systematic Review of Machine Learning and Big Data Applications. *Agronomy (Basel)*, *16*(5), 516. DOI: 10.3390/agronomy16050516

Dash, P. K. (2024). Remote sensing as a potential tool for advancing digital soil mapping. In *Remote Sensing of Soils* (pp. 357–370). Elsevier. DOI: 10.1016/B978-0-443-18773-5.00021-1

Delgado, J. A., Short, N. M.Jr, Roberts, D. P., & Vandenberg, B. (2019). Big data analysis for sustainable agriculture on a geospatial cloud framework. *Frontiers in Sustainable Food Systems*, *3*, 54. DOI: 10.3389/fsufs.2019.00054

Fan, Y., Wang, X., Funk, T., Rashid, I., Herman, B., Bompoti, N., Mahmud, M. D. S., Chrysochoou, M., Yang, M., Vadas, T. M., Lei, Y., & Li, B. (2022). A critical review for real-time continuous soil monitoring: Advantages, challenges, and perspectives. *Environmental Science & Technology*, *56*(19), 13546–13564. DOI: 10.1021/acs.est.2c03562 PMID: 36121207

Hamdouni, A. (2026). AI-Enabled Remote Sensing Assessment of Cultivated Land Quality and Sustainability Under Climate Stress: Evidence from Saudi Arabia. *Resources*, *15*(3), 44. DOI: 10.3390/resources15030044

Hosseini, F. S., Seo, M. B., Razavi-Termeh, S. V., Sadeghi-Niaraki, A., Jamshidi, M., & Choi, S. M. (2023). Geospatial artificial intelligence (GeoAI) and satellite imagery fusion for soil physical property predicting. *Sustainability (Basel), 15*(19), 14125. DOI: 10.3390/su151914125

Howari, F. (2025). AI–Remote Sensing for Soil Variability Mapping and Precision Agrochemical Management: A Comprehensive Review of Methods, Limitations, and Climate-Smart Applications. *Agrochemicals, 5*(1), 1. https://www.mdpi.com/2813-3145/5/1/1

Husain, A., & Bhowmick, S. (2026). Smart Geospatial Intelligence and AI for a Resilient Environment: A Review. *Advanced Geospatial Intelligence and AI for Environmental Resilience and Sustainable Development*, 417–432. https://link.springer.com/chapter/10.1007/978-3-032-03714-5_17

Karmaoui, A., El Jaafari, S., Chaachouay, H., & Hajji, L. (2023). A bibliometric review of geospatial analyses and artificial intelligence literature in agriculture. *GeoJournal, 88*(S1, Suppl 1), 343–360. DOI: 10.1007/s10708-023-10859-w

Laveglia, S., Altieri, G., Genovese, F., Matera, A., & Di Renzo, G. C. (2024). Advances in sustainable crop management: integrating precision agriculture and proximal sensing. *AgriEngineering, 6*(3), 3084–3120. https://www.mdpi.com/2624-7402/6/3/177

Mamatha, A., & Shalini, S. (2025). Artificial Intelligence-based Green Technologies for Efficient Agriculture. In *ITM Web of Conferences* (Vol. 79, p. 01027). EDP Sciences. DOI: 10.1051/itmconf/20257901027

Mondal, B. P., Sahoo, R. N., Das, B., Chakraborty, D., Kumar, N., & Raj, G. B. (2024). Digital Soil Mapping: concepts, methods, and applications-Remote sensing and GIS perspectives. In *Remote sensing of soils* (pp. 195–212). Elsevier. DOI: 10.1016/B978-0-443-18773-5.00018-1

Munir, A. (2026). Artificial Intelligence and Machine Learning for Smart and Sustainable Agriculture. *AI, 7*(1), 12. https://www.mdpi.com/2673-2688/7/1/12

Obasi, S. N., Aa, T. V., Obasi, C. C., Jokthan, G. E., Adjei, E. A., & Keyagha, E. R. (2024). Harnessing Artificial Intelligence for sustainable agriculture: A comprehensive review of African applications in spatial analysis and precision agriculture. *Big Data in Agriculture, 6*(1), 6–18. DOI: 10.26480/bda.01.2024.01.13

Padarian, J., Minasny, B., & McBratney, A. B. (2019). Using deep learning for digital soil mapping. *Soil (Göttingen), 5*(1), 79–89. DOI: 10.5194/soil-5-79-2019

Page, M. J., McKenzie, J. E., Bossuyt, P. M., Boutron, I., Hoffmann, T. C., Mulrow, C. D., Shamseer, L., Tetzlaff, J. M., Akl, E. A., Brennan, S. E., Chou, R., Glanville, J., Grimshaw, J. M., Hróbjartsson, A., Lalu, M. M., Li, T., Loder, E. W., Mayo-Wilson, E., McDonald, S., & Moher, D. (2021). The PRISMA 2020 statement: An updated guideline for reporting systematic reviews. *BMJ (Clinical Research Ed.), 372*, n71. Advance online publication. DOI: 10.1136/bmj.n71 PMID: 33782057

Pham, V., Weindorf, D. C., & Dang, T. (2021). Soil profile analysis using interactive visualizations, machine learning, and deep learning. *Computers and Electronics in Agriculture, 191*, 106539. DOI: 10.1016/j.compag.2021.106539

Postolache, S., Sebastião, P., Viegas, V., Postolache, O., & Cercas, F. (2022). IoT-based systems for soil nutrients assessment in horticulture. *Sensors (Basel), 23*(1), 403. DOI: 10.3390/s23010403 PMID: 36617000

Pranjal, P., & Soni, A. (2025). Next-Generation Geographical Information Systems: Integrating Artificial Intelligence, Geographic Information Systems, and Satellite Imagery for Mapping and Surveying. *Global Land-Ocean Geospatial Applications*, 216–233.

Pritee, K., & Garg, R. D. (2024). Geo-AI for urban planning. In *Spatial Intelligence for a Greener Planet* (pp. 115–130). CRC Press. DOI: 10.1201/9781032718323-9

Roy, D., Tamuk, P., Singh, A. P., & Mondal, S. (2025). Modern Technologies for Soil Health Monitoring and Management. In *Smart Farming, Smarter Solutions* (pp. 118–133). CRC Press. DOI: 10.1201/9781003536932-6

Senapaty, M. K., Ray, A., & Padhy, N. (2023). IoT-enabled soil nutrient analysis and crop recommendation model for precision agriculture. *Computers, 12*(3), 61. DOI: 10.3390/computers12030061

Sobhy, D. M., & Anandhi, A. (2025). Soil Nutrient Monitoring Technologies for Sustainable Agriculture: A Systematic Review. *Sustainability (Basel), 17*(18), 8477. DOI: 10.3390/su17188477

Tikariha, S., & Tripathi, D. (2025). Spatio-Temporal Modeling of Urbanization, Environmental Impact, and Ecosystem Functions Using a Unified Geo-AI Framework: A Review Paper. https://ijorarjournal.com/web/Download/2026-01-02-02-18-33-Paper%20Id%200651.pdf

Yousaf, A., Kayvanfar, V., Mazzoni, A., & Elomri, A. (2023). Artificial intelligence-based decision support systems in smart agriculture: Bibliometric analysis for operational insights and future directions. *Frontiers in Sustainable Food Systems, 6*, 1053921. DOI: 10.3389/fsufs.2022.1053921

Zhu, D., & Cao, G. (2023). Intelligent spatial prediction and interpolation methods. In *Handbook of geospatial artificial intelligence* (pp. 121–150). CRC Press. DOI: 10.1201/9781003308423-7

KEY TERMS AND DEFINITIONS

GeoAI and Agricultural Sustainability: GeoAI, or Geospatial Artificial Intelligence, helps make farming more sustainable by combining GIS, remote sensing, and AI to optimize resource use, increase yields, and improve environmental health.

GeoAI and Carbon Reduction: Geospatial Artificial Intelligence (GeoAI) leverages machine learning and geographic data to accelerate carbon emissions reduction. It is very important to monitor climate indicators, improve energy efficiency, and better manage the environment. Some important uses include mapping carbon sinks, identifying the best places for renewable energy, reducing transportation emissions, and monitoring land-use changes to reach net-zero goals.

GeoAI for Crop Monitoring: GeoAI (Geospatial Artificial Intelligence) transforms crop monitoring by leveraging satellite imagery, drone data, and machine learning to improve the accuracy and automation of agricultural management. It gives farmers real-time information on crop health, yield potential, disease identification, and soil analysis. This helps them make the best use of their resources and grow more crops in an environmentally friendly way.

GeoAI for Green Farm Sustainability: By fusing geographic information systems (GIS), remote sensing, and sophisticated machine learning, GeoAI (Geospatial Artificial Intelligence) is transforming agriculture and improving food security, resource efficiency, and sustainable agricultural methods. GeoAI enables high-precision, data-driven decisions that reduce waste and increase productivity by analyzing satellite imagery, drone data, and IoT sensor data, thereby promoting a "greener" agricultural industry.

GeoAI in Agriculture: Geospatial Artificial Intelligence (GeoAI) in agriculture is the use of AI, machine learning, and Geographic Information Systems (GIS) to analyze geographic data to make farming more efficient, sustainable, and accurate. GeoAI is like a "digital brain" for farms. It takes data from satellites, drones, and IoT devices and turns static photos into useful, real-time information that helps people make better decisions.

GeoAI in Environmental Impact: Using GeoAI for spatiotemporal study of land use...GeoAI (Geospatial Artificial Intelligence) uses AI and spatial data to monitor, model, and mitigate environmental impacts. It gives high-resolution analysis for sustainability. Some important uses include monitoring climate change indicators such as deforestation and rising sea levels, advancing renewable energy, protecting wildlife, and reducing urban hazards like heat islands.

GeoAI in Nutrient Management: Geospatial Artificial Intelligence (GeoAI) is transforming how farmers manage nutrients by leveraging machine learning, deep learning, and geospatial data analytics to identify optimal fertilizer application, improve crop health, and reduce environmental impact. GeoAI uses satellite images, drones, and sensors to provide useful information that lets farmers apply nutrients exactly where they need them, rather than blanket, inefficient methods.

GeoAI in Real-Time Dynamic Information: GeoAI (Geospatial Artificial Intelligence) is rapidly evolving from static, descriptive analysis to a real-time, dynamic, and predictive framework. GeoAI combines artificial intelligence (AI), such as machine learning and deep learning, with GIS, remote sensing, and the Internet of Things (IoT) to allow for real-time monitoring and analysis of moving, changing, and interacting spatial data.

Geospatial Artificial Intelligence: or GeoAI, uses AI, geographic data, science, and technology to analyze and understand complex location-based data. GeoAI automates the extraction, categorization, and detection of information from geographic data, such as photographs, videos, point clouds, and text, using machine learning, deep learning, and computer vision, integrated with Geographic Information Systems (GIS). GeoAI goes beyond mapping("where" objects are) to predictive spatial analytics

Soil Health Mapping: Using digital soil mapping (DSM), remote sensing, and machine learning, soil health mapping examines and visualizes how soil parameters such as organic carbon, texture, and pH vary across space. It helps identify degraded regions so that land management can be improved, agricultural output increased, and policy decisions on environmental sustainability made.

Chapter 14
The Guardian Agro-Intelligence System (GAIS):
AI-Driven IoT Framework for Real-Time Jawar and Pomegranate Disease Detection and Precision Treatment

Kutubuddin Sayyad Liyakat Kazi
http://orcid.org/0000-0001-5623-9211
Eudoxia Research University, USA

Heena Tajoddin Shaikh
Brahmdevdada Mane Institute of Technology, Solapur, India

ABSTRACT

Low-cost, edge-computing IoT nodes placed in agricultural areas are integrated into the GAIS architecture. High-resolution cameras and environmental sensors (which track temperature, humidity, and soil moisture) are used to strengthen these nodes. Data is continuously gathered and sent to a central cloud platform, mostly leaf photos. A specially trained KSK (Kutubuddin S. Kazi) Approach, which is ideal for identifying and categorizing dominant illnesses (such as wilt, Pomegranate bacterial blight, and Jawar leaf blight) is the fundamental intelligence. Following classification, the system makes use of a proprietary, rules-based expert system that quickly produces calibrated treatment protocols that take into account the disease stage, environmental factors, and viable fungicide/pesticide substitutes. According to preliminary testing, the GAIS greatly outperforms manual detection techniques in terms of speed and consistency, achieving a disease classification accuracy of over 96%.

1. INTRODUCTION

Maladies are an invisible enemy to the golden meals of the jawar, a food of millions of people, and the seed of the pomegranate, jewels, a treasury of antioxidants. These difficulties can adversely affect food security, livelihood, and harvests, and include the insidious wilts that kill jawar crops, fungal blights that annihilate the crimson splendor of pomegranates, etc. But what would happen when we are able to

DOI: 10.4018/979-8-3373-8037-7.ch014

give farmers a digital watchdog, a smart system, which would be able to diagnose these diseases at an early stage and offer direct treatment?

It is hard to imagine that in the world a farmer will be able to snap a picture of the spot-ridden leaf of a pomegranate or a hanging jawar stem with his smartphone. It is not a science fiction but a promise of a Jawar and Pomegranate Diseases Detection and Treatment Suggestion System. It is a clever technology which involves machine learning and artificial intelligence so that it can turn into a virtual agronomist in the fingertips of every farmer (Kazi K S,2025; Kutubuddin, 2024;).

Figure 1. Database with the information containing

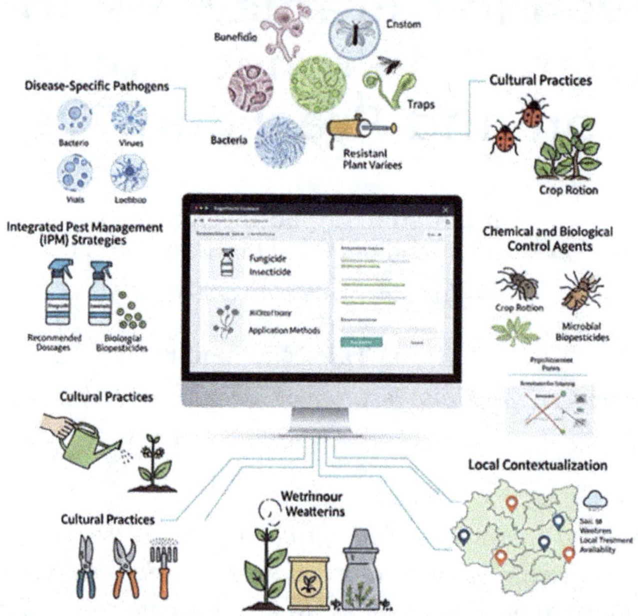

This system is based on a complex diagnostic engine. With trained data sets of images, healthy plants, and plants diseased with a variety of diseases, a powerful KSK (Kutubuddin S Kazi) Approach behaves like the appreciative eye. This AI model is trained to detect subtle visual indicators that can usually be hard to notice with the naked eye: the unusual discolouration of a jawar leaf signifying Downy Mildew, the typical lesions on a pomegranate fruit signifying Anthracnose, or the initial signs of stem borer infestation in jawar stalks.

It is extremely easy to the end-user. A farmer simply needs to:

- Take Clear pictures of the affected part of the plant (leaf, stem, fruit, or the plant itself).
- Upload: upload the image using an easy to use mobile application or web portal.
- Analyze: AI analysis of the image is then conducted by the system that compares features of the image to its large disease database.

In seconds, a diagnosis is given by the system which is the particular disease with a very high level of accuracy. This early diagnosis is very important, because in most cases the initial treatment is what determines the outcome of the treatment and the level of loss of crop.

Half the battle is the battle of detection. The real strength of this system consists in the fact it provides effective and personalized tips of treatment. After the identification of a disease the system retrieves an extensive database of agricultural knowledge. This database holds information (Figure 1) regarding:

- Pathogenic Agents: Pathogen Knowledge.
- IPM Strategies: Advocating the use of multi-pronged strategies in which a greater focus should be made on sustainable and environmentally friendly techniques.
- Chemical and Biological Control Agents: Recommendation of suitable fungicides, insecticides or bio-pesticides, their dose and mode of application.
- Cultural Practices: Recommending better irrigation and fertilization, pruning, and sanitation practices that would strengthen the health and vitality of plants.
- Local Contextualization: Importantly, the system may be modelled in such a way that it takes into account the local agricultural factors, soil type, climatic patterns, and the presence of a certain treatment in a given area. This will make sure that the recommendations are not only scientifically viable, but useful and available to the farmer.

As an example, when the system identifies the first symptoms of Ergot in jawar, it could be recommended:

Immediate Removal: It is recommended that the farmer should remove and destroy the infected heads to avoid further spread of the spores.

- Application of Fungicides: Prescription of particular fungicides that have been shown to be effective against Ergot, including timing of application (e.g. at the time of ear emergence) and application dose.
- Crop Rotation: Recommending the need to have jawar rotation with non-host crops in future seasons to interrupt the disease cycle.

In the same way, in case of a pomegranate with Alternaria Leaf Spot, the system may suggest:

- Sanitation: Promoting the elimination of leaves and diseased fruits that have fallen.
- Copper-based Fungicides:--Recommendation to use copper oxychloride or any broad-spectrum fungicides.
- Enhanced Air Movement: Recommendations regarding adequate spacing of trees and pruning to promote air flow thereby decreasing humidity which encourages the growth of fungi.

This system is not only a diagnostic tool but it is also an educational platform. It can:

- Present Visual Aids: Demonstrate to farmers the visual representation of the healthy plants versus the diseased ones, and the visual aids will increase the farmers knowledge and self-detection skills.

- Provide Best Practice Notifications: Notifications on coming disease seasons and preventives should be sent in advance.
- Promote Sharing of Knowledge: This can be used to potentially introduce farmers to agricultural experts or other farmers who are struggling on the same grounds.
- Contribute to Research: Anonymized data that is gathered by the system can be invaluable to monitor disease outbreaks as well as their propagation as well as, improve treatment plans as time progresses.

The Jawar and Pomegranate Diseases Detection and Treatment Suggestion System is a major stride towards the protection of such important crops. The power of AI can be used to empower farmers to enable them to fight diseases better, therefore, resulting in higher yields, better crops, more food security and more sustainable agriculture. The pomegranate in ruby-red, and the golden grain of jawar are now worthy to grow, and not be checked by any stern assault of sickness, and this wise mechanism is about to bring that vision to pass.

KSK Approach:

The study's results show that the KSK (Kutubuddin S Kazi) approach could be useful for finding diseases that harm jawar leaves. It is obvious that an ANN, a DT, and a K-NN algorithms may be used successfully in real-world situations since they are so accurate. The Suggested KSK approach is as shown in Figure 2. The study also shows the pros and cons for every algorithm, which can help people choose the one that works best for their needs (Liyakat,2025; Tamboli et al., 2024).

Figure 2. KSK Approach for leaf disease detection

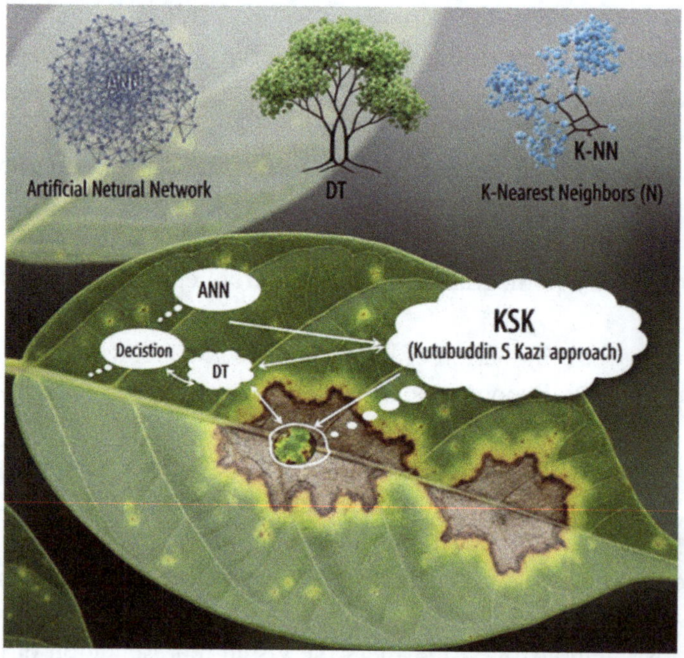

This study shows that artificial intelligence methods, especially ANN, can help find diseases that harm jawar leaf. The proposed approach can help farmers find contagions early, which will help them start the right management methods to reduce crop losses. Further research could explore the utilisation of advanced ML systems and the integration of sensors for realtime monitoring of diseases.

The KSK (Kutubuddin S Kazi) approach's findings indicate that artificial neural networks (ANN) outdone other algorithms in identifying jawar leaf illness, as exemplified. The artificial neural network (ANN) could see little patterns in information and handle non-linear relations because it was so complex. The method of decision-making that DT gave was clear, but it wasn't as accurate as ANN gave. K-NN did well, but ANN and DT did better.

The KSK approach utilizes the ANN with 3-Layerd structure, DT having C4.5 and KNN has K=3. These suggestions are for the better results. The three models are practicably stacked together in KSK Approach, wherein ANN offers deep and non-linear feature transformations; the C4.5 tree, tree offers rule-based guidance and the 3-NN further refines predictions based upon local similarity. They can be metametalearned on their outputs: with logistic regression or a shallow neural net, a composite classifier can be trained with greater accuracy, precision and recall than its components. Tuning the hidden-layer size, pruning confidence, and distance measurements through careful cross-validation takes care of ensuring that each of the components is working at its optimum level to provide a harmonized high-performing solution to the challenging classification problems.

Below is info of three-layered ANN built—complete with the numbers, the functions, the training regimen, and the gathering of images that form its education. Table 1 represents the Layers of ANN and Activations Functions employed in each layer. The employed ANN is 3 layered structure and the purpose of each layer in proposed system is explained in Table 1.

Table 1. ANN architecture

Layer	Purpose	Number of Neurons	Activation Function
Input (Floor 0)	Flattened 64 × 64 × 3 RGB image –received (a total of **12 288** pixels)	12288 (no activation)	—
Hidden 1 (Floor 1)	Extracts low-level patterns – edges, colour blobs, the faint speckle of pomegranate arils or the smooth curve of Jawar skin	**256**	**ReLU** (Rectified Linear Unit)
Hidden 2 (Floor 2)	Consolidates those patterns into mid-level concepts – "seed clusters", "fleshy sheen", "stem curvature"	**64**	**ReLU**
Output (Floor 3)	Decides which leaf/fruit is in-front of camera	2 (one per class)	**Softmax** (produces probability distribution)

Why these numbers?

- The structure usages a wide-then-narrow architecture from 256 → 64hidden units. This enables high-dimensional feature extractions for detecting the broad aspects of features, followed by dimensionality reduction focused on sparse, high discriminative features.
- For classification between Jawar and Pomegranate, 2-Neuron defense layer combined with Softwax activation Function is employed. This configuration effectively outputs a normalized probability distribution, providing a specific confidence score for each of the two classes.

Table 2 explains the Training algorithm employed, Learning rate, Epochs, batch size of each learning, and why the algorithm is selected and its foundation.

Table 2. Learning engine

Element	Choice	Rationale
Training Algorithm	**Adam** (Adaptive Moment Estimation)	Adam adapts the learning rate for each weight, handling the diverse gradients that appear when the network swings between the glossy red of a pomegranate and the matte green-yellow of a Jawar.
Learning Rate (η)	**0.001** (default Adam)	Fast enough to see progress in the first few epochs, yet gentle enough to avoid overshooting the subtle boundary between the two fruit textures.
Number of Epochs	**30**	Empirically, validation loss stabilises around epoch 25; a few extra passes give the model a safety margin without over-fitting.
Batch Size	**32**	A sweet spot for GPU memory on a modest workstation, providing smooth gradient estimates while keeping the training loop lively.

Table 3 represents the dataset collected in this study, the data size, class distribution, image resolutions and augmentation.

Table 3. Dataset collection style

Metric	Value
Dataset Size	5422 images total (Collected by Author's) at Solapur, Maharashtra.
Class Distribution	**1772 Jawar** images – **3650 Pomegranate** images
Image Resolution	64 × 64 × 3 (RGB) – down-sampled to keep the network lightweight
Augmentation	Random horizontal flips, slight rotations (±15°), brightness jitter (±10 %) – to mimic the natural variance in orchard lighting and fruit orientation

A two-week field data collection campaign was conducted to acquire high-resolution imagery of *Jawar* (sorghum) and pomegranates by author(s). Raw images were captured using a smartphone camera (200 × 200 pixel format) and subsequently processed via cropping and resizing to establish a standardized input dimension. To ensure high variance and feature representation, the dataset includes, but is not limited to, the following environmental and structural conditions:

- *Jawar* was imaged both in-situ (on the vine) and post-harvest, exhibiting surface variations ranging from natural orchard soil accumulation to polished, clean surfaces.
- Pomegranates were captured in multiple states, including whole fruit, cross-sections (halved), and exposed, glistening arils.

To prevent bias in the Softmax activation function toward higher-frequency classes, a balanced class distribution was rigorously maintained in the dataset. This balancing enables the optimization process to learn uniform decision boundaries, ensuring reliable classification performance across all fruit types and conditions.

The results of our study indicate that KSK(Kutubuddin S Kazi) Approach has an effective technique for detecting diseases that impact jawar leaf. Because an ANN, a DT, and a K-NN procedures can be very accurate, it makes sense to think that those methods might be used to make illness detection methods which are both consistent and useful.

2. LITERARURE SURVEY

From the sun-drenched fields of Jawar (Sorghum), whose golden kernels form a lifeline for millions, to the copses where the ruby-red arils of Pomegranate promise both nutrition and delight, agriculture is a constant dance with nature. Yet, prowling in the shadows of growth and abundance is a silent, unescapable threat: plant diseases. These microscopic opponents can destroy yields, compromise quality, and shatter the livelihoods of farmers. The key to protection these vital crops lies in early, accurate, and innovative disease detection – a field rapidly evolving from the farmer's intense judgment to the precision of artificial intelligence (Abouelmagd et al., 2024; Bissadu et al., 2025; Liu et al., 2023; Mincuzzi et al., 2022).

Imagine a Jawar field, lush and vibrant, suddenly submitting to the insidious grip of silky mold or smut, and turning promising stalks into shrunken shells. Or a Pomegranate orchard, its fruits just ripening, developing the revealing black lacerations of bacterial disfigurement (oily spot) or yielding to anthracnose. The window for effective intervention is often narrow. Late detection means wider spread, heavier pesticide use, or, worse, complete crop loss. Therefore, the methods we service to reveal these hidden enemies are not just scientific protocols; they are the guardians of food security and rural affluence as shown in Figure 3.

Figure 3. Different methods for diseases detection

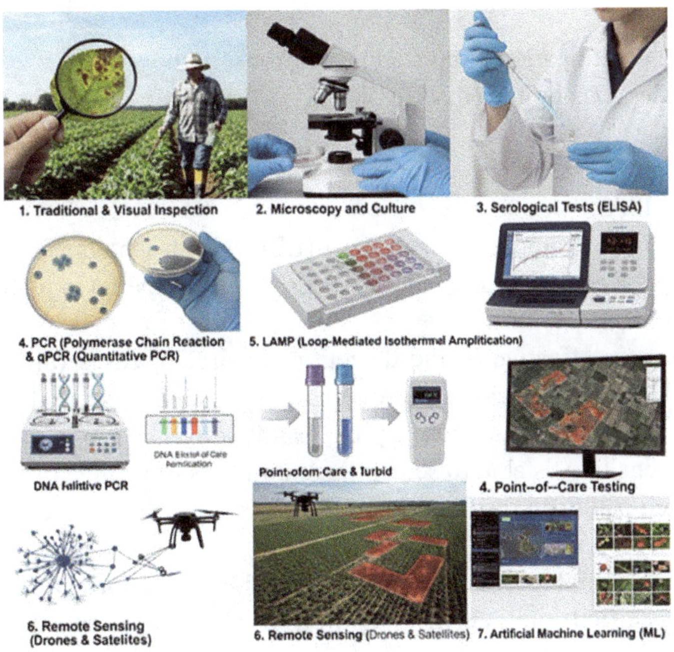

The First Line of Traditional and Visual Inspection in the Farmer.

The careful husband has been the foremost fortification. They can easily see subtle changes because they have known the crop lifecycle, and they have experienced eye:

- Jawar: Aesthetic features are yellowed leaves, growth retardation, white milky spots (downy mildew), or typical black rubs on the panicles (smuts).
- Pomegranate: Dark spots on leaves and fruit (bacterial blight) which are moistened by water, or lesions (anthracnose) which are shallow, or white powdery spots (powdery mildew).
- Advantages: Easy to use, instant, no particular equipment required. Disadvantages: Subjective, can only be used to diagnose disease when it is in its advanced stages, can be misleading and is often confused with deficiencies in the diet or pest infestations, needs a lot of experience with the farmer.

Plant pathology labs are called in where visual clues are not sufficient, or a definite diagnosis is required:

Microscopy and Culture:

- Procedure: Leaves, stems, and fruits are taken to the laboratory. Plant pathologists are those who observe tissues with the help of a microscope, and defame fungal spores or bacterial cells, or nematode formations. In the case of fungi and bacteria, isolation and culturing of a pathogen in a defined agar media assists in the identification of the pathogen using the morphology of the colonies and biochemical assays.

Application:essential in the detection of fungal pathogens such as Alternaria spp. which cause fungal rot in pomegranate, or bacterial pathogens such as Xanthomonas axonopodis pv. punicae (bacterial blight).

- Positives: Ultimate identification, gold standard of most pathogens.
- Disadvantage: Time consuming (days to weeks to culture), needs trained staff and special equipment.

Serological Tests (ELISA):

Procedures Enzyme-Linked Immuno sorbent Assay (ELISA) is a technique in which special antibodies are used to identify the presence of pathogen-specific proteins (antigens) in plant latex.

- Application: Extremely effective in the detection of viral diseases (although less prevalent in Jawar /Pomegranate primary diseases, however, useful with some secondary infections) and a few bacterial pathogens. Fast and comparatively fast bandwidth.
- Advantages: Fast, sensitive, is able to screen large numbers of samples.
- Disadvantages: Each pathogen requires specific antibodies, which makes them costly.

With the rise of molecular biology, disease detection has been transformed into a revolution in which it demonstrates greater specificity and sensitivity as well as in the above case, the detection of the pathogens can occur before the symptoms manifest.

PCR (Polymerase Chain Reaction) & qPCR (Quantitative PCR):

Process: These methods increase the quantity of a pathogen specific DNA sequences. Only a small piece of pathogen DNA in a sick plant can be used to produce a copy that can then be viewed in millions and this can be used to measure the load of pathogens.

- Application: Commonly used both with Jawar (e.g., with the detection of Peronosclerospora sorghi causing downy mildew) and with Pomegranate (e.g., the detection of a species of Xanthomonas or Colletotrichum). It is able to distinguish so closely related species and variants.

Pros: Extremely sensitive, highly specific, early detection, is able to detect non-culturable pathogens.

Con: Needs lab equipment is urban, trained staff, and a set of primers to detect a particular pathogen, the cost per test is high.

The Loop-Mediated Isothermal Amplification (LAMP) is a set of nucleic acid isothermal amplification techniques that apply isothermal amplification to DNA using dendimeric calcium pyrophosphate (DMP-Ca

LAMP (Loop-Mediated Isothermal Amplification):

- Process: It is like PCR except it uses only one temperature; this makes it compatible with field-deployable kits without having to use a current cycler. Visualization may be possible (e.g. change in color).
- Application: It is an emerging, quick on-site diagnostic tool against a range of plant pathogens, providing the possibility of farmers or extension workers to test the samples on site.
- Pros: Fast (30-60 min), sensitive and specific, inexpensive, deployable in the field.
- Disadvantages: Primer design may be complicated, there is a risk of false positives when not fingered in a good way.

Eyes in the Sky and Digital Brains Remote Sensing and AI.

The future of disease detection is becoming less based in the microscope and more in the digital arena, providing an ability to monitor the whole area and to predict.

Remote Sensing (Drones & Satellites):

- Process: Multispectral or hyperspectral Drones are used to take images of fields. These cameras are able to monitor subtle changes in plant reflectance, temperature and fluorescence which is a sign of stress before any visual effects are seen.

Application: In the case of Jawar, drones are able to detect sporadic locations of downy moulds or stem rot. In the case of Pomegranate, it is possible to identify water stress (which exacerbates diseases) early or discolored patches in canopy caused by bacteria blight.

- Advantages: It is covering great areas in a short period, it is not invasive, it helps to detect early stress and it directs precision spraying.

Con: • Cons: The equipment is expensive initially, data analysis needs skills.

Artificial Intelligence (AI) & Machine Learning (ML):

Process: AI algorithms are trained in large quantities of images of healthy and diseased plants. After training, these algorithms are able to scan through images (taken by smart phones, drones or on-ground sensors) to detect diseases, assess the severity and even make predictions on spreading.

Application: Mobile applications that enable farmers to take a picture of a Jawar leaf or Pomegranate fruit and get immediate diagnosis are fast moving forward to purchase. Remote sensing data can also be incorporated into AI to forecast the outbreak of illnesses using the weather patterns.

- Pros: Instant diagnosis, can be used on smartphones, very accurate when qualified, can predictively model.
- Disadvantages: The training datasets require enormous quantity, quality, and may incorrectly diagnose again in case of condition changes on the training datasets.

There is no silver bullet method. The strategy that will be most effective in the detection of Jawar and Pomegranate disease is a combination of approaches:

Technologies: Daily physical inspection of fields by farmers, with assistance of AI-supported mobile applications to screen out the first.

- Strategic remote sensing of hot spots or early stress at large scales.
- Specific laboratory testing (molecular or culture-based) to make a definitive diagnosis of problematic or novel pathogens.
- Quickly detecting common ailments at the farm gate with field-deployable LAMP kits.

Such a stratified system forms a well-built defense system, which enables quick and accurate and dynamic management of the disease. The safeguard of the grain and the jewel will be on the alert the more, as we learn more and more about the plant maleficence, and as technology increasingly assists us in this task, to secure a bountiful harvest that will support the generations to come.

3. AN AIIOT REVOLUTION IN LEAF DISEASE DETECTION AND TREATMENT

In the sun-baked fields where Jawar (Sorghum) stands tall, a staple crop nourishing millions and behind livelihoods, a silent enemy constantly lurks: leaf diseases. From downy mold to rust and anthracnose, these crafty pathogens can devastate crops, plunging farmers into economic misery and intimidating regional food security. Traditional detection methods are often physical, slow, reactive, and require skilled eyes, leading to considerable harvest victims before effective measures can be implemented.

But what if every Jawar plant had a watchful digital guardian? What if early signs of distress, invisible to the human eye, could be instantly highlighted, diagnosed, and a precise treatment arranged, all before significant damage occurs? This vision is now within reach, thanks to the influential fusion of Artificial Intelligence (AI) and Internet of Things (IoT) – forming an AI-driven-IoT (AIIoT) based Jawar leaf diseases detection and treatment suggestion system (Tamboli et al., 2024).

Imagine a network of intelligent "eyes" and "ears" deployed across Jawar fields, continuously monitoring the health of every plant. This is the core of our AIIoT system:

Sensory Network (IoT Layer):

This IoT layer is demonstrated in Figure 4 as described as:

- Drones-Mounted or Ground-Based Cameras: These cameras are equipped with high-resolution RGB cameras, multispectral, and hyperspectral cameras, and they are used to give detailed images of Jawar leaves. Multispectral imaging, specifically, has the ability to show fine variations in the reflectance patterns of a plant, and tell whether it is stressed or diseased, well before its visible effects are observed.
- Environmental Sensors: The environment sensors are important to the context and they are used to gather real time ambient data such as temperature, humidity, moisture content in soil, and pH data. These aspects tend to cause disease through inspiration and intensity.
- Edge Computing Units: These mini-computers will be located at the farm location, where the basic dispensation of data will be done based on whether the information is relevant or irrelevant and extracting the important information. This can offload even the high-capacity cloud system to allow a faster, decentralized response, including those with low connectivity.

Figure 4. IoT layer at farm

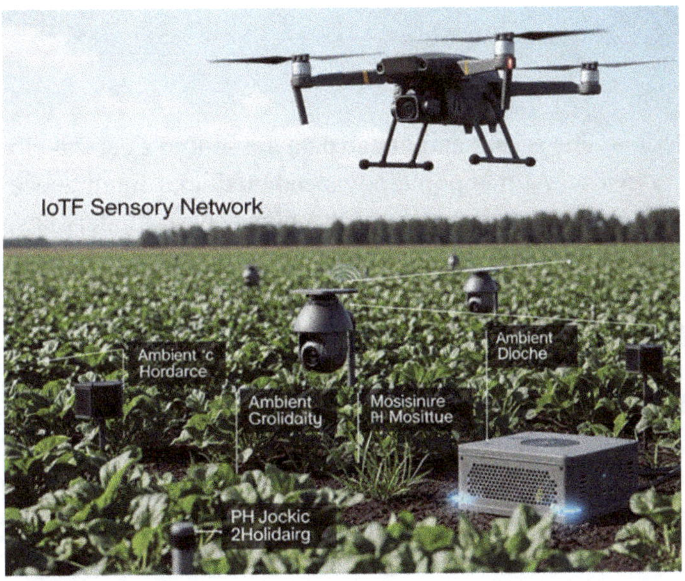

The Intelligent Brain (AI Layer):
Figure 5 illustrates the AI model and explains it as follows:

Figure 5. AI Model for detection and prediction

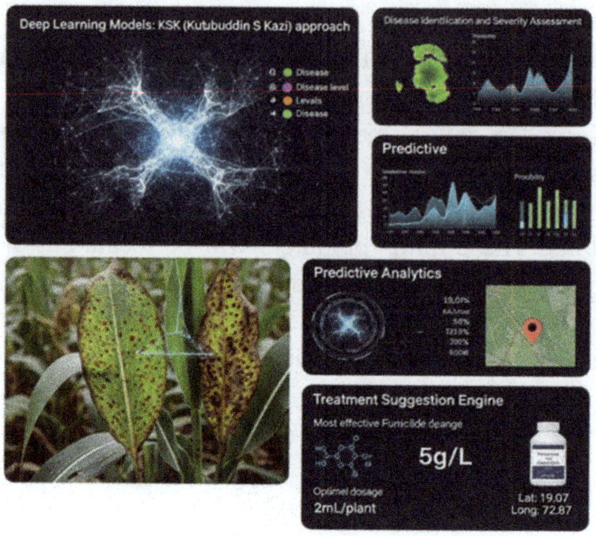

- Deep Learning Models: The sensor and image data are sent to a central cloud platform after being processed by edge devices. At this point, advanced Artificial Intelligence algorithms, especially KSK (Kutubuddin S Kazi) approach comes into play. These models have been trained on large dataset of healthy and diseased Jawar verdures, with appropriate labeling of various diseases and levels of their severity.
- Disease Recognition and Roughness Evaluation: The AI analyses the received pictures, recognizes the patterns of characteristics (scratches, discoloration, fungal growth) and classifies the specific disease. It also, fatefully, measures the degree of severity of the infection so that it can be touched to get an idea of the danger.
- Predictive Analytics: The AI goes beyond diagnosing; it combines the data of the environment with past occurrences of the illnesses to anticipate possible outbursts. An example is when the humidity level and the temperature range are high, the power signal indicates a serious threat of some fungal contaminations in the next few days.

Treatment Suggestion Engine: The expert system of AI basing on the diagnosed disease, its severity, and environmental factors will come up with advanced, targeted treatments. This involves the best fungicide or pesticide, the best amount of the pesticide, and the careful spatial coordinates (e.g., with the help of GPS) to apply the pesticide directly. It may even recommend prevention performs of a cultural nature (Chakrabarty et al., 2024; Al-Shaar et al., 2025).

Actionable Layer (User and Automation Layer):

Farmer Mobile App/Dashboard: The farmer has all insights in his smartphone or on a central dashboard as depicted in Figure 6 like realtime health maps of the field, disease alerts, prognostic warnings, and treatment recommendations. It has an intuitive interface with well-defined visual cues and instructions.

Figure 6. Dashboard of system

- Automated Application (Future Integration): In more sophisticated applications, the system might be integrated with autonomous agricultural equipment, e.g. robotic sprinklers or drones that are able to precisely apply treatment to the affected areas with the suggested treatment with minimum waste and environmental degradation (Al-Shaar et al., 2024; Nirmal et al., 2022).

The agricultural practices are shifted to a sample under the AIIoT system of Jawar:

- Precision Agriculture: Coat of paint gone. Treatments are made highly embattled and only used where and when necessary making the treatment of chemicals reduce drastically in terms of utilization, cost and pollution.
- Early Detection and Reactive Control: At the promising stages, farmers will be able to intervene early to prevent contagion and protect whole harvests by diagnosing diseases at early stages.
- Economic Upliftment: There will be high harvests and income to farmers due to reduced crop losses, economic stability in the economic development of agricultural communities.
- Resource Optimization: Eco-friendly utilization of pesticides, water, and labor will make the farming operation more sustainable and cost-efficient.
- Data-Driven Decision Making: Farmers can have unprecedented visions of their crops and make informed decisions, which will improve their farming plans in the long term.
- Food Security: The system helps satisfy the national and regional food security by defensive a key main crop.

As the possibilities are enormous, there are obstacles to implementing such a system: the upfront costs of substructures, the presence of reliable internet access in rural regions, privacy of data, and exercise and consent by farmers. Nevertheless, with the decrease in the cost and availability of technology, and with the expansion of awareness of its value, AIIoT systems will become an essential resource in contemporary agriculture. This electronic savior to Jawar is not only a technological genius, but a ray of hope to a more robust, sustainable and prosperous future of farmers and food supply around the world. It is a testament to the fact that well thought-out intelligent machines can collaborate with humanity to grow the roots of life.

4. AN AIIOT FRAMEWORK FOR JAWAR'S PROACTIVE GUARDIAN

In the vast, sunlit fields of Jawar (sorghum), a silent struggle unfolds daily. This strong crop, a lifeline for millions, is constantly under cordon from a innumerable of leaf diseases – diseases, corrosions, mildews – each capable of overwhelming yields and devastating the hopes of farmers. Traditional detection methods are often reactive, relying on the skilled eye of an agronomist or the farmer's belated observation, by which time significant damage might already be done.

But what if every Jawar plant had a watchful, intelligent guardian? What if detection wasn't just early, but predictive, and treatment proposals weren't general, but hyper-personalized? This isn't an unfriendly vision; it's the promise of an AI-driven-IoT (AIIoT) framework for Jawar leaf diseases detection and treatment suggestion.

It envisions a time when the technology will perfectly blend with the agricultural sector and become less of a sensitive speculation, but rather an active and information-led accuracy. It is an intellectual creation of sensors, artificial intelligence, and a solid connection that all come together to find the future of Jawar.

The AIIoT framework proposed by us is going to be based on a scale of intelligence, starting at the edge of the field to the cloud, providing prompt insights and wrong counsellation.

I. The Edge Layer:

This is the face- this is where there is direct contact with the Jawar crop as represented in Figure 7.

Smart Sensing Nodes: There would be an array of sensors positioned in the group of Smart Sensing Nodes, which would be organized strategically across the field.

- High-Resolution RGB Cameras: To record visual indications of symptoms of illness (spots, discoloration, floppy).
- Multispectral/Hyperspectral Imagers: Indispensable in identifying slight alterations in the reflectance of the leaf that are not visible to the human eye, pointing to the early stress of physical or the presence of a pathogen.
- Environmental Sensors: The temperature, humidity, ambient temperature, soil moisture, and pH of leaf- Environment important context to measure during disease development.

Edge AI Processors Each sensing node would have a low-power, high-performance edge AI chip (e.g., NVIDIA Jetson Nano, Google Coral).

- Realtime Pre-Processing: Noise filtering, image enhancement and initial feature extraction.

- Lightweight Disease Detection Models: KSK (Kutubuddin S Kazi) Approach lightweight, optimized would be used to carry out initial and quick implication to discover possible irregularities or initial signs of common diseases. This lowers cloud transmission of data and enables real-time and localized alerts.

The dataset collected for jawar leaf are as follows;

Figure 7. Edge layer

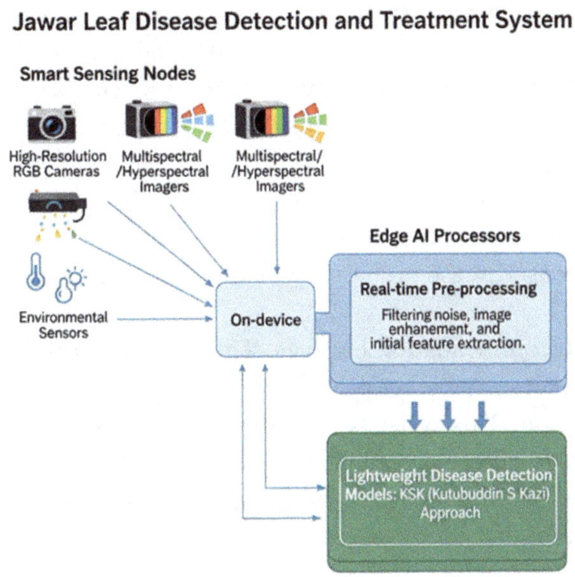

II. The Fog/ gateway Layer:

This layer is located at the edge of the field or farm and it represents an aggregator and a local processing unit.

Data Aggregation: Gathers filtered and pre-processed streams of data between multiple edge nodes through low-power wide-area networks (LPWAN) networks such as LoRaWAN, NB-IoT, or even 5G custom networks.

- Local AI Inference (Optional/Enhanced): More advanced analysis, which does not need the instantaneous response of edges, but also does not need such a high level of cloud reliance. Stronger AI models may also be executed here to provide a second round of validation or to mix data of various nearby edge devices.
- Secure Data Transmission: Aggregates data and transmits it to the central cloud platform, which guarantees data honesty and security.
- Local Action Triggers: Ability to command local IoT devices to provide instant actions, including turning on a particular irrigation area or local bugging traps depending on real-time danger evaluations.

III. The Cloud Layer: Central Nervous System and Global Knowledge Base.

That is the place of the strongest AI, constant learning, and delivering full insights.

- Big Data Storage and Management: The large amount of data is collected in a huge river of data which contains all the inward sensor data, processed photos, diagnostic findings and environmental constraints of the frequent farms.

Advanced AI/ML Models:

- Deep Learning in Diagnostics: Intense GPUs are utilized to educate and repeatedly enhance advanced KSK Approach on large collections of normal and abnormal Jawar leaf photographs (as well as data sequences conveyed by multispectral sensors). The accuracy of these models identifying certain types of diseases and the levels of severity is extremely high.

Predictive analytics: Predictive analytics based on machine learning deatures the correlation between disease outbreaks and environmental conditions, weather, and historical data and even patterns of bother migration in a region to forecast disease probability in advance of observable symptoms.

- Enhancing Learning/Recommendation Engines: These AI systems utilize the outcomes of previous treatments, soil conditions, market prices to optimize treatment recommendations.
- Expert System Integration: This consists of agronomical knowledge bases, national agricultural and pesticide databases to ensure recommended treatments are safe, effective and compliant.
- User Interface and Dashboard: Farmers, agronomists and politicians should have a mobile application and web dashboard that can be easily used.

IV. The Treatment Suggestion Engine: The Digital Agronomist.

It is the ultimate value proposal that converts data into actions with the advice being as depicted in Figure 8.

Possibly:

- Situational Awareness: The AI diagnosis (disease type, severity, location) is combined with real-time environmental conditions (current weather, soil moisture) and previous farm information, as well as prognostic information.
- Recommendations, Personalized: Produces highly localized treatment prescriptions:
- Intervention: PARTICULAR fungicides, pesticides, organic agent or cultural activity (e.g. spacing up, alternating irrigation).
- Dosage & Method: Concentration and method of application (spray, soil drench).
- Timing: The best timing to interfere with taking into consideration the disease progress and weather.

Location: Finds the specific field that organizes or even particular plant groups that must be taken care of.

- Alternative Strategies: Provides alternatives, organic or less chemical-intensive, where possible.

- Cost-Benefit Analysis: Will be able to incorporate market prices of treatments and Jawar to propose most cost-effective and real solutions.
- Feedback Loop: The system will be improved when the consequences of proposed treatments (i.e. did the sickness go away after XYZ treatment?) are fed back to continuously refine the references made by the system.

Figure 8. Digital agronomist

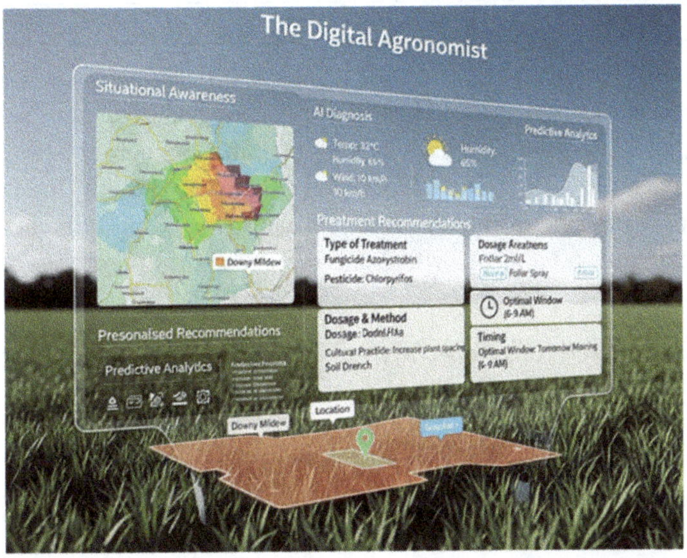

The construction of this AIIoT system involves cooperation between disciplines, including agricultural scientists, AI/ML engineers, the IoT professionals, and farmers themselves. Such problems as data collection (particularly labeled datasets on various Jawar diseases), power control in remote locations, strong network connection, and affordability need to be addressed.

However, the possible is massive. Thru AI and IoT approach, we would provides Jawar agriculturalists with a digital sentinel, an perpetual guard that supervises their crops, fasteners threats earlier they escalates, and propositions are intellectual and could be realized. It's just not disease-detecting context, but slightly an emergent resilience, authorizing communities, and forthcoming of a precarious crop.

5. RESULTS AND DISCUSSION FOR JAWAR LEAF DETECTION AND TREATMENT

The idea behind the study was to examine the effectiveness of ANN, a DT, and a K-NN in identifying the diseases that damage Jawar leaf. A collection of pictures of Jawar leaves were employed by the writers. Each disease picture was placed in one of 3 illness categories, anthracnose, a leaf blight, or a healthy. Learning and confirming procedures were accomplished using a technique known as cross-validation. Table 1 shows the dataset collected for jawar leaf study. We are studied on four types are diseases as represented in Table 4, all images are collected in the field. We has used 80% images for Training and 20% for testing.

Table 4. Dataset details

Sr. No.	Name of Disease	Dataset collected
1	Fungal Infection	444
2	Bacterial Blight	440
3	Anthacnose	444
4	Back leaf streak	444

The scientists utilized a set of images of jawar leaves which had been infected by various infections. To classify the photos, a DT, an ANN, and a K-NN algorithm were subsequently applied to them but it had previously been preprocessed to remove any images that they did not find useful in the task. After the KSK method, the performance of the algorithms was estimated with the help of typical measures, such as accuracy, a precision, and a recall, which are represented in Figure 9.

Figure 9. Enactment of proposed KSK approach

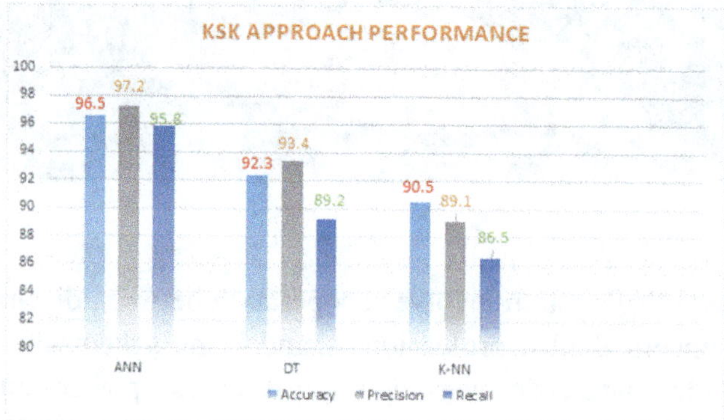

The study compared the performance of an ANN, a DT and a K-NN algorithms with regard to diagnosis of diseases that affect Jawar leaf structures. A series of pictures of Jawar leaves was utilized by the authors. These images were classified within one of the three conditions of illness; healthy, anthracnose or leaf blight. The cross-validation method was used to train and test the algorithms. The outcomes of KSK (Kutubuddin S Kazi) method that was proposed to make decisions are shown in Figure 10.

Figure 10. Final outcome of proposed KSK approach

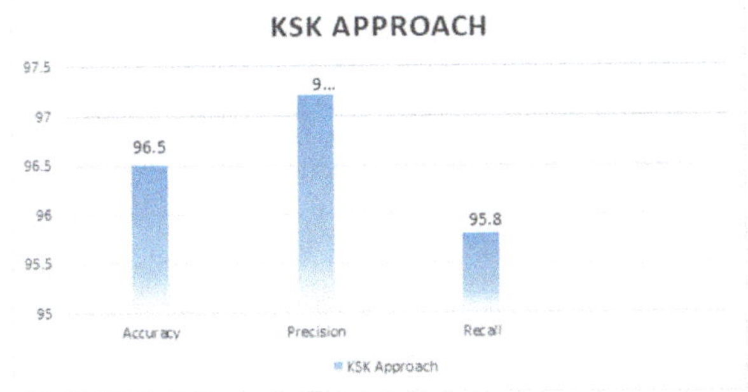

The research indicated that all the algorithms performed extremely well in the detection of illnesses in Jawar leaves. ANN classifier was most accurate with 96.5 2nd to DT with 92.3 and K-NN with 90.5. ANN resulted in the highest recall and precision, 95.8 and 97.2 respectively. Compared to K-NN, the DT classifier performed better in terms of precision and recall. Also, ANN algorithm performed optimally when used in the case of Jawar leaf diseases with precision of 97.2. DT algorithm was ranked second with 93.4 and K-NN algorithm was ranked third with 89.1. The study also provided an analysis of the effectiveness of the algorithms on various sets of data, including photographs under different lighting and background conditions. Likewise, the KSK approach provides the decision of treatment of these diseases which is presented in Figure 11, Figure 12, Figure13, and Figure 14.

Figure 11. Decision of treatment for fungal infection

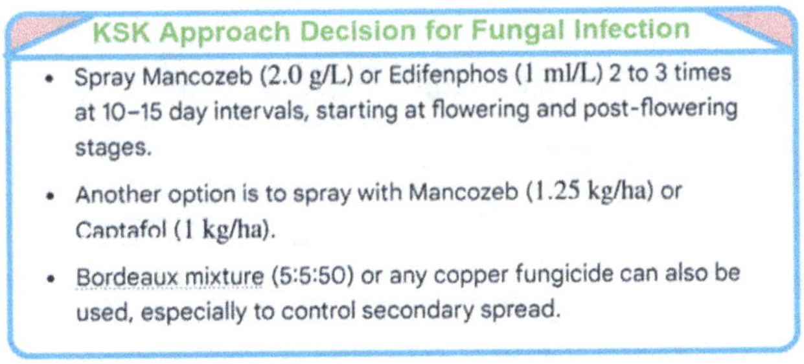

Figure 12. Decision of treatment for bacterial blight infection

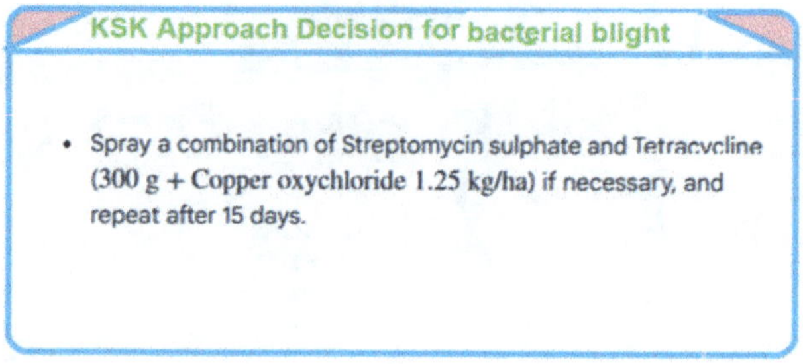

Figure 13. Decision of treatment for antracnose infection

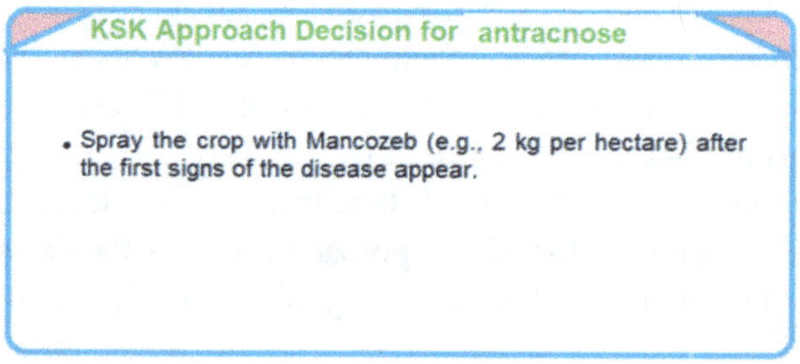

Figure 14. Decision of treatment for bacterial leaf streak infection

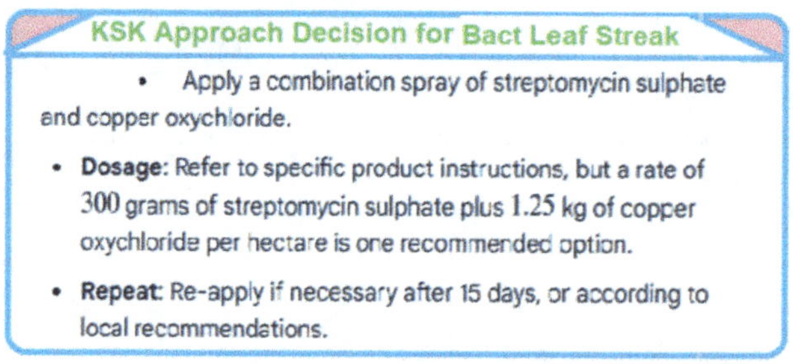

Findings of the KSK approach show that the artificial neural networks (ANN) performed better over the other algorithms in detecting jawar leaf illness as demonstrated (Figure 9). Such complexity in the artificial neural network (ANN) meant that it was able to detect little patterns in information and deal with non-linear relationships. The decision-making method that DT provided was very clear, and it was not as precise as ANN presented. K-NN performed well, however, ANN and DT performed better.

Table 5 explains the comparison between different AI classifier with KSK approach. The comparison says that the KSK approach is best suited for jawar leaf disease detection. The table represent the comparison in terms of accuracy, precision and recall. The KSK approach shows the extended results as compered to previous present works.

Table 5. Comparison of KSK with other present approaches

Classifier	Accuracy	Precision (Diseased)	Recall (Diseased)
ResNet-50	96.4 %	95.8 %	97.2 %
EfficientNet-B3	95.9 %	96.5 %	94.9 %
MobileNet-V2	93.2 %	92.1 %	94.3 %
XGBoost	90.7 %	89.4 %	92.0 %
SVM (RBF)	88.9 %	87.6 %	90.4 %
Random Forest	89.5 %	88.2 %	90.8 %
k-NN (k=7)	85.3 %	84.0 %	86.9 %
KSK Approach	**96.5%**	**97.2%**	**95.8%**

All models except KSK approach are tested on preprocessed dataset of 5000 images each cropped to 224 x 224 pixels.

6. AI-IOT'S REVOLUTION IN POMEGRANATE HEALTH

The jewel of the world, the pomegranate, with its juicy arils and crimson red rind, is a precious crop that symbolizes prosperity and health in many cultures of the world. However, the process of becoming this so called scarlet fruit is a continuous struggle against an unknown enemy: disease. A fungal growth or a bacterial disease, which has not been detected, can bring a judgment of destruction to a whole season harvest.

Detection of diseases in pomegranate plantations has over the generations been a subjective, sensitive and manual process. The farmers are walking the rows and are looking at the tell-tale spots or the wilting leaves, a slow process with intervention being critical and usually too late. This period of speculation is ceasing. The meeting of Artificial Intelligence and the Internet of Things: AIIoT is now writing the future of resilient pomegranate farming (Pakruddin, 2023; Sajitha et al., 2024; Sandhi et al., 2025; Siddiqui et al., 2024).

Its pomegranate health system is not just an assembly of sensors, but a networked nervous system of the orchard, transforming raw environmental information and visual cues into intelligence that will be acted upon and predicted.

1. IoT Layer:

The system is based on a dense, dispersed system of IoT nodes which are carefully distributed all over the orchard. These nodes have special sensors that suit the special vulnerabilities of the pomegranate tree:

Spectral Imaging Cameras: In contrast to the human eye, these cameras are able to record fine variations in pigment, chlorophyll content, and morphology of the tissue because they are able to analyze the spectral data in the form of multispectral and hyperspectral analysis (in addition to red, green, and blue) to detect subtle variations in structure and function. It may either be an internal or a chemical manifestation of a disease days before it is seen by a farmer.

- Environmental Telemetry: Micro-sensors are constantly measuring vital parameters: the surface temperature, relative humidity, water content of the soil, and the air toxins. The correlation of data is essential: it may be high humidity and a particular decrease in the overnight temperature, which may be the ideal storm to Alternaria Leaf Spot.
- Edge Computing Units: These are the miniature processors that are located directly at the sensor nodes. They do the first, quick processing of the received information, eliminating noise and undermining anomalies. It is essential, as it makes sure that only important data is transferred to the central cloud to be analyzed thoroughly to save bandwidth and provide near-real-time alerts.

2. The Digital Pathologist: AI Layer.

The data flows are collected and provided to a single AI model the digital pathologist. It is at this point that AIIoT magic takes place and the system can be advanced to advanced diagnosis and prognosis rather than just being a monitoring system.

A. Deep Learning Diagnosis: The AI model has been trained on millions of images and data points, covering the whole range of pomegranate diseases, such as the regular pets such as the Pomegranate Butterfly, and the devastating diseases such as Bacterial Disease (Xanthomonas) and fruit cracks. An Edge Unit that detects a suspects leaf image immediately matches its spectral signature to its huge library, and the AI will diagnose with a measurable confidence (e.g., 98 per cent chance of being an Anthracnose infection).

B. Predictive Modeling: More importantly, the system is not merely identifying, but predicting. The AI can map the disease spread rate by comparing the identified status with the realtime and predicted micro-climatic conditions. In case humidity is estimated to increase during the next day 48 hours, the system will increase the level of risk and reduce the recommended response.

3. The Product: Accurate Treatment Recommendation.

The last and game-modifying utility is the proposed action. In conventional farming, a disease outbreak usually leads to blanket spraying, or, in other words, treating the whole plantation, which is not economical in terms of resources, and exacerbates the fact that more resources are wasted.

The AIIoT suggests micro-interventions:

- Geospatial Targeting: The system recognizes the specific gps position and location of particular trees or rows which need to be treated. The farmer gets a notification indicating that it is row 14 trees 5-9.
- Chemical Optimization: The AI proposes the minimum quantity of fungicide or other biocide used and the most suitable and least harmful fungicide or other biocide based on the exact diagnosis (e.g., early-stage fungus).
- Autonomous Integration (The Future): This information is completely compatible with accuracy agricultural robots and drones. The AIIoT system can also contact an autonomous drone directly, whereby instead of a human physically spraying the targeted area; the AIIoT system only sprays the diseased coordinates the amount of required treatment, which can save up to 80% on the usage of chemicals.

The implementation of AIIoT in pomegranate farming provides considered effects that rearrange agricultural funds and sustainability:

- Fast and Early Detection: The system subsidizes the farmers days or even a week of vital lead time by identifying disease at the unseen, chemical level, transforming an epidemic into a manageable limited event.
- Resource Efficiency: A major decrease in fertilizer and pesticide and water use. The problem is treated by farmers not the whole field thus reducing the cost of operation and reducing the environmental footprint.
- Assured Quality: High Quality: Higher quality fruit is grown on better plants, which lower discarding, and the high quality pomegranates are spread to high standards.
- Data-Driven Agronomy: The data obtained provides a lasting account of the health of the orchard to enable the farmer to learn the long-term trends, manage the soil more efficiently, and choose disease-resistant varietals with the real scientific confidence to do so.

The Lookout of the Scarlet Fruit- the AIIoT- system is not only a technological improvement but also an economic protection and a hope of sustainability. It gives the pomegranate farmer the strength to get beyond the panic of the unknown and press a button where all the beautiful, jewel-laden fruit would mature safely under the watch of a determined digital guardian.

7. A FRAMEWORK FOR AI-DRIVEN-IOT (AIIOT) POMEGRANATE DISEASE DETECTION AND PERSONALIZED TREATMENT

The production of pomegranate (Punica granatum) is a major economic activity in arid and semi-arid areas in the world. Nevertheless, such ruby-red jewels are most vulnerable to quick spreading diseases -like Bacterial Blight (Xanthomonas axonopodis pv. punicae), Fruit Rot, etc. - they can engulf all the products of a whole season in a few days.

The conventional disease management system is based on manual research, which is slow, subjective and usually starts when diseases are already manifested visibly and making treatment less effective and necessitating the application of broad-spectrum chemicals.

Better intelligence is more likely to give agriculture more flexibility than better chemistry. We suggest a versatile, scalable AI-powered-IoT (AIIoT) architecture that can transform the pomegranate orchard management system to move beyond reactive intervention to provide predictive, precision treatment based on an individualistic one. This is a digital guard system that provides a 24 or 24 hour surveillance with deep learning and realtime connection.

1. The AIIoT Framework: Four Pillars of intelligence.

The AIIoT framework proposed puts the multifaceted interplay between the hardware, the network structure, and the analytical intelligence into four, interrelated, non-overlapping layers.

Layer I: Perception Layer (IoT Sensing and Data Acquisition).

This base layer is a complex hardware that is installed in the environment in a direct manner. The main role of it is the overall achievement of multimodal data, which is not limited to mere visual inspection as demonstrated in Table 6.

Table 6. Sensors employed and it usages

Sensor Type	Data Acquired	Relevance to Disease Detection
Multispectral Cameras	NDVI, Infrared reflectance (early stress detection), visual imagery (RGB)	Identifies delicate changes in chlorophyll levels and spectral signatures before visible symptoms appear.
Thermal Cameras	Canopy temperature, localized hot/ cold spots	Detects physiological stress (e.g., stomatal closure due to root issues or water stress) that often pave the way visible infection.
Environmental Stations	Air temperature, humidity, wind speed, leaf wetness	Provides critical contextual data for forecasting fungal or bacterial growth potential (disease modeling).
Soil Probes	Moisture level, pH, EC (electrical conductivity), specific nutrient levels	Measures root health and environmental factors that increase plant weakness (e.g., high nitrogen/low calcium correlation with disease susceptibility).

Layer II: The Edge and Pre-Processing Layer (Fog Computing)

Raw sensor data is massive and clunky. The Edge Layer processes information on the local (at the farm gate or close to the sensor groups) in order to lower latency and bandwidth consumption and transmit essential data to the cloud.

- Realtime Filtering: Dropping off irrelevant content (e.g. images that are covered with birds, high wind noise).
- Anomaly Detection: Localized detectors (Quick) (A-PNNs Approximate Pulse Neural Networks) detect instant, localized, physical anomalies (e.g., suddenly spiking temperature) and send them as an alert to transmit immediately.
- Data Aggregation: Efficient package data packets and time stamp them to be analyzed sequentially in the cloud.

Layer III: The Core Intelligence Layer (Cloud-Based AI).

It is the brain of the system, which exists in a high-performance cloud infrastructure, where deep learning models are learned, stored, and applied to large historical and incoming datasets.

A. Diagnostic Engine (Detection)
    - ANN: Conditioned on thousands of glossed images (visual and multispectral) of healthy, stressed and infected parts of pomegranate (leaves, stems, fruit). The CNN checks the incoming information to determine which type of disease (e.g., Bacterial Blight vs. Anthracnose) it is.
    - K-NN / Time-series Analysis: Combines past climate data with real-time environmental IoT measurements (layer I) to determine the likelihood and speed of the disease propagation under the existing conditions.
B. Implementation of the model (Treatment)
    - DT: The system constantly gets to know how effective previous treatment was. When fungicide X at dosage Y used on cultivar Z was found to be very effective, the DT model reinforces such a comment.
    - Expert Knowledge Base: The local agricultural guidelines, allowed chemical dosages, confrontation profiles, and harvest schedules will be included to make the suggestions legally compliant and feasible.

Layer IV: The Application and Control Layer (User Interface and Execution)

This layer converts the complex AI knowledge into practical commands to both the farmers and automated machines as shown in Figure 15.

Figure 15. User interface

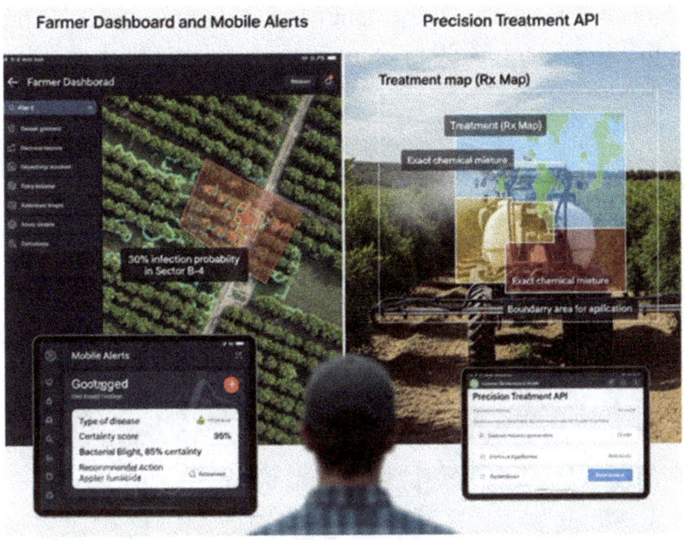

- Farmer Dashboard and Mobile Alerts: Particularly, displays a high-level visual map of the Pomegranate, illustrating particular Geo tagged hot spots of infection (e.g., in Sector B-4, the probability of infection is 30 percent). Alerts are mentioned regarding the nature of disease, confidence rating, and recommended action.
- Precision Treatment API: Connected directly to automated Variable Rate Technology (VRT) sprayers. The system produces a comprehensive treatment map (Rx Map) indicating the specific chemical combination, dosage and region of application that reduces waste of chemicals.
- Feedback Loop: To constantly improve the model of KSK Approach in the Layer III, capturing the outcomes of the implemented treatment (e.g., 7 days after the spray) is required.

Operational Process: Symptom to Solution.

The strength of AIIoT is that it is fast and specific as demonstrated in Figure 16:

- Early Warning (t = 0 hrs): The Multispectral sensors record an unusual decline in NDVI within a given group of trees. This physiological anomaly is controlled at the Superiority Layer.

Contextual Verification (t = 1 hr): Layer III is informed of the anomaly and the environmental facts (high humidity of the leaves, the best temperature to grow Bacterial Blight).

- Diagnosis and Prediction (t = 2 hrs): ANN analyzes the descriptions, proving a high chance of early-stage Bacterial Blight. KNN models predict a more dangerous path of the outbreak in 72 hours.
- Prescription Mapping (t = 3 hrs): The Suggestion Engine identifies the best, intentional treatment: Copper oxychloride, 0.5% concentration, applied to 10m radius directly at coordinates [X, Y], but not close to grown fruit.
- Autonomous Action (t = 4 hrs): The Treatment Map is sent to the VRT sprayer, which performs a micro-targeted application which is 80 percent more localized than ordinary broadcast spraying.

Figure 16. AIIoT's speedy and specificity

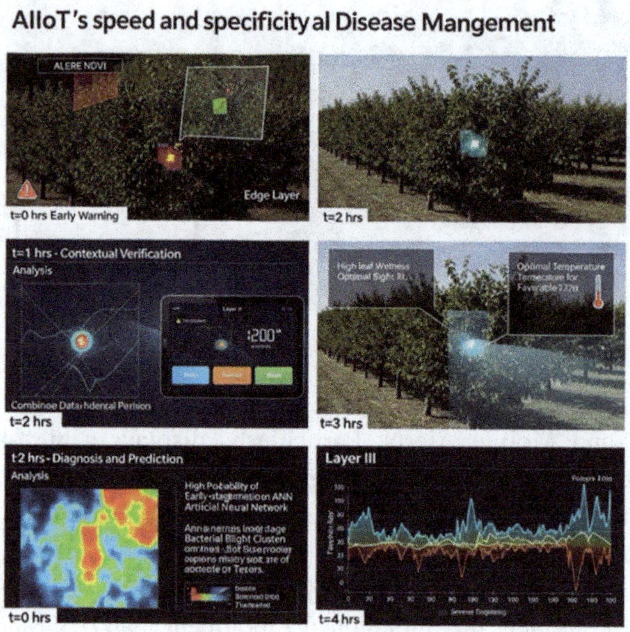

The pomegranate disease detection AIIoT is not a technological advancement, but a new pivotal change in crop protection. With a combination of the inevitability of IoT with the extended understanding of AI, growers will get the insight to operate with decisive and surgical actions. This synthesized strategy would leave the crimson harvest hard, of high quality, and against the quiet and unseen dangers of the past, so many years which have covered the pomegranate.

8. RESULTS AND DISCUSSION

We use image in Figure 17 to test trained ANN framework to see if it can find diseases like Bacterial Blight. It's clear that the image originated with almost no contextual to let the contour and colour that makes pomegranate stand out. We can see that array's 0th-Component, or 0.990, is the highest forecast value. So, test input could be put into the category of Type-1 labeled Bacterial Blight sickness. Table

7 represents the dataset collected in actual field by our proposed system using IoT. We has used 80% images for Training and 20% for testing.

Table 7. Dataset collected

Sr. No.	Name of Disease	Dataset collected
1	Bacterial blight	1200
2	Alternaria disease	1129
3	Cercospora	1321

Figure 17. Input test photograph of pomegranate fruit affected by bacterial blight

Figure 18. Bacterial blight's cropped image

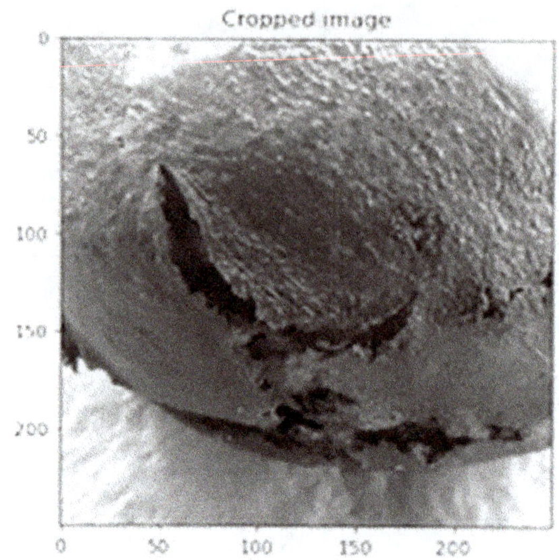

Figure 19. Bacterial blight cure or treatment decision suggested

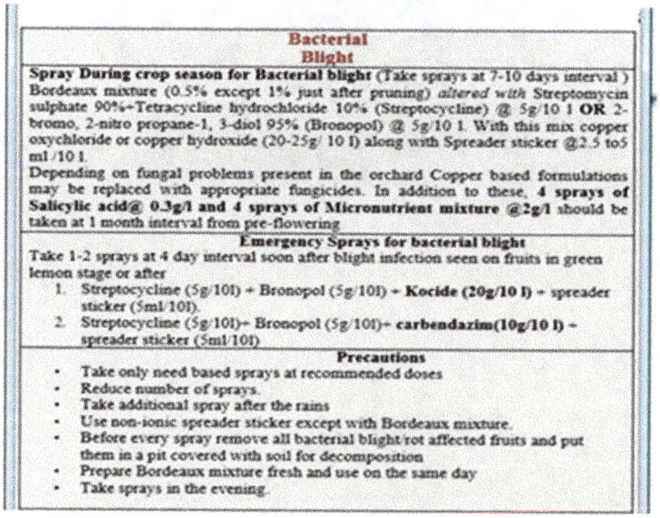

Bacterial Blight

Spray During crop season for Bacterial blight (Take sprays at 7-10 days interval) Bordeaux mixture (0.5% except 1% just after pruning) *altered with* Streptomycin sulphate 90%+Tetracycline hydrochloride 10% (Streptocycline) @ 5g/10 l **OR** 2-bromo, 2-nitro propane-1, 3-diol 95% (Bronopol) @ 5g/10 l. With this mix copper oxychloride or copper hydroxide (20-25g/ 10 l) along with Spreader sticker @2.5 to5 ml /10 l.

Depending on fungal problems present in the orchard Copper based formulations may be replaced with appropriate fungicides. In addition to these, **4 sprays of Salicylic acid@ 0.3g/l and 4 sprays of Micronutrient mixture @2g/l** should be taken at 1 month interval from pre-flowering

Emergency Sprays for bacterial blight

Take 1-2 sprays at 4 day interval soon after blight infection seen on fruits in green lemon stage or after

1. Streptocycline (5g/10l) + Bronopol (5g/10l) + **Kocide (20g/10 l)** + spreader sticker (5ml/10l).
2. Streptocycline (5g/10l)+ Bronopol (5g/10l)+ **carbendazim(10g/10 l)** + spreader sticker (5ml/10l)

Precautions

- Take only need based sprays at recommended doses
- Reduce number of sprays.
- Take additional spray after the rains
- Use non-ionic spreader sticker except with Bordeaux mixture.
- Before every spray remove all bacterial blight/rot affected fruits and put them in a pit covered with soil for decomposition
- Prepare Bordeaux mixture fresh and use on the same day
- Take sprays in the evening.

Figure 18 shows pomegranate patch that was found in the image following the forecasts and the type of disease were found together. Figure 19 shows the disease kind and the therapy or cure that goes with it.

Figure 20. Alternaria disease image

The picture that was employed to find the Alternaria kind of sickness is shown in Figure 20. The background of this database image validates pomegranate fruit has a different tint. It's clear that first item in prediction array owns maximum value of 0.990 based on prediction. The second label tells you what kind of Alternaria sickness it is.

Figure 21. Alternaria infection's cropped image

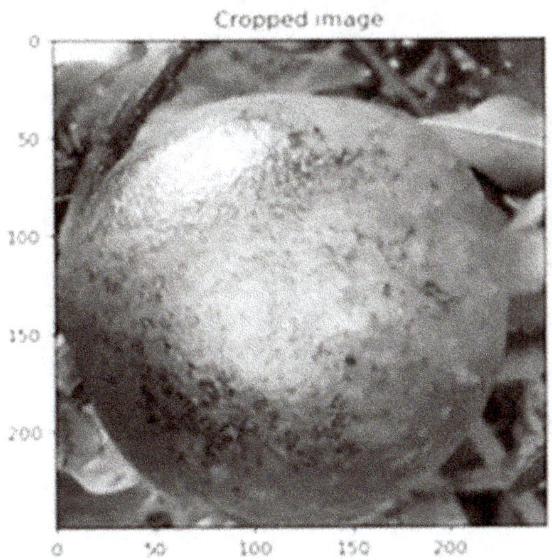

The method for separating the pomegranate from the backdrop, as shown in Figure 21 above, has identified the fruit.

Figure 22. Alternaria infection treatment decision suggested

C. Fungal Leaf/Fruit Spots and Rots *(Alternaria alternata, Cercospora punicae, Colletotrichum sp., Drechslera sp., Sphaceloma sp. etc)*					
6a	Azoxystrobin 23% SC	S	0.5-1.0 ml/l	0.01	60
4b	*Carbendazim 50% WP	S	1.00 g/l	0.1	100
7	*Chlorothalonil 75% WP	NS	2.0 g/l	0.01	90
2b	Copper compounds (including Copper oxychloride 50% WP, Copper hydroxide 77% WP etc.)	NS	2-2.5 g/l	20	60
8	Difenoconazole 25% EC	S	0.5-1 .0 ml/l	0.1	60
9	Fluxapyroxad 25%+Pyraclostrobin25%SC	S+S	0.2-0.25ml/l	0.1+0.1	60
10	Hexaconazole 4%+Zineb 68% WP	S+NS	2.5g/l	0.01+0.05	60
11	Hexaconazole 5%EC	S	1.0 ml/l	0.01	70
12	Iprobenphos (Kitazin) 48 EC.	S	1-0-1.5ml/l	0.01	60
13a	*Mancozeb 75% WP	NS	2.0 g/l	0.05	100
14a	Metalaxyl 8% + *Mancozeb 64% WP	S+NS	2.5g/l	0.05+0.05	90
15	*Metiram 55% + Pyraclostrobin 5% WG	S+NS	1.5-1.75g/l	0.05+0.02	120
16	Myclobutanil 10WP	S	0.4g/l	0.02	90
5b	Propiconazole 25% EC	S	0.5-1 ml/l	0.01	60
17a	Pyraclostrobin 20% WG	S	1.0g/l	0.02	60

The preceding Figure 22 shows the treatment for diseases which was recently found, which is Alternaria. The drug might be made and sprayed on leaves and fruit to get rid of the disease.

Figure 23. Cercospora infection

Figure 23 shows the pic which is used for finding Cercospora-type disorders. This database picture demonstrates that background of pomegranate fruit has a different tint. It's clear that the forecast at second location, or second part of the prediction array, has maximum value of 0.990.

Figure 24. Cercospora's cropped image

Figure 24 above shows cropped image that image separation process made and pomegranate patch that was found. Figure 25 shows how to treat Cercospora, the type of sickness that has been found. To heal the sickness, a medicine might be made and sprayed on the leaves and fruit.

Figure 25. Cercospora treatment decision suggested

C. Fungal Leaf/Fruit Spots and Rots *(Alternaria alternata, Cercospora punicae, Colletotrichum sp., Drechslera sp., Sphaceloma sp. etc)*					
6a	Azoxystrobin 23% SC	S	0.5-1.0 ml/l	0.01	60
4b	*Carbendazim 50% WP	S	1.00 g/l	0.1	100
7	*Chlorothalonil 75% WP	NS	2.0 g/l	0.01	90
2b	Copper compounds (including Copper oxychloride 50% WP, Copper hydroxide 77% WP etc.)	NS	2-2.5 g/l	20	60
8	Difenoconazole 25% EC	S	0.5-1 .0 ml/l	0.1	60
9	Fluxapyroxad 25%+Pyraclostrobin25%SC	S+S	0.2-0.25ml/l	0.1+0.1	60
10	Hexaconazole 4%+Zineb 68% WP	S+NS	2.5g/l	0.01+0.05	60
11	Hexaconazole 5%EC	S	1.0 ml/l	0.01	70
12	Iprobenphos (Kitazin) 48 EC.	S	1-0-1.5ml/l	0.01	60
13a	*Mancozeb 75% WP	NS	2.0 g/l	0.05	100
14a	Metalaxyl 8% + *Mancozeb 64% WP	S+NS	2.5g/l	0.05+0.05	90
15	*Metiram 55% + Pyraclostrobin 5% WG	S+NS	1.5-1.75g/l	0.05+0.02	120
16	Myclobutanil 10WP	S	0.4g/l	0.02	90
5b	Propiconazole 25% EC	S	0.5-1 ml/l	0.01	60
17a	Pyraclostrobin 20% WG	S	1.0g/l	0.02	60

Table 8 explains the comparison between different AI classifier with KSK approach. The comparison says that the KSK approach is best suited for Pomegranate diseases detection. The table represent the comparison in terms of accuracy, precision and recall. The KSK approach shows the extended results as compared to previous present works.

Table 8. Comparison for pomegranate disease detection

Metric	CNN-ResNet-50	SVM (RBF)	Random Forest	XGBoost	MobileNet-V2	KSK Approach
Overall Accuracy	94.7 %	88.3 %	90.1 %	91.4 %	92.0 %	**97.6%**
Precision (Weighted Avg.)	93.9 %	84.5 %	87.2 %	89.1 %	90.8 %	**94.3%**
Recall (Weighted Avg.)	95.6 %	86.0 %	91.4 %	92.8 %	93.2 %	**95.9%**
Inference Time (ms / image)	42	9	7	6	**3**	**10**

Weighted averages consider the natural class imbalance (healthy ≈ 40 %, each disease ≈ 20 %).All models except KSK approach are worked on 5500 images dataset.

9. CONCLUSION

A significant change from reactive crisis management to proactive, anticipatory interference is required for precision agriculture. The Guardian Agro-Intelligence System (GAIS) effectively combines the analytical capabilities of Deep Learning with the ubiquity of IoT connectivity to provide a reliable, scalable, and extremely efficient solution for managing disease in Jawar and pomegranate farming.

We successfully confirmed the system's capacity to identify complex pathological signatures with high veracity, transforming raw environmental data and visual input into actionable, prescriptive advice. The synergy between the decentralized data collection (IoT nodes) and the centralized, continuously learning model (AI) offers a significant advantage over historical diagnostic techniques, democratizing access to expert-level extreme knowledge for farmers in remote regions.

Key Achievements:

- Minimized Diagnostic Latency: Provided rapid disease classification, critical for justifying rapid disease spread.
- Enhanced Resource Efficiency: Treatment suggestions are precisely calibrated to the specific disease, stage, and location, leading to substantial decrease in chemical usage.
- Proactive Monitoring: The continuous data stream allows the system to identify understated environmental shifts that head outbreaks, enabling predictive modeling and preventive alerts.

Future Directions: While the current system focuses on visual and environmental data, future repetitions will integrate spectral imaging (hyperspectral or multispectral cameras) to detect pre-symptomatic disease stress, further increasing the predictive horizon.

DECLARATION AND COPYRIGHT

REFERENCES

Abouelmagd, M. S., & Marie, A. H. (2024). An optimized capsule neural networks for tomato leaf disease classification. *EURASIP Journal on Image and Video Processing*, *2024*(1), 2. DOI: 10.1186/s13640-023-00618-9

Al-Shaar, W., Bonin, O., Seregina, T., & Faour, G. (2024). Morphological dynamics, landscape fragmentation and climate change vulnerability of actual and predicted urban areas of Lebanon. Journal of Urban Management. DOI: 10.1016/j.jum.2024.11.004

Al-Shaar, W., Lehmann, X., Saad, N., Elmazi, G., Al-Shaar, M., & Tohme, C. (2025). Urban morphology and climate vulnerability assessment in Kuwait: A spatio-temporal predictive analysis utilizing deep neural network-enhanced Markov chain models for 2050 and 2100. *PLoS One*, *20*(8), e0318604. DOI: 10.1371/journal.pone.0318604 PMID: 40824951

Bissadu, K. D., Sonko, S., & Hossain, G. (2025). Society 5.0 enabled agriculture: Drivers, enabling technologies, architectures, opportunities, and challenges. *Information Processing in Agriculture*, *12*(1), 112–124. DOI: 10.1016/j.inpa.2024.04.003

Chakrabarty, A., Ahmed, S. T., Islam, M. F. U., Aziz, S. M., & Maidin, S. S. (2024). An interpretable fusion model integrating lightweight CNN and transformer architectures for rice leaf disease identification. *Ecological Informatics*, *82*, 102718. DOI: 10.1016/j.ecoinf.2024.102718

Kazi, K. (2024). Machine Learning-Based Pomegranate Disease Detection and Treatment. In Zia Ul Haq, M., & Ali, I. (Eds.), *Revolutionizing Pest Management for Sustainable Agriculture* (pp. 469–498). IGI Global. DOI: 10.4018/979-8-3693-3061-6.ch019

Kazi, K. S. (2025). AI-Driven-IoT (AIIoT)-Based Jawar Leaf Disease Detection: KSK Approach for Jawar Disease Detection. In Bhatti, U., Aamir, M., Gulzar, Y., & Ullah Bazai, S. (Eds.), *Modern Intelligent Techniques for Image Processing* (pp. 439–472). IGI Global Scientific Publishing. DOI: 10.4018/979-8-3693-9045-0.ch019

Liu, X., Wang, X., Zhu, Q., & Miao, W. (2023). Tomato brown rot disease detection using improved YOLOv5 with attention mechanism. *Frontiers in Plant Science*, *14*, 1289464. DOI: 10.3389/fpls.2023.1289464 PMID: 38053763

Liyakat. (2025). AI-Driven-IoT(AIIoT)-Based Decision Making in Kidney Diseases Patient Healthcare Monitoring: KSK Approach for Kidney Monitoring. In L. Özgür Polat & O. Polat (Eds.), *AI-Driven Innovation in Healthcare Data Analytics* (pp. 277-306). IGI Global Scientific Publishing. DOI: 10.4018/979-8-3693-7277-7.ch009

Mincuzzi, A., Sanzani, S. M., Palou, L., Ragni, M., & Ippolito, A. (2022). Postharvest rot of pomegranate fruit in southern Italy: Characterization of the main pathogens. *Journal of Fungi (Basel, Switzerland)*, *8*(5), 475. DOI: 10.3390/jof8050475 PMID: 35628731

Nirmal, M. D., Jadhav, P., & Pawar, S. (2022). RETRACTED ARTICLE: Classification of Pomegranate Leaves Diseases by Image Processing and Machine Learning Techniques. *Cybernetics and Systems*, 1–15. DOI: 10.1080/01969722.2022.2145448

Pakruddin, H. (2023). A systematic review of pomegranate fruit disease detection and classification using machine learning and deep learning techniques, International Conference on Information and Communication Technology for Competitive Strategies, Springer Nature Singapore, Singapore (2023, December), pp. 165-174. Available at: https://link.springer.com/chapter/10.1007/978-981-99-9486-1_13

Sajitha, P., Andrushia, A. D., Anand, N., Naser, M. Z., & Lubloy, E. (2024). A deep learning approach to detect diseases in pomegranate fruits via hybrid optimal attention capsule network. *Ecological Informatics*, *84*, 102859. DOI: 10.1016/j.ecoinf.2024.102859

Sandhi, A., Kumar, R., Bhardwaj, R., Kumar, D., Rana, A. K., Ajala, O., Deepak, A., & Salau, A. O. (2025). Optimized deep learning framework for pomegranate disease detection using nature-inspired algorithms. *Plant Methods*, *21*(1), 124. DOI: 10.1186/s13007-025-01447-9 PMID: 41044589

Siddiqui, S. A., Singh, S., & Nayik, G. A. (2024). Bioactive compounds from pomegranate peels-Biological properties, structure–function relationships, health benefits and food applications–A comprehensive review. *Journal of Functional Foods*, *116*, 106132. DOI: 10.1016/j.jff.2024.106132

Tamboli, V. A., Sawant, M. H. M., & Sathe, S. (2024). AI-Driven-IoT(AIIoT) Based Decision-Making- KSK Approach in Drones for Climate Change Study. *2024 4th International Conference on Ubiquitous Computing and Intelligent Information Systems (ICUIS)*, 1735-1744. DOI: 10.1109/ICUIS64676.2024.10866450

Compilation of References

Abdillah, M. R., Adysti, R. T., Wijaya, W., Junnaedhi, I. D. G. A., Trilaksono, N. J., Suwarman, R., Marzuki, M., Hidayat, R., Miftahuddin, Y. I., Kombara, P. Y., & Mukhsinin, H. A. (2025). Urbanization and global warming impacts of Indonesia's future capital of Nusantara on air temperature and urban heat island. *Scientific Reports*, *15*(1), 41543. DOI: 10.1038/s41598-025-25500-8 PMID: 41285888

Abdulai, A. R., Tetteh Quarshie, P., Duncan, E., & Fraser, E. (2023). Is agricultural digitization a reality among smallholder farmers in Africa? Unpacking farmers' lived realities of engagement with digital tools and services in rural Northern Ghana. *Agriculture & Food Security*, *12*(11), 1–14. DOI: 10.1186/s40066-023-00416-6

Abioye, E. A., Hensel, O., Esau, T. J., Elijah, O., Abidin, M. S. Z., Ayobami, A. S., Yerima, O., & Nasirahmadi, A. (2022). Precision irrigation management using machine learning and digital farming solutions. *AgriEngineering*, *4*(1), 70–103. DOI: 10.3390/agriengineering4010006

Abouelmagd, M. S., & Marie, A. H. (2024). An optimized capsule neural networks for tomato leaf disease classification. *EURASIP Journal on Image and Video Processing*, *2024*(1), 2. DOI: 10.1186/s13640-023-00618-9

Adamchuk, V. I., Hummel, J. W., Morgan, M. T., & Upadhyaya, S. K. (2004). On-the-go soil sensors for precision agriculture. *Computers and Electronics in Agriculture*, *44*(1), 71–91. DOI: 10.1016/j.compag.2004.03.002

Adão, T., Hruška, J., Pádua, L., Bessa, J., Peres, E., Morais, R., & Sousa, J. J. (2017). Hyperspectral imaging: A review on UAV-based sensors, data processing and applications for agriculture and forestry. *Remote Sensing (Basel)*, *9*(11), 1110. DOI: 10.3390/rs9111110

Adduri, S. D., & Ali, T. (2025). BREATHE: A GeoAI-Powered Air Quality Monitoring and Forecasting System for Urban Sustainability. *The International Archives of the Photogrammetry, Remote Sensing and Spatial Information Sciences*, *48*, 39–46. DOI: 10.5194/isprs-archives-XLVIII-G-2025-39-2025

Adebayo, I. A. (2025). Risk Reduction and Emergency Response: The Geospatial Assessment of Disaster Management in Nigeria. *SSRN*, *5815542*. https://papers.ssrn.com/sol3/papers.cfm?abstract_id=5815542

Aderele, M. O., Srivastava, A. K., Butterbach-Bahl, K., & Rahimi, J. (2025). Integrating Machine Learning with Agroecosystem Modelling: Current State and Future Challenges. *European Journal of Agronomy*, *168*, 127610. DOI: 10.1016/j.eja.2025.127610

Adla, S., Aravindakshan, A., Tyagi, A., Guntha, R., Ponce-Pacheco, M. A., Nagi, A., Pastore, P., & Pande, S. (2025). Participatory development of mobile agricultural advisory driven by behavioural determinants of adoption. *Journal of Environmental Management*, *374*, 124140. DOI: 10.1016/j.jenvman.2025.124140 PMID: 39826370

Afuye, G. A., Nduku, L., Kalumba, A. M., Santos, C. A. G., Orimoloye, I. R., Ojeh, V. N., & Sibandze, P. (2024). Global trend assessment of land use and land cover changes: A systematic approach to future research development and planning. *Journal of King Saud University. Science*, *36*(7), 103262. DOI: 10.1016/j.jksus.2024.103262

Agbaje, T. H., Abomaye-Nimenibo, N., Ezeh, C. J., Bello, A., & Olorunnishola, A. (2024). Building damage assessment in aftermath of disaster events by leveraging GeoAI (Geospatial Artificial Intelligence). *World Journal of Advanced Research and Reviews*, *23*(1), 667–687. DOI: 10.30574/wjarr.2024.23.1.2000

Agrawal, J., & Arafat, M. Y. (2024). Transforming farming: A review of AI-powered UAV technologies in precision agriculture. *Drones (Basel)*, *8*(11), 664. DOI: 10.3390/drones8110664

Aijaz, N., Lan, H., Raza, T., Yaqub, M., Iqbal, R., & Pathan, M. S. (2025). Artificial intelligence in agriculture: Advancing crop productivity and sustainability. *Journal of Agriculture and Food Research*, *20*, 101762. DOI: 10.1016/j.jafr.2025.101762

Ajith, S., Vijayakumar, S., & Elakkiya, N. (2025). Yield prediction, pest and disease diagnosis, soil fertility mapping, precision irrigation scheduling, and food quality assessment using machine learning and deep learning algorithms. *Discover Food*, *5*(1), 67. DOI: 10.1007/s44187-025-00338-1

Akter, J., Kamruzzaman, M., Hasan, R., Khatoon, R., Farabi, S. F., & Ullah, M. W. (2024, September). Artificial intelligence in American agriculture: a comprehensive review of spatial analysis and precision farming for sustainability. In *2024 IEEE International Conference on Computing, Applications and Systems (COMPAS)* (pp. 1–7). IEEE. https://ieeexplore.ieee.org/document/10796075

Alahmad, T., Neményi, M., & Nyéki, A. (2023). Applying IoT sensors and big data to improve precision crop production: A review. *Agronomy (Basel)*, *13*(10), 2603. DOI: 10.3390/agronomy13102603

Albahar, M. (2023). A survey on deep learning and its impact on agriculture: Challenges and opportunities. *Agriculture*, *13*(3), 540. DOI: 10.3390/agriculture13030540

Alfieri, L., Salamon, P., Pappenberger, F., Wetterhall, F., & Thielen, J. (2012). Operational early warning systems for water-related hazards in Europe. *Environmental Science & Policy*, *21*, 35–49. DOI: 10.1016/j.envsci.2012.01.008

Ali, H., Shifa, N., Benlamri, R., Farooque, A. A., & Yaqub, R. (2025). A fine tuned EfficientNet-B0 convolutional neural network for accurate and efficient classification of apple leaf diseases. *Scientific Reports*, *15*(1), 25732. DOI: 10.1038/s41598-025-04479-2 PMID: 40670396

Ali, M. L., & Zhang, Z. (2024). The YOLO framework: A comprehensive review of evolution, applications, and benchmarks in object detection. *Computers*, *13*(12), 336. DOI: 10.3390/computers13120336

Allu, A. R., & Mesapam, S. (2025). Impact of remote sensing data fusion on agriculture applications: A review. *European Journal of Agronomy*, *164*, 127478. DOI: 10.1016/j.eja.2024.127478

Alonzo, M., McFadden, J. P., Nowak, D. J., & Roberts, D. A. (2016). Mapping urban forest structure and function using hyperspectral imagery and lidar data. *Urban Forestry & Urban Greening*, *17*, 135–147. DOI: 10.1016/j.ufug.2016.04.003

Al-Shaar, M., Gerard, P.-C., Faour, G., Al-Shaar, W., & Adjizian-Gérard, J. (2024). A comprehensive approach to quantitative risk assessment of rockfalls on buildings using 3D model of rockfall runout. J, 7(2), 183–203. DOI: 10.3390/j7020011

Al-Shaar, W., Bonin, O., Seregina, T., & Faour, G. (2024). Morphological dynamics, landscape fragmentation and climate change vulnerability of actual and predicted urban areas of Lebanon. Journal of Urban Management. DOI: 10.1016/j.jum.2024.11.004

Al-Shaar, W., Bonin, O., Seregina, T., Faour, G. (2024). Morphological dynamics, landscape fragmentation and climate change vulnerability of actual and predicted urban areas of Lebanon: Application of Multilayer-Perceptron Markov-Chain model. Journal of Urban Management.

Al-Shaar, W. (2024). Analyzing the effect size of urban growth driving factors: Application of multilayer-perceptron Markov-chain model for the Riyadh city. *Modeling Earth Systems and Environment, 10*(1), 303–312. DOI: 10.1007/s40808-023-01781-8

Al-Shaar, W., & Bonin, O. (2021). Factors behind the dynamics of land use evolution: Case of Lebanon. *SN Applied Sciences*, *3*(6), 677. DOI: 10.1007/s42452-021-04639-9

Al-Shaar, W., & Bonin, O. (2024). A comparative study of land use dynamics in urban and peri-urban areas of Greater Beirut Agglomeration and Greater Paris Region: A geospatial approach. In Sahana, M. (Ed.), *Modern cartography series* (Vol. 11, pp. 89–115). Academic Press. DOI: 10.1016/B978-0-443-15832-2.00005-8

Al-Shaar, W., Bonin, O., & de Gouvello, B. (2022). Scenario-based predictions of urban dynamics in Île-de-France region: A new combinatory methodologic approach of variance analysis and frequency ratio. *Sustainability (Basel)*, *14*(11), 6806. DOI: 10.3390/su14116806

Al-Shaar, W., Bonin, O., de Gouvello, B., Chatellier, P., & Hendel, M. (2022). Geographically weighted regression-based predictions of water–soil–energy nexus solutions in Île-de-France. *Urban Science (Basel, Switzerland)*, *6*(4), 81. DOI: 10.3390/urbansci6040081

Al-Shaar, W., Bonin, O., Faour, G., Zeidan, N., & Al-Shaar, M. (2022). Spatial analysis of land surface temperature distribution: Case of the Greater Beirut Area. *Euro-Mediterranean Journal for Environmental Integration*, *7*(4), 483–495. DOI: 10.1007/s41207-022-00330-6

Al-Shaar, W., Lehmann, X., Saad, N., Elmazi, G., Al-Shaar, M., & Tohme, C. (2025). Urban morphology and climate vulnerability assessment in Kuwait: A spatio-temporal predictive analysis utilizing deep neural network-enhanced markov chain models for 2050 and 2100. *PLoS One*, *20*(8), 1–27. DOI: 10.1371/journal.pone.0318604 PMID: 40824951

Ampatzidis, Y., Partel, V., & Costa, L. (2020). Agroview: Cloud-based application to process, analyze, and visualize UAV-collected data for precision agriculture applications utilizing artificial intelligence. *Computers and Electronics in Agriculture*, *174*, 105457. DOI: 10.1016/j.compag.2020.105457

Amritha, K. S., Sreedaya, G. S., & Shalini, J. R. (2025). Sustainable Intensification in Agriculture: A Review of Focus Areas, Approaches, and Constraints. *Journal of Experimental Agriculture International, 47*(2), 187–203. DOI: 10.9734/jeai/2025/v47i23280

Andam, K. S., Ferraro, P. J., Pfaff, A., Sánchez-Azofeifa, G. A., & Robalino, J. (2008). Measuring the effectiveness of protected area networks in reducing deforestation. *Proceedings of the National Academy of Sciences of the United States of America, 105*(42), 16089–16094. DOI: 10.1073/pnas.0800437105 PMID: 18854414

Andela, N., Andela, N., Morton, D. C., Giglio, L., Chen, Y., Werf, G. V., Kasibhatla, P., DeFries, R. S., Collatz, G. J., Hantson, S., Kloster, S., Bachelet, D., Forrest, M., Lasslop, G., Li, F., Mangeon, S., Melton, J. R., Yue, C., & Randerson, J. T. (2017). A human-driven decline in global burned area. *Science, 356*(6345), 1356–1362. DOI: 10.1126/science.aal4108 PMID: 28663495

Anderson, M. C., Kustas, W. P., Alfieri, J. G., Gao, F., Hain, C. R., Prueger, J. H., Evett, S. R., Colaizzi, P. D., Howell, T., & Chávez, J. L. (2012). Mapping daily evapotranspiration at Landsat spatial scales during the BEAREX'08 field campaign. *Advances in Water Resources, 50*, 162–177. DOI: 10.1016/j.advwatres.2012.06.005

An, H., Ouyang, C., & Chen, X. (2025). Real-time estimation of SMAP soil moisture in mountainous areas and its impact on rainfall-runoff simulation. *Journal of Hydrology (Amsterdam), 133487*. Advance online publication. DOI: 10.1016/j.jhydrol.2025.133487

Appiah, R. (2020). Community-based participatory research in rural African contexts: Ethico-cultural considerations and lessons from Ghana. *Public Health Review, 41*(27), 1–13. DOI: 10.1186/s40985-020-00145-2 PMID: 33292760

Arneth, A., Olsson, L., Cowie, A., Erb, K.-H., Hurlbert, M., Kurz, W. A., Mirzabaev, A., & Rounsevell, M. D. A. (2021). Restoring degraded lands. *Annual Review of Environment and Resources, 46*(1), 569–599. DOI: 10.1146/annurev-environ-012320-054809

Arora, N. K. (2019). Impact of climate change on agricultural production and its sustainable solutions. *Environmental Sustainability, 2*(2), 95–96. DOI: 10.1007/s42398-019-00078-w

Arulnathan, V., Li, E., & Pelletier, N. (2026). Implementation of a Participatory Design Approach to the Development of a Sustainability Decision Support Tool for Canadian Egg Farmers. *Sustainability (Basel), 18*(1), 69. DOI: 10.3390/su18010069

Aslan, M. F., Sabanci, K., & Aslan, B. (2024). Artificial intelligence techniques in crop yield estimation based on Sentinel-2 data: A comprehensive survey. *Sustainability (Basel), 16*(18), 8277. DOI: 10.3390/su16188277

Asner, G. P., Martin, R. E., Knapp, D. E., Tupayachi, R., Anderson, C. B., Sinca, F., Vaughn, N. R., & Llactayo, W. (2017). Airborne laser-guided imaging spectroscopy to map forest trait diversity and guide conservation. *Science, 355*(6323), 385–389. DOI: 10.1126/science.aaj1987 PMID: 28126815

Aspinall, R., & Staiano, M. (2017). A conceptual model for land system dynamics as a coupled human–environment system. *Land (Basel), 6*(4), 81. DOI: 10.3390/land6040081

Astutik, S., Ahimbisibwe, V., Hintz, K. S., Purwanto, P., & Humaedi, M. A. (2025). Medicinal Plant Production System Management in Rural Java, Indonesia: Views of Local Actors from a Participatory Rural Appraisal Approach. *Forest and Society*, *9*(1), 20–47. DOI: 10.24259/fs.v9i1.31352

Astutik, S., Pretzsch, J., Kimengsi, J. N., & Kapp, G. (2023). Medicinal plants production systems in rural Indonesia: Management practices and performance insights. *Forest Policy and Economics*, *153*(March), 1–17. DOI: 10.1016/j.forpol.2023.102972

Athesan, A. I., & Ghazi, M. F. E. M. (2024). Innovative dashboard designs for real-time smart farming data visualization. *National EngiTech Digest, 1*(1). Retrieved from https://jktss.puo.edu.my/jurnalpuo/index.php/NED/article/view/143

Atik, C., & Urdu, D. (2025). *Bridging the Interoperability Gap in Digital Agriculture: Insights from Policy*. Practice, and Standards.

Atzberger, C. (2013). Advances in remote sensing of agriculture: Context description, existing operational monitoring systems and major information needs. *Remote Sensing (Basel)*, *5*(2), 949–981. DOI: 10.3390/rs5020949

Audebert, N., Le Saux, B., & Lefèvre, S. (2019). Deep learning for classification of hyperspectral data: A comparative review. *IEEE Geoscience and Remote Sensing Magazine*, *7*(2), 159–173. DOI: 10.1109/MGRS.2019.2912563

Auernhammer, H. (2001). Precision farming—The environmental challenge. *Computers and Electronics in Agriculture*, *30*(1–3), 31–43. DOI: 10.1016/S0168-1699(00)00153-8

Avinash, T. (2018). Mid-day meal scheme under food security: A study with special reference to upper primary schools of rural Bhadravthi taluk. *International Journal of Research in Social Sciences*, *8*(1). https://indianjournals.com/article/ijrss-8-1-065

Awada, H., Di Prima, S., Sirca, C., Giadrossich, F., Marras, S., Spano, D., & Pirastru, M. (2021). Daily actual evapotranspiration estimation in a Mediterranean ecosystem from Landsat observations using SEBAL approach. *Forests*, *12*(2), 189. DOI: 10.3390/f12020189

Ayadi, R., Forouheshfar, Y., & Moghadas, O. (2025). Enhancing System Resilience to Climate Change through Artificial Intelligence: A Systematic Literature Review. *Frontiers in Climate*, *7*, 1585331. DOI: 10.3389/fclim.2025.1585331

Ayesha Barvin, P., & Sampradeepraj, T. (2023). Crop recommendation systems based on soil and environmental factors using graph convolution neural network: A systematic literature review. *Engineering Proceedings*, *58*(1), 97. DOI: 10.3390/ecsa-10-16010

Azevedo, J. A., Chapman, L., & Muller, C. L. (2016). Quantifying the daytime and night-time urban heat island in Birmingham, UK: A comparison of satellite derived land surface temperature and high-resolution air temperature observations. *Remote Sensing (Basel)*, *8*(2), 153. DOI: 10.3390/rs8020153

Babar, M. I., & Akan, B. (2024). Smart irrigation systems in IoT for sustainable water management. *Journal of Environmental Management*, *316*, 115347.

Baccini, A., Goetz, S. J., Walker, W. S., Laporte, N. T., Sun, M., Sulla-Menashe, D., Hackler, J., Beck, P. S. A., Dubayah, R., Friedl, M. A., Samanta, S., & Houghton, R. A. (2012). Estimated carbon dioxide emissions from tropical deforestation improved by carbon-density maps. *Nature Climate Change*, *2*(3), 182–185. DOI: 10.1038/nclimate1354

Bajpai, N., Sachs, J. D., & Volavka, N. H. (2004). Reaching the Millennium Development Goals in South Asia. *Academic Commons*. Columbia University Libraries. https://academiccommons.columbia.edu/doi/10.7916/D8BV7PFH

Banks, S., Armstrong, A., Carter, K., Graham, H., Hayward, P., Henry, A., Holland, T., Holmes, C., Lee, A., McNulty, A., Moore, N., Nayling, N., Stokoe, A., & Strachan, A. (2013). Everyday ethics in community-based participatory research. *Contemporary Social Science*, *8*(3), 263–277. DOI: 10.1080/21582041.2013.769618

Baralla, G., Pinna, A., Tonelli, R., Marchesi, M., & Ibba, S. (2021). Ensuring transparency and traceability of food local products: A blockchain application to a Smart Tourism Region. *Concurrency and Computation*, *33*(1), e5857. DOI: 10.1002/cpe.5857

Barbedo, J. G. A., da Silva, M. S., Cazzolato, M. T., Valem, L. P., Tinós, R., Romero, R. A. F., Murta, L. O.Junior, & Holanda, A. (2026). Explainability and privacy in AI-enabled crop monitoring: Trends and future directions in soybean research. *Computers and Electronics in Agriculture*, *243*, 111392. DOI: 10.1016/j.compag.2025.111392

Barbosa Júnior, M. R., Moreira, B. R., Carreira, V. S., Brito Filho, A. L., Trentin, C., Souza, F. L. P., Tedesco, D., Setiyono, T., Flores, J. P., Ampatzidis, Y., Silva, R. P., & Shiratsuchi, L. S. (2024). Precision agriculture in the United States: A comprehensive meta-review inspiring further research, innovation, and adoption. *Computers and Electronics in Agriculture*, *221*, 108993. DOI: 10.1016/j.compag.2024.108993

Barroso-Barroso, C., Vega-Muñoz, A., Maradiaga-López, J., Salazar-Sepúlveda, G., & Carabantes-Silva, R. (2026). Smart Farming and the SDGs: Emerging Research Patterns and Sustainability Implications. *Agriculture*, *16*(1), 81. DOI: 10.3390/agriculture16010081

Bastida, E. M., Tseng, T.-S., McKeever, C., & Jr, L. J. (2010). Ethics and Community-Based Participatory Research: Perspectives From the Field. *Health Promotion Practice*, *11*(1), 16–20. DOI: 10.1177/1524839909352841 PMID: 20038649

Bastin, J., Finegold, Y., Garcia, C. A., Mollicone, D., Rezende, M., Routh, D., Zohner, C. M., & Crowther, T. W. (2019). The global tree restoration potential. *Science*, *365*(6448), 76–79. DOI: 10.1126/science.aax0848 PMID: 31273120

Bayar, J., Ali, N., Cao, Z., Ren, Y., & Dong, Y. (2025). Artificial intelligence of things (AIoT) for precision agriculture: Applications in smart irrigation, nutrient and disease management. *Smart Agricultural Technology*, *12*, 101629. DOI: 10.1016/j.atech.2025.101629

Bazrafkan, A., Igathinathane, C., Bandillo, N., & Flores, P. (2025). Optimizing integration techniques for UAS and satellite image data in precision agriculture — A review. *Frontiers in Remote Sensing*, *6*, 1622884. Advance online publication. DOI: 10.3389/frsen.2025.1622884

Becker-Reshef, I., Justice, C., Barker, B., Humber, M.L., Rembold, F., Bonifaçio, R., Zappacosta, M., Budde, M.E., Magadzire, T., Shitote, C., Pound, J., Constantino, A., Nakalembe, C., Mwangi, K., Sobue, S., Newby, T.S., Whitcraft, A.K., Jarvis, I.A., Verdin, J.P. (2020). Strengthening agricultural decisions in countries at risk of food insecurity: The GEOGLAM Crop Monitor for Early Warning. Remote Sensing of Environment.

Becker-Reshef, I., Vermote, E. F., Lindeman, M. A., & Justice, C. O. (2010). A generalized regression-based model for forecasting winter wheat yields in Kansas and Ukraine using MODIS data. *Remote Sensing of Environment, 114*(6), 1312–1323. DOI: 10.1016/j.rse.2010.01.010

Bégué, A., Arvor, D., Bellon, B., Betbeder, J., De Abelleyra, D., Ferraz, R. P., Lebourgeois, V., Lelong, C., Simões, M., & Verón, S. R. (2018). Remote sensing and cropping practices: A review. *Remote Sensing, 10*(1), 99. DOI: 10.3390/rs10010099

Behrenfeld, M. J., O'Malley, R. T., Siegel, D. A., McClain, C. R., Sarmiento, J. L., Feldman, G. C., Milligan, A. J., Falkowski, P. G., Letelier, R. M., & Boss, E. S. (2006). Climate-driven trends in contemporary ocean productivity. *Nature, 444*(7120), 752–755. DOI: 10.1038/nature05317 PMID: 17151666

Berger, C., Voltersen, M., Hese, S., Walde, I., & Schmullius, C. (2013). Robust Extraction of Urban Land Cover Information From HSR Multi-Spectral and LiDAR Data. *IEEE Journal of Selected Topics in Applied Earth Observations and Remote Sensing*, *6*(5), 1–16. DOI: 10.1109/JSTARS.2013.2252329

Berisha, F., Mooney, P., Pourzolfaghar, Z., & Helfert, M. (2026). Shared yet owned: The dual path of data ownership in agriculture: A systematic review. *Data & Policy*, *8*, e3. DOI: 10.1017/dap.2025.10050

Bernard, F., Manolis, I., Barat, I., Bolea Alamañac, A., Such Taboada, M., Mingorance, P., Ciapponi, A., Cardone, T., Fernandez Nunez, I., Garcia, A., Hallibert, P., Hammar, A., Henriot, P., Merodio Codinachs, D., Patti, S., Prado Montes, P., Skrzypek, P., Steenari, D., Weixler, S., & Gandullo Holgueras, D. (2023). The Copernicus land surface temperature monitoring (LSTM) mission: Design, technology and status. *Remote Sensing.*

Bhagat, P. R., Naz, F., & Magda, R. (2022). Artificial intelligence solutions enabling sustainable agriculture: A bibliometric analysis. *PLoS One*, *17*(6), e0268989. DOI: 10.1371/journal.pone.0268989 PMID: 35679287

Bhagwan, P. V., Anjaiah, T., Ravali, C., Chary, D. S., Zamani, A. T., Ullah, S., Rebouh, N. Y., & Tariq, A. (2025). Delineation and evaluation of management zones for site-specific nutrient management using a geostatistical and fuzzy C mean cluster approach. *Scientific Reports*, *15*(1), 20991. DOI: 10.1038/s41598-025-07283-0 PMID: 40594516

Bhalla, G. S., Hazell, P., & Kerr, J. (2001). *Prospects for India's Cereal Supply and Demand to 2020, Food, Agriculture and the Environment, Discussion Paper 29.* International Food Security Institute. https://ideas.repec.org/p/fpr/2020dp/29.html

Bhunia, G. S., & Shit, P. K. (2021). *Geospatial Technology for Multi-hazard Risk Assessment.* Advances in Geographic Information Science.

Birner, R., Daum, T., & Pray, C. (2021). Who drives the digital revolution in agriculture? A review of supply-side trends, players and challenges. *Applied Economic Perspectives and Policy*, *43*(4), 1260–1285. DOI: 10.1002/aepp.13145

Bissadu, K. D., Sonko, S., & Hossain, G. (2025). Society 5.0 enabled agriculture: Drivers, enabling technologies, architectures, opportunities, and challenges. *Information Processing in Agriculture*, *12*(1), 112–124. DOI: 10.1016/j.inpa.2024.04.003

Blaschke, T. (2010). Object based image analysis for remote sensing. *ISPRS Journal of Photogrammetry and Remote Sensing*, *65*(1), 2–16. DOI: 10.1016/j.isprsjprs.2009.06.004

Blondeau-Patissier, D., Gower, J. F., Dekker, A. G., Phinn, S. R., & Brando, V. E. (2014). A review of ocean color remote sensing methods and statistical techniques for the detection, mapping and analysis of phytoplankton blooms in coastal and open oceans. *Progress in Oceanography*, *123*, 123–144. DOI: 10.1016/j.pocean.2013.12.008

Bolton, D. K., & Friedl, M. A. (2013). Forecasting crop yield using remotely sensed vegetation indices and crop phenology metrics. *Agricultural and Forest Meteorology*, *173*, 74–84. DOI: 10.1016/j.agrformet.2013.01.007

Bordogna, G., Kliment, T., Frigerio, L., Brivio, P. A., Crema, A., Stroppiana, D., Boschetti, M., & Sterlacchini, S. (2016). A Spatial Data Infrastructure Integrating Multisource Heterogeneous Geospatial Data and Time Series: A Study Case in Agriculture. *ISPRS International Journal of Geo-Information*, *5*(5), 73. DOI: 10.3390/ijgi5050073

Boussetta, S., Balsamo, G., Dutra, E., Beljaars, A., & Albergel, C. (2015). Assimilation of surface albedo and vegetation states from satellite observations and their impact on numerical weather prediction. *Remote Sensing of Environment*, *163*, 111–126. DOI: 10.1016/j.rse.2015.03.009

Boutayeb, A., Lahsen-Cherif, I., & El Khadimi, A. (2025). When Machine Learning Meets Geospatial Data: A Comprehensive GeoAI Review. *IEEE Journal of Selected Topics in Applied Earth Observations and Remote Sensing*, *18*, 13135–13191. Advance online publication. DOI: 10.1109/JSTARS.2025.3568715

Boyce, D. G., Lewis, M. R., & Worm, B. (2010). Global phytoplankton decline over the past century. *Nature*, *466*(7306), 591–596. DOI: 10.1038/nature09268 PMID: 20671703

Bradley, B. A. (2013). Remote detection of invasive plants: A review of spectral, textural and phenological approaches. *Biological Invasions*. Advance online publication. DOI: 10.1007/s10530-013-0578-9

Brandt, P., Kvakić, M., Butterbach-Bahl, K., & Rufino, M. C. (2017). How to target climate-smart agriculture? Concept and application of the consensus-driven decision support framework "targetCSA.". *Agricultural Systems*, *151*, 234–245. DOI: 10.1016/j.agsy.2015.12.011

Breunig, F. M., Galvão, L. S., Dalagnol, R., Dauve, C. E., Parraga, A., Santi, A. L., Della Flora, D. P., & Chen, S. (2020). Delineation of management zones in agricultural fields using cover–crop biomass estimates from PlanetScope data. *International Journal of Applied Earth Observation and Geoinformation*, *85*, 102004. DOI: 10.1016/j.jag.2019.102004

Brinkhoff, J., Hornbuckle, J., & Dowling, T. (2018). Multisensor capacitance probes for simultaneously monitoring rice field soil-water-crop-ambient conditions. *Sensors (Basel), 18*(1), 53. DOI: 10.3390/s18010053 PMID: 29278397

Brito, M. M., Evers, M., & Almoradie, A. (2017). Participatory flood vulnerability assessment: A multi-criteria approach. *Hydrology and Earth System Sciences, 22*(1), 373–390. DOI: 10.5194/hess-22-373-2018

Bronson, K. (2019). Looking through a responsible innovation lens at uneven engagements with digital farming. *NJAS Wageningen Journal of Life Sciences, 90–91*(1), 1–6. DOI: 10.1016/j.njas.2019.03.001

Bronson, K., & Knezevic, I. (2016). Big data in food and agriculture. *Big Data & Society, 3*(1). Advance online publication. DOI: 10.1177/2053951716648174

Budah, A. G., Habeeb, T., & Faiza, S. (2025). Integrating GeoAI and deep learning for sustainable agriculture: Sentinel-2 based crop mapping and yield prediction for selected crops in Sokoto, Nigeria. Journal of Geography. *Environment and Earth Science International, 29*(12), 200–212. DOI: 10.9734/jgeesi/2025/v29i12992

Buitenwerf, R., Rose, L., & Higgins, S. I. (2015). Three decades of multi-dimensional change in global leaf phenology. *Nature Climate Change, 5*(4), 364–368. DOI: 10.1038/nclimate2533

Burabe, B. I., Dutse, F., Shuaibu, H., Ahmed, B., & Yusuf, H. O. (2024). Influence of Socio-Economics and Institutional Factors on Adoption of Agricultural Information Technologies of Irrigation Rice Farmers in Northeastern Nigeria. *Journal of Arid Agriculture, 25*(3), 14–21. https://jaaunimaid.ng/index.php/home/article/view/37

Burks, T. F., Lee, W. S., & Pydipati, R. (2006). citrus disease detection based on discriminant analysis and color texture features. Agriculture Electronics and Computers. DOI: 10.13031/2013.21733

Cabernard, L., Pfister, S., & Hellweg, S. (2024). Biodiversity impacts of recent land-use change driven by increases in agri-food imports. *Nature Sustainability, 7*(11), 1512–1524. DOI: 10.1038/s41893-024-01433-4

Cahyadi, E. R., Hidayati, N., Zahra, N., & Arif, C. (2024). Integrating Circular Economy Principles into Agri-Food Supply Chain Management: A Systematic Literature Review. *Sustainability (Basel), 16*(16), 1–17. DOI: 10.3390/su16167165

Calvino-Cancela, M., Mendez-Rial, R., Reguera-Salgado, J., & Martin-Herrero, J. (2014). Alien plant monitoring with ultralight airborne imaging spectroscopy. *PLoS One, 9*(7), e102381. DOI: 10.1371/journal.pone.0102381 PMID: 25010601

Canfield, M., & Ntambirweki, B. (2024). Datafying African agriculture: From data governance to farmers' rights. *Development, 67*(1-2), 5–13. DOI: 10.1057/s41301-024-00405-7

Carolan, M. (2018). "Smart" farming techniques as political ontology: Access, sovereignty and the performance of neoliberal and not-so-neoliberal worlds. *Sociologia Ruralis, 58*(4), 745–764. DOI: 10.1111/soru.12202

Carroll, S. R., Garba, I., Figueroa-Rodríguez, O. L., Holbrook, J., Lovett, R., Materechera, S., Parsons, M., Raseroka, K., Rodriguez-Lonebear, D., Rowe, R., Sara, R., Walker, J. D., Anderson, J., & Hudson, M. (2020). The CARE principles for Indigenous data governance. *Data Science Journal, 19*, 43. Advance online publication. DOI: 10.5334/dsj-2020-043

Cartolano, A., Cuzzocrea, A., & Pilato, G. (2024). Analyzing and assessing explainable AI models for smart agriculture environments. *Multimedia Tools and Applications, 83*(12), 37225–37246. DOI: 10.1007/s11042-023-17978-z

Castellanos, S. A. S. (2023). *A policy proposal for agricultural data governance* (Doctoral dissertation, Portland State University). DOI: 10.15760/etd.3567

Castelluccio, M., Poggi, G., Sansone, C., & Verdoliva, L. (2015). Land use classification in remote sensing images by convolutional neural networks. https://arxiv.org/abs/1508.00092

Cerjak, M., Medici, M., Faletar, I., Sundeep, J. V., & Canavari, M. (2025). Adoption of mobile-based agricultural extension services: Evidence from South India. *Journal of Rural Studies, 120*, 103851. DOI: 10.1016/j.jrurstud.2025.103851

Cesco, S., Sambo, P., Borin, M., Basso, B., Orzes, G., & Mazzetto, F. (2023). Smart agriculture and digital twins: Applications and challenges in a vision of sustainability. *European Journal of Agronomy, 146*, 126809. DOI: 10.1016/j.eja.2023.126809

Chakrabarty, A., Ahmed, S. T., Islam, M. F. U., Aziz, S. M., & Maidin, S. S. (2024). An interpretable fusion model integrating lightweight CNN and transformer architectures for rice leaf disease identification. *Ecological Informatics, 82*, 102718. DOI: 10.1016/j.ecoinf.2024.102718

Chakraborty, T., & Lee, X. (2019). A simplified urban-extent algorithm to characterize surface urban heat islands on a global scale and examine vegetation control on their spatiotemporal variability. *International Journal of Applied Earth Observation and Geoinformation, 74*, 269–280. DOI: 10.1016/j.jag.2018.09.015

Chamorro-Padial, J., García, R., & Gil, R. (2024). A systematic review of open data in agriculture. *Computers and Electronics in Agriculture, 219*, 108775. DOI: 10.1016/j.compag.2024.108775

Chan, S., Weitz, N., Persson, Å., & Trimmer, C. (2018). SDG 12: Responsible consumption and production. A review of research needs. Technical annex to the Formas report *Forskning för Agenda 2030*. https://www.sei.org/wp-content/uploads/2018/11/review-sdg12-research-needs-final.pdf

Chao, X., Sun, G., Zhao, H., Li, M., & He, D. (2020). Identification of apple tree leaf diseases based on deep learning models. *Symmetry, 12*(7), 1065. DOI: 10.3390/sym12071065

Chen, R. (2021). *Mapping data governance legal frameworks around the world: Findings from the global data regulation diagnostic*. https://ideas.repec.org/p/wbk/wbrwps/9615.html

Cheng, K. S., Ling, J., Lin, T., Liu, Y., Shen, Y., & Kono, Y. (2021). Quantifying Uncertainty in Land-Use/Land-Cover Classification Accuracy: A Stochastic Simulation Approach. *Frontiers in Environmental Science, 9*, 628214. Advance online publication. DOI: 10.3389/fenvs.2021.628214

Cheng, K., Zhang, C., Fan, Y., Diao, H., & Xia, S. (2025). AI-based approaches for wetland mapping and classification: A review of current practices and future perspectives. *Artificial Intelligence in Geosciences*, *6*(2), 100165. DOI: 10.1016/j.aiig.2025.100165

Cheng, X., Sun, Y., Zhang, W., Wang, Y., Cao, X., & Wang, Y. (2023). Application of deep learning in multitemporal remote sensing image classification. *Remote Sensing (Basel)*, *15*(15), 3859. DOI: 10.3390/rs15153859

Chen, H., Yang, L., & Wu, Q. (2023). Enhancing Land Cover Mapping and Monitoring: An Interactive and Explainable Machine Learning Approach Using Google Earth Engine. *Remote Sensing (Basel)*, *15*(18), 4585. DOI: 10.3390/rs15184585

Chen, J., Liao, W., & Yu, C. (2021). Route optimization for cold chain logistics of front warehouses based on traffic congestion and carbon emission. *Computers & Industrial Engineering*, *161*, 107663. DOI: 10.1016/j.cie.2021.107663

Chen, X. (2025). The role of modern agricultural technologies in improving agricultural productivity and land use efficiency. *Frontiers in Plant Science*, *16*(September), 1–19. DOI: 10.3389/fpls.2025.1675657 PMID: 41036393

Chen, X., Zhang, H., & Wong, C. U. I. (2025). Dynamic monitoring and precision fertilization decision system for agricultural soil nutrients using UAV remote sensing and GIS. *Agriculture*, *15*(15), 1627. DOI: 10.3390/agriculture15151627

Chen, Y., Jiao, T., Song, J., He, G., & Jin, Z. (2024). AI-enabled animal behavior analysis with high usability: A case study on open-field experiments. *Applied Sciences (Basel, Switzerland)*, *14*(11), 4583. DOI: 10.3390/app14114583

Chi, M., Plaza, A., Benediktsson, J. A., Sun, Z., Shen, J., & Zhu, Y. (2016). Big data for remote sensing: Challenges and opportunities. *Proceedings of the IEEE*, *104*(11), 2207–2219. DOI: 10.1109/JPROC.2016.2598228

Chini, M., Pelich, R., Pulvirenti, L., Pierdicca, N., Hostache, R., & Matgen, P. (2019). Sentinel-1 InSAR coherence to detect floodwater in urban areas: Houston and Hurricane Harvey as a test case. *Remote Sensing (Basel)*, *11*(2), 107. DOI: 10.3390/rs11020107

Choi, Y. (2023). GeoAI: Integration of Artificial Intelligence, Machine Learning, and Deep Learning with GIS. *Applied Sciences (Basel, Switzerland)*, *13*(6), 3895. DOI: 10.3390/app13063895

Choudhury, B. (2018). Balancing soft and hard law for business and human rights. *The International and Comparative Law Quarterly*, *67*(4), 961–986. DOI: 10.1017/S0020589318000155

Chowdhury, S., Paul, C., Hamidov, A., Bayer, L., Arndt, M., McPherson, J., & Helming, K. (2025). Developing a generic DPSIR framework for land use and soil management: A systems approach to maximise soil health. *European Journal of Soil Science*, *76*(6), e70230. DOI: 10.1111/ejss.70230

Christofi, D., Mettas, C., Evagorou, E., Stylianou, N., Eliades, M., Theocharidis, C., Chatzipavlis, A., Hasiotis, T., & Hadjimitsis, D. (2025). A Review of Open Remote Sensing Data with GIS, AI, and UAV Support for Shoreline Detection and Coastal Erosion Monitoring. *Applied Sciences (Basel, Switzerland), 15*(9), 4771. DOI: 10.3390/app15094771

Chukwuma, E. C., Afolabi, O. O. D., Okonkwo, C. C., & Okonkwo, C. E. (2024). Application of geospatial technology and decision model in the development of an improved food security index. *Scientific Reports, 14*(1), 30204. DOI: 10.1038/s41598-024-77845-1 PMID: 39632976

Claverie, M., Ju, J., Masek, J. G., Dungan, J. L., Vermote, E. F., Roger, J. C., Skakun, S. V., & Justice, C. (2018). The harmonized Landsat and Sentinel-2 surface reflectance data set. *Remote Sensing of Environment, 219*, 145–161. DOI: 10.1016/j.rse.2018.09.002

Cockx, K., Voorde, T. V., & Canters, F. (2014). Quantifying uncertainty in remote sensing-based urban land-use mapping. *International Journal of Applied Earth Observation and Geoinformation, 31*, 154–166. DOI: 10.1016/j.jag.2014.03.016

Colpaert, A. (2022). Satellite and UAV platforms, remote sensing for geographic information systems. *Sensors (Basel), 22*(12), 4564. DOI: 10.3390/s22124564 PMID: 35746346

Comiso, J. C., Meier, W. N., & Gersten, R. (2017). Variability and trends in the Arctic sea ice cover: Results from different techniques. *Journal of Geophysical Research. Oceans, 122*(8), 6883–6900. Advance online publication. DOI: 10.1002/2017JC012768

Coppin, P., Jonckheere, I., Nackaerts, K., Muys, B., & Lambin, E. (2004). Review ArticleDigital change detection methods in ecosystem monitoring: A review. *International Journal of Remote Sensing, 25*(9), 1565–1596. DOI: 10.1080/0143116031000101675

Corallo, A., Latino, M. E., Pizzi, R., Spennato, A., Fortunato, L., & Menegoli, M. (2019, September). Human factor in smart labelling: How enhance food awareness in consumers. In *2019 6th International Conference on Frontiers of Industrial Engineering (ICFIE)* (pp. 89-93). IEEE. https://ieeexplore.ieee.org/abstract/document/8907769/

Cornwall, A., & Jewkes, R. (1995). What is participatory research? *Social Science & Medicine, 41*(12), 1667–1676. DOI: 10.1016/0277-9536(95)00127-S PMID: 8746866

Costa, D. G. (2026). The risks of GenAI in shaping our cities. *Cities (London, England), 173*, 107025. DOI: 10.1016/j.cities.2026.107025

Costa, F., Frecassetti, S., Rossini, M., & Portioli-Staudacher, A. (2023). Industry 4.0 digital technologies enhancing sustainability: Applications and barriers from the agricultural industry in an emerging economy. *Journal of Cleaner Production, 408*, 137208. DOI: 10.1016/j.jclepro.2023.137208

Cravero, A., Sepúlveda, S., Gutiérrez, F., & Muñoz, L. (2026). From Precision Agriculture to Intelligent Agricultural Ecosystems: A Systematic Review of Machine Learning and Big Data Applications. *Agronomy (Basel), 16*(5), 516. DOI: 10.3390/agronomy16050516

Cressie, N., Calder, C. A., Clark, J. S., Hoef, J. M. V., & Wikle, C. K. (2009). Accounting for uncertainty in ecological analysis: The strengths and limitations of hierarchical statistical modeling. *Ecological Applications : A Publication of the Ecological Society of America, 19*(3), 553–570. DOI: 10.1890/07-0744.1 PMID: 19425416

Dai, Y., Shangguan, W., Wei, N., Xin, Q., Yuan, H., Zhang, S., Liu, S., Lu, X., Wang, D., & Yan, F. (2019). A review of the global soil property maps for Earth system models. *Soil (Göttingen), 5*(2), 137–158. DOI: 10.5194/soil-5-137-2019

Dai, Z., Liu, H., Le, Q. V., & Tan, M. (2021). CoAtNet: Attention convolution with all data sizes. Innovations in Neural Information Processing Systems (NeurIPS). https://arxiv.org/abs/2106.04803

Dara, R., Hazrati Fard, S. M., & Kaur, J. (2022). Recommendations for ethical and responsible use of artificial intelligence in digital agriculture. *Frontiers in Artificial Intelligence, 5*, 1–11. DOI: 10.3389/frai.2022.884192 PMID: 35968036

Dash, P. K. (2024). Remote sensing as a potential tool for advancing digital soil mapping. In *Remote Sensing of Soils* (pp. 357–370). Elsevier. DOI: 10.1016/B978-0-443-18773-5.00021-1

Das, N., & Edgaonkar, A. (2025). Economic valuation of ecosystem services at the watershed scale for strategic land-use planning. *Environmental Conservation, 53*(1), 30–39. DOI: 10.1017/S0376892925100271

Dattatreya, S., Khan, A. N., Jena, K., & Chatterjee, G. (2023). Conventional to modern methods of soil npk sensing: A review. *IEEE Sensors Journal, 24*(3), 2367–2380. DOI: 10.1109/JSEN.2023.3334243

de Bruin, S., Dengerink, J., & van Vliet, J. (2021). Urbanisation as driver of food system transformation and opportunities for rural livelihoods. *Food Security, 13*(4), 781–798. DOI: 10.1007/s12571-021-01182-8 PMID: 34221191

De Caires, S. A., St. Martin, C., Atwell, M. A., Kaya, F., Wuddivira, G. A., & Wuddivira, M. N. (2025). Advancing soil mapping and management using geostatistics and integrated machine learning and remote sensing techniques: A synoptic review. *Discover Soil, 2*(1), 53. DOI: 10.1007/s44378-025-00082-z

de la Cruz, L. G. (2024). *Optimizing sustainable agriculture with EO-derived data for soil health assessment* (Doctoral dissertation, Central European University). https://www.etd.ceu.edu/2024/de-la-cruz_leizel.pdf

De la Rosa, D., & Anaya-Romero, M. (2010). MicroLEIS DSS: For planning agro-ecological soil use and management systems. In *Decision Support Systems in Agriculture, Food and the Environment: Trends, Applications and Advances* (pp. 339–376). IGI Global. https://doi.org/DOI: 10.4018/978-1-61520-881-4.ch016

de Smith, M. J., Goodchild, M. F., & Longley, P. A. (2018). *Geospatial analysis: A comprehensive guide to principles, techniques, and software tools* (6th ed.). Drumlin Security. https://www.spatialanalysisonline.com

Deichmann, U., Goyal, A., & Mishra, D. (2016). Will digital technologies transform agriculture in developing countries? *Agricultural Economics, 47*(S1), 21–33. DOI: 10.1111/agec.12300

Delgado, J. A., Short, N. M.Jr, Roberts, D. P., & Vandenberg, B. (2019). Big data analysis for sustainable agriculture on a geospatial cloud framework. *Frontiers in Sustainable Food Systems*, *3*, 54. DOI: 10.3389/fsufs.2019.00054

Dembani, R., Karvelas, I., Akbar, N. A., Rizou, S., Tegolo, D., & Fountas, S. (2025). Agricultural data privacy and federated learning: A review of challenges and opportunities. *Computers and Electronics in Agriculture*, *232*, 110048. DOI: 10.1016/j.compag.2025.110048

Demissie, W. A., Sebastiani, L., & Rossetto, R. (2026). Integration of artificial intelligence and remote sensing for crop yield prediction and crop growth parameter estimation in Mediterranean agroecosystems: Methodologies, emerging technologies, research gaps, and future directions. *European Journal of Agronomy*, *127894*(173), 127894. DOI: 10.1016/j.eja.2025.127894

Deputy, K. V., Passi, K., & Jain, C. K. (2023). Crop Disease Detection Using Deep Learning Techniques on Images. *Journal of Computational Science*, *19*(12), 1438–1449. DOI: 10.3844/jcssp.2023.1438.1449

Deressu, T. F., Bojer, A. K., Debelee, T. G., Negera, W. G., Nadarajah, S., & Gebissa, K. W. (2025). Enhancing land use and land cover classification with deep learning-based satellite imagery segmentation. *International Journal of Applied Earth Observation and Geoinformation*, *144*, 104839. DOI: 10.1016/j.jag.2025.104839

Devare, M., Arnaud, E., Antezana, E., & King, B. (2023). Governing agricultural data: Challenges and recommendations. In *Towards responsible plant data linkage: Data challenges for agricultural research and development* (pp. 201–222). Springer International Publishing. DOI: 10.1007/978-3-031-13276-6_11

DeVries, B., Decuyper, M., Verbesselt, J., Zeileis, A., Herold, M., & Joseph, S. (2015). Tracking disturbance-regrowth dynamics in tropical forests using structural change detection and Landsat time series. *Remote Sensing of Environment*, *169*, 320–334. DOI: 10.1016/j.rse.2015.08.020

Dhamija, G., Ojha, M., & Roychowdhury, P. (2021). Hunger and health: The impact of household food insecurity on child malnutrition in India. *The Journal of Development Studies*, *58*(6), 1181–1210. DOI: 10.1080/00220388.2022.2029419

Dhanaraju, M., Chenniappan, P., Ramalingam, K., Pazhanivelan, S., & Kaliaperumal, R. (2022). Smart farming: Internet of Things (IoT)-based sustainable agriculture. *Agriculture*, *12*(10), 1745. DOI: 10.3390/agriculture12101745

Ding, H., Tian, J., Yu, W., Wilson, D. I., Young, B. R., Cui, X., & Li, W. (2023). The application of artificial intelligence and big data in the food industry. *Foods*, *12*(24), 4511. DOI: 10.3390/foods12244511 PMID: 38137314

Domingues, T., Brandão, T., & Ferreira, J. C. (2022). Machine learning for detection and prediction of crop diseases and pests: A comprehensive survey. *Agriculture*, *12*(9), 1350. DOI: 10.3390/agriculture12091350

Dong, J., Metternicht, G., Hostert, P., Fensholt, R., & Roy Chowdhury, R. (2019). Remote sensing and geospatial technologies in support of a normative land system science: Status and prospects. *Current Opinion in Environmental Sustainability*, *38*, 44–52. DOI: 10.1016/j.cosust.2019.05.003

Dong, Y., Huang, W., Li, H., & Han, L. (2025). Editorial: Advancements in Agricultural Monitoring with AI Enhanced Remote Sensing Techniques. *Frontiers in Remote Sensing*, *6*, 1664060. DOI: 10.3389/frsen.2025.1664060

Dosovitskiy, A., Beyer, L., Kolesnikov, A., Weissenborn, D., Zhai, X., Unterthiner, T., . . . Houlsby, N. (2020). An image is worth 16x16 words: Transformers for image recognition at scale. *arXiv preprint arXiv:2010.11929*

Dosovitskiy, A., Beyer, L., & Kolesnikov, A. (2021). An image is worth 16x16 words: Transformers for image recognition at scale. *International Conference on Learning Representations (ICLR 2021)*. DOI: 10.48550/arXiv.2010.11929

Drago, E., Campardelli, R., Pettinato, M., & Perego, P. (2020). Innovations in smart packaging concepts for food: An extensive review. foods, 9(11), 1628. https://www.mdpi.com/2304-8158/9/11/1628

Dritsas, E., & Trigka, M. (2025). Remote sensing and geospatial analysis in the big data era: A survey. *Remote Sensing (Basel)*, *17*(3), 550. DOI: 10.3390/rs17030550

Drusch, M., Del Bello, U., Carlier, S., Colin, O., Fernandez, V., Gascon, F., Hoersch, B., Isola, C., Laberinti, P., Martimort, P., Meygret, A., Spoto, F., Sy, O., Marchese, F., & Bargellini, P. (2012). Sentinel-2: ESA's optical high-resolution mission for GMES operational services. *Remote Sensing of Environment*, *120*, 25–36. DOI: 10.1016/j.rse.2011.11.026

Du, A., Doan, A.-D., Law, Y. W., & Chin, T.-J. (2024). Domain adaptation for satellite-borne multispectral cloud detection. *Remote Sensing (Basel)*, *16*(18), 3469. DOI: 10.3390/rs16183469

Duncan, B. N., Prados, A. I., Lamsal, L. N., Liu, Y., Streets, D. G., Gupta, P., Hilsenrath, E., Kahn, R. A., Nielsen, J. E., Beyersdorf, A. J., Burton, S. P., Fiore, A. M., Fishman, J., Henze, D. K., Hostetler, C. A., Krotkov, N. A., Lee, P., Lin, M., Pawson, S., & Ziemba, L. D. (2014). Satellite Data of Atmospheric Pollution for U.S. Air Quality Applications: Examples of Applications, Summary of Data End-User Resources, Answers to FAQs, and Common Mistakes to Avoid. *Atmospheric Environment*, *94*, 647–662. DOI: 10.1016/j.atmosenv.2014.05.061

Duncan, E., Ruder, S.-L., Kc, K. B., Wittman, H., & Fraser, E. (2026). The political economy of agricultural data governance: Farmer perspectives on trust in data sharing. *Journal of Rural Studies*, *123*, 104076. DOI: 10.1016/j.jrurstud.2026.104076

Dwyer, J. L., Roy, D. P., Sauer, B., Jenkerson, C. B., Zhang, H. K., & Lymburner, L. (2018). Analysis ready data: Enabling analysis of the Landsat archive. *Remote Sensing (Basel)*, *10*(9), 1363. DOI: 10.3390/rs10091363

Dyson, T., & Hanchate, A. (2000). India's Demographic and Food Prospects: State Level Analysis. *Economic and Political Weekly, 35*. https://www.jstor.org/stable/i399209

Eckebil, P. P. T., Mintah, F., Bürgi, M., Akinyemi, F. O., Sonwa, D. J., & Speranza, C. I. (2025). Tropical ecosystem multifunctionality assessment and insights for sustainable land management: A systematic literature review using the Drivers-Pressures-State-Impacts-Responses framework. *Frontiers in Forests and Global Change*, *8*, 1623266. DOI: 10.3389/ffgc.2025.1623266

Edler, J., Blind, K., Kroll, H., & Schubert, T. (2023). Technology sovereignty as an emerging frame for innovation policy: Defining rationales, ends and means. *Research Policy, 52*(5), 104765. DOI: 10.1016/j.respol.2023.104765

Ehlers, M.-H., Finger, R., El Benni, N., Gocht, A., Sørensen, C. A. G., Gusset, M., Pfeifer, C., Poppe, K., Regan, Á., Rose, D. C., Wolfert, S., & Huber, R. (2022). Scenarios for European agricultural policy-making in the era of digitalisation. *Agricultural Systems, 196*, 103318. DOI: 10.1016/j.agsy.2021.103318

Ehlers, M.-H., Huber, R., & Finger, R. (2021). Agricultural policy in the era of digitalisation. *Food Policy, 100*, 102019. DOI: 10.1016/j.foodpol.2020.102019

Eisfelder, C., Boemke, B., Gessner, U., Sogno, P., Alemu, G., Hailu, R., & Huth, J. (2024). Cropland and crop type classification with Sentinel-1 and Sentinel-2 time series using Google Earth Engine for agricultural monitoring in Ethiopia. *Remote Sensing (Basel), 16*(5), 866. DOI: 10.3390/rs16050866

El Jarroudi, M., Kouadio, L., Delfosse, P., Bock, C. H., Mahlein, A. K., Fettweis, X., Mercatoris, B., Adams, F., Lenné, J. M., & Hamdioui, S. (2024). Leveraging edge artificial intelligence for sustainable agriculture. *Nature Sustainability, 7*(7), 846–854. DOI: 10.1038/s41893-024-01352-4

Elbasi, E., Mostafa, N., Zaki, C., AlArnaout, Z., Topcu, A. E., & Saker, L. (2024). Optimizing Analysis Techniques through Agricultural Data AI-Powered Decision-Making Processes. *Applied Sciences (Basel, Switzerland), 14*(17), 8018. DOI: 10.3390/app14178018

Eli-Chukwu, N. C. (2019). Applications of Artificial Intelligence in Agriculture: A Review. *Eng. Technol. Appl. Sci. Res., 9*(4), 4377–4383. DOI: 10.48084/etasr.2756

Ellur, R., Ananthakumar, M. A., & Desai, K. (2025). Prediction and mapping of soil texture at high spatial resolution in a canal irrigated region using machine learning. Journal of Geography. *Environment and Earth Science International, 29*(6), 114–125. DOI: 10.9734/jgeesi/2025/v29i6911

Epasto, S. (2025). Spatializing Trust: A GeoAI-Based Model for Mapping Digital Trust Ecosystems in Mediterranean Smart Regions. *ISPRS International Journal of Geo-Information, 14*(12), 491. DOI: 10.3390/ijgi14120491

Escobar-Naranjo, J., Caiza, G., Ayala, P., Jordan, E., Garcia, C. A., & Garcia, M. V. (2023). Autonomous navigation of robots: Optimization with DQN. *Applied Sciences (Basel, Switzerland), 13*(12), 7202. DOI: 10.3390/app13127202

Eshetie, G. G., Alemie, B. K., & Wubie, A. M. (2024). Geospatial technologies in support of responsible land tenure governance: A systematic review. *Geomatica, 76*(2), 100014. DOI: 10.1016/j.geomat.2024.100014

Espinel, R., Herrera-Franco, G., Rivadeneira García, J. L., & Escandón-Panchana, P. (2024). Artificial Intelligence in Agricultural Mapping. *Agriculture, 14*(7), 1071. DOI: 10.3390/agriculture14071071

Estacio, I., Sianipar, C. P. M., Onitsuka, K., Basu, M., & Hoshino, S. (2023). A statistical model of land use/cover change integrating logistic and linear models: An application to agricultural abandonment. *International Journal of Applied Earth Observation and Geoinformation, 120*, 103339. DOI: 10.1016/j.jag.2023.103339

Estoque, R. C., Murayama, Y., & Myint, S. W. (2017). Effects of landscape composition and pattern on land surface temperature: An urban heat island study in the megacities of Southeast Asia. *The Science of the Total Environment*, *577*, 349–359. Advance online publication. DOI: 10.1016/j.scitotenv.2016.10.195 PMID: 27832866

Everett, J. R., Staskowski, R. J., & Jengo, C. M. (2002). Remote sensing and GIS enable future exploration success.

Fabiani, S., Vanino, S., Napoli, R., Zajíček, A., Duffková, R., Evangelou, E., & Nino, P. (2020). Assessment of the economic and environmental sustainability of Variable Rate Technology (VRT) application in different wheat intensive European agricultural areas. A Water energy food nexus approach. *Environmental Science & Policy*, *114*, 366–376. DOI: 10.1016/j.envsci.2020.08.019

Fairbairn, M., & Kish, Z. (2023). Setting data free: The politics of open data for food and agriculture. *New Media & Society*, *25*(8), 1935–1959. DOI: 10.1177/14614448231174520

Fan, Y., Wang, X., Funk, T., Rashid, I., Herman, B., Bompoti, N., Mahmud, M. D. S., Chrysochoou, M., Yang, M., Vadas, T. M., Lei, Y., & Li, B. (2022). A critical review for real-time continuous soil monitoring: Advantages, challenges, and perspectives. *Environmental Science & Technology*, *56*(19), 13546–13564. DOI: 10.1021/acs.est.2c03562 PMID: 36121207

FAO (Food and Agriculture Organization of the United Nations). (2023). Land degradation assessment and its impact on agricultural productivity. https://www.fao.org/land-water/land/land-assessment/en/

FAO (Food and Agriculture Organization of the United Nations). (2025). The state of the world's land and water resources for food and agriculture 2025: Managing systems at risk. Rome, Italy: https://www.fao.org/publications/fao-flagship-publications/the-state-of-the-worlds-land-and-water-resources-for-food-and-agriculture/en

FAO, Food and Agriculture Organization of the United Nations. (2013). Climate-smart agriculture: Policies, practices and financing for food security, adaptation and mitigation. FAO. https://www.fao.org/3/i3325e/i3325e.pdf

FAO, Food and Agriculture Organization of the United Nations. (2018). Sustainable food and agriculture: Guidance for policymakers. FAO. https://www.fao.org/3/CA2096EN/ca2096en.pdf

FAO. (2018). *Scaling up agroecology initiative: Transforming food and agricultural systems in support of the SDGs*. https://www.fao.org/3/I9049EN/i9049en.pdf

FAO. (2021). *The State of Food and Agriculture 2021: Making agrifood systems more resilient to shocks and stresses*. DOI: 10.4060/cb7351en

FAO. (2022). Leveraging automation in agriculture for transforming agrifood systems. *Food and Agriculture Organization of the United Nations*.

FAO. (2023). *Promoting the digitalization of small and medium-sized agrifood enterprises in Asia and the Pacific*. DOI: 10.4060/cc8826en

FAO. (2025). *Agro-informatics*. https://www.fao.org/agroinformatics/en

Farhadi, M., Abbaspour-Gilandeh, Y., Mahmoudi, A., & Mari Maja, J. (2020). An integrated system of artificial intelligence and signal processing techniques for the sorting and grading of nuts. *Applied Sciences (Basel, Switzerland), 10*(9), 3315. DOI: 10.3390/app10093315

Farhan, S. M., Yin, J., Chen, Z., & Memon, M. S. (2024). A comprehensive review of LiDAR applications in crop management for precision agriculture. *Sensors (Basel), 24*(16), 5409. DOI: 10.3390/s24165409 PMID: 39205103

Farooq, M. S., Javid, R., Riaz, S., & Atal, Z. (2022). IoT based smart greenhouse framework and control strategies for sustainable agriculture. *IEEE Access : Practical Innovations, Open Solutions, 10*, 99394–99420. DOI: 10.1109/ACCESS.2022.3204066

Farooq, M. S., Riaz, S., Abid, A., Abid, K., & Naeem, M. A. (2019). A Survey on the Role of IoT in Agriculture for the Implementation of Smart Farming. *IEEE Access : Practical Innovations, Open Solutions, 7*, 156237–156271. DOI: 10.1109/ACCESS.2019.2949703

Feddema, J. J., Oleson, K. W., Bonan, G. B., Mearns, L. O., Buja, L. E., Meehl, G. A., & Washington, W. M. (2005). The importance of land-cover change in simulating future climates. *Science, 310*(5754), 1674–1678. DOI: 10.1126/science.1118160 PMID: 16339443

Fei, S., Hassan, M. A., Xiao, Y., Su, X., Chen, Z., Cheng, Q., Duan, F., Chen, R., & Ma, Y. (2023). UAV-based multi-sensor data fusion and machine learning algorithm for yield prediction in wheat. *Precision Agriculture, 24*(1), 187–212. DOI: 10.1007/s11119-022-09938-8 PMID: 35967193

Ferentinos, K. P. (2018). Plant disease diagnosis and detection Deep learning model. *Computers and Electronics in Agriculture*. Advance online publication. DOI: 10.1016/j.compag.2018.01.009

Fernandez, C. M., Alves, J., Gaspar, P. D., & Lima, T. M. (2021). Fostering awareness on environmentally sustainable technological solutions for the post-harvest food supply chain. *Processes (Basel, Switzerland), 9*(9), 1611. DOI: 10.3390/pr9091611

Fick, S. E., & Hijmans, R. J. (2017). WorldClim 2: New 1-km spatial resolution climate surfaces for global land areas. *International Journal of Climatology, 37*(12), 4302–4315. Advance online publication. DOI: 10.1002/joc.5086

Fiedler, J. L., & Yadav, S. (2017). How can we better capture food away from home? Lessons from India's linking person-level meal and household-level food data. *Food Policy, 72*, 81–93. DOI: 10.1016/j.foodpol.2017.08.015 PMID: 29093609

Filintas, A., Gougoulias, N., Kourgialas, N., & Hatzichristou, E. (2023). Management zones delineation, correct and incorrect application analysis in a coriander field using precision agriculture, soil chemical, granular and hydraulic analyses, fuzzy k-means zoning, factor analysis and geostatistics. *Water (Basel), 15*(18), 3278. DOI: 10.3390/w15183278

Finck, M., & Pallas, F. (2020). They who must not be identified—Distinguishing personal from non-personal data under the GDPR. *International Data Privacy Law, 10*(1), 11–36. DOI: 10.1093/idpl/ipz026

Foley, J. A., DeFries, R., Asner, G. P., Barford, C., Bonan, G., Carpenter, S. R., Chapin, F. S., Coe, M. T., Daily, G. C., Gibbs, H. K., Helkowski, J. H., Holloway, T., Howard, E. A., Kucharik, C. J., Monfreda, C., Patz, J. A., Prentice, I. C., Ramankutty, N., & Snyder, P. K. (2005). Global consequences of land use. *Science, 309*(5734), 570–574. DOI: 10.1126/science.1111772 PMID: 16040698

Foley, J. A., Ramankutty, N., Brauman, K. A., Cassidy, E. S., Gerber, J. S., Johnston, M., Mueller, N. D., O'Connell, C., Ray, D. K., West, P. C., Balzer, C., Bennett, E. M., Carpenter, S. R., Hill, J., Monfreda, C., Polasky, S., Rockström, J., Sheehan, J., Siebert, S., & Zaks, D. P. M. (2011). Solutions for a Cultivated Planet. *Nature, 478*(7369), 337–342. DOI: 10.1038/nature10452 PMID: 21993620

Food and Agriculture Organization (FAO). (2022). The State of Food and Agriculture 2022: Leveraging automation in agriculture for transforming agrifood systems. *FAO*. DOI: 10.4060/cb9479en

Food and Agriculture Organization of the United Nations. (2021). *AGROVOC semantic data interoperability on food and agriculture*. DOI: 10.4060/cb2838en

Fritz, S., See, L., Perger, C., McCallum, I., Schill, C., Schepaschenko, D., Duerauer, M., Karner, M., Dresel, C., Laso-Bayas, J.-C., Lesiv, M., Moorthy, I., Salk, C. F., Danylo, O., Sturn, T., Albrecht, F., You, L., Kraxner, F., & Obersteiner, M. (2017). A global dataset of crowdsourced land cover and land use reference data. *Scientific Data, 4*(1), 170075. DOI: 10.1038/sdata.2017.75 PMID: 28608851

Fu, Y., Zhu, Z., Liu, L., Zhan, W., He, T., Shen, H., & Ao, Z. (2024). Remote sensing time series analysis: A review of data and applications. *Journal of Remote Sensing, 4*, 0285. DOI: 10.34133/remotesensing.0285

Fuglie, K., Gautam, M., Goyal, A., & Maloney, W. F. (2020). *Harvesting prosperity: Technology and productivity growth in agriculture*. World Bank. DOI: 10.1596/978-1-4648-1393-1

Gagliardi, G., Lupia, M., Cario, G., Cicchello Gaccio, F., D'Angelo, V., Cosma, A. I. M., & Casavola, A. (2021). An internet of things solution for smartagriculture. *Agronomy (Basel), 11*(11), 2140. DOI: 10.3390/agronomy11112140

Gaitani, N., Spanou, A., Saliari, M., Synnefa, A., Vassilakopoulou, K., Papadopoulou, K., Pavlou, K., Santamouris, M., Papaioannou, M., & Lagoudaki, A. (2011). Improving the microclimate in urban areas: A case study in the centre of Athens. *Building Services Engineering Research and Technology, 32*(1), 53–71. DOI: 10.1177/0143624410394518

Gans Combe, C., & Camaréna, S. (2025). Data sovereignty and valuation model for sustainable agriculture innovation and equity. *npj Sustainable Agriculture, 3*, 61. DOI: 10.1038/s44264-025-00102-z

Gao, J., Gujarati, K., Hegde, M., Arra, P., Gupta, S., & Buch, N. (2025). Integration of UAV and remote sensing data for early diagnosis and severity mapping of diseases in maize crop through deep learning and reinforcement learning. *Remote Sensing (Basel), 17*(20), 3427. DOI: 10.3390/rs17203427

Garcia-Ruiz, F., Sankaran, S., Maja, J. M., Lee, W. S., Rasmussen, J., & Ehsani, R. (2013). Comparison of two aerial imaging platforms for identification of Huanglongbing-infected citrus trees. *Computers and Electronics in Agriculture, 91*, 106–115. DOI: 10.1016/j.compag.2012.12.002

Garonna, I., Jong, R., & Schaepman, M. E. (2016). Variability and evolution of global land surface phenology over the past three decades (1982–2012). *Global Change Biology*, *22*(4), 1456–1468. DOI: 10.1111/gcb.13168 PMID: 26924776

Gascon, F. (2018). Sentinel-2 for Agricultural Monitoring. IGARSS 2018 - 2018 IEEE International Geoscience and Remote Sensing Symposium, 8166-8168, DOI: 10.1109/IGARSS.2018.8517942

Gaur, S., & Singh, R. (2023). A comprehensive review on land use/land cover (LULC) change modeling for urban development: Current status and future prospects. *Sustainability (Basel)*, *15*(2), 903. DOI: 10.3390/su15020903

Gebbers, R., & Adamchuk, V. I. (2010). Precision agriculture and food security. *Science*, *327*(5967), 828–831. DOI: 10.1126/science.1183899 PMID: 20150492

Geissdoerfer, M., Savaget, P., Bocken, N. M. P., & Hultink, E. J. (2017). The Circular Economy - A New Sustainability Paradigm? *Journal of Cleaner Production*, *143*, 757–768. DOI: 10.1016/j.jclepro.2016.12.048

Gelagay, H. S. (2017). Geospatial Data Sharing Barriers across Organisations and the Possible Solution for Ethiopia. *International Journal of Spatial Data Infrastructures Research*, *12*, 62–84. DOI: 10.2902/1725-0463.2017.12.art4

Geldmann, J., Manica, A., Burgess, N. D., Coad, L., & Balmford, A. (2019). A global-level assessment of the effectiveness of protected areas at resisting anthropogenic pressures. *Proceedings of the National Academy of Sciences of the United States of America*, *116*(46), 23209–23215. DOI: 10.1073/pnas.1908221116 PMID: 31659036

George, N. A., & McKay, F. H. (2019). The public distribution system and food security in India. *International Journal of Environmental Research and Public Health*, *16*(17), 3221. DOI: 10.3390/ijerph16173221 PMID: 31484374

Getahun, S., Kefale, H., & Gelaye, Y. (2024). Application of precision agriculture technologies for sustainable crop production and environmental sustainability: A systematic review. *TheScientificWorldJournal*, *2126734*(1). Advance online publication. DOI: 10.1155/2024/2126734 PMID: 39421732

Gevaert, C. M. (2022). Explainable AI for earth observation: A review including societal and regulatory perspectives. *International Journal of Applied Earth Observation and Geoinformation*, *112*, 102869. DOI: 10.1016/j.jag.2022.102869

Ge, Y., Thomasson, J. A., & Sui, R. (2011). Remote sensing of soil properties in precision agriculture: A review. *Frontiers of Earth Science*, *5*(3), 229–238. DOI: 10.1007/s11707-011-0175-0

Ghisellini, P., Cialani, C., & Ulgiati, S. (2016). A review on circular economy: The expected transition to a balanced interplay of environmental and economic systems. *Journal of Cleaner Production*, *114*, 11–32. DOI: 10.1016/j.jclepro.2015.09.007

Ghulam, S. T., & Abushammala, H. (2023). Challenges and opportunities in the management of electronic waste and its impact on human health and environment. *Sustainability (Basel)*, *15*(3), 1837. DOI: 10.3390/su15031837

Giuliani, G., Mazzetti, P., Santoro, M., Nativi, S., Van Bemmelen, J., Colangeli, G., & Lehmann, A. (2020). Knowledge generation using satellite earth observations to support sustainable development goals (SDG): A use case on land degradation. *International Journal of Applied Earth Observation and Geoinformation*, *88*, 102068. DOI: 10.1016/j.jag.2020.102068

Godfray, H. C. J., Beddington, J. R., Crute, I. R., Haddad, L., Lawrence, D., Muir, J. F., Pretty, J., Robinson, S., Thomas, S. M., & Toulmin, C. (2010). Food Security: The Challenge of Feeding 9 Billion People. *Science*, *327*(5967), 812–818. DOI: 10.1126/science.1185383 PMID: 20110467

Gogoi, N. K., Deka, B., & Bora, L. C. (2018). Remote sensing and its use in detection and monitoring plant diseases: A review. *Agricultural Reviews (Karnal)*, *39*(4). Advance online publication. DOI: 10.18805/ag.R-1835

Gokool, S., Mahomed, M., Brewer, K., Naiken, V., Clulow, A., Sibanda, M., & Mabhaudhi, T. (2024). Crop mapping in smallholder farms using unmanned aerial vehicle imagery and geospatial cloud computing infrastructure. *Heliyon*, *10*(5), e26913. DOI: 10.1016/j.heliyon.2024.e26913 PMID: 38455562

Golden, C. D., Koehn, J. Z., Shepon, A., Passarelli, S., Free, C. M., Viana, D. F., & Thilsted, S. H. (2021). Aquatic foods to nourish nations. *Nature*, *598*(7880), 315–320. DOI: 10.1038/s41586-021-03917-1 PMID: 34526720

Gong, Z., Gao, F., Chang, X., Hu, T., & Li, Y. (2024). A review of interactions between irrigation and evapotranspiration. *Ecological Indicators*, *169*, 112870. DOI: 10.1016/j.ecolind.2024.112870

Goodfellow, I., Pouget-Abadie, J., Mirza, M., Xu, B., Warde-Farley, D., Ozair, S., Courville, A., & Bengio, Y. (2020). Generative adversarial networks. *Communications of the ACM*, *63*(11), 139–144. DOI: 10.1145/3422622

Gorelick, N., Hancher, M., Dixon, M., Ilyushchenko, S., Thau, D., & Moore, R. (2017). Google Earth Engine: Planetary-scale geospatial analysis for everyone. *Remote Sensing of Environment*, *202*, 18–27. DOI: 10.1016/j.rse.2017.06.031

Goyal, A., Aggarwal, R., Mahajan, S., & Saran, S. (2025). Automated GUI-based system for in-season crop classification and acreage estimation using multi-temporal SAR data and machine learning. *Archives of Agriculture Research and Technology*, *6*, 1082. DOI: 10.54026/AART/1082

Griscom, B. W., Adams, J., Ellis, P. W., Houghton, R. A., Lomax, G., Miteva, D. A., Schlesinger, W. H., Shoch, D. T., Siikamäki, J., Smith, P., Woodbury, P. B., Zganjar, C., Blackman, A., Campari, J. S., Conant, R. T., Delgado, C., Elias, P., Gopalakrishna, T., Hamsik, M. R., & Fargione, J. E. (2017). Natural climate solutions. *Proceedings of the National Academy of Sciences of the United States of America*, *114*(44), 11645–11650. DOI: 10.1073/pnas.1710465114 PMID: 29078344

Guanter, L., Kaufmann, H., Segl, K., Foerster, S., Rogass, C., Chabrillat, S., Küster, T., Hollstein, A., Rossner, G., Chlebek, C., Straif, C., Fischer, S., Schrader, S., Storch, T., Heiden, U., Müller, A., Bachmann, M., Mühle, H., Müller, R., & Sang, B. (2015). The EnMAP Spaceborne Imaging Spectroscopy Mission for Earth Observation. *Remote Sensing (Basel)*, *7*(7), 8830–8857. DOI: 10.3390/rs70708830

Guebsi, R., Mami, S., & Chokmani, K. (2024). Drones in precision agriculture: A comprehensive review of applications, technologies, and challenges. *Drones (Basel)*, *8*(11), 686. DOI: 10.3390/drones8110686

Guimarães, N., Sousa, J. J., Couto, P., Bento, A., & Pádua, L. (2024). Combining UAV-based multispectral and thermal infrared data with regression modeling and SHAP analysis for predicting stomatal conductance in almond orchards. *Remote Sensing (Basel), 16*(13), 2467. DOI: 10.3390/rs16132467

Gumbi, N., Gumbi, L., & Twinomurinzi, H. (2023). Towards Sustainable Digital Agriculture for Smallholder Farmers: A Systematic Literature Review. *Sustainability (Basel), 15*(16), 1–20. DOI: 10.3390/su151612530

Gupta, A., & Singh, A. (2023). Agri-gnn: A novel genotypic-topological graph neural network framework built on graphsage for optimized yield prediction. *arXiv preprint arXiv:2310.13037*

Gupta, G., & Kumar Pal, S. (2025). Applications of AI in precision agriculture. *Discover Agriculture, 3*(61), 1–19. DOI: 10.1007/s44279-025-00220-9

Hadj-Salah, A., Verdier, R., Caron, C., Picard, M., & Capelle, M. (2019). Schedule Earth observation satellites with deep reinforcement learning. arXiv preprint arXiv:1911.05696. https://arxiv.org/abs/1911.05696

Hafeez, A., Husain, M. A., Singh, S. P., Chauhan, A., Khan, M. T., Kumar, N., & Soni, S. K. (2023). Implementation of drone technology for farm monitoring & pesticide spraying: A review. *Information Processing in Agriculture, 10*(2), 192–203. DOI: 10.1016/j.inpa.2022.02.002

Hagolle, O., Huc, M., Villa Pascual, D., & Dedieu, G. (2015). A multi-temporal and multi-spectral method to estimate aerosol optical thickness over land, for the atmospheric correction of FormoSat-2, Landsat, VENμS and Sentinel-2 images. *Remote Sensing (Basel), 7*(3), 2668–2691. DOI: 10.3390/rs70302668

Haley, J., Conroy, D., & Errmann, A. (2024). *Shaping agricultural policy: Societal considerations of digitalization solutions. SSRN*. DOI: 10.2139/ssrn.4797293

Hamdouni, A. (2026). AI-Enabled Remote Sensing Assessment of Cultivated Land Quality and Sustainability Under Climate Stress: Evidence from Saudi Arabia. *Resources, 15*(3), 44. DOI: 10.3390/resources15030044

Hamidi, M., Homayouni, S., Safari, A., & Hasani, H. (2024). Deep learning-based crop-type mapping using SAR and optical data fusion. *International Journal of Applied Earth Observation and Geoinformation, 129*, 103860. DOI: 10.1016/j.jag.2024.103860

Hamilton, S. E., & Casey, D. (2016). Creation of a high spatio-temporal resolution global database of continuous mangrove forest cover for the 21st century (CGMFC-21). *Global Ecology and Biogeography, 25*(6), 729–738. DOI: 10.1111/geb.12449

Han, H., Liu, Z., Li, J., & Zeng, Z. (2024). Challenges in remote sensing–based climate and crop monitoring: Navigating the complexities using AI. *Journal of Cloud Computing (Heidelberg, Germany), 13*(1), 34. DOI: 10.1186/s13677-023-00583-8

Hansen, M. C., Krylov, A., Tyukavina, A., Potapov, P. V., Turubanova, S., Zutta, B., Ifo, S., Margono, B., Stolle, F., & Moore, R. (2016). Humid tropical forest disturbance alerts using Landsat data. *Environmental Research Letters, 11*(3), 034008. DOI: 10.1088/1748-9326/11/3/034008

Hansen, M. C., Potapov, P. V., Moore, R., Hancher, M., Turubanova, S. A., Tyukavina, A., Thau, D., Stehman, S. V., Goetz, S. J., Loveland, T. R., Kommareddy, A., Egorov, A., Chini, L., Justice, C. O., & Townshend, J. R. G. (2013). High-resolution global maps of 21st-century forest cover change. *Science*, *342*(6160), 850–853. DOI: 10.1126/science.1244693 PMID: 24233722

Hardie, M. (2020). Review of novel and emerging proximal soil moisture sensors for use in agriculture. *Sensors (Basel)*, *20*(23), 6934. DOI: 10.3390/s20236934 PMID: 33291606

Haregeweyn, N., Tsunekawa, A., Tsubo, M., Almaw Fenta, A., Ebabu, K., Vanmaercke, M., Borrelli, P., Panagos, P., Berihun, M. L., Langendoen, E. J., Nigussie, Z., Setargie, T. A., Maurice, B. N., Minichil, T., Elias, A., Sun, J., & Poesen, J. (2023). Progress and challenges in sustainable land management initiatives: A global review. *The Science of the Total Environment*, *858*(Part 3), 160027. DOI: 10.1016/j.scitotenv.2022.160027 PMID: 36356757

Harris, I. C., Osborn, T. J., Jones, P. D., & Lister, D. H. (2020). Version 4 of the CRU TS monthly high-resolution gridded multivariate climate dataset. *Scientific Data*, *7*(1), 7. DOI: 10.1038/s41597-020-0453-3 PMID: 32246091

He, K. S., Bradley, B. A., Cord, A. F., Rocchini, D., Tuanmu, M., Schmidtlein, S., Turner, W., Wegmann, M., & Pettorelli, N. (2015). Will remote sensing shape the next generation of species distribution models? *Remote Sensing in Ecology and Conservation*, *1*(1), 4–18. Advance online publication. DOI: 10.1002/rse2.7

He, K., Gkioxari, G., Dollár, P., & Girshick, R. (2017). Mask r-cnn. In *Proceedings of the IEEE international conference on computer vision* (pp. 2961-2969).

He, K., Zhang, X., Ren, S., & Sun, J. (2016). Deep residual learning for image recognition. In *Proceedings of the IEEE conference on computer vision and pattern recognition* (pp. 770–778). https://arxiv.org/abs/1512.03385

Helmold, M. (2020). Lean management on the upstream (supply side). In *Supply management research* (pp. 91–106). Springer., DOI: 10.1007/978-3-030-46981-8_10

Hemalatha, S., & Jayachandran, J. J. B. (2024). A multitask learning-based vision transformer for plant disease localization and classification. *International Journal of Computational Intelligence Systems*, *17*(1), 188. DOI: 10.1007/s44196-024-00597-3

Hengl, T., Mendes de Jesus, J., Heuvelink, G. B., Ruiperez González, M., Kilibarda, M., Blagotić, A., Wei, S., Wright, M. N., Geng, X., Bauer-Marschallinger, B., Guevara, M., Vargas, R., MacMillan, R. A., Batjes, N. H., Leenaars, J. G., Ribeiro, E., Wheeler, I., Mantel, S., & Kempen, B. (2017). SoilGrids250m: Global gridded soil information based on machine learning. *PLoS One*, *12*(2), 12. DOI: 10.1371/journal.pone.0169748 PMID: 28207752

Herath, H. M. B. S., & Nayanajith, J. (2025). Land use changes and sustainability: A review of geospatial, AI and socioeconomic approaches. Journal of Geography. *Environment and Earth Science International*, *29*(11), 144–158. DOI: 10.9734/jgeesi/2025/v29i11973

Hidrobo, M., Kosec, K., Gartaula, H. N., Van Campenhout, B., & Carrillo, L. (2024). Making complementary agricultural resources, technologies, and services more gender-responsive. *Global Food Security, 42*(June), 1–14. DOI: 10.1016/j.gfs.2024.100778

Hilmi, Y. S., Tóth, J., Gabnai, Z., Király, G., & Temesi, Á. (2024). Farmers' resilience to climate change through the circular economy and sustainable agriculture: A review from developed and developing countries. *Renewable Agriculture and Food Systems, 39*, 1–11. DOI: 10.1017/S1742170524000097

Hislop, S., Jones, S., Soto-Berelov, M., Skidmore, A., Haywood, A., & Nguyen, T. H. (2018). Using Landsat Spectral Indices in Time-Series to Assess Wildfire Disturbance and Recovery. *Remote Sensing (Basel), 10*(3), 460. DOI: 10.3390/rs10030460

HLPE. (2020). Food security and nutrition: building a global narrative towards 2030. High-Level Panel of Experts on Food Security and Nutrition of the Committee. https://openknowledge.fao.org/server/api/core/bitstreams/8357b6eb-8010-4254-814a-1493faaf4a93/content

Ho, J., Abbeel, P., & Jain, A. (2020). denoising probabilistic models of diffusion. Advancements in Neural Information Processing Systems (NeurIPS). https://doi.org//arXiv.2006.11239DOI: 10.48550

Holzinger, A., Fister, I., Fister, I., Kaul, H.-P., & Asseng, S. (2024). Human-Centered AI in Smart Farming : Toward Agriculture 5.0. *IEEE Access : Practical Innovations, Open Solutions, 12*(April), 62199–62214. DOI: 10.1109/ACCESS.2024.3395532

Hosseini, F. S., Seo, M. B., Razavi-Termeh, S. V., Sadeghi-Niaraki, A., Jamshidi, M., & Choi, S.-M. (2023). Geospatial artificial intelligence (GeoAI) and satellite imagery fusion for soil physical property predicting. *Sustainability (Basel), 15*(19), 14125. DOI: 10.3390/su151914125

Howari, F. (2025). AI–Remote Sensing for Soil Variability Mapping and Precision Agrochemical Management: A Comprehensive Review of Methods, Limitations, and Climate-Smart Applications. *Agrochemicals, 5*(1), 1. https://www.mdpi.com/2813-3145/5/1/1

Huang, X., Wang, S., & Song, Y. (2025). Digital Divide and GeoAI. In X. Huang, S. Wang, J. Wilson, & P. Kedron (Eds.), GeoAI and Human Geography (pp. 361-371). Springer Geography, Springer, Cham. DOI: 10.1007/978-3-031-87421-5_25

Huang, G., Liu, Z., Van Der Maaten, L., & Weinberger, K. Q. (2017). Densely connected convolutional networks. In *Proceedings of the IEEE conference on computer vision and pattern recognition* (pp. 4700-4708).

Huang, X., Ziniti, B., Torbick, N., & Ducey, M. J. (2018). Assessment of Forest above Ground Biomass Estimation Using Multi-Temporal C-band Sentinel-1 and Polarimetric L-band PALSAR-2 Data. *Remote Sensing (Basel), 10*(9), 1424. DOI: 10.3390/rs10091424

Hu, C., Ru, Y., Fang, S., Rong, Z., Zhou, H., Yan, X., & Liu, M. (2025). Orchard variable rate spraying method and experimental study based on multidimensional prescription maps. *Computers and Electronics in Agriculture, 235*, 110379. DOI: 10.1016/j.compag.2025.110379

Huete, A., Didan, K., Miura, T., Rodriguez, E. P., Gao, X., & Ferreira, L. G. (2002). Overview of the radiometric and biophysical performance of the MODIS vegetation indices. *Remote Sensing of Environment, 83*(1–2), 195–213. DOI: 10.1016/S0034-4257(02)00096-2

Huffman, G. J., Bolvin, D. T., Braithwaite, D., Hsu, K., Joyce, R., Kidd, C., Nelkin, E. J., Sorooshian, S., Tan, J., & Xie, P. (2019). Algorithm theoretical basis document (ATBD) version 06 NASA global precipitation measurement (GPM) integrated multi-satellitE retrievals for GPM (IMERG). National Aeronautics and Space Administration. https://pmm.nasa.gov/sites/default/files/document_files/IMERG_ATBD_V06.pdf

Hunt, E. R.Jr, & Daughtry, C. S. (2018). What good are unmanned aircraft systems for agricultural remote sensing and precision agriculture? *International Journal of Remote Sensing, 39*(15–16), 5345–5376. DOI: 10.1080/01431161.2017.1410300

Husain, A., & Bhowmick, S. (2026). Smart Geospatial Intelligence and AI for a Resilient Environment: A Review. *Advanced Geospatial Intelligence and AI for Environmental Resilience and Sustainable Development*, 417–432. https://link.springer.com/chapter/10.1007/978-3-032-03714-5_17

Husaini, A. M., & Sohail, M. (2024). Agrochemical-free genetically modified and genome-edited crops: Towards achieving the United Nations sustainable development goals and a'greener'green revolution. *Journal of Biotechnology, 389*, 68–77. DOI: 10.1016/j.jbiotec.2024.04.015 PMID: 38663518

Hu, X., Li, L., Huang, J., Zeng, Y., Zhang, S., Su, Y., Hong, Y., & Hong, Z. (2024). Radar Vegetation Indices for Monitoring Surface Vegetation: Developments, Challenges, and Trends. *The Science of the Total Environment, 945*, 173974. DOI: 10.1016/j.scitotenv.2024.173974 PMID: 38897467

Ijaware, V. A. (2024). GIS and artificial intelligence application in smart forest ecosystem sustainability evaluation of Olokemeji Forest Reserve, Ogun State, Nigeria. *American Journal of Geospatial Technology, 3*(1), 9–16. DOI: 10.54536/ajgt.v3i1.2621

Ines, A. V., Das, N. N., Hansen, J. W., & Njoku, E. G. (2013). Assimilation of remotely sensed soil moisture and vegetation with a crop simulation model for maize yield prediction. *Remote Sensing of Environment, 138*, 149–164. DOI: 10.1016/j.rse.2013.07.018

Intellecap Advisory Services. (2024). *Unlocking data sharing in agriculture*. https://www.intellecap.com/publications/unlocking-data-sharing-in-agriculture/

IPCC. (2022). *Climate Change 2022: Impacts, Adaptation and Vulnerability*. Cambridge University Press. DOI: 10.1017/9781009325844

Islam, M. R., Oliullah, K., Kabir, M. M., Alom, M., & Mridha, M. F. (2023). Machine learning enabled IoT system for soil nutrients monitoring and crop recommendation. *Journal of Agriculture and Food Research, 14*, 100880. DOI: 10.1016/j.jafr.2023.100880

Isola, P., Zhu, J. Y., Zhou, T., & Efros, A. A. (2017). Image-to-image translation with conditional adversarial networks. In *Proceedings of the IEEE conference on computer vision and pattern recognition* (pp. 1125-1134).

Jafarzadeh, S., Yildiz, Z., Yildiz, P., Strachowski, P., Forough, M., Esmaeili, Y., & Abdollahi, M. (2024). Advanced technologies in biodegradable packaging using intelligent sensing to fight food waste. *International Journal of Biological Macromolecules, 261*, 129647. DOI: 10.1016/j.ijbiomac.2024.129647 PMID: 38281527

Jagan Mohan, R. N. V., Rayanoothala, P. S., & Sree, R. P. (2025). Next-gen agriculture: Integrating AI and XAI for precision crop yield predictions. *Frontiers in Plant Science, 15*, 1451607. DOI: 10.3389/fpls.2024.1451607 PMID: 39845494

Jahangeer, J., Joshi, P., Kapoor, A., & Tang, Z. (2025). A review of AI-driven Google Earth Engine applications in surface water monitoring, assessment, and management. *Discover Geoscience, 3*(1), 140. DOI: 10.1007/s44288-025-00255-x PMID: 41018550

Janowicz, K., Gao, S., McKenzie, G., Hu, Y., & Bhaduri, B. (2020). GeoAI: Spatially explicit artificial intelligence for geographic knowledge discovery. *ABACUS, 16*(2), 1–25. DOI: 10.1080/13658816.2019.1684500

Janssen, M., Brous, P., Estevez, E., Barbosa, L. S., & Janowski, T. (2020). Data governance: Organizing data for trustworthy artificial intelligence. *Government Information Quarterly, 37*(3), 101493. DOI: 10.1016/j.giq.2020.101493

Jaramillo, J., Muchugu, E., Vega, F. E., Davis, A. P., Borgemeister, C., & Chabi-Olaye, A. (2011). Some Like It Hot: The Influence and Implications of Climate Change on Coffee Berry Borer (Hypothenemus hampei) and Coffee Production in East Africa. *PLoS One, 6*(9), 6. DOI: 10.1371/journal.pone.0024528 PMID: 21935419

Javadi, S. H., Guerrero, A., & Mouazen, A. M. (2022). Clustering and smoothing pipeline for management zone delineation using proximal and remote sensing. *Sensors (Basel), 22*(2), 645. DOI: 10.3390/s22020645 PMID: 35062606

Jena, R. K., Bandyopadhyay, S., Pradhan, U. K., Moharana, P. C., Kumar, N., Sharma, G. K., Roy, P. D., Ghosh, D., Ray, P., Padua, S., Ramachandran, S., Das, B., Singh, S. K., Ray, S. K., Alsuhaibani, A. M., Gaber, A., & Hossain, A. (2022). Geospatial modelling for delineation of crop management zones using local terrain attributes and soil properties. *Remote Sensing (Basel), 14*(9), 2101. DOI: 10.3390/rs14092101

Jennewein, J. S., Davis, B. W., Seehaver-Eagen, S., Nicolette, J., Pittman, J., Hively, W. D., Goldsmith, A., Hidalgo, C., Reberg-Horton, C., & Mirsky, S. B. (2025). Multi-sensor proximal remote sensing for cover crop biomass estimation at high and moderate spatial resolutions. *Smart Agricultural Technology, 12*, 101201. DOI: 10.1016/j.atech.2025.101201

Jetz, W., Cavender-Bares, J. M., Pavlick, R. P., Schimel, D. S., Davis, F. W., Asner, G. P., Guralnick, R. P., Kattge, J., Latimer, A. M., Moorcroft, P. R., Schaepman, M. E., Schildhauer, M., Schneider, F. D., Schrodt, F., Stahl, U., & Ustin, S. L. (2016). Monitoring plant functional diversity from space. *Nature Plants*, 2. PMID: 27249357

Jha, K., Doshi, A., Patel, P., & Shah, M. (2019). A comprehensive review on automation in agriculture using artificial intelligence. *Artificial Intelligence in Agriculture, 2*, 1–12. DOI: 10.1016/j.aiia.2019.05.004

Jha, P. (2007). Some Aspects of the Well Being of India's Agricultural Labour in the Context of Contemporary Agrarian Crisis". *The Indian Journal of Labour Economics : The Quarterly Journal of the Indian Society of Labour Economics, 2007*. https://www.macroscan.org/anl/feb07/Agrarian_Crisis.pdf

Jiang, C., Guo, X., Li, Y., Lai, N., Peng, L., & Geng, Q. (2025). Multimodal deep learning models in precision agriculture: Cotton yield prediction based on unmanned aerial vehicle imagery and meteorological data. *Agronomy (Basel)*, *15*(5), 1217. DOI: 10.3390/agronomy15051217

Jiang, S. (2025). Big data sharing: A comprehensive survey. *Data*, *10*(11), 182. DOI: 10.3390/data10110182

Jin, N., Tao, B., Ren, W., Feng, M., Sun, R., He, L., Zhuang, W., & Yu, Q. (2016). Mapping irrigated and rainfed wheat areas using multi-temporal satellite data. *Remote Sensing (Basel)*, *8*(3), 207. DOI: 10.3390/rs8030207

Joshi, K., Kumar, S., Bhatla, S., Kaushik, V., Anandaram, H., & Kumar, A. (2023). Automation intervention for monitoring agriculture farming scale: A review. In *2023 14th International Conference on Computing Communication and Networking Technologies (ICCCNT)* (pp. 1–6). IEEE. DOI: 10.1109/ICCCNT56998.2023.10307391

Joshi, A., Pradhan, B., Gite, S., & Chakraborty, S. (2023). Remote-sensing data and deep-learning techniques in crop mapping and yield prediction: A systematic review. *Remote Sensing (Basel)*, *15*(8), 2014. Advance online publication. DOI: 10.3390/rs15082014

Joshi, N., Baumann, M., Ehammer, A., Fensholt, R., Grogan, K., Hostert, P., Jepsen, M. R., Kuemmerle, T., Meyfroidt, P., Mitchard, E. T. A., Reiche, J., Ryan, C. M., & Waske, B. (2016). A Review of the Application of Optical and Radar Remote Sensing Data Fusion to Land Use Mapping and Monitoring. *Remote Sensing (Basel)*, *8*(1), 70. DOI: 10.3390/rs8010070

Kabato, W., Hailegnaw, N., Mutum, L., & Molnar, Z. (2025). Managing soil health for climate resilience and crop productivity in a changing environment. *The Science of the Total Environment*, *1000*, 180460. DOI: 10.1016/j.scitotenv.2025.180460 PMID: 40946612

Kabisch, N., Selsam, P., Kirsten, T., Lausch, A., & Bumberger, J. (2019). A multi-sensor and multi-temporal remote sensing approach to detect land cover change dynamics in heterogeneous urban landscapes. *Ecological Indicators*, *99*, 273–282. DOI: 10.1016/j.ecolind.2018.12.033

Kaku, K., Aso, N., & Takiguchi, F. (2015). Space-based response to the 2011 Great East Japan Earthquake: Lessons learnt from JAXA's support using earth observation satellites. *International Journal of Disaster Risk Reduction*, *12*, 134–153. DOI: 10.1016/j.ijdrr.2014.12.009

Kalogiannidis, S., Kalfas, D., Giannarakis, G., & Paschalidou, M. (2023). Integration of water resources management strategies in land use planning towards environmental conservation. *Sustainability (Basel)*, *15*(21), 15242. DOI: 10.3390/su152115242

Kamilaris, A., & Prenafeta-Boldú, F. (2018). Deep learning in agriculture. *Computers and Electronics in Agriculture*, *147*, 70–90. Advance online publication. DOI: 10.1016/j.compag.2018.02.016

Karmakar, P., Teng, S. W., Murshed, M., Pang, S., Li, Y., & Lin, H. (2024). Crop monitoring by multimodal remote sensing: A review. *Remote Sensing Applications: Society and Environment, 33*, 101093. DOI: 10.1016/j.rsase.2023.101093

Karmaoui, A., El Jaafari, S., Chaachouay, H., & Hajji, L. (2023). A bibliometric review of geospatial analyses and artificial intelligence literature in agriculture. *GeoJournal, 88*(1), 343–360. DOI: 10.1007/s10708-023-10859-w

Karpatne, A., Atluri, G., Faghmous, J. H., Steinbach, M., Banerjee, A., Ganguly, A., Shekhar, S., Samatova, N., & Kumar, V. (2017). Theory-guided data science: A new paradigm for scientific discovery from data. *IEEE Transactions on Knowledge and Data Engineering, 29*(10), 2318–2331. Advance online publication. DOI: 10.1109/TKDE.2017.2720168

Kausika, B. B., & van Altena, V. (2025). GeoAI in Topographic Mapping: Navigating the Future of Opportunities and Risks. *ISPRS International Journal of Geo-Information, 14*(8), 313. DOI: 10.3390/ijgi14080313

Kazi, K. (2024). Machine Learning-Based Pomegranate Disease Detection and Treatment. In Zia Ul Haq, M., & Ali, I. (Eds.), *Revolutionizing Pest Management for Sustainable Agriculture* (pp. 469–498). IGI Global. DOI: 10.4018/979-8-3693-3061-6.ch019

Kazi, K. S. (2025). AI-Driven-IoT (AIIoT)-Based Jawar Leaf Disease Detection: KSK Approach for Jawar Disease Detection. In Bhatti, U., Aamir, M., Gulzar, Y., & Ullah Bazai, S. (Eds.), *Modern Intelligent Techniques for Image Processing* (pp. 439–472). IGI Global Scientific Publishing. DOI: 10.4018/979-8-3693-9045-0.ch019

Kendall, A., Olson, C. M., & Frongillo, E. A.Jr. (1995). Validation of the Radimer/Cornell measures of hunger and food insecurity. *The Journal of Nutrition, 125*(11), 2793–2801. DOI: 10.1093/jn/125.11.2793 PMID: 7472659

Kennedy, R. E., Yang, Z., & Cohen, W. B. (2010). Detecting trends in forest disturbance and recovery using yearly Landsat time series: 1. LandTrendr—Temporal segmentation algorithms. *Remote Sensing of Environment, 114*(12), 2897–2910. DOI: 10.1016/j.rse.2010.07.008

Keshava, N., & Mustard, J. F. (2002). Spectral unmixing. *IEEE Signal Processing Magazine, 19*(1), 44–57. DOI: 10.1109/79.974727

Kgaphola, M. J., Ramoelo, A., Odindi, J., Mwenge Kahinda, J.-M., Seetal, A., & Musvoto, C. (2023). Social–ecological system understanding of land degradation in response to land use and cover changes in the Greater Sekhukhune District Municipality. *Sustainability (Basel), 15*(4), 3850. DOI: 10.3390/su15043850

Khan, S. H., Hayat, M., Naseer, M., & Khan, F. S. (2021). A vision survey of the transformers. Surveys of ACM Computing. DOI: 10.1145/3505244

Khan, A., Sohail, A., Zahoora, U., & Qureshi, A. S. (2021). A survey of the recent architectures of deep convolutional neural networks. *Artificial Intelligence Review, 53*(8), 5455–5516. DOI: 10.1007/s10462-020-09825-6

Khanal, S., Kc, K., Fulton, J. P., Shearer, S., & Ozkan, E. (2020). Remote sensing in agriculture—Accomplishments, limitations, and opportunities. *Remote Sensing (Basel)*, *12*(22), 3783. DOI: 10.3390/rs12223783

Khaspuria, G., Khandelwal, A., Agarwal, M., Bafna, M., Yadav, R., & Yadav, A. (2024). Adoption of precision agriculture technologies among farmers: A comprehensive review. *Journal of Scientific Research and Reports*, *30*(7), 671–686. DOI: 10.9734/jsrr/2024/v30i72180

Khatri, V., & Brown, C. V. (2010). Designing data governance. *Communications of the ACM*, *53*(1), 148–152. DOI: 10.1145/1629175.1629210

Khattri, P., Sayers, R., Singh, K. K., Slapikas, R., Tamang, C. B., Tamang, D., Sagara, B., & BenYishay, A. (2025). Geospatial impact evaluation of a low-cost agricultural intervention for enhancing environmental resilience. *International Journal of Applied Earth Observation and Geoinformation*, *141*, 104657. DOI: 10.1016/j.jag.2025.104657 PMID: 40599674

Khofiyah, N. A., Nugraha, I., & Windyatri, H. (2025). An extended TAM approach to understanding behavioral and institutional drivers of drone technology adoption in Indonesia's green economy. *Journal of Advances in Information and Industrial Technology*, *7*(2), 227–240. DOI: 10.52435/jaiit.v7i2.732

Kibblewhite, M. G., Ritz, K., & Swift, M. J. (2008). Soil health in agricultural systems. *Philosophical Transactions of the Royal Society of London. Series B, Biological Sciences*, *363*(1492), 685–701. DOI: 10.1098/rstb.2007.2178 PMID: 17785275

Kingma, D. P., & Ba, J. (2014). Adam: A method for stochastic optimization. *arXiv preprint arXiv:1412.6980.*

Klerkx, L., Jakku, E., & Labarthe, P. (2019). A review of social science on digital agriculture, smart farming and agriculture 4.0: New contributions and a future research agenda. *NJAS Wageningen Journal of Life Sciences*, *90–91*(October), 1–16. DOI: 10.1016/j.njas.2019.100315

Kogan, F., Guo, W., & Yang, W. (2019). Drought and food security prediction from NOAA new generation of operational satellites. *Geomatics, Natural Hazards & Risk*, *10*(1), 651–666. DOI: 10.1080/19475705.2018.1541257

Koldasbayeva, D., Tregubova, P., Gasanov, M., Zaytsev, A., Petrovskaia, A., & Burnaev, E. (2024). Challenges in Data-Driven Geospatial Modeling for Environmental Research and Practice. *Nature Communications*, *15*(1), 10700. DOI: 10.1038/s41467-024-55240-8 PMID: 39702456

Kopp, S., Becker, P., Doshi, A., Wright, D. J., Zhang, K., & Xu, H. (2019). Achieving the Full Vision of Earth Observation Data Cubes. *Data*, *4*(3), 94. DOI: 10.3390/data4030094

Korhonen, J., Honkasalo, A., & Seppälä, J. (2018). Circular Economy: The Concept and its Limitations. *Ecological Economics*, *143*, 37–46. DOI: 10.1016/j.ecolecon.2017.06.041

Kornelsen, K. C., & Coulibaly, P. (2013). Advances in soil moisture retrieval from synthetic aperture radar and hydrological applications. *Journal of Hydrology (Amsterdam)*, *476*, 460–489. DOI: 10.1016/j.jhydrol.2012.10.044

Kotwal, J., Kashyap, R., Shafi, P. M., & Kimbahune, V. (2024). Enhanced leaf disease detection: UNet for segmentation and optimized EfficientNet for disease classification. *Software Impacts*, *22*, 100701. DOI: 10.1016/j.simpa.2024.100701

Kouadio, L., El Jarroudi, M., Belabess, Z., Laasli, S. E., Roni, M. Z. K., Amine, I. D. I., Mokhtari, N., Mokrini, F., Junk, J., & Lahlali, R. (2023). A review on UAV-based applications for plant disease detection and monitoring. *Remote Sensing (Basel)*, *15*(17), 4273. DOI: 10.3390/rs15174273

Kpodo, J., & Nejadhashemi, A. P. (2025). Navigating challenges and opportunities in developing smart agricultural extension platforms: Multi-media data mining techniques. *Artificial Intelligence in Agriculture*, *15*(3), 426–448. DOI: 10.1016/j.aiia.2025.04.001

Krizhevsky, A., Sutskever, I., & Hinton, G. E. (2012). Imagenet classification with deep convolutional neural networks. *Advances in Neural Information Processing Systems, 25*. https://proceedings.neurips.cc/paper/2012/file/c399862d3b9d6b76c8436e924a68c45b-Paper.pdf

Krizhevsky, A., Sutskever, I., & Hinton, G. E. (2012). Imagenet classification with deep convolutional neural networks. *Advances in Neural Information Processing Systems*, 25.

Kross, A., Kaur, G., & Jaeger, J. A. G. (2022). A geospatial framework for the assessment and monitoring of environmental impacts of agriculture. *Environmental Impact Assessment Review*, *97*, 106851. DOI: 10.1016/j.eiar.2022.106851

Kroupová, Z. Ž., Aulová, R., Rumánková, L., Bajan, B., Čechura, L., Šimek, P., & Jarolímek, J. (2025). Drivers and Barriers to Precision Agriculture Technology and Digitalisation Adoption: Meta-Analysis of Decision Choice Models. *Precision Agriculture*, *26*(1), 17. DOI: 10.1007/s11119-024-10213-1

Kshetri, N. (2021). Blockchain and sustainable supply chain management in developing countries. *International Journal of Information Management*, *60*, 102376. DOI: 10.1016/j.ijinfomgt.2021.102376

Kulp, S. A., & Strauss, B. H. (2019). New elevation data triple estimates of global vulnerability to sea-level rise and coastal flooding. *Nature Communications*, *10*(1), 4844. DOI: 10.1038/s41467-019-12808-z PMID: 31664024

Kumagai, N. H., Molinos, J. G., Yamano, H., Takao, S., Fujii, M., & Yamanaka, Y. (2018). Ocean currents and herbivory drive macroalgae-to-coral community shift under climate warming. *Proceedings of the National Academy of Sciences of the United States of America*, *115*(36), 8990–8995. DOI: 10.1073/pnas.1716826115 PMID: 30126981

Kumar, A. S. (2022). Capacity development in geospatial technology – An overview. ISPRS Annals of the Photogrammetry, Remote Sensing and Spatial Information Sciences, 5, 9–16. DOI: 10.5194/isprs-annals-V-5-2022-9-2022

Kumar, A., Bantilan, M. C. S., Kumar, P., Kumar, S., & Jee, S. (2012). Food Security in India: Trends, Patterns, and Determinants. *Indian Journal of Agricultural Economics*, *67*(3). https://oar.icrisat.org/6590/1/IJAgriEco_67_3_445-463_2012.pdf

Kumari, K., Nafchi, A. M., Mirzaee, S., & Abdalla, A. (2025). Ai-driven future farming: Achieving climate-smart and sustainable agriculture. *AgriEngineering.*, *7*(3), 89. DOI: 10.3390/agriengineering7030089

Kumari, S., Ali, N., Dagati, M., & Dong, Y. (2025). IoT-enabled soil moisture and conductivity monitoring under controlled and field fertigation systems. *AgriEngineering, 7*(7), 207. DOI: 10.3390/agriengineering7070207

Kumar, S., Yadav, A., & Sharma, D. K. (2020). Deep Learning and Computer Vision in Smart Agriculture. In Pradeep, N., Kautish, S., Nirmala, C., Goyal, V., & Abdellatif, S. (Eds.), *Modern Techniques for Agricultural Disease Management and Crop Yield Prediction* (pp. 66–88). IGI Global Scientific Publishing. DOI: 10.4018/978-1-5225-9632-5.ch004

Kussul, N., Lavreniuk, M., Skakun, S., & Shelestov, A. (2017). Deep learning classification of land cover and crop types using remote sensing data. *IEEE Geoscience and Remote Sensing Letters, 14*(5), 778–782. DOI: 10.1109/LGRS.2017.2681128

Kwok, R., & Cunningham, G. F. (2015). Variability of Arctic sea ice thickness and volume from CryoSat-2. *Philosophical Transactions. Series A, Mathematical, Physical, and Engineering Sciences, 373*(2045), 20140157. Advance online publication. DOI: 10.1098/rsta.2014.0157 PMID: 26032317

Lacerda, L. N., Snider, J. L., Cohen, Y., Liakos, V., Gobbo, S., & Vellidis, G. (2022). Using UAV-based thermal imagery to detect crop water status variability in cotton. *Smart Agricultural Technology, 2*, 100029. DOI: 10.1016/j.atech.2021.100029

Laird, S. A., & Wynberg, R. P. (2018). *Digital sequence information on genetic resources*. Convention on Biological Diversity. https://www.cbd.int/dsi-gr/2017-2018#peerreview

Lambin, E. F., & Meyfroidt, P. (2011). Global land use change, economic globalization, and the looming land scarcity. *Proceedings of the National Academy of Sciences of the United States of America, 108*(9), 3465–3472. DOI: 10.1073/pnas.1100480108 PMID: 21321211

Lanjouw, P., & Rinku, M. (2009). Poverty decline, agricultural wages and non-farm employment in rural India: 1983-2004, Policy Research Working Paper Series 4858. The World Bank. https://openknowledge.worldbank.org/entities/publication/3ee5b391-f80d-5568-95a2-0c8b21c4582e

Lausch, A., Blaschke, T., Haase, D., Herzog, F., Syrbe, R.-U., Tischendorf, L., & Walz, U. (2015). Understanding and quantifying landscape structure—A review on relevant process characteristics, data models and landscape metrics. *Ecological Modelling, 295*, 31–41. DOI: 10.1016/j.ecolmodel.2014.08.018

Laveglia, S., Altieri, G., Genovese, F., Matera, A., & Di Renzo, G. C. (2024). Advances in sustainable crop management: integrating precision agriculture and proximal sensing. *AgriEngineering, 6*(3), 3084–3120. https://www.mdpi.com/2624-7402/6/3/177

Lee, C.-H., Chen, K.-Y., & Liu, L.-Y. D. (2024). Effect of texture feature distribution on agriculture field type classification with multitemporal UAV RGB images. *Remote Sensing (Basel), 16*(7), 1221. DOI: 10.3390/rs16071221

Lefsky, M. A., Cohen, W. B., Parker, G. G., & Harding, D. J. (2002). Lidar remote sensing for ecosystem studies: Lidar, an emerging remote sensing technology that directly measures the three-dimensional distribution of plant canopies, can accurately estimate vegetation structural attributes and should be of particular interest to forest, landscape, and global ecologists. *Bioscience, 52*(1), 19–30. DOI: 10.1641/0006-3568(2002)052[0019:LRSFES]2.0.CO;2

Leng, J., Jiang, P., & Xu, K. (2020). Blockchain-empowered sustainable manufacturing and product lifecycle management in Industry 4.0: A survey. *Renewable & Sustainable Energy Reviews*, *132*, 110112. DOI: 10.1016/j.rser.2020.110112

Lewis, A., Oliver, S., Lymburner, L., Evans, B. J., Wyborn, L., Mueller, N., Raevksi, G., Hooke, J., Woodcock, R., Sixsmith, J., Wu, W., Tan, P., Li, F., Killough, B. D., Minchin, S., Roberts, D. O., Ayers, D., Bala, B., Dwyer, J. L., & Wang, L. (2017). The Australian Geoscience Data Cube - foundations and lessons learned. *Remote Sensing of Environment*, *202*, 276–292. DOI: 10.1016/j.rse.2017.03.015

Li, W. (2025). Artificial intelligence in Earth science: A GeoAI perspective. *Journal of Geophysical Research: Machine Learning and Computation*, *2*(3), e2025JH000691. https://doi.org/DOI: 10.1029/2025JH000691

Liang, J., & Jiang, W. (2023). A ResNet50-DPA model for tomato leaf disease identification. *Frontiers in Plant Science*, *14*, 1258658. DOI: 10.3389/fpls.2023.1258658 PMID: 37908831

Li, D., Liao, W., Rigden, A. J., Liu, X., Wang, D., Malyshev, S., & Shevliakova, E. (2019). Urban heat island: Aerodynamics or imperviousness? *Science Advances*, *5*(4), eaau4299. Advance online publication. DOI: 10.1126/sciadv.aau4299 PMID: 30949572

Li, J., Hong, D., Gao, L., Yao, J., Zheng, K., Zhang, B., & Chanussot, J. (2022). Deep learning in multimodal remote sensing data fusion: A comprehensive review. *International Journal of Applied Earth Observation and Geoinformation*, *112*, 102926. DOI: 10.1016/j.jag.2022.102926

Li, J., & Roy, D. P. (2017). A global analysis of Sentinel-2A, Sentinel-2B and Landsat-8 data revisit intervals and implications for terrestrial monitoring. *Remote Sensing (Basel)*, *9*(9), 902. DOI: 10.3390/rs9090902

Li, K., Zhao, W., Chen, J., Zhang, L., Hu, D., & Wang, Q. (2023). Predicting crop growth patterns with spatial–temporal deep feature exploration for early mapping. *Remote Sensing (Basel)*, *15*(13), 3285. DOI: 10.3390/rs15133285

Li, M., Imou, K., Wakabayashi, K., & Yokoyama, S. (2009). Review of research on agricultural vehicle autonomous guidance. *International Journal of Agricultural and Biological Engineering*, *2*(3), 1–16. https://ijabe.org/index.php/ijabe/article/view/160/82

Li, M., Zhu, Z., Deng, J., Zhang, J., & Li, Y. (2025). Geospatial explainable AI uncovers eco-environmental effects and its driving mechanisms—Evidence from the Poyang Lake Region, China. *Land (Basel)*, *14*(7), 1361. DOI: 10.3390/land14071361

Lim, B., Arık, S. Ö., Loeff, N., & Pfister, T. (2021). Temporal fusion transformers for interpretable multi-horizon time series forecasting. *International Journal of Forecasting*, *37*(4), 1748–1764. DOI: 10.1016/j.ijforecast.2021.03.012

Lin, T. Y., Goyal, P., Girshick, R., He, K., & Dollár, P. (2017). Focal loss for dense object detection. In *Proceedings of the IEEE international conference on computer vision* (pp. 2980-2988).

Lipper, L., Thornton, P. K., Campbell, B. M., Baedeker, T., Braimoh, A. K., Bwalya, M., Caron, P., Cattaneo, A., Garrity, D. P., Henry, K., Hottle, R. D., Jackson, L. E., Jarvis, A., Kossam, F., Mann, W., McCarthy, N., Meybeck, A., Neufeldt, H., Remington, T., & Torquebiau, E. F. (2014). Climate-smart agriculture for food security. *Nature Climate Change*, *4*(12), 1068–1072. DOI: 10.1038/nclimate2437

Li, Q., Tian, J., & Tian, Q. (2023). Deep learning application for crop classification via multi-temporal remote sensing images. *Agriculture*, *13*(4), 906. DOI: 10.3390/agriculture13040906

Li, S., Huang, J., Xiao, G., Huang, H., Sun, Z., & Li, X. (2024). Improved winter wheat yield estimation by combining remote sensing data, machine learning, and phenological metrics. *Remote Sensing (Basel)*, *16*(17), 3217. DOI: 10.3390/rs16173217

Littlefield, C. E., Mcrae, B. H., Michalak, J. L., Lawler, J. J., & Carroll, C. (2017). Connecting today's climates to future climate analogs to facilitate movement of species under climate change. *Conservation Biology : The Journal of the Society for Conservation Biology*, *31*(6), 31. DOI: 10.1111/cobi.12938 PMID: 28339121

Liu, X., Chen, J., & Wen, Q. (2023). A survey on graph classification and link prediction based on gnn. *arXiv preprint arXiv:2307.00865*

Liu, J., Bowman, K. W., Schimel, D. S., Parazoo, N. C., Jiang, Z., Lee, M., Bloom, A. A., Wunch, D., Frankenberg, C., Sun, Y., O'Dell, C. W., Gurney, K. R., Menemenlis, D., Gierach, M., Crisp, D., & Eldering, A. (2017). Contrasting carbon cycle responses of the tropical continents to the 2015–2016 El Niño. *Science*, *358*(6360), eaam5690. Advance online publication. DOI: 10.1126/science.aam5690 PMID: 29026011

Liu, J., Iryadi, R., & Böhlen, M. (2023). Agroforestry in the Alas Mertajati of Bali, Indonesia. A Case Study in Sustainable Small-Scale Farming Practices and GeoAI. *International Archives of the Photogrammetry, Remote Sensing and Spatial Information Sciences - ISPRS Archives*, *48*(4/W7), 101–108. DOI: 10.5194/isprs-archives-XLVIII-4-W7-2023-101-2023

Liu, J., & Wang, X. (2021). Plant diseases and pests detection based on deep learning: A review. *Plant Methods*, *17*(1), 22. DOI: 10.1186/s13007-021-00722-9 PMID: 33627131

Liu, M., Wei, H., Dong, X., Wang, X.-C., Zhao, B., & Zhang, Y. (2022). Integrating land use, ecosystem service, and human well-being: A systematic review. *Sustainability (Basel)*, *14*(11), 6926. DOI: 10.3390/su14116926

Liu, P., & Biljecki, F. (2022). A review of spatially-explicit GeoAI applications in urban geography. *International Journal of Applied Earth Observation and Geoinformation*, *112*, 102936. DOI: 10.1016/j.jag.2022.102936

Liu, R., Sun, Y., Li, M., Zhang, M., Zhang, Z., Li, H., & Yang, W. (2022). Development and application experiments of a grain yield monitoring system. *Computers and Electronics in Agriculture*, *195*, 106851. DOI: 10.1016/j.compag.2022.106851

Liu, X., Huang, Y., Xu, X., Li, X., Li, X., Ciais, P., Lin, P., Gong, K., Ziegler, A. D., Chen, A., Gong, P., Chen, J., Hu, G., Chen, Y., Wang, S., Wu, Q., Huang, K., Estes, L. D., & Zeng, Z. (2020). High-spatiotemporal-resolution mapping of global urban change from 1985 to 2015. *Nature Sustainability*, *3*(7), 1–7. DOI: 10.1038/s41893-020-0521-x

Liu, X., Wang, X., Zhu, Q., & Miao, W. (2023). Tomato brown rot disease detection using improved YOLOv5 with attention mechanism. *Frontiers in Plant Science*, *14*, 1289464. DOI: 10.3389/fpls.2023.1289464 PMID: 38053763

Liu, Y., Fan, Y., Yue, J., Ma, Y., Yang, F., Fan, J., Chen, R., Bian, M., & Feng, H. (2025). UAV-based remote sensing estimation of above-ground biomass in different crops: A review. *International Journal of Applied Earth Observation and Geoinformation*, *144*, 104938. DOI: 10.1016/j.jag.2025.104938

Liu, Y., Zhang, S., Zhang, J., Tang, L., & Bai, Y. (2021). Assessment and comparison of six machine learning models in estimating evapotranspiration over croplands using remote sensing and meteorological factors. *Remote Sensing (Basel)*, *13*(19), 3838. DOI: 10.3390/rs13193838

Liu, Z., Wang, S., Zhang, Y., Feng, Y., Liu, J., & Zhu, H. (2023). Artificial intelligence in food safety: A decade review and bibliometric analysis. *Foods*, *12*(6), 1242. DOI: 10.3390/foods12061242 PMID: 36981168

Li, W. (2020). *Deep learning in geospatial*. ISPRS Journal. DOI: 10.1016/j.isprsjprs.2020.04.012

Li, W., Hsu, C. Y., Wang, S., & Kedron, P. (2024). GeoAI Reproducibility and Replicability: A Computational and Spatial Perspective. *Annals of the American Association of Geographers*, *114*(9), 2085–2103. DOI: 10.1080/24694452.2024.2373787

Li, X., Pan, Y.-C., Ge, Z.-Q., & Zhao, C.-J. (2007). Delineation and scale effect of precision agriculture management zones using yield monitor data over four years. *Agricultural Sciences in China*, *6*(2), 180–188. DOI: 10.1016/S1671-2927(07)60033-9

Li, Y., Deng, G., Zhao, H., Liu, B., Liu, C., Qian, W., & Qiao, X. (2026, January). A review of UAV remote sensing technology applications in common gramineous crops. *Information Processing in Agriculture*. Advance online publication. DOI: 10.1016/j.inpa.2026.01.009

Liyakat. (2025). AI-Driven-IoT(AIIoT)-Based Decision Making in Kidney Diseases Patient Healthcare Monitoring: KSK Approach for Kidney Monitoring. In L. Özgür Polat & O. Polat (Eds.), *AI-Driven Innovation in Healthcare Data Analytics* (pp. 277-306). IGI Global Scientific Publishing. DOI: 10.4018/979-8-3693-7277-7.ch009

Li, Z., Chen, G., & Zhang, T. (2020). A CNN-transformer hybrid approach for crop classification using multitemporal multisensor images. *IEEE Journal of Selected Topics in Applied Earth Observations and Remote Sensing*, *13*, 847–858. DOI: 10.1109/JSTARS.2020.2971763

Lobell, D., Schlenker, W., & Costa-Roberts, J. (2011). Climate Trends and Global Crop Production Since 1980. *Science*, *333*(6042), 616–620. DOI: 10.1126/science.1204531 PMID: 21551030

Löbmann, M. T., Maring, L., Prokop, G., Brils, J., Bender, J., Bispo, A., & Helming, K. (2022). Systems knowledge for sustainable soil and land management. *The Science of the Total Environment, 822*, 153389. DOI: 10.1016/j.scitotenv.2022.153389 PMID: 35104520

Loosvelt, L., Peters, J., Skriver, H., Lievens, H., Van Coillie, F. M. B., De Baets, B., & Verhoest, N. E. C. (2012). Random forests as a tool for estimating uncertainty at pixel-level in SAR image classification. *International Journal of Applied Earth Observation and Geoinformation, 19*, 173–184. DOI: 10.1016/j.jag.2012.05.011

Lou, S., Hu, R. Q., Liu, Y., Zhang, W., & Yang, S. Q. (2022). The formulation of irrigation and nitrogen application strategies under multi-dimensional soil fertility targets based on preference neural network. *Scientific Reports, 12*(1), 20918. DOI: 10.1038/s41598-022-25133-1 PMID: 36463318

Lowe, D. (2004). *SIFT features*. IJCV. DOI: 10.1023/B:VISI.0000029664.99615.94

Luijendijk, A., Hagenaars, G., Ranasinghe, R., Baart, F., Donchyts, G., & Aarninkhof, S. (2018). The state of the world's beaches. *Scientific Reports*, •••, 8.

Lu, J., Tan, L., & Jiang, H. (2021). Review on convolutional neural network (CNN) applied to plant leaf disease classification. *Agriculture, 11*(8), 707. DOI: 10.3390/agriculture11080707

Lulla, K. P., Duane Nellis, M., & Rundquist, B. C. (2013). The Landsat 8 is ready for geospatial science and technology researchers and practitioners. *Geocarto International, 28*(3), 191–191. DOI: 10.1080/10106049.2013.812346

Lu, Y., Yi, S., Zeng, N., Liu, Y., & Zhang, Y. (2020). Data augmentation of the rice disease detection using CycleGAN. *Computers and Electronics in Agriculture*. Advance online publication. DOI: 10.1016/j.compag.2020.105434

Maestrini, B., & Basso, B. (2021). Subfield crop yields and temporal stability in thousands of US Midwest fields. *Precision Agriculture, 22*(6), 1749–1767. DOI: 10.1007/s11119-021-09810-1 PMID: 34744492

Mafenya, K. (2024). Determining the uptake of ICT, GIS, RS, and 4IR technologies in improving food security in the agricultural sector in Gauteng Province, South Africa (Master's thesis, University of South Africa). Retrieved from https://search.proquest.com/openview/9f7b2f5d25bd961acb89d3bbce9dc2c8/1

Magdas, D. A., Hategan, A. R., David, M., & Berghian-Grosan, C. (2025). The Journey of Artificial Intelligence in Food Authentication: From Label Attribute to Fraud Detection. *Foods, 14*(10), 1808. DOI: 10.3390/foods14101808 PMID: 40428587

Mahajan, S. (2024). An irrigation system: Design, implementation, and benefits. In *Smart agri-tech: Robotics, AI, and Internet of Things (IoT) in agriculture* (pp. 309–328). https://doi.org/DOI: 10.1002/9781394302994.ch11

Mahendra, D., & Alakh, N. S. (2010). Food Security in India: Performance, Challenges and Policies Oxfam India Working Paper Series, September 2010. https://policy-practice.oxfam.org/resources/food-security-in-india-performance-challenges-and-policies-346637/

Mahlein, A. K., Steiner, U., Hillnhütter, C., Dehne, H. W., & Oerke, E. C. (2012). Hyperspectral imaging for small-scale analysis of symptoms caused by different sugar beet diseases. *Plant Methods*, *8*(1), 3. DOI: 10.1186/1746-4811-8-3 PMID: 22273513

Mai, G., Xie, Y., Jia, X., Lao, N., Rao, J., Zhu, Q., Liu, Z., Chiang, Y.-Y., & Jiao, J. (2025). Towards the next generation of geospatial artificial intelligence. *International Journal of Applied Earth Observation and Geoinformation*, *136*, 104368. DOI: 10.1016/j.jag.2025.104368

Maimaitijiang, M., Sagan, V., Sidike, P., Daloye, A. M., Erkbol, H., & Fritschi, F. B. (2020). Crop monitoring using satellite/UAV data fusion and machine learning. *Remote Sensing (Basel)*, *12*(9), 1357. DOI: 10.3390/rs12091357

Ma, J., Rabin, S. S., Anthoni, P., Bayer, A. D., Nyawira, S. S., Olin, S., Xia, L., & Arneth, A. (2022). Assessing the impacts of agricultural managements on soil carbon stocks, nitrogen loss, and crop production – a modelling study in eastern Africa. *Biogeosciences*, *19*(8), 2145–2169. DOI: 10.5194/bg-19-2145-2022

Ma, L., Liu, Y., Zhang, X., Ye, Y., Yin, G., & Johnson, B. A. (2019). Deep learning in remote sensing applications: A meta-analysis and review. *ISPRS Journal of Photogrammetry and Remote Sensing*, *152*, 166–177. DOI: 10.1016/j.isprsjprs.2019.04.015

Mamatha, A., & Shalini, S. (2025). Artificial Intelligence-based Green Technologies for Efficient Agriculture. In *ITM Web of Conferences* (Vol. 79, p. 01027). EDP Sciences. DOI: 10.1051/itmconf/20257901027

Mandal, D., Patra, S., Sharma, N. K., Alam, N. M., Jana, C., & Lal, R. (2023). Impacts of soil erosion on soil quality and agricultural sustainability in the north-western Himalayan region of India. *Sustainability (Basel)*, *15*(6), 5430. DOI: 10.3390/su15065430

Mansoor, S., Iqbal, S., Popescu, S. M., Kim, S. L., Chung, Y. S., & Baek, J. H. (2025). Integration of smart sensors and IOT in precision agriculture: Trends, challenges and future prospectives. *Frontiers in Plant Science*, *16*, 1587869. DOI: 10.3389/fpls.2025.1587869 PMID: 40438737

Manzoor, F., Wei, L., Siraj, M., Lu, X., & Qiyang, G. (2025). Digital agriculture technology adoption in low and middle-income countries—A review of contemporary literature. *Frontiers in Sustainable Food Systems*, *9*(September), 1–11. DOI: 10.3389/fsufs.2025.1621851

Marasinghe, R., Yigitcanlar, T., Mayere, S., Washington, T., & Limb, M. (2024). Towards responsible urban geospatial AI: Insights from the white and grey literatures. *Journal of Geovisualization and Spatial Analysis*, *8*(2), 24. DOI: 10.1007/s41651-024-00184-2

Markham, B. L., Storey, J. C., & Morfitt, R. A. (2015). Landsat-8 Sensor Characterization and Calibration. *Remote Sensing (Basel)*, *7*(3), 2279–2282. DOI: 10.3390/rs70302279

Marshall, E., Cherian, D., Henderson, S., Scheick, J., & Forster, R. (2025). Cloud-native geospatial data cube workflows with open-source tools. *The Journal of Open Source Education*, *8*(89), 267. DOI: 10.21105/jose.00267

Martinez-Harms, M. J., Bryan, B. A., Balvanera, P., Law, E. A., Rhodes, J. R., Possingham, H. P., & Wilson, K. A. (2015). Making decisions for managing ecosystem services. *Biological Conservation, 184*, 229–238. DOI: 10.1016/j.biocon.2015.01.024

Mashala, M. J., Dube, T., Ayisi, K. K., & Ramudzuli, M. R. (2023a). Using the Google Earth Engine cloud-computing platform to assess the long-term spatial temporal dynamics of land use and land cover within the Letaba watershed, South Africa. *Geocarto International, 38*(1), 2252781. DOI: 10.1080/10106049.2023.2252781

Mashala, M. J., Dube, T., Mudereri, B. T., Ayisi, K. K., & Ramudzuli, M. R. (2023b). A Systematic Review on Advancements in Remote Sensing for Assessing and Monitoring Land Use and Land Cover Changes Impacts on Surface Water Resources in Semi-Arid Tropical Environments. *Remote Sensing (Basel), 15*(16), 3926. DOI: 10.3390/rs15163926

Masolele, R. N., De Sy, V., Herold, M., Marcos, D., Verbesselt, J., Gieseke, F., Mullissa, A. G., & Martius, C. (2021). Spatial and temporal deep learning methods for deriving land-use following deforestation: A pan-tropical case study using Landsat time series. *Remote Sensing of Environment, 264*, 112600. DOI: 10.1016/j.rse.2021.112600

Mattijssen, T. J. M., Hennen, W., Buijs, A. E., De Dooij, P., Van Lammeren, R., & Walet, L. (2024). Urban greening co-creation: Participatory spatial modelling to bridge data-driven and citizen-centred approaches. *Urban Forestry & Urban Greening, 94*, 128257. DOI: 10.1016/j.ufug.2024.128257

Mavisclara, O. A., Esekie, J. O., Atoyebi, T. O., Atumah, P. E., Akadiri, O. O., Jimoh, R. A., & Oshobugie, I. (2025). GeoAI at the forefront of climate action: Mapping mitigation and adaptation with artificial intelligence. *Global Journal of Engineering and Technology Advances, 24*(2), 217–234. DOI: 10.30574/gjeta.2025.24.2.0248

Maxwell, A. E., Warner, T. A., & Fang, F. (2018). Implementation of machine-learning classification in remote sensing: An applied review. *International Journal of Remote Sensing, 39*(9), 2784–2817. DOI: 10.1080/01431161.2018.1433343

Ma, Y., Chen, S., Ermon, S., & Lobell, D. B. (2024). Transfer Learning in Environmental Remote Sensing. *Remote Sensing of Environment, 301*, 113924. DOI: 10.1016/j.rse.2023.113924

Maynard, J. J., Yeboah, E., Owusu, S., Buenemann, M., Neff, J. C., & Herrick, J. E. (2023). Accuracy of regional-to-global soil maps for on-farm decision-making: Are soil maps "good enough"? *Soil (Göttingen), 9*(1), 277–300. DOI: 10.5194/soil-9-277-2023

McBratney, A., Whelan, B., Ancev, T., & Bouma, J. (2005). Future directions of precision agriculture. *Precision Agriculture, 6*(1), 7–23. DOI: 10.1007/s11119-005-0681-8

Mccabe, M. F., Rodell, M., Alsdorf, D. E., Miralles, D. G., Uijlenhoet, R., Wagner, W., Lucieer, A., Houborg, R., Verhoest, N. E., Franz, T. E., Shi, J., Gao, H., & Wood, E. F. (2017). The Future of Earth Observation in Hydrology. *Hydrology and Earth System Sciences, 21*(7), 3879–3914. DOI: 10.5194/hess-21-3879-2017 PMID: 30233123

McKay, F. H., Sims, A., & Pligt, P. (2023). Measuring food insecurity in India: A systematic review of the current evidence. *Current Nutrition Reports*, *12*(2), 358–367. DOI: 10.1007/s13668-023-00470-3 PMID: 37022635

Meghraoui, K., Sebari, I., Pilz, J., Ait El Kadi, K., & Bensiali, S. (2024). Applied deep learning-based crop yield prediction: A systematic analysis of current developments and potential challenges. *Technologies*, *12*(4), 43. DOI: 10.3390/technologies12040043

Mehmood, M., Shahzad, A., Zafar, B., Shabbir, A., & Ali, N. (2022). Remote sensing image classification: A comprehensive review and applications. *Mathematical Problems in Engineering*, *5880959*, 1–24. Advance online publication. DOI: 10.1155/2022/5880959

Mekasha, A., Gérard, B., Tesfaye, K., Nigatu, L., & Duncan, A. J. (2014). Inter-connection between land use/land cover change and herders'/farmers' livestock feed resource management strategies: A case study from three Ethiopian eco-environments. *Agriculture, Ecosystems & Environment*, *188*, 150–162. DOI: 10.1016/j.agee.2014.02.022

Melichar, M., Didan, K., Barreto-Muñoz, A., Duberstein, J. N., Jiménez Hernández, E., Crimmins, T., & Nagler, P. L. (2023). Random forest classification of multitemporal Landsat 8 spectral data and phenology metrics for land cover mapping in the Sonoran and Mojave deserts. *Remote Sensing (Basel)*, *15*(5), 1266. DOI: 10.3390/rs15051266

Melo-Velasco, J., Wilson, K., Hendrickson, M., & Valdivia, C. (2025). Understanding the social implications of digital agricultural technologies. *Journal of Agriculture, Food Systems, and Community Development*, 1–16. Advance online publication. DOI: 10.5304/jafscd.2025.143.010

Mena, F., Pathak, D., Najjar, H., Sanchez, C., Helber, P., Bischke, B., Habelitz, P., Miranda, M., Siddamsetty, J., Nuske, M., Charfuelan, M., Arenas, D., Vollmer, M., & Dengel, A. (2025). Adaptive fusion of multi-modal remote sensing data for optimal sub-field crop yield prediction. *Remote Sensing of Environment*, *318*, 114547. DOI: 10.1016/j.rse.2024.114547

Mendes, J., Pinho, T. M., dos Santos, F. N., Sousa, J. J., Peres, E., Boaventura-Cunha, J., Cunha, M., & Morais, R. (2020). Smartphone applications targeting precision agriculture practices—A systematic review. *Agronomy (Basel)*, *10*(6), 855. DOI: 10.3390/agronomy10060855

Mendoza-Bernal, J., González-Vidal, A., & Skarmeta, A. F. (2024). A convolutional neural network approach for image-based anomaly detection in smart agriculture. *Expert Systems with Applications*, *247*, 123210. Advance online publication. DOI: 10.1016/j.eswa.2024.123210

Mentaschi, L., Vousdoukas, M. I., Pekel, J. F., Voukouvalas, E., & Feyen, L. (2018). Global long-term observations of coastal erosion and accretion. *Scientific Reports*, *8*(1), 12876. DOI: 10.1038/s41598-018-30904-w PMID: 30150698

Messina, G., & Modica, G. (2020). Applications of UAV thermal imagery in precision agriculture: State of the art and future research outlook. *Remote Sensing (Basel)*, *12*(9), 1491. DOI: 10.3390/rs12091491

Michalk, D. L., Kemp, D. R., Badgery, W. B., Wu, J., Zhang, Y., & Thomassin, P. J. (2019). Sustainability and future food security—A global perspective for livestock production. *Land Degradation & Development*, *30*(5), 561–573. DOI: 10.1002/ldr.3217

Micheli, M., Gevaert, C. M., Carman, M., Craglia, M., Daemen, E., Ibrahim, R. E., Kotsev, A., Mohamed-Ghouse, Z., Schade, S., Schneider, I., Shanley, L. A., Tartaro, A., & Vespe, M. (2022). AI Ethics and Data Governance in the Geospatial Domain of Digital Earth. *Big Data & Society*, *9*(2). Advance online publication. DOI: 10.1177/20539517221138767

Miller, T., Mikiciuk, G., Durlik, I., Mikiciuk, M., Łobodzińska, A., & Śnieg, M. (2025). The IoT and AI in agriculture: The time is now—A systematic review of smart sensing technologies. *Sensors (Basel)*, *25*(12), 3583. DOI: 10.3390/s25123583 PMID: 40573470

Mincuzzi, A., Sanzani, S. M., Palou, L., Ragni, M., & Ippolito, A. (2022). Postharvest rot of pomegranate fruit in southern Italy: Characterization of the main pathogens. *Journal of Fungi (Basel, Switzerland)*, *8*(5), 475. DOI: 10.3390/jof8050475 PMID: 35628731

Mirzaei, A., Bagheri, H., & Khosravi, I. (2023). Enhancing crop classification accuracy through synthetic SAR-optical data generation using deep learning. *ISPRS International Journal of Geo-Information*, *12*(11), 450. DOI: 10.3390/ijgi12110450

Mirzaeitalarposhti, R., Shafizadeh-Moghadam, H., Taghizadeh-Mehrjardi, R., & Demyan, M. S. (2022). Digital soil texture mapping and spatial transferability of machine learning models using Sentinel-1, Sentinel-2, and terrain-derived covariates. *Remote Sensing (Basel)*, *14*(23), 5909. DOI: 10.3390/rs14235909

Mitchell, A. L., Williams, M. L., Tapley, I., & Milne, A. K. (2012). Interoperability of multi-frequency SAR data for forest information extraction in support of national MRV systems. *2012 IEEE International Geoscience and Remote Sensing Symposium*, 3166-3169. DOI: 10.1109/IGARSS.2012.6350752

Mittelstadt, B. D., Allo, P., Taddeo, M., Wachter, S., & Floridi, L. (2016). The ethics of algorithms: Mapping the debate. *Big Data & Society*, *3*(2). Advance online publication. DOI: 10.1177/2053951716679679

Mkhitaryan, K., Sanamyan, A., Mnatsakanyan, M., Kirakosyan, E., & Ratner, S. (2025). Integrating AI and Geospatial technologies for sustainable smart city development: A case study of Yerevan. *Urban Science (Basel, Switzerland)*, *9*(10), 389. DOI: 10.3390/urbansci9100389

Mohamed, E. S., Belal, A. A., Abd-Elmabod, S. K., El-Shirbeny, M. A., Gad, A., & Zahran, M. B. (2021). Smart farming for improving agricultural management. *The Egyptian Journal of Remote Sensing and Space Sciences*, *24*(3), 971–981. DOI: 10.1016/j.ejrs.2021.08.007

Mohanty, S. P., Salathé, M., & Hughes, D. P. (2016). Image-based deep learning plant disease diagnosis. *Frontiers in Plant Science*. Advance online publication. DOI: 10.3389/fpls.2016.01419 PMID: 27713752

Mondal, B. P., Sahoo, R. N., Das, B., Chakraborty, D., Kumar, N., & Raj, G. B. (2024). Digital Soil Mapping: concepts, methods, and applications-Remote sensing and GIS perspectives. In *Remote sensing of soils* (pp. 195–212). Elsevier. DOI: 10.1016/B978-0-443-18773-5.00018-1

Monteiro, A., Santos, S., & Gonçalves, P. (2021). Precision agriculture for crop and livestock farming - Brief review. *Animals (Basel)*, *11*(8), 2345. DOI: 10.3390/ani11082345 PMID: 34438802

Montesino Pouzols, F., Toivonen, T., Di Minin, E., Kukkala, A. S., Kullberg, P., Kuusterä, J., Lehtomäki, J., Tenkanen, H., Verburg, P. H., & Moilanen, A. (2014). Global protected area expansion is compromised by projected land-use and parochialism. *Nature, 516*(7531), 383–386. DOI: 10.1038/nature14032 PMID: 25494203

Moomen, A.-W., Yevugah, L. L., Boakye, L., Osei, J. D., & Muthoni, F. (2024). Review of applications of remote sensing towards sustainable agriculture in the northern savannah regions of Ghana. *Agriculture, 14*(4), 546. DOI: 10.3390/agriculture14040546

Morales-Ona, A., Camberato, J., Nielsen, R., Paul, S., & Quinn, D. (2025). Integration of satellite, UAV, soil, and topographic data for assessing corn nitrogen uptake at early vegetative growth stages. *Precision Agriculture, 26*(6), 97. DOI: 10.1007/s11119-025-10293-7

Moravec, D., Komárek, J., López-Cuervo Medina, S., & Molina, I. (2021). Effect of atmospheric corrections on NDVI: Intercomparability of Landsat 8, Sentinel-2, and UAV sensors. *Remote Sensing (Basel), 13*(18), 3550. DOI: 10.3390/rs13183550

Moumane, A., Elmotawakkil, A., Hasan, M. M., Kranjčić, N., Batchi, M., Karkouri, J. A., & Youssef, Y. M. (2025). Integrating GIS, remote sensing, and machine learning to optimise sustainable groundwater recharge in arid Mediterranean landscapes: A case study from the Middle Draa Valley, Morocco. *Water (Basel), 17*(15), 2336. DOI: 10.3390/w17152336

Mulla, D. J. (2013). Twenty five years of remote sensing in precision agriculture: Key advances and remaining knowledge gaps. *Biosystems Engineering, 97*(4), 66–85. DOI: 10.1016/j.biosystemseng.2012.08.009

Mulya, S. P., Hudalah, D., & Prilandita, N. (2024). Developing a new methodology for determining typologies of peri-urban agriculture: A case study of the Jakarta Bandung Mega Urban Region Indonesia. *Land Use Policy, 146*(May), 1–19. DOI: 10.1016/j.landusepol.2024.107310

Mulya, S. P., Putro, H. P. H., & Hudalah, D. (2023). Review of peri-urban agriculture as a regional ecosystem service. *Geography and Sustainability, 4*(3), 244–254. DOI: 10.1016/j.geosus.2023.06.001

Munir, A. (2026). Artificial Intelligence and Machine Learning for Smart and Sustainable Agriculture. *AI, 7*(1), 12. https://www.mdpi.com/2673-2688/7/1/12

Muñoz-Sabater, J., Dutra, E., Agustí-Panareda, A., Albergel, C., Arduini, G., Balsamo, G., Boussetta, S., Choulga, M., Harrigan, S., Hersbach, H., Martens, B., Miralles, D. G., Piles, M., Rodríguez-Fernández, N. J., Zsoter, E., Buontempo, C., & Thépaut, J.-N. (2021). ERA5-Land: A state-of-the-art global reanalysis dataset for land applications. *Earth System Science Data, 13*(9), 4349–4383. DOI: 10.5194/essd-13-4349-2021

Munteanu, A. (2024). Data Cubes and Cloud-Native Environments for Earth Observation: An Overview. Scalable Computing. *Practice and Experience, 25*(6), 5745–5759. DOI: 10.12694/scpe.v25i6.4999

Muromba, P., Keeni, M., & Fuyuki, K. (2025). A systematic review of mobile agricultural service applications for smallholder farmers in sub-Saharan Africa: Perspectives from the technology acceptance model. *Agriculture & Food Security, 14*(1), 34. DOI: 10.1186/s40066-025-00563-y

Muruganandhan, S., Hanish, H., Vinish, H., Thamilmani, R., Kumarasamy, P., Vani, H., & Sivasubramanian, K. (2025). Measuring the application of innovative digital technology in the retail sector for sustainable business. In *Transforming business through digital sustainability models* (pp. 1–22). IGI Global. DOI: 10.4018/979-8-3373-0608-7.ch014

Muruganantham, P., Wibowo, S., Grandhi, S., Samrat, N. H., & Islam, N. (2022). A systematic literature review on crop yield prediction with deep learning and remote sensing. *Remote Sensing (Basel)*, *14*(9), 1990. DOI: 10.3390/rs14091990

Musyaffa, M. S. I., Yudistira, N., Rahman, M. A., Basori, A. H., Mansur, A. B., & Batoro, J. (2024). IndoHerb: Indonesia medicinal plants recognition using transfer learning and deep learning. *Heliyon*, *10*(23), 1–16. DOI: 10.1016/j.heliyon.2024.e40606 PMID: 39660181

Nagasubramanian, K., Jones, S., Singh, A. K., Sarkar, S., Singh, A., & Ganapathysubramanian, B. (2019). Plant disease identification using explainable 3D deep learning on hyperspectral images. *Plant Methods*, *15*(1), 98. DOI: 10.1186/s13007-019-0479-8 PMID: 31452674

Nagendra, H., Lucas, R. M., Honrado, J. P., Jongman, R. H., Tarantino, C., Adamo, M., & Mairota, P. (2013). Remote sensing for conservation monitoring: Assessing protected areas, habitat extent, habitat condition, species diversity, and threats. *Ecological Indicators*, *33*, 45–59. DOI: 10.1016/j.ecolind.2012.09.014

Narain, J. P. (2016). Public health challenges in India: Seizing the opportunities. *Indian Journal of Community Medicine : Official Publication of Indian Association of Preventive & Social Medicine*, *41*(2), 85. DOI: 10.4103/0970-0218.177507 PMID: 27051080

National Academies of Sciences, Engineering, and Medicine. (2018). Thriving on our changing planet: A decadal strategy for Earth observation from space. The National Academies Press. DOI: 10.17226/24938

Neethirajan, S. (2020). Transforming the adaptation physiology of farm animals through sensors. *Animals (Basel)*, *10*(9), 1512. DOI: 10.3390/ani10091512 PMID: 32859060

Nenkam, A. M., Wadoux, A. M. J.-C., Minasny, B., Silatsa, F. B. T., Yemefack, M., Ugbaje, S. U., Akpa, S., Van Zijl, G., Bouasria, A., Bouslihim, Y., Chabala, L. M., Ali, A., & McBratney, A. B. (2024). Applications and challenges of digital soil mapping in Africa. *Geoderma*, *449*, 117007. DOI: 10.1016/j.geoderma.2024.117007

Nerem, R. S., Beckley, B., Fasullo, J. T., Hamlington, B. D., Masters, D., & Mitchum, G. T. (2018). Climate-change–driven accelerated sea-level rise detected in the altimeter era. *Proceedings of the National Academy of Sciences of the United States of America*, *115*(9), 2022–2025. DOI: 10.1073/pnas.1717312115 PMID: 29440401

Neupane, J., & Guo, W. (2019). Agronomic basis and strategies for precision water management: A review. *Agronomy (Basel)*, *9*(2), 87. DOI: 10.3390/agronomy9020087

Neyra, J. M. V., Schmitt, V. G. H., & Cequea, M. M. (2025). Sustainable circular strategies and short supply chains among smallholder producers: The case of Peru's Agroferias Campesinas. *Agricultural and Food Economics*, *13*(77), 1–24. DOI: 10.1186/s40100-025-00423-0

Ng, A. H., Ge, L., Li, X., & Zhang, K. (2012). Monitoring ground deformation in Beijing, China with persistent scatterer SAR interferometry. *Journal of Geodesy, 86*(6), 375–392. DOI: 10.1007/s00190-011-0525-4

Ngugi, H. N., Akinyelu, A. A., & Ezugwu, A. E. (2024). Machine learning and deep learning for crop disease diagnosis: Performance analysis and review. *Agronomy (Basel), 14*(12), 3001. DOI: 10.3390/agronomy14123001

Nikolaou, G., Neocleous, D., Christou, A., Kitta, E., & Katsoulas, N. (2020). Implementing sustainable irrigation in water-scarce regions under the impact of climate change. *Agronomy (Basel), 10*(8), 1120. DOI: 10.3390/agronomy10081120

Nikolić, N., Mattivi, P., Pappalardo, S. E., De Marchi, M., & Masin, R. (2025). From detection to action: Creating operational prescription maps for weed management using low-cost UAVs. *Italian Journal of Agronomy, 20*(3), 100047. DOI: 10.1016/j.ijagro.2025.100047

Nirmal, M. D., Jadhav, P., & Pawar, S. (2022). RETRACTED ARTICLE: Classification of Pomegranate Leaves Diseases by Image Processing and Machine Learning Techniques. *Cybernetics and Systems*, 1–15. DOI: 10.1080/01969722.2022.2145448

Nitin, & Bal Gupta, S. (2023). Artificial Intelligence in Smart Agriculture: Applications and Challenges. *Current Applied Science and Technology, 24*(2), e0254427. DOI: 10.55003/cast.2023.254427

Nogueira, V. B., Teixeira, D. G., de Lima, I. A. C. N., Moreira, M. V. C., de Oliveira, B. S. C., Pedrosa, I. M. B., de Queiroz, J. W., & Jeronimo, S. M. B. (2022). Towards an inclusive digital literacy: An experimental intervention study in a rural area of Brazil. *Education and Information Technologies, 27*(2), 2807–2834. DOI: 10.1007/s10639-021-10711-z PMID: 34493924

Nourani, M., King, J. T., & Ragan, E. D. (2020). The role of domain expertise in user trust and the impact of first impressions with intelligent systems. Proceedings of the AAAI Conference on Human Computation and Crowdsourcing, 8, 112–121. https://ojs.aaai.org/index.php/HCOMP/article/view/7469

Nouri, H., Beecham, S., Kazemi, F., & Hassanli, A. (2013). A review of ET measurement techniques for estimating the water requirements of urban landscape vegetation. *Urban Water Journal, 10*(4), 247–259. DOI: 10.1080/1573062X.2012.726360

Noyan, M. A. (2022). Uncovering bias in the PlantVillage dataset. *arXiv preprint arXiv:2206.04374*

Nxumalo, G. S., & Chauke, H. (2025). Challenges and opportunities in smallholder agriculture digitization in South Africa. *Frontiers in Sustainable Food Systems, 9*, 1583224. DOI: 10.3389/fsufs.2025.1583224

Nyamekye, A. B., Dewulf, A., Van Slobbe, E., Termeer, K., & Pinto, C. (2018). Governance arrangements and adaptive decision-making in rice farming systems in Northern Ghana. *NJAS Wageningen Journal of Life Sciences, 86*(1), 39–50. DOI: 10.1016/j.njas.2018.07.004

Obasi, S. N., Aa, T. V., Obasi, C. C., Jokthan, G. E., Adjei, E. A., & Keyagha, E. R. (2024). Harnessing Artificial Intelligence for sustainable agriculture: A comprehensive review of African applications in spatial analysis and precision agriculture. *Big Data in Agriculture, 6*(1), 6–18. DOI: 10.26480/bda.01.2024.01.13

Obubu, J. P., Odong, R., Alamerew, T., Fetahi, T., & Mengistou, S. (2022). Application of DPSIR model to identify the drivers and impacts of land use and land cover changes and climate change on land, water, and livelihoods in the L. Kyoga basin: Implications for sustainable management. *Environmental Systems Research, 11*(1), 11. DOI: 10.1186/s40068-022-00254-8 PMID: 35607482

OECD. (2019). *Enhancing access to and sharing of data: Reconciling risks and benefits for data re-use across societies*. OECD Publishing. DOI: 10.1787/276aaca8-

Oguamanam, C., de Beer, J., & Ubalijoro, É. (2023). *Ownership, control, and governance of the benefits of data for food and agriculture*. https://www.oguamanam.com/publications/2023-3-31/ownership-control-and-governance-of-the-benefits-of-data-for-food-and-agriculture

Ogungbuyi, M. G., Mohammed, C., Fischer, A. M., Turner, D., Whitehead, J., & Harrison, M. T. (2024). Integration of Drone and Satellite Imagery Improves Agricultural Management Agility. *Remote Sensing (Basel), 16*(24), 4688. DOI: 10.3390/rs16244688

Oğuztürk, G. E. (2025). AI-driven irrigation systems for sustainable water management: A systematic review and meta-analytical insights. *Smart Agricultural Technology, 11*, 100982. DOI: 10.1016/j.atech.2025.100982

Okori, P., Munthali, W., Msere, H., Charlie, H., Chitaya, S., Sichali, F., Chilumpha, E., Chirwa, T., Seetha, A., Chinyamuyamu, B., Monyo, E., Siambi, M., & Chirwa, R. (2022). Improving efficiency of knowledge and technology diffusion using community seed banks and farmer-to-farmer extension: Experiences from Malawi. *Agriculture & Food Security, 11*(38), 1–14. DOI: 10.1186/s40066-022-00375-4

Oliver, M. A., & Webster, R. (2015). *Basic steps in geostatistics: The variogram and kriging*. Springer., DOI: 10.1007/978-3-319-15865-5

Olofsson, P., Foody, G. M., Herold, M., Stehman, S. V., Woodcock, C. E., & Wulder, M. A. (2014). Good practices for estimating area and assessing accuracy of land change. *Remote Sensing of Environment, 148*, 42–57. DOI: 10.1016/j.rse.2014.02.015

Olsen, T. L. (2026). Artificial Intelligence in Agricultural Supply Chains: Practice and Prospects. In Cohen, M. C., & Dai, T. (Eds.), *AI in Supply Chains. Springer Series in Supply Chain Management* (Vol. 27). Springer. 8_18.DOI: 10.1007/978-3-032-07054-8_18

Olson, C. M. (1999). Nutrition and health outcomes associated with food insecurity and hunger. *The Journal of Nutrition, 129*(2), 521S–524S. DOI: 10.1093/jn/129.2.521S PMID: 10064322

Oluoch, I. (2024). Crossing boundaries: The ethics of AI and geographic information technologies. *ISPRS International Journal of Geo-Information, 13*(3), 87. DOI: 10.3390/ijgi13030087

Omia, E., Bae, H., Park, E., Kim, M. S., Baek, I., Kabenge, I., & Cho, B.-K. (2023). Remote Sensing in Field Crop Monitoring: A Comprehensive Review of Sensor Systems, Data Analyses and Recent Advances. *Remote Sensing (Basel), 15*(2), 354. DOI: 10.3390/rs15020354

Omotayo, A. O., Adediran, S. A., Omotoso, A. B., Olagunju, K. O., & Omotayo, O. P. (2025). Artificial Intelligence in Agriculture: Ethics, Impact Possibilities, and Pathways for Policy. Computers and Electronics in Agriculture, 239(Part A), 110927. DOI: 10.1016/j.compag.2025.110927

Orellana, E., Gonzalez, T., Álvarez, A., Fernández-Campusano, C., Muñoz, M., & Carrasco, R. (2025). Precision farming: Exploring the challenges and opportunities for smallholder farmers in Chile. *Agriculture, 15*(24), 2555. DOI: 10.3390/agriculture15242555

Padarian, J., Minasny, B., & McBratney, A. B. (2019). Using deep learning for digital soil mapping. *Soil (Göttingen), 5*(1), 79–89. DOI: 10.5194/soil-5-79-2019

Page, M. J., McKenzie, J. E., Bossuyt, P. M., Boutron, I., Hoffmann, T. C., Mulrow, C. D., Shamseer, L., Tetzlaff, J. M., Akl, E. A., Brennan, S. E., Chou, R., Glanville, J., Grimshaw, J. M., Hróbjartsson, A., Lalu, M. M., Li, T., Loder, E. W., Mayo-Wilson, E., McDonald, S., & Moher, D. (2021). The PRISMA 2020 statement: An updated guideline for reporting systematic reviews. *BMJ (Clinical Research Ed.), 372*, n71. Advance online publication. DOI: 10.1136/bmj.n71 PMID: 33782057

Pakruddin, H. (2023). A systematic review of pomegranate fruit disease detection and classification using machine learning and deep learning techniques, International Conference on Information and Communication Technology for Competitive Strategies, Springer Nature Singapore, Singapore (2023, December), pp. 165-174. Available at: https://link.springer.com/chapter/10.1007/978-981-99-9486-1_13

Pallottino, F., Antonucci, F., Costa, C., Bisaglia, C., Figorilli, S., & Menesatti, P. (2019). Optoelectronic proximal sensing vehicle-mounted technologies in precision agriculture: A review. *Computers and Electronics in Agriculture, 162*, 859–873. DOI: 10.1016/j.compag.2019.05.034

Panda, S. S., Siddique, A., Terrill, T. H., Mahapatra, A. K., Morgan, E., Pech-Cervantes, A. A., & Van Wyk, J. A. (2025). Decision support system for *Lespedeza cuneata* production and quality evaluation: A WebGIS dashboard approach to precision agriculture. *Frontiers in Plant Science, 16*, 1520163. DOI: 10.3389/fpls.2025.1520163 PMID: 40747527

Pandey, D., & Misra, R. (2024). Towards sustainable agriculture: Harnessing AI for global food security. *Artificial Intelligence in Agriculture., 12*, 72–84. DOI: 10.1016/j.aiia.2024.04.003

Pandit, V., Sravya, M., Ravali, C., Fiaz, S., Tariq, A., & Chatterjee, S. (2026). Precision agriculture in the era of technological advancement. In Fiaz, S., Nadeem, M. A., Baloch, F. S., & Chung, Y. S. (Eds.), *Artificial intelligence and data sciences for precision agriculture* (pp. 159–186). Springer. DOI: 10.1007/978-3-032-12770-9_8

Pantazi, X.-E., Moshou, D., & Bravo, C. (2016). Active learning system for weed species recognition based on hyperspectral sensing. *Biosystems Engineering, 146*, 193–202. DOI: 10.1016/j.biosystemseng.2016.01.014

Papadakis, E., Adams, B., Gao, S., Martins, B., Baryannis, G., & Ristea, A. (2022). Explainable artificial intelligence in the spatial domain (X-GeoAI). *Transactions in GIS, 26*(6), 2413–2414. DOI: 10.1111/tgis.12996

Paszke, A., Gross, S., Massa, F., Lerer, A., Bradbury, J., Chanan, G., & Chintala, S. (2019). Pytorch: An imperative style, high-performance deep learning library. *Advances in Neural Information Processing Systems*, 32.

Pathak, D., Miranda, M., Mena, F., Sanchez, C., Helber, P., Bischke, B., Habelitz, P., Najjar, H., Siddamsetty, J., Arenas, D., Vollmer, M., Charfuelan, M., Nuske, M., & Dengel, A. (2023). Predicting crop yield with machine learning: An extensive analysis of input modalities and models on a field and sub-field level. In IGARSS 2023 - 2023 IEEE International Geoscience and Remote Sensing Symposium (pp. 2767–2770). Pasadena, CA, USA. DOI: 10.1109/IGARSS52108.2023.10282318

Pathak, J., Mahanta, T. G., Arora, P., Kalita, D., & Kaur, G. (2020). Malnutrition and household food insecurity in children attending Anganwadi centres in a district of North East India. *Indian Journal of Community Medicine : Official Publication of Indian Association of Preventive & Social Medicine*, *45*(4), 405. DOI: 10.4103/ijcm.IJCM_428_19 PMID: 33623190

Patrício, D. I., & Rieder, R. (2018). Computer vision and artificial intelligence in precision agriculture for grain crops: A systematic review. *Computers and Electronics in Agriculture*, *153*, 69–81. DOI: 10.1016/j.compag.2018.08.001

Pawlak, K., & Kołodziejczak, M. (2020). The role of agriculture in ensuring food security in developing countries: Considerations in the context of the problem of sustainable food production. *Sustainability (Basel)*, *12*(13), 5488. DOI: 10.3390/su12135488

Pedersen, S. M., & Lind, K. M. (2017). Precision agriculture – From mapping to site-specific application. In Pedersen, S. M., & Lind, K. M. (Eds.), *Precision agriculture: Technology and economic perspectives. Progress in Precision Agriculture* (pp. 1–18). Springer. DOI: 10.1007/978-3-319-68715-5_1

Peeters, A., Cohen, Y., Bahat, I., Ohana-Levi, N., Goldshtein, E., Netzer, Y., Tenreiro, T. R., Alchanatis, V., & Ben-Gal, A. (2024). A spatial machine-learning model for predicting crop water stress index for precision irrigation of vineyards. *Computers and Electronics in Agriculture*, *227*(Part 2), 109578. DOI: 10.1016/j.compag.2024.109578

Peng, S., Piao, S., Ciais, P., Friedlingstein, P., Ottle, C., Breon, F., Nan, H., Zhou, L., & Myneni, R. B. (2012). Surface Urban Heat Island Across 419 Global Big Cities. *Environmental Science & Technology*, *46*(2), 696–703. DOI: 10.1021/es2030438 PMID: 22142232

Petraki, D., Gazoulis, I., Kokkini, M., Danaskos, M., Kanatas, P., Rekkas, A., & Travlos, I. (2025). Digital tools and decision support systems in agroecology: Benefits, challenges, and practical implementations. *Agronomy (Basel)*, *15*(1), 236. DOI: 10.3390/agronomy15010236

Petrovics, D., & Giezen, M. (2022). Planning for sustainable urban food systems: An analysis of the up-scaling potential of vertical farming. *Journal of Environmental Planning and Management*, *65*(5), 785–808. DOI: 10.1080/09640568.2021.1903404

Pettorelli, N., Laurance, W. F., O'Brien, T. G., Wegmann, M., Nagendra, H., & Turner, W. (2014). Satellite remote sensing for applied ecologists: Opportunities and challenges. *Journal of Applied Ecology*, *51*(4), 839–848. DOI: 10.1111/1365-2664.12261

Pham-Duc, B., Nguyen, H., Phan, H., & Tran-Anh, Q. (2023). Trends and Applications of Google Earth Engine in Remote Sensing and Earth Science Research: A Bibliometric Analysis Using Scopus Database. *Earth Science Informatics*, *16*(3), 2355–2371. DOI: 10.1007/s12145-023-01035-2

Pham, V., Weindorf, D. C., & Dang, T. (2021). Soil profile analysis using interactive visualizations, machine learning, and deep learning. *Computers and Electronics in Agriculture, 191*, 106539. DOI: 10.1016/j.compag.2021.106539

Piao, S., Liu, Q., Chen, A., Janssens, I. A., Fu, Y., Dai, J., Liu, L., Lian, X., Shen, M., & Zhu, X. (2019). Plant phenology and global climate change: Current progresses and challenges. *Global Change Biology, 25*(6), 1922–1940. DOI: 10.1111/gcb.14619 PMID: 30884039

Piao, S., Nan, H., Huntingford, C., Ciais, P., Friedlingstein, P., Sitch, S. A., Peng, S., Ahlström, A., Canadell, J. G., Cong, N., Levis, S., Levy, P. E., Liu, L., Lomas, M. R., Mao, J., Myneni, R. B., Peylin, P., Poulter, B. I., Shi, X., & Chen, A. (2014). Evidence for a weakening relationship between interannual temperature variability and northern vegetation activity. *Nature Communications, 5*(1), 5018. Advance online publication. DOI: 10.1038/ncomms6018 PMID: 25318638

Pierce, F. J., & Nowak, P. (1999). *Aspects of precision agriculture, Advances in agronomy* (Vol. 67). Academic Press. DOI: 10.1016/S0065-2113(08)60513-1

Pierdicca, R., & Paolanti, M. (2022). GeoAI: A Review of Artificial Intelligence Approaches for the Interpretation of Complex Geomatics Data. *Geoscientific Instrumentation, Methods and Data Systems, 11*(1), 195–218. DOI: 10.5194/gi-11-195-2022

Pokhariyal, S., Patel, N. R., & Govind, A. (2023). Machine learning-driven remote sensing applications for agriculture in India—A systematic review. *Agronomy (Basel), 13*(9), 2302. DOI: 10.3390/agronomy13092302

Polasky, S., Nelson, E., Pennington, D., & Johnson, K. A. (2011). The impact of land-use change on ecosystem services, biodiversity and returns to landowners: A case study in the State of Minnesota. *Environmental and Resource Economics, 48*(2), 219–242. DOI: 10.1007/s10640-010-9407-0

Postolache, S., Sebastião, P., Viegas, V., Postolache, O., & Cercas, F. (2022). IoT-based systems for soil nutrients assessment in horticulture. *Sensors (Basel), 23*(1), 403. DOI: 10.3390/s23010403 PMID: 36617000

Poulter, B., MacBean, N., Hartley, A., Khlystova, I., Arino, O., Betts, R., Bontemps, S., Boettcher, M., Brockmann, C., Defourny, P., Hagemann, S., Herold, M., Kirches, G., Lamarche, C., Lederer, D., Ottlé, C., Peters, M., & Peylin, P. (2015). Plant functional type classification for earth system models: Results from the European Space Agency's Land Cover Climate Change Initiative. *Geoscientific Model Development, 8*(7), 2315–2328. DOI: 10.5194/gmd-8-2315-2015

Pourmohammadi, P., Strager, M. P., Dougherty, M. J., & Adjeroh, D. A. (2021). Analysis of land development drivers using geographically weighted ridge regression. *Remote Sensing (Basel), 13*(7), 1307. DOI: 10.3390/rs13071307

Pramanik, M. K., Dash, P., & Behal, D. (2021). Improving outcomes for socioeconomic variables with coastal vulnerability index under significant sea-level rise: An approach from Mumbai coasts. *Environment, Development and Sustainability, 23*(9), 13819–13853. DOI: 10.1007/s10668-021-01239-w

Pramanik, M., Khanna, M., Singh, M., Singh, D. K., Sudhishri, S., Bhatia, A., & Ranjan, R. (2022). Automation of soil moisture sensor-based basin irrigation system. *Smart Agricultural Technology, 2*, 100032. DOI: 10.1016/j.atech.2021.100032

Pranjal, P., & Soni, A. (2025). Next-Generation Geographical Information Systems: Integrating Artificial Intelligence, Geographic Information Systems, and Satellite Imagery for Mapping and Surveying. *Global Land-Ocean Geospatial Applications*, 216–233.

Prăvălie, R. (2021). Exploring the multiple land degradation pathways across the planet. *Earth-Science Reviews, 220*, 103689. DOI: 10.1016/j.earscirev.2021.103689

Pritee, K., & Garg, R. D. (2024). Geo-AI for urban planning. In *Spatial Intelligence for a Greener Planet* (pp. 115–130). CRC Press. DOI: 10.1201/9781032718323-9

Priyadharshini, V. M., Aldehim, G., Negm, N., & Subathradevi, S. (2025). Integrating geospatial techniques and machine learning for assessing soil erosion and associated geomorphic risks. *Journal of South American Earth Sciences, 156*, 105463. DOI: 10.1016/j.jsames.2025.105463

Qazi, S., Khawaja, B. A., & Farooq, Q. U. (2022). IoT-equipped and AI-enabled next generation smart agriculture: A critical review, current challenges and future trends. *IEEE Access : Practical Innovations, Open Solutions, 10*, 21219–21235. DOI: 10.1109/ACCESS.2022.3152544

Qian, J., Yang, J., Sun, W., Zhao, L., Shi, L., Shi, H., Liao, L., Dang, C., & Dou, Q. (2025). Soil organic carbon estimation and transfer framework in agricultural areas based on spatiotemporal constraint strategy combined with active and passive remote sensing. *Remote Sensing (Basel), 17*(2), 333. DOI: 10.3390/rs17020333

Quaife, T., Lewis, P., De Kauwe, M., Williams, M., Law, B. E., Disney, M., & Bowyer, P. (2008). Assimilating canopy reflectance data into an ecosystem model with an Ensemble Kalman Filter. *Remote Sensing of Environment, 112*(4), 1347–1364. DOI: 10.1016/j.rse.2007.05.020

Quisumbing, A., Cole, S., Elias, M., Faas, S., Galiè, A., Malapit, H., Meinzen-Dick, R., Myers, E., Seymour, G., & Twyman, J. (2023). Measuring Women's Empowerment in Agriculture: Innovations and evidence. *Global Food Security, 38*, 1–15. DOI: 10.1016/j.gfs.2023.100707 PMID: 37752898

Radhakrishna, R., & Ray, S. (2005). *Handbook of Poverty in India: Perspective, Policies and Programmes*. Oxford University Press. https://www.semanticscholar.org/paper/Handbook-of-poverty-in-India-%3A-perspectives%2C-and-Radhakrishna-Ray/17854e567f800a453a9be2b317a947bd1a821acf

Radočaj, D., Obhođaš, J., Jurišić, M., & Gašparović, M. (2020). Global Open Data Remote Sensing Satellite Missions for Land Monitoring and Conservation: A Review. *Land (Basel), 9*(11), 402. DOI: 10.3390/land9110402

Rajeswari, S., Savithiri, D., Usharani, S., Sujatha, P., Kasthuri, R., & Sivasubramanian, K. (2025). Economic Analysis of Trend in Agricultural Price: A Statistical Picture with Reference to India. In Hamdan, A., & Braendle, U. (Eds.), *Harnessing AI, Machine Learning, and IoT for Intelligent Business. Studies in Systems, Decision and Control* (Vol. 555). Springer. DOI: 10.1007/978-3-031-67890-5_50

Rani, S., Mishra, A. K., Kataria, A., Mallik, S., & Qin, H. (2023). Machine learning-based optimal crop selection system in smart agriculture. *Scientific Reports*, *13*(1), 15997. DOI: 10.1038/s41598-023-42356-y PMID: 37749111

Rautela, G., Ali, M. K., Prabhakaran, D., Narayan, K. V., Tandon, N., Mohan, V., & Jaacks, L. M. (2020). Prevalence and correlates of household food insecurity in Delhi and Chennai. *India's. Food Security*, *12*(2), 391–404. DOI: 10.1007/s12571-020-01015-0 PMID: 33456633

Ray, S., & Ray, I. A. (2011). Role of the effectiveness of the public distribution system in assuring food security in India: An appraisal. *Journal of Economics and Sustainable Development*, *2*(4). https://iiste.org/Journals/index.php/JEDS/article/view/359

Rebecca, S., Ulfia, A. C., & Thomas, C. (2023). The application of artificial intelligence models for food security. *Revista de Agricultura (Piracicaba)*, *13*(10). https://www.mdpi.com/2077-0472/13/10/2037

Reddy, B. R., Kalnoor, G., Devashish, M., & Reddy, P. S. K. (2025). Deep Learning Based Mobile Application for Automated Plant Disease Detection. *IEEE Access : Practical Innovations, Open Solutions*, *13*, 107917–107925. Advance online publication. DOI: 10.1109/ACCESS.2025.3581099

Redmon, J., Divvala, S., Girshick, R., & Farhadi, A. (2016). You only look once: Unified, real-time object detection. In *Proceedings of the IEEE conference on computer vision and pattern recognition* (pp. 779-788). DOI: 10.1109/CVPR.2016.91

Regulation (EU) 2016/679 of the European Parliament and of the Council of 27 April 2016 on the protection of natural persons with regard to the processing of personal data and on the free movement of such data (General Data Protection Regulation). (2016). *Official Journal of the European Union*. https://eur-lex.europa.eu/eli/reg/2016/679/oj/eng

Reiche, J., Lucas, R., Mitchell, A. L., Verbesselt, J., Hoekman, D. H., Haarpaintner, J., Kellndorfer, J. M., Rosenqvist, A., Lehmann, E. A., Woodcock, C. E., Seifert, F. M., & Herold, M. (2016). Combining satellite data for better tropical forest monitoring. *Nature Climate Change*, *6*(2), 120–122. DOI: 10.1038/nclimate2919

Reichstein, M., Camps-Valls, G., Stevens, B., Jung, M., Denzler, J., Carvalhais, N., & Prabhat, F. (2019). Deep learning and process understanding for data-driven Earth system science. *Nature*, *566*(7743), 195–204. DOI: 10.1038/s41586-019-0912-1 PMID: 30760912

Reitano, M., Segovia, M. S., & Nayga, R. S.Jr. (2025). A systematic review on the impact of Artificial Intelligence in the agri-food supply chain. *Food Policy*, *102983*(137), 102983. DOI: 10.1016/j.foodpol.2025.102983

Rhee, J., Im, J., & Carbone, G. J. (2010). Monitoring agricultural drought for arid and humid regions using multi sensor remote sensing data. *Remote Sensing of Environment*, *114*(12), 2875–2887. DOI: 10.1016/j.rse.2010.07.005

Ribeiro, V., De Souza Mendes, F., Singh, C., & Heilmayr, R. (2024). Leveraging remote sensing for transparency and accountability in Amazonian commodity supply chains. *One Earth*, *7*(11), 1935–1940. DOI: 10.1016/j.oneear.2024.10.011

Richter, K.-F., & Scheider, S. (2023). Current Topics and Challenges in GeoAI. *Künstliche Intelligenz*, *37*(1), 11–16. DOI: 10.1007/s13218-022-00796-0

Ridwan, M.A., Radzi, N.A., Ahmad, W.S., Ahmad, W.S., Mustafa, I.S., Mustafa, I.S., Din, N.M., Jalil, Y.E., Jalil, Y.E., Isa, A.M., Isa, A.M., Othman, N.S., Zaki, W.M. (2018). Applications of Landsat-8 Data: a Survey. International Journal of Engineering and Technology.

Rolandi, S., Brunori, G., Bacco, M., & Scotti, I. (2021). The digitalization of agriculture and rural areas: Towards a taxonomy of the impacts. *Sustainability (Basel)*, *13*(9), 5172. DOI: 10.3390/su13095172

Romero, K. F., & Heenkenda, M. K. (2024). Developing site-specific prescription maps for sugarcane weed control using high-spatial-resolution images and light detection and ranging (LiDAR). *Land (Basel)*, *13*(11), 1751. DOI: 10.3390/land13111751

Rosier, J. F., Taubenböck, H., Verburg, P. H., & van Vliet, J. (2022). Fusing Earth observation and socioeconomic data to increase the transferability of large-scale urban land use classification. *Remote Sensing of Environment*, *278*, 113076. DOI: 10.1016/j.rse.2022.113076

Rotz, S., Duncan, E., Small, M., Botschner, J., Dara, R., Mosby, I., Reed, M., & Fraser, E. D. G. (2019). The politics of digital agricultural technologies: A preliminary review. *Sociologia Ruralis*, *59*(2), 203–229. DOI: 10.1111/soru.12233

Rouš, R., Peller, J., Polder, G., Hageraats, S., Ruigrok, T., & Blok, P. M. (2023). Apple scab detection in orchards using deep learning on colour and multispectral images. *arXiv preprint arXiv:2302.08818.*

Roussel, C., Böhm, K., & Reiterer, A. (2025). Q-GGXAI: A framework for quality-aware explainable AI in geospatial analysis. *Journal of Photogrammetry, Remote Sensing and Geoinformation Science*. Advance online publication. DOI: 10.1007/s41064-025-00360-z

Roy, D., Tamuk, P., Singh, A. P., & Mondal, S. (2025). Modern Technologies for Soil Health Monitoring and Management. In *Smart Farming, Smarter Solutions* (pp. 118–133). CRC Press. DOI: 10.1201/9781003536932-6

Roy, S., Hoque, A., Saikia, P., & Padhiary, M. (2025). Climate-Smart Agriculture: AI-Based Solutions for Enhancing Crop Resilience and Reducing Environmental Impact. *Asian Research Journal of Agriculture*, *18*(1), 291–310. DOI: 10.9734/arja/2025/v18i1665

Rozenstein, O., Cohen, Y., Alchanatis, V., Behrendt, K., Bonfil, D. J., Eshel, G., Harari, A., Harris, W. E., Klapp, I., Laor, Y., Linker, R., Paz-Kagan, T., Peets, S., Rutter, S. M., Salzer, Y., & Lowenberg-DeBoer, J. (2024). Data-driven agriculture and sustainable farming: Friends or foes? *Precision Agriculture*, *25*(2), 520–531. DOI: 10.1007/s11119-023-10061-5

Ruder, S. L., Faxon, H. O., Orzel, E. C., Devkota, R., & Bronson, K. (2026). Reviewing the Evidence on Precision Agriculture and Environmental Sustainability. NPJ Sustainable Agriculture, 4, 9. DOI: 10.1038/s44264-026-00128-x

Ruf, C. S., Chew, C., Lang, T., Morris, M. G., Nave, K., Ridley, A., & Balasubramaniam, R. (2018). A new paradigm in Earth environmental monitoring with the CYGNSS small satellite constellation. *Scientific Reports*, *8*(1), 8782. DOI: 10.1038/s41598-018-27127-4 PMID: 29884899

Running, S. W., Justice, C. O., Salomonson, V. V., Hall, D. K., Barker, J. L., Kaufmann, Y. J., Strahler, A. H., Huete, A. R., Muller, J., Vanderbilt, V. C., Wan, Z., Teillet, P. M., & Carneggie, D. M. (1994). Terrestrial remote sensing science and algorithms planned for EOS/MODIS. *International Journal of Remote Sensing*, *15*(17), 3587–3620. DOI: 10.1080/01431169408954346

Rußwurm, M., & Körner, M. (2018). Multi-temporal land cover classification with sequential recurrent encoders. *ISPRS International Journal of Geo-Information*, *7*(4), 129. DOI: 10.3390/ijgi7040129

Ryo, M. (2022). Explainable Artificial Intelligence and Interpretable Machine Learning for Agricultural Data Analysis. *Artificial Intelligence in Agriculture*, *6*, 257–265. DOI: 10.1016/j.aiia.2022.11.003

Sabbir, M., Ruman, A., Azim, N.A., & Mahmud, S., Parvez, Miah, T. (2025). Soil Moisture Retrieval from Soil Moisture Active Passive (SMAP) Satellite: A Validation and Error Analysis Using Ground-Based Observation Over Bangladesh. *European Journal of Applied Sciences*.

Saha, S., Kucher, O. D., Utkina, A. O., & Rebouh, N. Y. (2025). Precision agriculture for improving crop yield predictions: A literature review. *Frontiers in Agronomy*, *7*, 1566201. DOI: 10.3389/fagro.2025.1566201

Saikanth, D., Ragini, M., Tripathi, G., Rajesh, K., Giri, A., Pandey, S. K., & Verma, L. K. (2024). The impact of emerging technologies on sustainable agriculture and rural development. International Journal of Environment and Climate Change, 14(1), 253-263. https://hal.science/hal-05135258/

Sajitha, P., Andrushia, A. D., Anand, N., Naser, M. Z., & Lubloy, E. (2024). A deep learning approach to detect diseases in pomegranate fruits via hybrid optimal attention capsule network. *Ecological Informatics*, *84*, 102859. DOI: 10.1016/j.ecoinf.2024.102859

Sakhaee, A., Gebauer, A., Ließ, M., & Don, A. (2022). Spatial prediction of organic carbon in German agricultural topsoil using machine learning algorithms. *Soil (Göttingen)*, *8*(2), 587–604. DOI: 10.5194/soil-8-587-2022

Salah, K., Nizamuddin, N., Jayaraman, R., & Omar, M. (2019). Blockchain-based soybean traceability in agricultural supply chain. *IEEE Access : Practical Innovations, Open Solutions*, *7*, 73295–73305. DOI: 10.1109/ACCESS.2019.2918000

Salman, A. G., Kanigoro, B., & Heryadi, Y. (2015, October). Weather forecasting using deep learning techniques. In *2015 international conference on advanced computer science and information systems (ICACSIS)* (pp. 281-285). IEEE. DOI: 10.1109/ICACSIS.2015.7415154

Salman, Z., Muhammad, A., & Han, D. (2025). Plant disease classification in the wild using vision transformers and mixture of experts. *Frontiers in Plant Science*, *16*, 1522985. DOI: 10.3389/fpls.2025.1522985 PMID: 40606487

Salvati, L., Venezian Scarascia, M. E., Zitti, M., Ferrara, A., Urbano, V., Sciortino, M., & Giupponi, C. (2009). The integrated assessment of land degradation. *Italian Journal of Agronomy*, *4*(3), 77–90. DOI: 10.4081/ija.2009.3.77

Sánchez-Arnau, E., Ferrer-Sapena, A., & Cárcel-Mas, M. C. (2026). Sustainability indicators and data management for small-scale agricultural systems: The case of water resources. *Regional Environmental Change*, *26*(2), 90. DOI: 10.1007/s10113-026-02577-z

Sandhi, A., Kumar, R., Bhardwaj, R., Kumar, D., Rana, A. K., Ajala, O., Deepak, A., & Salau, A. O. (2025). Optimized deep learning framework for pomegranate disease detection using nature-inspired algorithms. *Plant Methods*, *21*(1), 124. DOI: 10.1186/s13007-025-01447-9 PMID: 41044589

Sangeetha, C., Moond, V., Rajesh, G. M., Damor, J. S., Pandey, S. K., Kumar, P., & Singh, B. (2024). Remote sensing and geographic information systems for precision agriculture: A review. *International Journal of Environment and Climate Change*, *14*(2), 287–309. DOI: 10.9734/ijecc/2024/v14i23945

Sankaran, S., Khot, L. R., Zúñiga Espinoza, C., Jarolmasjed, S., Sathuvalli, V. R., Vandemark, G. J., Miklas, P. N., Carter, A. H., Pumphrey, M. O., Knowles, N. R., & Pavek, M. J. (2015). Low-altitude, high-resolution aerial imaging systems for row and field crop phenotyping: A review. *European Journal of Agronomy*, *70*, 112–123. DOI: 10.1016/j.eja.2015.07.004

Santos, R., Bertolami, O., Castanho, E., Silva, P., Costa, A., Guerra, A. G. C., Arantes, M., Martin, M., Figueiredo, P., Cecilio, C. M., Castelão, I., Azevedo, L. F., Faria, J., Silva, H., Fontes, J., Prendergast, S., Tieppo, M., Pereira, E., Miranda, T., . . . Payne, C. (2022). The AEROS ocean observation mission and its CubeSat pathfinder. The 4S Symposium 2022.

Santos, M., Moreira, H., Cabral, J. A., Gabriel, R., Teixeira, A., Bastos, R., & Aires, A. (2022). Contribution of Home Gardens to Sustainable Development: Perspectives from A Supported Opinion Essay. *International Journal of Environmental Research and Public Health*, *19*(20), 1–26. DOI: 10.3390/ijerph192013715 PMID: 36294295

Sanyé-Mengual, E., Gasperi, D., Michelon, N., Orsini, F., Ponchia, G., & Gianquinto, G. (2018). Eco-efficiency assessment and food security potential of home gardening: A case study in Padua, Italy. *Sustainability (Basel)*, *10*(7), 1–25. DOI: 10.3390/su10072124

Sapna, G., Gauttam, H., Chauhan, V., Pattanaik, K. K., Trivedi, A., & Ghosh, H. (2026). A comprehensive review of edge computing empowered smart agriculture: Trends, opportunities, and future directions. *Computers and Electronics in Agriculture*, *241*, 111252. DOI: 10.1016/j.compag.2025.111252

Sarif, M. O., & Gupta, R. D. (2022). Evaluation of seasonal ecological vulnerability using LULC and thermal state dynamics using Landsat and MODIS data: A case study of Prayagraj City, India (1987–2018). *Environmental Science and Pollution Research International*, *29*(51), 77502–77535. DOI: 10.1007/s11356-022-21225-7 PMID: 35676584

Satapathy, T., Dietrich, J., & Ramadas, M. (2024). Agricultural drought monitoring and early warning at the regional scale using a remote sensing-based combined index. *Environmental Monitoring and Assessment*, *196*(11), 1132. DOI: 10.1007/s10661-024-13265-y PMID: 39476296

Savithiri, D., Kasthuri, R., Rajeswari, P., Vinothina, V., & Sivasubramanian, K. (2026). Self-help group model for financial sustainability of rural women in India: An explanatory study. In B. S. Awwad (Ed.), Artificial intelligence and technology: Systems management, decisions and control for sustainability in the digital age (Studies in Systems, Decision and Control, Vol. 645, pp. 969–976). Springer, Cham. DOI: 10.1007/978-3-032-10016-0_77

Sawyer, G., Mamais, E., & Papadakis, D. (2022). The Six Dimensions of Value Associated to the Use of Copernicus Sentinel Data: Key Findings from the Sentinel Benefits Study. *Frontiers in Environmental Science*, *10*, 804862. DOI: 10.3389/fenvs.2022.804862

Saxena, A., Amin, A., Mohan, S. B., & Mohan, P. (2020). Food insecurity in tribal high-migration communities in Rajasthan. *India Food and Nutrition Bulletin*, *41*(4), 513–518. https://pubmed.ncbi.nlm.nih.gov/33143470/

Saxena, N. C. (2011). Hunger, Under-Nutrition and Food Security in India. Working Paper 44, Chronic Poverty Research Centre, Indian Institute of Public Administration, New Delhi. https://assets.publishing.service.gov.uk/media/57a08ac540f0b652dd0008cc/CPRC-IIPA44.pdf

Saxena, N. C. (2018). Poverty, chronic poverty, and poverty dynamics. *Springer*. https://link.springer.com/book/10.1007/978-981-13-0677-8

Schauberger, B., Jägermeyr, J., & Gornott, C. (2020). A systematic review of local to regional yield forecasting approaches and frequently used data resources. *European Journal of Agronomy*, *120*, 126153. DOI: 10.1016/j.eja.2020.126153

Scheuchl, B., Mouginot, J., Rignot, E., Morlighem, M., & Khazendar, A. (2016). Grounding line retreat of Pope, Smith, and Kohler Glaciers, West Antarctica, measured with Sentinel-1a radar interferometry data. *Geophysical Research Letters*, *43*(16), 8572–8579. DOI: 10.1002/2016GL069287

Schirpke, U., Kohler, M., Leitinger, G., Fontana, V., Tasser, E., & Tappeiner, U. (2017). Future impacts of changing land-use and climate on ecosystem services of mountain grassland and their resilience. *Ecosystem Services*, *26*, 79–94. DOI: 10.1016/j.ecoser.2017.06.008 PMID: 31110934

Schroeder, W., Oliva, P., Giglio, L., & Csiszar, I. (2014). The New VIIRS 375 m active fire detection data product: Algorithm description and initial assessment. *Remote Sensing of Environment*, *143*, 85–96. DOI: 10.1016/j.rse.2013.12.008

Schuster, J., Mittermayer, M., Maidl, F. X., Nätscher, L., & Hülsbergen, K.-J. (2023). Spatial variability of soil properties, nitrogen balance and nitrate leaching using digital methods on heterogeneous arable fields in southern Germany. *Precision Agriculture*, *24*(2), 647–676. DOI: 10.1007/s11119-022-09967-3

Schwilch, G., Bestelmeyer, B., Bunning, S., Critchley, W., Herrick, J., Kellner, K., Liniger, H. P., Nachtergaele, F., Ritsema, C. J., Schuster, B., Tabo, R., van Lynden, G., & Winslow, M. (2011). Experiences in monitoring and assessment of sustainable land management. *Land Degradation & Development*, *22*(2), 214–225. DOI: 10.1002/ldr.1040

See, L., Lesiv, M., & Schepaschenko, D. (2024). Integrating remote sensing and geospatial big data for land cover and land use mapping and monitoring. *Land (Basel)*, *13*(6), 769. DOI: 10.3390/land13060769

Seidl, R., Thom, D., Kautz, M., Martin-Benito, D., Peltoniemi, M., Vacchiano, G., Wild, J., Ascoli, D., Petr, M., Honkaniemi, J., Lexer, M. J., Trotsiuk, V., Mairota, P., Svoboda, M., Fabrika, M., Nagel, T. A., & Reyer, C. P. (2017). Forest disturbances under climate change. *Nature Climate Change*, *7*(6), 395–402. DOI: 10.1038/nclimate3303 PMID: 28861124

Seifert, E., Seifert, S., Vogt, H., Drew, D., Van Aardt, J., Kunneke, A., & Seifert, T. (2019). Influence of drone altitude, image overlap, and optical sensor resolution on multi-view reconstruction of forest images. *Remote Sensing (Basel)*, *11*(10), 1252. DOI: 10.3390/rs11101252

Sen, A. (2023). *Overview of the imaging spectrometer instrument for NASA's Surface Biology and Geology Mission.* Sensors, Systems, and Next-Generation Satellites XXVII.

Senanayake, I. P., Pathira Arachchilage, K. R. L., Yeo, I.-Y., Khaki, M., Han, S.-C., & Dahlhaus, P. G. (2024). Spatial Downscaling of Satellite-Based Soil Moisture Products Using Machine Learning Techniques: A Review. *Remote Sensing (Basel)*, *16*(12), 2067. DOI: 10.3390/rs16122067

Senapaty, M. K., Ray, A., & Padhy, N. (2023). IoT-enabled soil nutrient analysis and crop recommendation model for precision agriculture. *Computers*, *12*(3), 61. DOI: 10.3390/computers12030061

Sertel, E., & Huang, X. (2025). Advancing human geography with GeoAI: Transforming land use and land cover analysis. In X. Huang, S. Wang, J. Wilson, & P. Kedron (Eds.), GeoAI and human geography (pp. 373–386). Springer Geography, Springer, Cham. DOI: 10.1007/978-3-031-87421-5_26

Shafique, K., Khawaja, B. A., Sabir, F., Qazi, S., & Mustaqim, M. (2020). Internet of things (IoT) for next-generation smart systems: A review of current challenges, future trends and prospects for emerging 5G-IoT scenarios. *IEEE Access : Practical Innovations, Open Solutions*, *8*, 23022–23040. DOI: 10.1109/ACCESS.2020.2970118

Shafi, U., Mumtaz, R., García-Nieto, J., Hassan, S. A., Raza Zaidi, S. A., & Iqbal, N. (2019). Precision agriculture techniques and practices: From considerations to applications. *Sensors (Basel)*, *19*(17), 3796. DOI: 10.3390/s19173796 PMID: 31480709

Shahab, H., Naeem, M., Iqbal, M., Aqeel, M., & Ullah, S. S. (2025). IoT-driven smart agricultural technology for real-time soil and crop optimization. *Smart Agricultural Technology*, *10*, 100847. DOI: 10.1016/j.atech.2025.100847

Shaikh, F. K., Karim, S., Zeadally, S., & Nebhen, J. (2022). Recent trends in internet-of-things-enabled sensor technologies for smart agriculture. *IEEE Internet of Things Journal*, *9*(23), 23583–23598. DOI: 10.1109/JIOT.2022.3210154

Shalwee, D., Dhupper, R., Kumari, M., Gupta, A. K., & Kumar, D. (2025). Implementing Multi-Criteria Decision-Making Approaches to Evaluate and Map Drought Vulnerability with Geospatial Artificial Intelligence (Geo-AI) Based Investigations. *Discover Applied Sciences*, *7*(10), 1185. DOI: 10.1007/s42452-025-07189-6

Sharma, H., & Kaur, S. (2025). Deep learning for sustainable development across climate, energy, agriculture and urban systems. *Discover Sustainability*, *6*(1), 1408. DOI: 10.1007/s43621-025-02186-6

Sharma, M., Kumar, A., Muthuraman, S., & Chaturvedi, P. (2026). Deep learning techniques for crop classification in complex agricultural landscapes. *Scientific Reports, 16*(1), 8831. DOI: 10.1038/s41598-026-37806-2 PMID: 41820421

Sharma, P., Sharma, P., & Thakur, N. (2024). Sustainable farming practices and soil health: A pathway to achieving SDGs and future prospects. *Discover Sustainability*, *5*(1), 250. DOI: 10.1007/s43621-024-00447-4

Shawon, S. M., Ema, F. B., Mahi, A. K., Fahima Lokman Niha, F. L., & Zubair, H. T. (2025). Crop yield prediction using machine learning: An extensive and systematic literature review. *Smart Agricultural Technology*, *10*, 100718. DOI: 10.1016/j.atech.2024.100718

Shen, J., Wang, H., & Tareque, H. (2025). Toward resilience in broadacre agriculture: A methodological review of remote sensing in crop productivity, phenology, and environmental stress detection. *Remote Sensing (Basel)*, *17*(23), 3886. DOI: 10.3390/rs17233886

Shiferaw, N., Habte, L., & Waleed, M. (2025). Land use dynamics and their impact on hydrology and water quality of a river catchment: A comprehensive analysis and future scenario. *Environmental Science and Pollution Research International*, *32*(7), 4124–4136. DOI: 10.1007/s11356-025-35946-y PMID: 39862375

Shirzaei, M., & Bürgmann, R. (2018). Global climate change and local land subsidence exacerbate inundation risk to the San Francisco Bay Area. *Science Advances*, *4*(3), eaap9234. Advance online publication. DOI: 10.1126/sciadv.aap9234 PMID: 29536042

Shi, W., Cao, J., Zhang, Q., Li, Y., & Xu, L. (2016). Edge computing: Vision and challenges. *IEEE Internet of Things Journal*, *3*(5), 637–646. DOI: 10.1109/JIOT.2016.2579198

Shorten, C., & Khoshgoftaar, T. (2019). Data augmentation survey. *Journal of Big Data*. Advance online publication. DOI: 10.1186/s40537-019-0197-0

Siddiqui, S. A., Singh, S., & Nayik, G. A. (2024). Bioactive compounds from pomegranate peels-Biological properties, structure–function relationships, health benefits and food applications–A comprehensive review. *Journal of Functional Foods*, *116*, 106132. DOI: 10.1016/j.jff.2024.106132

Sieberth, T., Wackrow, R., & Chandler, J. H. (2016). Automatic detection of blurred images in UAV image sets. *ISPRS Journal of Photogrammetry and Remote Sensing*, *122*, 1–16. DOI: 10.1016/j.isprsjprs.2016.09.010

Siegel, D. A., Buesseler, K. O., Doney, S. C., Sailley, S. F., Behrenfeld, M. J., & Boyd, P. W. (2014). Global assessment of ocean carbon export by combining satellite observations and food-web models. *Global Biogeochemical Cycles*, *28*(3), 181–196. DOI: 10.1002/2013GB004743

Simard, M., Fatoyinbo, L., Smetanka, C., Rivera-Monroy, V. H., Castañeda-Moya, E., Thomas, N., & Van der Stocken, T. (2019). Mangrove canopy height globally related to precipitation, temperature and cyclone frequency. *Nature Geoscience*, *12*(1), 40–45. DOI: 10.1038/s41561-018-0279-1

Simón Sánchez, A. M., González-Piqueras, J., de la Ossa, L., & Calera, A. (2022). Convolutional neural networks for agricultural land use classification from Sentinel-2 image time series. *Remote Sensing (Basel)*, *14*(21), 5373. DOI: 10.3390/rs14215373

Simonyan, K., & Zisserman, A. (2014). Very deep convolutional networks for large-scale image recognition. *arXiv preprint arXiv:1409.1556*. https://arxiv.org/abs/1409.1556

Simonyan, K., & Zisserman, A. (2015). VGG16. *ICLR*. arXiv.1409.1556

Sinclair, R., Nodi, S., & Kabir, M. A. (2024). Evaluating mobile applications for estimating soil properties: Quality of current apps, limitations and future directions. *Computers and Electronics in Agriculture*, *216*, 108527. DOI: 10.1016/j.compag.2023.108527

Singh, A. (1989). Digital change detection techniques using remotely sensed data. *International Journal of Remote Sensing*, *10*(6), 989–1003. DOI: 10.1080/01431168908903939

Singh, T., Mukhopadhyay, R., Ravita, , Ks, S., Deep, N., Tarannum, , Chaudhary, S., Paramjita, D., Goyal, A., & Singh, G. (2025). Drone-assisted agronomy: Monitoring, spraying and precision application in the field. *International Journal of Research in Agronomy*, *8*(6), 717–727. DOI: 10.33545/2618060X.2025.v8.i6i.3094

Sishodia, R. P., Ray, R. L., & Singh, S. K. (2020). Applications of Remote Sensing in Precision Agriculture: A Review. *Remote Sensing (Basel)*, *12*(19), 3136. DOI: 10.3390/rs12193136

Sivasubramanian, K. (2025). Assessing the impact of public-private partnership on the progress of renewable energy start-up industries for sustainable development. In *Navigating strategic partnerships for sustainable startup growth* (pp. 1–20). IGI Global. DOI: 10.4018/979-8-3693-4066-0.ch005

Sivasubramanian, K., Acharjee, S., Adarsh, R., Kumarasamy, P., Thamilmani, R., & Vani, H. (2025). Application of AI technology in the electronic commerce industry for sustainable business growth in India. In *AI and digital technologies transforming global industries* (pp. 1–28). IGI Global. DOI: 10.4018/979-8-3373-6097-3.ch001

Skirving, W. J., Heron, S. F., Marsh, B. L., Liu, G., De La Cour, J. L., Geiger, E. F., & Eakin, C. M. (2019). The relentless march of mass coral bleaching: A global perspective of changing heat stress. *Coral Reefs*, *38*(4), 547–557. DOI: 10.1007/s00338-019-01799-4

Smith, G. (2021). *The power of farm data to enhance lending and crop insurance services*. U.S. Farmers & Ranchers in Action. https://usfarmersandranchers.org/stories/economic-sustainability/the-power-of-farm-data-to-enhance-lending-and-crop-insurance-services/

Sobhy, D. M., & Anandhi, A. (2025). Soil Nutrient Monitoring Technologies for Sustainable Agriculture: A Systematic Review. *Sustainability (Basel)*, *17*(18), 8477. DOI: 10.3390/su17188477

Sohl, T. L., Wimberly, M. C., Radeloff, V. C., Theobald, D. M., & Sleeter, B. M. (2016). Divergent projections of future land use in the United States arising from different models and scenarios. *Ecological Modelling*, *337*, 281–297. DOI: 10.1016/j.ecolmodel.2016.07.016

Soleimani, A., Abbaspour-Fard, M. H., Sapkota, R., & Sanaeifar, A. (2026). Artificial intelligence in agricultural machinery & robotics: A systematic review and future outlook. *Applied Soft Computing*, *194*, 114889. DOI: 10.1016/j.asoc.2026.114889

Somvanshi, S., Kumar, D., & Kumari, M. (2025). Trends of recent data, AI/ML approaches for geospatial AI in Earth observation towards sustainable development goals. *ITU Journal : ICT Discoveries*, *6*(1), 11–28. DOI: 10.52953/CDIE7940

Sondarjee, M. (2024). Coloniality of epistemic power in international practices: NGO inclusion in World Bank policymaking. *Global Society*, *38*(3), 328–350. DOI: 10.1080/13600826.2023.2236129

Song, J., Meng, C., & Ermon, S. (2022). Low-light image and obstructed picture restoration using denoising diffusion models. Image Processing Transactions of the IEEE. DOI: 10.1109/TIP.2022.3148289

Songsom, V., Koedsin, W., Ritchie, R. J., & Huete, A. (2019). Mangrove phenology and environmental drivers derived from remote sensing in southern Thailand. *Remote Sensing (Basel)*, *11*(8), 955. DOI: 10.3390/rs11080955

Song, Y., Biljecki, F., Camps-Valls, G., & Atkinson, P. M. (2025). GeoAI: Beyond mapping earth and cities through explainability, adaptability, and sustainability. *iScience*, *29*(1), 114407. DOI: 10.1016/j.isci.2025.114407 PMID: 41536980

Song, Y., Kalacska, M., Gašparović, M., Yao, J., & Najibi, N. (2023). Advances in geocomputation and geospatial artificial intelligence (GeoAI) for mapping. *International Journal of Applied Earth Observation and Geoinformation*, *120*, 103300. DOI: 10.1016/j.jag.2023.103300

Subeesh, A., & Naveen Chauhan, N. (2026). Agricultural digital twin for smart farming: A review. *Green Technologies and Sustainability*, *4*(2), 100299. Advance online publication. DOI: 10.1016/j.grets.2025.100299

Sudha, S. P., & Loret, J. B. S. (2026). A review on machine learning-based precision agriculture techniques for crop farming monitoring with IOT. *Discover Environment*, *4*(1), 10. DOI: 10.1007/s44274-025-00305-8

Sullivan, C. S., Gemtou, M., Anastasiou, E., & Fountas, S. (2024). Building Trust: A Systematic Review of the Drivers and Barriers of Agricultural Data Sharing. *Smart Agricultural Technology*, *8*, 100477. DOI: 10.1016/j.atech.2024.100477

Suneetha, C., Kumar, L. S., Sreenivas, K., & Mitran, T. (2025). Deep learning-driven soil texture classifier using Landsat 8 images. *Remote Sensing Applications: Society and Environment*, *38*, 101568. DOI: 10.1016/j.rsase.2025.101568

Syaban, A. S. N., & Appiah-Opoku, S. (2024). Unveiling the complexities of land use transition in Indonesia's new capital city IKN Nusantara: A multidimensional conflict analysis. *Land (Basel)*, *13*(5), 606. DOI: 10.3390/land13050606

Symeonakis, E., Caccetta, P., Koukoulas, S., Furby, S., & Karathanasis, N. (2012). Multi-temporal land-cover classification and change analysis with conditional probability networks: The case of Lesvos Island (Greece). *International Journal of Remote Sensing*, *33*(13), 4075–4093. DOI: 10.1080/01431161.2011.640961

Szeliski, R. (2022). *Computer Vision: Algorithms and Applications* (2nd ed.). Springer., DOI: 10.1007/978-3-030-34372-9

Tadono, T., Ishida, H., Oda, F., Naito, S., Minakawa, K., & Iwamoto, H. (2014). Precise global DEM generation by ALOS PRISM. *ISPRS Annals of the Photogrammetry, Remote Sensing and Spatial Information Sciences, II-4*, 71–76. DOI: 10.5194/isprsannals-II-4-71-2014

Talaviya, T., Shah, D., Patel, N., Yagnik, H., & Shah, M. (2020). Implementation of artificial intelligence in agriculture for optimisation of irrigation and application of pesticides and herbicides. *Artificial Intelligence in Agriculture, 4*, 58–73. DOI: 10.1016/j.aiia.2020.04.002

Talukder, A., Kiess, L., Huq, N., de Pee, S., Hill, I. D., & Bloem, M. W. (2000). Increasing the Production and Consumption of Vitamin-A Rich Fruits and Vegetables: Lessons Learned in taking the Bangladesh Homestead Gardening Programme to a National Scale. *Food and Nutrition Bulletin, 21*(2), 165–172. DOI: 10.1177/156482650002100210

Tamboli, V. A., Sawant, M. H. M., & Sathe, S. (2024). AI-Driven-IoT(AIIoT) Based Decision-Making- KSK Approach in Drones for Climate Change Study. *2024 4th International Conference on Ubiquitous Computing and Intelligent Information Systems (ICUIS)*, 1735-1744. DOI: 10.1109/ICUIS64676.2024.10866450

Tamilkodi, R., Leelavathy, N., Sujatha, B., & Kumar, K. P. (2025). *Innovative AI approaches to enhance land use prediction and management*. https://doi.org/DOI: 10.21203/rs.3.rs-6352625/v1

Tamiminia, H., Salehi, B., Mahdianpari, M., Quackenbush, L., Adeli, S., & Brisco, B. (2020). Google Earth Engine for geo-big data applications: A meta-analysis and systematic review. *ISPRS Journal of Photogrammetry and Remote Sensing, 164*, 152–170. DOI: 10.1016/j.isprsjprs.2020.04.001

Tang, R., Peng, Z., Liu, M., Li, Z. L., Jiang, Y., Hu, Y., & Fisher, J. B. (2024). Spatial-temporal patterns of land surface evapotranspiration from global products. *Remote Sensing of Environment, 304*, 114066. DOI: 10.1016/j.rse.2024.114066

Tan, Y.-C., Duarte, L., & Teodoro, A. C. (2024). Comparative Study of Random Forest and Support Vector Machine for Land Cover Classification and Post-Wildfire Change Detection. *Land (Basel), 13*(11), 1878. DOI: 10.3390/land13111878

Taylor, L. (2017). What is data justice? The case for connecting digital rights and freedoms globally. *Big Data & Society, 4*(2). DOI: 10.1177/2053951717736335

Tefera, G. W., Ray, R. L., Jackson, R., Deegala, B., & Akenzua, O. (2026). Integrating machine learning models with ground sensors to enhance soil moisture prediction in agroecosystems of Texas. *Computers and Electronics in Agriculture, 242*, 111358. DOI: 10.1016/j.compag.2025.111358

Tellman, B., Sullivan, J. A., Kuhn, C., Kettner, A. J., Doyle, C. S., Brakenridge, G. R., Erickson, T. A., & Slayback, D. A. (2021). Satellite imaging reveals increased proportion of population exposed to floods. *Nature, 596*(7870), 80–86. DOI: 10.1038/s41586-021-03695-w PMID: 34349288

Thakuri, S., Adhikari, M., Tashi, S., Acharya, P. R., Chauhan, R., Maharjan, B., & Shrestha, M. L. (2025). Integrating geospatial technologies in climate-smart agriculture planning and management: A review focused on South Asia. APN Science Bulletin, https://www.apn-gcr.org/bulletin/wp-content/uploads/2025/10/8-CBA2020-13MY-Thakuri.pdf

Thapa, R. B., Matin, M. A., & Bajracharya, B. (2019). Capacity Building Approach and Application: Utilization of Earth Observation Data and Geospatial Information Technology in the Hindu Kush Himalaya. *Frontiers in Environmental Science*, *7*, 165. DOI: 10.3389/fenvs.2019.00165

Thorp, K. R., Gore, M. A., Andrade-Sanchez, P., Carmo-Silva, A. E., Welch, S. M., White, J. W., & French, A. N. (2015). Proximal hyperspectral sensing and data analysis approaches for field-based plant phenomics. *Computers and Electronics in Agriculture*, *118*, 225–236. DOI: 10.1016/j.compag.2015.09.005

Tian, X., Li, J., Zhang, F., Zhang, H., & Jiang, M. (2024). Forest aboveground biomass estimation using multisource remote sensing data and deep learning algorithms: A case study over Hangzhou area in China. *Remote Sensing (Basel)*, *16*(6), 1074. DOI: 10.3390/rs16061074

Tikariha, S., & Tripathi, D. (2025). Spatio-Temporal Modeling of Urbanization, Environmental Impact, and Ecosystem Functions Using a Unified Geo-AI Framework: A Review Paper. https://ijorarjournal.com/web/Download/2026-01-02-02-18-33-Paper%20Id%200651.pdf

Tikhonova, M. V., & Buzylev, A. V. (2024). Application of DSS in agroecological assessment of the potential of use of chernozem soils in the conditions of the Penza region. In *BIO Web of Conferences* (Vol. 82, p. 02004). EDP Sciences. DOI: 10.1051/bioconf/20248202004

Tobler, W. R. (1970). A computer movie simulating urban growth in the Detroit region. *Economic geography, 46*(sup1), 234-240. .DOI: 10.2307/143141

Togneri, R., dos Santos, D. F., Camponogara, G., Nagano, H., Custódio, G., Prati, R., Fernandes, S., & Kamienski, C. (2022). Soil moisture forecast for smart irrigation: The primetime for machine learning. *Expert Systems with Applications*, *207*, 117653. DOI: 10.1016/j.eswa.2022.117653

Tonnang, H. E., Sokame, B. M., Abdel-Rahman, E. M., & Dubois, T. (2022). Measuring and modelling crop yield losses due to invasive insect pests under climate change. *Current Opinion in Insect Science*, *50*, 100873. DOI: 10.1016/j.cois.2022.100873 PMID: 35051620

Topouzelis, K., & Singha, S. (2017). Oil spill detection using space-borne Sentinel-1 SAR imagery. *Oil spill science and technology*, 387-394.

Toutin, T. (2004). Review article: Geometric processing of remote sensing images: Models, algorithms and methods. *International Journal of Remote Sensing*, *25*(10), 1893–1924. DOI: 10.1080/0143116031000101611

Transon, J., d'Andrimont, R., Maugnard, A., & Defourny, P. (2018). Survey of Hyperspectral Earth Observation Applications from Space in the Sentinel-2 Context. *Remote Sensing (Basel)*, *10*(2), 157. DOI: 10.3390/rs10020157

Tsolakidis, D., Gymnopoulos, L. P., & Dimitropoulos, K. (2024, August). Artificial intelligence and machine learning technologies for personalized nutrition: a review. In Informatics (Vol. 11, No. 3, p. 62). MDPI. https://www.mdpi.com/2227-9709/11/3/62

Tsouros, D. C., Bibi, S., & Sarigiannidis, P. G. (2019). A review on UAV-based applications forprecision agriculture. *Information (Basel), 10*(11), 349. DOI: 10.3390/info10110349

Tufail, R., Ahmad, A., Javed, M. A., & Ahmad, S. R. (2022). A machine learning approach for accurate crop type mapping using combined SAR and optical time series data. *Advances in Space Research, 69*(1), 331–346. DOI: 10.1016/j.asr.2021.09.019

Tugrul, B., Elfatimi, E., & Eryigit, R. (2022). Convolutional neural networks in detection of plant leaf diseases. *Revista de Agricultura (Piracicaba), 12*(8), 1192. https://www.mdpi.com/2077-0472/12/8/1192

Tzachor, A. (2021). Barriers to AI Adoption in Indian Agriculture: An Initial Inquiry. *International Journal of Innovation in the Digital Economy, 12*(3), 30–44. DOI: 10.4018/IJIDE.2021070103

Ubina, N. A., & Cheng, S. C. (2022). A review of unmanned system technologies with its application to aquaculture farm monitoring and management. *Drones (Basel), 6*(1), 12. DOI: 10.3390/drones6010012

Ullah, S., Qiao, X., & Abbas, M. (2024). Addressing the impact of land use and land cover changes on land surface temperature using machine learning algorithms. *Scientific Reports, 14*(1), 18746. DOI: 10.1038/s41598-024-68492-7 PMID: 39138301

Uma Maheswari, V., Balasubramanian, G., Sugatha Kumari, B., Ramya, C., Thamilmani, R., & Sivasubramanian, K. (2025). Business Strategies and Employment Growth of Retail Industry: An Economic Perspective in India. In Hamdan, A. (Ed.), *Achieving Sustainable Business through AI, Technology Education and Computer Science. Studies in Big Data* (Vol. 158). Springer. DOI: 10.1007/978-3-031-70855-8_41

United Nations Department of Economic and Social Affairs (UNDESA). (2024). *World Population Prospects 2024: Summary of Results*. Stylus Publishing, LLC. https://population.un.org/wpp/assets/Files/WPP2024_Summary-of-Results.pdf

United Nations. (2015). Transforming our world: The 2030 Agenda for Sustainable Development. United Nations. https://sdgs.un.org/2030agenda

Upadhyay, R. P., & Palanivel, C. (2011). Challenges in achieving food security in India. *Iranian Journal of Public Health, 40*(4), 31–36. https://pmc.ncbi.nlm.nih.gov/articles/PMC3481742/ PMID: 23113100

Urdánigo-Zambrano, J., Torres, B., De-Pablos-Heredero, C., Herrera-Feijoo, R., & García, A. (2026). Open data reuse in agricultural, livestock, and environmental systems: A global scoping review with a case analysis of Ecuador. *Land (Basel), 15*(1), 13. DOI: 10.3390/land15010013

Usman, M., Liedl, R., Shahid, M. A., & Abbas, A. (2015). Land use/land cover classification and its change detection using multi-temporal MODIS NDVI data. *Journal of Geographical Sciences, 25*(12), 1479–1506. DOI: 10.1007/s11442-015-1247-y

Ustin, S. L., DiPietro, D., Olmstead, K. S., Underwood, E. C., & Scheer, G. J. (2002). Hyperspectral remote sensing for invasive species detection and mapping. *IEEE International Geoscience and Remote Sensing Symposium*, 3, 1658-1660. DOI: 10.1109/IGARSS.2002.1026212

Uyar, N. (2025). Index-driven soil loss mapping across environmental scenarios: Insights from a remote sensing approach. *Sustainability (Basel), 17*(17), 7913. DOI: 10.3390/su17177913

Vali, A., Comai, S., & Matteucci, M. (2020). Deep learning for land use and land cover classification based on hyperspectral and multispectral Earth observation data: A review. *Remote Sensing (Basel), 12*(15), 2495. DOI: 10.3390/rs12152495

Valle, D. R., Izbicki, R., & Leite, R. V. (2023). Quantifying uncertainty in land-use land-cover classification using conformal statistics. *Remote Sensing of Environment, 295*, 113682. DOI: 10.1016/j.rse.2023.113682

van der Burg, S., Bogaardt, M.-J., & Wolfert, S. (2019). Ethics of smart farming: Current questions and directions for responsible innovation towards the future. *NJAS Wageningen Journal of Life Sciences, 90–91*(1), 1–10. DOI: 10.1016/j.njas.2019.01.001

Van Dijk, M., Morley, T., Rau, M. L., & Saghai, Y. (2021). A meta-analysis of projected global food demand and population at risk of hunger for the period 2010–2050. *Nature Food, 2*(7), 494–501. DOI: 10.1038/s43016-021-00322-9 PMID: 37117684

Vanitha, V., Dwibedi, R. K., Painadthu, B., Venugopal, K., & Thomas, A. (2021). IOT based Soil Moisture Measuring System for Indian Agriculture. *Journal of Physics: Conference Series, 1964*(6), 062054. Advance online publication. DOI: 10.1088/1742-6596/1964/6/062054

Vatistas, C., Avgoustaki, D. D., & Bartzanas, T. (2022). A systematic literature review on controlled-environment agriculture: How vertical farms and greenhouses can influence the sustainability and footprint of urban microclimate with local food production. *Atmosphere (Basel), 13*(8), 1258. DOI: 10.3390/atmos13081258

Veach, V., Moilanen, A., & Di Minin, E. (2017). Threats from Urban Expansion, Agricultural Transformation and Forest Loss on Global Conservation Priority Areas. *PLoS One, 12*(11), e0188397. DOI: 10.1371/journal.pone.0188397 PMID: 29182662

Verbesselt, J., Hyndman, R., Newnham, G., & Culvenor, D. (2010). Detecting trend and seasonal changes in satellite image time series. *Remote Sensing of Environment, 114*(1), 106–115. DOI: 10.1016/j.rse.2009.08.014

Verburg, P. H., Crossman, N., Ellis, E. C., Heinimann, A., Hostert, P., Mertz, O., Nagendra, H., Sikor, T., Erb, K.-H., Golubiewski, N., Grau, R., Grove, M., Konaté, S., Meyfroidt, P., Parker, D. C., Roy Chowdhury, R., Shibata, H., Thomson, A., & Zhen, L. (2015). Land system science and sustainable development of the earth system: A global land project perspective. *Anthropocene, 12*, 29–41. DOI: 10.1016/j.ancene.2015.09.004

Verburg, P., Soepboer, W., Veldkamp, A., Limpiada, R., Espaldon, V., & Mastura, S. S. A. (2002). Modeling the spatial dynamics of regional land use: The CLUE-S model. *Environmental Management, 30*(3), 391–405. DOI: 10.1007/s00267-002-2630-x PMID: 12148073

Verrelst, J., Camps-Valls, G., Muñoz-Marí, J., Rivera, J. P., Veroustraete, F., Clevers, J. G. P. W., & Moreno, J. (2015). Optical remote sensing and the retrieval of terrestrial vegetation bio-geophysical properties – A review. *ISPRS Journal of Photogrammetry and Remote Sensing*, *108*, 273–290. DOI: 10.1016/j.isprsjprs.2015.05.005

Vidican, R., Mălina , A., Ranta, O., Moldovan, C., Marian, O., Ghe e, A., Ghi e, C. R., Popovici, F., & Cătunescu, G. M. (2023). Using remote sensing vegetation indices for the discrimination and monitoring of agricultural crops: A critical review. *Agronomy (Basel)*, *13*(12), 3040. DOI: 10.3390/agronomy13123040

Vinuesa, R., Azizpour, H., Leite, I., Balaam, M., Dignum, V., Domisch, S., Felländer, A., Langhans, S. D., Tegmark, M., & Nerini, F. F. (2020). The role of artificial intelligence in achieving the Sustainable Development Goals. *Nature Communications*, *11*(1), 233. DOI: 10.1038/s41467-019-14108-y PMID: 31932590

Vissarapu, S. (2025). AI in agriculture: Cloud-powered precision farming with real-time data analytics. *Journal of Computer Science and Technology Studies*, *7*(5), 298–306. DOI: 10.32996/jcsts.2025.7.5.37

Vitale, A., & Lamonaca, F. (2025). Advancing built-up area monitoring through multi-temporal satellite data fusion and machine learning-based geospatial analysis. *Remote Sensing (Basel)*, *17*(11), 1830. DOI: 10.3390/rs17111830

von Keyserlingk, J., Thieken, A. H., & Paton, E. N. (2023). Approaches to assess land degradation risk: A synthesis. *Ecology and Society*, *28*(1), 53. DOI: 10.5751/ES-13951-280153

VoPham, T., Hart, J. E., Laden, F., & Chiang, Y.-Y.VoPham. (2018). Emerging Trends in Geospatial Artificial Intelligence (GeoAI): Potential Applications for Environmental Epidemiology. *Environmental Health : A Global Access Science Source*, *17*(1), 40. DOI: 10.1186/s12940-018-0386-x PMID: 29665858

Vos, K., Splinter, K. D., Harley, M. D., Simmons, J. A., & Turner, I. L. (2019). CoastSat: A Google Earth Engine-enabled Python toolkit to extract shorelines from publicly available satellite imagery. *Environmental Modelling & Software*, *122*, 122. DOI: 10.1016/j.envsoft.2019.104528

Vousdoukas, M. I., Mentaschi, L., Voukouvalas, E., Verlaan, M., Jevrejeva, S., Jackson, L. P., & Feyen, L. (2018). Global probabilistic projections of extreme sea levels show intensification of coastal flood hazard. *Nature Communications*, *9*(1), 2360. DOI: 10.1038/s41467-018-04692-w PMID: 29915265

Walaszczyk, A., & Galińska, B. (2020). Food origin traceability from a consumer's perspective. *Sustainability (Basel)*, *12*(5), 1872. DOI: 10.3390/su12051872

Wang, C., Xia, Y., Xia, L., Wang, Q., & Gu, L. (2025). Dual discriminator GAN-based synthetic crop disease image generation for precise crop disease identification. *Plant Methods*, *21*(1), 46. DOI: 10.1186/s13007-025-01361-0 PMID: 40159478

Wang, J., Bretz, M., Dewan, M. A. A., & Delavar, M. A. (2022). Machine learning in modelling land-use and land cover-change (LULCC): Current status, challenges and prospects. *The Science of the Total Environment*, *822*, 153559. DOI: 10.1016/j.scitotenv.2022.153559 PMID: 35114222

Wang, J., Wang, Y., Li, G., & Qi, Z. (2024). Integration of remote sensing and machine learning for precision agriculture: A comprehensive perspective on applications. *Agronomy (Basel), 14*(9), 1975. DOI: 10.3390/agronomy14091975

Wang, S., Huang, X., Liu, P., Zhang, M., Biljecki, F., Hu, T., Fu, X., Liu, L., Liu, X., Wang, R., Huang, Y., Yan, J., Jiang, J., Chukwu, M., Naghedi, S. R., Hemmati, M., Shao, Y., Jia, N., Xiao, Z., & Bao, S. (2024). Mapping the landscape and roadmap of geospatial artificial intelligence (GeoAI) in quantitative human geography: An extensive systematic review. *International Journal of Applied Earth Observation and Geoinformation, 128*, 103734. DOI: 10.1016/j.jag.2024.103734

Wang, X. (2022). Managing land carrying capacity: Key to achieving sustainable production systems for food security. *Land (Basel), 11*(4), 484. DOI: 10.3390/land11040484

Wang, X., Feng, Y., Wang, Y., Zhu, H., Song, D., Shen, C., & Luo, Y. (2025). Enhancing optical non-destructive methods for food quality and safety assessments with machine learning techniques: A survey. *Journal of Agriculture and Food Research, 19*, 101734. DOI: 10.1016/j.jafr.2025.101734

Wang, X., Zeng, H., Yang, X., Shu, J., Wu, Q., Que, Y., Yang, X., Yi, X., Khalil, I., & Zomaya, A. Y. (2025). Remote sensing revolutionizing agriculture: Toward a new frontier. *Future Generation Computer Systems, 166*, 107691. DOI: 10.1016/j.future.2024.107691

Wang, Y., Wang, L., Han, W., & Yan, J. (2024). Digital twin of Earth: A novel information framework for managing a sustainable Earth. *The Innovation Geoscience, 2*(4), 100092. DOI: 10.59717/j.xinn-geo.2024.100092

Waqas, M. A., Humphries, U. W., Hlaing, P. T., Dechpichai, P., & Wangwongchai, A. (2025). Applications of machine learning and deep learning in agriculture: A comprehensive review. *Green Technologies and Sustainability, 3*(3), 100199. Advance online publication. DOI: 10.1016/j.grets.2025.100199

Weill, P., & Ross, J. W. (2004). *IT governance: How top performers manage IT decision rights for superior results*. Harvard Business Press.

Weiss, M., Jacob, F., & Duveiller, G. (2020). Remote sensing for agricultural applications: A meta-review. *Remote Sensing of Environment, 236*, 111402. DOI: 10.1016/j.rse.2019.111402

Wen, J., Xi, J., Pan, Y., Wang, S., Fan, Z., & Fu, W. (2024). Spatial Conservation Prioritization for Land in Megacity Facing Climate Change and Biodiversity Loss. *Sustainability (Basel), 16*(4), 1392. DOI: 10.3390/su16041392

Wilkinson, M. D., Dumontier, M., Aalbersberg, I. J., Appleton, G., Axton, M., Baak, A., Blomberg, N., Boiten, J.-W., da Silva Santos, L. B., Bourne, P. E., Bouwman, J., Brookes, A. J., Clark, T., Crosas, M., Dillo, I., Dumon, O., Edmunds, S., Evelo, C. T., Finkers, R., & Mons, B. (2016). The FAIR Guiding Principles for scientific data management and stewardship. *Scientific Data, 3*(1), 1–9. DOI: 10.1038/sdata.2016.18 PMID: 26978244

William & Kumar, S. (2024). The role of digital technologies in enhancing agricultural economic efficiency. *Stallion Journal for Multidisciplinary Associated Research Studies, 3*(5), 255–263. DOI: 10.55544/sjmars.3.5.19

Wirjodirdjo, B., Fudhla, A. F., Singgih, M. L., & Ridho, H. (2025). Driving forces for sustainable development of smallholder sugarcane agribusiness: A system dynamics approach to circular economy strategies. *Agricultural and Resource Economics: International Scientific E-Journal, 11*(2), 32–64. DOI: 10.51599/are.2025.11.02.02

Wiseman, L., Sanderson, J., Zhang, A., & Jakku, E. (2019). Farmers and their data: An examination of farmers' reluctance to share their data through the lens of the laws impacting smart farming. *NJAS Wageningen Journal of Life Sciences, 90–91*(1), 1–10. DOI: 10.1016/j.njas.2019.04.007

Wolfert, S., Ge, L., Verdouw, C., & Bogaardt, M. J. (2017). Big data in smart farming: A review. *Agricultural Systems, 153*, 69–80. DOI: 10.1016/j.agsy.2017.01.023

World Bank. (2019). *Future of food: Harnessing digital technologies to improve food system outcomes.* https://www.worldbank.org/en/topic/agriculture/publication/future-of-food-harnessing-digital-technologies-to-improve-food-system-outcomes

World Food Summit. (1996). Food and Agriculture Organization of the United Nations; Rome, Italy: 1996. Rome's declaration on the World Food Security and World Food Summit Plan of Action. https://www.fao.org/4/w3613e/w3613e00.htm

Wu, B., Zhang, M., Zeng, H., Tian, F., Potgieter, A. B., Qin, X., Yan, N., Chang, S., Zhao, Y., Dong, Q., Boken, V., Plotnikov, D., Guo, H., Wu, F., Zhao, H., Deronde, B., Tits, L., & Loupian, E. (2023). Challenges and Opportunities in Remote Sensing-Based Crop Monitoring: A Review. *National Science Review, 10*(4), nwac290. Advance online publication. DOI: 10.1093/nsr/nwac290 PMID: 36960224

Wu, J. (2004). Effects of changing scale on landscape pattern analysis: Scaling relations. *Landscape Ecology, 19*(2), 125–138. DOI: 10.1023/B:LAND.0000021711.40074.ae

Wu, J., & Li, H. (2006). Concepts of scale and scaling. In Wu, J., Jones, K. B., Li, H., & Loucks, O. L. (Eds.), *Scaling and uncertainty analysis in ecology: Methods and applications* (pp. 3–15). Springer. DOI: 10.1007/1-4020-4663-4_1

Wulder, M. A., Masek, J. G., Cohen, W. B., Loveland, T. R., & Woodcock, C. E. (2012). Opening the archive: How free data has enabled the science and monitoring promise of Landsat. *Remote Sensing of Environment, 122*, 2–10. DOI: 10.1016/j.rse.2012.01.010

Wulder, M. A., Roy, D. P., Radeloff, V. C., Loveland, T. R., Anderson, M. C., Johnson, D. M., Healey, S. P., Zhu, Z., Scambos, T. A., Pahlevan, N., Hansen, M. C., Gorelick, N., Crawford, C. J., Masek, J. G., Hermosilla, T., White, J. C., Belward, A. S., Schaaf, C., Woodcock, C. E., & Cook, B. D. (2022). Fifty years of Landsat science and impacts. *Remote Sensing of Environment, 280*, 113195. Advance online publication. DOI: 10.1016/j.rse.2022.113195

Xiao, J., Aggarwal, A. K., Duc, N. H., Arya, A., Rage, U. K., & Avtar, R. (2023). A review of remote sensing image spatiotemporal fusion: Challenges, applications, and recent trends. *Remote Sensing Applications: Society and Environment, 32*, 101005. DOI: 10.1016/j.rsase.2023.101005

Xie, M., Jean, N., Burke, M., Lobell, D., & Ermon, S. (2016). Transfer learning from deep features for remote sensing and poverty mapping. In *Proceedings of the Thirtieth AAAI Conference on Artificial Intelligence (AAAI-16)* (pp. 3929–3935). AAAI Press. DOI: 10.1609/aaai.v30i1.9906

Xing, Y., Liu, X., & Wang, X. (2026). Integrating UAVs, satellite remote sensing, and machine learning in precision agriculture: Pathways to sustainable food production, resource efficiency, and scalable innovation. *Frontiers in Agronomy*, *7*, 1670380. DOI: 10.3389/fagro.2025.1670380

Xu, C., Du, X., Fan, X., Giuliani, G., Hu, Z., Wang, W., Liu, J., Wang, T., Yan, Z., Zhu, J., Jiang, T., & Guo, H. (2022a). Cloud-Based Storage and Computing for Remote Sensing Big Data: A Technical Review. *International Journal of Digital Earth*, *15*(1), 1417–1445. DOI: 10.1080/17538947.2022.2115567

Xu, C., Du, X., Jian, H., Dong, Y., Qin, W., Mu, H., Yan, Z., Zhu, J., & Fan, X. (2022b). Analyzing Large-Scale Data Cubes with User-Defined Algorithms: A Cloud-Native Approach. *International Journal of Applied Earth Observation and Geoinformation*, *109*, 102784. DOI: 10.1016/j.jag.2022.102784

Xu, J., Quackenbush, L. J., Volk, T. A., & Im, J. (2020). Forest and crop leaf area index estimation using remote sensing: Research trends and future directions. *Remote Sensing (Basel)*, *12*(18), 2934. DOI: 10.3390/rs12182934

Xu, J., Su, Q., Li, X., Ma, J., Song, W., Zhang, L., & Su, X. (2024). A spatial downscaling framework for SMAP soil moisture based on a stacking strategy. *Remote Sensing (Basel)*, *16*(1), 200. DOI: 10.3390/rs16010200

Yang, C. (2018). High resolution satellite imaging sensors for precision agriculture. *Frontiers of Agricultural Science and Engineering*, *5*(0), 393–405. DOI: 10.15302/J-FASE-2018226

Yang, C., Suh, C. P. C., & Fritz, B. K. (2026). Effects of flight and processing parameters on UAS image-based point clouds for plant height estimation. *Remote Sensing (Basel)*, *18*(2), 360. DOI: 10.3390/rs18020360

Yange, T. S., Egbunu, C. O., Rufai, M. A., Onyekwere, O., Abdulrahman, A. A., & Abdulkadri, I. (2020). Using prescriptive analytics for the determination of optimal crop yield. *International Journal of Data Science and Analytics*, *6*(3), 72–82. DOI: 10.11648/j.ijdsa.20200603.11

Yang, H., Li, S., Chen, J., Zhang, X., & Xu, S. (2017). The Standardization and Harmonization of Land Cover Classification Systems Towards Harmonized Datasets: A Review. *ISPRS International Journal of Geo-Information*, *6*(5), 154. DOI: 10.3390/ijgi6050154

Yang, L., Driscol, J., Sarigai, S., Wu, Q., Chen, H., & Lippitt, C. D. (2022). Google Earth Engine and artificial intelligence (AI): A comprehensive review. *Remote Sensing (Basel)*, *14*(14), 3253. DOI: 10.3390/rs14143253

Yang, X., Chen, J., Lu, X., Liu, H., Liu, Y., Bai, X., Qian, L., & Zhang, Z. (2025). Advances in UAV remote sensing for monitoring crop water and nutrient status: Modeling methods, influencing factors, and challenges. *Plants*, *14*(16), 2544. DOI: 10.3390/plants14162544 PMID: 40872167

Yang, Y., Zhang, S., Yang, J., Chang, L., Bu, K., & Xing, X. (2014). A review of historical reconstruction methods of land use/land cover. *Journal of Geographical Sciences*, *24*(4), 746–766. DOI: 10.1007/s11442-014-1117-z

Yan, W. Y., Shaker, A., & El-Ashmawy, N. (2015). Urban land cover classification using airborne LiDAR data: A review. *Remote Sensing of Environment*, *158*, 295–310. DOI: 10.1016/j.rse.2014.11.001

Yao, W., Liu, C., Liu, Y., Zheng, Q., Wang, J., Yu, H., Chen, C., & Guo, S. (2025). Unmanned aerial vehicle payload technology applications in agriculture and other low-altitude scenarios: A review. *Frontiers in Plant Science*, *16*, 1–27. DOI: 10.3389/fpls.2025.1721484 PMID: 41459473

Yenkikar, A., Mishra, V. P., Bali, M., & Ara, T. (2025). An Explainable AI-Based Hybrid Machine Learning Model for Interpretability and Enhanced Crop Yield Prediction. *MethodsX*, *15*, 103442. DOI: 10.1016/j.mex.2025.103442 PMID: 40612261

Yin, H., Brandão, A.Jr, Buchner, J., Helmers, D., Iuliano, B., Kimambo, N. E., Lewińska, K. E., Razenkova, E., Rizayeva, A., Rogova, N., Spawn, S. A., Xie, Y., & Radeloff, V. C. (2020). Monitoring cropland abandonment with Landsat time series. *Remote Sensing of Environment*, *246*, 111873. DOI: 10.1016/j.rse.2020.111873

You, J., Li, X., Low, M., Lobell, D., & Ermon, S. (2017, February). Deep gaussian process for crop yield prediction based on remote sensing data. *Proceedings of the AAAI Conference on Artificial Intelligence*, *31*(1). DOI: 10.1609/aaai.v31i1.11172

Yousaf, A., Kayvanfar, V., Mazzoni, A., & Elomri, A. (2023). Artificial intelligence-based decision support systems in smart agriculture: Bibliometric analysis for operational insights and future directions. *Frontiers in Sustainable Food Systems*, *6*, 1053921. DOI: 10.3389/fsufs.2022.1053921

Yu, L., Du, Z., Li, X., Zheng, J., Zhao, Q., Wu, H., & Huang, X. (2025). Enhancing global agricultural monitoring system for climate-smart agriculture. Climate Smart Agriculture, 2(1), 100037. https://www.sciencedirect.com/science/article/pii/S2950409024000376

Yu, H., Kong, B., Hou, Y., Xu, X., Chen, T., & Liu, X. (2022). A critical review on applications of hyperspectral remote sensing in crop monitoring. *Experimental Agriculture*, *58*, e26. DOI: 10.1017/S0014479722000278

Yu, J., Zhao, L., Liu, Y., Chang, Q., & Wang, N. (2025). Automatic crop type mapping based on crop-wise indicative features. *International Journal of Applied Earth Observation and Geoinformation*, *139*, 104554. DOI: 10.1016/j.jag.2025.104554

Yu, P., Teng, F., Zhu, W., Shen, C., Chen, Z., & Song, J. (2025). Cloud–edge–device collaborative computing in smart agriculture: Architectures, applications, and future perspectives. *Frontiers in Plant Science*, *16*, 1668545. DOI: 10.3389/fpls.2025.1668545 PMID: 41164258

Zaabar, N., Niculescu, S., Mihoubi, M.K. (2021). Assessment of combining convolutional neural networks and object-based image analysis to land cover classification using sentinel 2 satellite imagery (TENES REGION, ALGERIA).

Zaidenberg, D. A., Sebastianelli, A., Spiller, D., & Ullo, S. L. (2021). Advantages and Bottlenecks of Quantum Machine Learning for Remote Sensing. 2021 IEEE International Geoscience and Remote Sensing Symposium IGARSS, 5680-5683.

Zarco-Tejada, P. J., Camino, C., Beck, P. S. A., Calderon, R., Hornero, A., Hernández-Clemente, R., Kattenborn, T., Montes-Borrego, M., Susca, L., Morelli, M., Gonzalez-Dugo, V., North, P. R. J., Landa, B. B., Boscia, D., Saponari, M., & Navas-Cortes, J. A. (2018). Previsual symptoms of Xylella fastidiosa infection revealed in spectral plant-trait alterations. *Nature Plants*, *4*(7), 432–439. DOI: 10.1038/s41477-018-0189-7 PMID: 29942047

Zeng, L., Wardlow, B. D., Wang, R., Shan, J., Tadesse, T., Hayes, M. J., & Li, D. (2016). A hybrid approach for detecting corn and soybean phenology with time-series MODIS data. *Remote Sensing of Environment*, *181*, 237–250. DOI: 10.1016/j.rse.2016.03.039

Zerga, B. (2025). Integrated watershed management: A review. *Discover Sustainability*, *6*(1), 657. DOI: 10.1007/s43621-025-01352-0

Zhang, B., Liu, L., Zhang, Y., Wei, B., Gong, D., & Li, L. (2024). Spatial Consistency and Accuracy Analysis of Multi-Source Land Cover Products on the Southeastern Tibetan Plateau, China. *Remote Sensing (Basel)*, *16*(17), 3219. DOI: 10.3390/rs16173219

Zhang, B., Tian, H., Ren, W., Tao, B., Lu, C., Yang, J., Banger, K., & Pan, S. (2016). Methane emissions from global rice fields: Magnitude, spatiotemporal patterns, and environmental controls. *Global Biogeochemical Cycles*, *30*(9), 1246–1263. DOI: 10.1002/2016GB005381

Zhang, C., & Kovacs, J. M. (2012). The application of small unmanned aerial systems for precision agriculture: A review. *Precision Agriculture*, *13*(6), 693–712. DOI: 10.1007/s11119-012-9274-5

Zhang, J., Guan, K., Peng, B., Pan, M., Zhou, W., Jiang, C., Kimm, H., Franz, T. E., Grant, R. F., Yang, Y., Rudnick, D. R., Heeren, D. M., Suyker, A. E., Bauerle, W. L., & Miner, G. L. (2021). Sustainable irrigation based on co-regulation of soil water supply and atmospheric evaporative demand. *Nature Communications*, *12*(1), 5549. DOI: 10.1038/s41467-021-25254-7 PMID: 34545076

Zhang, J., Xu, S., Zhao, Y., Sun, J., Xu, S., & Zhang, X. (2022). Aerial orthoimage generation for UAV remote sensing [Review]. *Information Fusion*, *89*, 91–120. DOI: 10.1016/j.inffus.2022.08.007

Zhang, N., Wang, M., & Wang, N. (2002). Precision agriculture, a worldwide overview. *Computers and Electronics in Agriculture*, *36*(2–3), 113–132. DOI: 10.1016/S0168-1699(02)00096-0

Zhang, R., Feng, Y., Li, Y., & Zheng, K. (2024). Can different information channels promote farmers' adoption of agricultural green production technologies? Empirical insights from Sichuan Province. *PLoS One*, *19*(8), e0308398. DOI: 10.1371/journal.pone.0308398 PMID: 39116048

Zhang, R., Wu, X., Li, J., Zhao, P., Zhang, Q., Wuri, L., Zhang, D., Zhang, Z., & Yang, L. (2025). A bibliometric review of deep learning in crop monitoring: Trends, challenges, and future perspectives. *Frontiers in Artificial Intelligence*, *8*, 1636898. DOI: 10.3389/frai.2025.1636898 PMID: 41041084

Zhang, X., Cao, Z., & Dong, W. (2020). Overview of edge computing in the agricultural internet of things: Key technologies, applications, challenges. *IEEE Access : Practical Innovations, Open Solutions*, *8*, 141748–141761. DOI: 10.1109/ACCESS.2020.3013005

Zhang, X., Feng, G., & Sun, X. (2024). Advanced technologies of soil moisture monitoring in precision agriculture: A review. *Journal of Agriculture and Food Research, 18*, 101473. DOI: 10.1016/j.jafr.2024.101473

Zhang, X., Shi, L., Jia, X., Seielstad, G., & Helgason, C. (2010). Zone mapping application for precision-farming: A decision support tool for variable rate application. *Precision Agriculture, 11*(2), 103–114. DOI: 10.1007/s11119-009-9130-4

Zhang, X., Sun, Y., & Sun, Y. (2022). Research on cold chain logistics traceability system of fresh agricultural products based on blockchain. *Computational Intelligence and Neuroscience, 2022*(1), 1957957. DOI: 10.1155/2022/1957957 PMID: 35154298

Zhang, Y., Wang, S., & Ji, G. (2019). *Time-series analysis RNN-based time-series analysis to monitor plant disease progression*. IEEE. DOI: 10.1109/ACCESS.2019.2931752

Zhao, T., Wang, S., Ouyang, C., Chen, M., Liu, C., Zhang, J., Yu, L., Wang, F., Xie, Y., Li, J., Wang, F., Grunwald, S., Wong, B. M., Zhang, F., Qian, Z., Xu, Y., Yu, C., Han, W., Sun, T., & Wang, L. (2024). Artificial intelligence for geoscience: Progress, challenges, and perspectives. *Innovation (Cambridge (Mass.), 5*(5), 100691. DOI: 10.1016/j.xinn.2024.100691 PMID: 39285902

Zha, X., Jia, S., Han, Y., Zhu, W., & Lv, A. (2025). Enhancing soil moisture prediction in drought-prone agricultural regions using remote sensing and machine learning approaches. *Remote Sensing (Basel), 17*(2), 181. DOI: 10.3390/rs17020181

Zheng, W., Li, K., & Liu, X. (2025). GeoAI for land use observations, analysis, and forecasting. *Land (Basel), 14*(10), 2058. DOI: 10.3390/land14102058

Zhou, K., Ming, D., Lv, X., Fang, J., & Wang, M. (2019). CNN-based land cover classification combining stratified segmentation and fusion of point cloud and very high-spatial resolution remote sensing image data. *Remote Sensing (Basel), 11*(17), 2065. DOI: 10.3390/rs11172065

Zhou, Z., Zheng, F., Wen, J., Chen, Y., Li, X., Liu, Y., Wang, Y., & Chen, W. (2023). A user-driven sampling model for large-scale geographical point data visualization via convolutional neural networks. *IEEE Transactions on Human-Machine Systems, 53*(5), 885–894. DOI: 10.1109/THMS.2023.3296692

Zhu, D., & Cao, G. (2023). Intelligent spatial prediction and interpolation methods. In *Handbook of geospatial artificial intelligence* (pp. 121–150). CRC Press. DOI: 10.1201/9781003308423-7

Zhu, X. X., Tuia, D., Mou, L., Xia, G.-S., Zhang, L., Xu, F., & Fraundorfer, F. (2017). Deep learning in remote sensing: A comprehensive review and list of resources. *IEEE Geoscience and Remote Sensing Magazine, 5*(4), 8–36. DOI: 10.1109/MGRS.2017.2762307

Zhu, X., Guo, R., Liu, T., & Xu, K. (2021). Crop yield prediction based on agrometeorological indexes and remote sensing data. *Remote Sensing (Basel), 13*(10), 2016. DOI: 10.3390/rs13102016

Zhu, Z., Piao, S., Myneni, R. B., Huang, M., Zeng, Z., Canadell, J. G., Ciais, P., Ciais, P., Sitch, S. A., Friedlingstein, P., Arneth, A., Cao, C., Cheng, L., Kato, E., Koven, C. D., Li, Y., Lian, X., Liu, Y., Liu, R., & Zeng, N. (2016). Greening of the Earth and its drivers. *Nature Climate Change, 6*(8), 791–795. DOI: 10.1038/nclimate3004

Zhu, Z., Wang, S., & Woodcock, C. E. (2015). Improvement and expansion of the Fmask algorithm: Cloud, cloud shadow, and snow detection for Landsats 4–7, 8, and Sentinel-2 images. *Remote Sensing of Environment, 159*, 269–277. DOI: 10.1016/j.rse.2014.12.014

Zhu, Z., & Woodcock, C. E. (2014). Continuous change detection and classification of land cover using all available Landsat data. *Remote Sensing of Environment, 144*, 152–171. DOI: 10.1016/j.rse.2014.01.011

Zhu, Z., Wulder, M. A., Roy, D. P., Woodcock, C. E., Hansen, M. C., Radeloff, V. C., Healey, S. P., Schaaf, C., Hostert, P., Strobl, P., Pekel, J.-F., Lymburner, L., Pahlevan, N., & Scambos, T. A. (2019). Benefits of the Free and Open Landsat Data Policy. *Remote Sensing of Environment, 224*, 382–385. DOI: 10.1016/j.rse.2019.02.016

Ziadat, F. M., Zdruli, P., Christiansen, S., Caon, L., Abdel Monem, M., & Fetsi, T. (2022). An overview of land degradation and sustainable land management in the Near East and North Africa. *Sustainable Agriculture Research, 11*(1), 11–24. DOI: 10.5539/sar.v11n1p11

Zink, M., Bachmann, M., Brautigam, B., Fritz, T., Hajnsek, I., Moreira, A., Wessel, B., & Krieger, G. (2014). TanDEM-X: The New Global DEM Takes Shape. *IEEE Geoscience and Remote Sensing Magazine, 2*(2), 8–23. DOI: 10.1109/MGRS.2014.2318895

About the Contributors

Walid Al-Shaar earned his PhD in Geography (Urban Planning) from Université Saint Joseph de Beyrouth and PhD in Civil Engineering (Transport) from L'école doctorale des sciences et technologies de L'université Libanaise (2018-2021), with a research mobility period at Université Sorbonne Nouvelle (Paris 3) through Erasmus+. He holds a Master's degree in Highway Transportation and Traffic Engineering from Lebanese University and is a certified Project Management Professional (PMP). Served as Assistant Professor of Civil Engineering at the American University of the Middle East, Kuwait, Dr. AL-SHAAR currently holds the position of Research Director in Novum Research and Innovation center in Beirut, Lebanon as as well as the position of Associate Researcher (Data Scientist) at CNRS Lebanon's National Center for Remote Sensing, where he specializes in satellite image analysis and climate change impact detection. His extensive research portfolio encompasses urban planning, Autonomous Mobility, transportation engineering, geospatial analysis, land use dynamics, and climate vulnerability assessment. Dr. Al-Shaar has authored over 25 peer-reviewed research articles published in prestigious international journals, including Scientific Reports, and various Springer journals, along with two book chapters. He serves as an Editorial Board Member for PLOS One journal and maintains active GitHub repositories sharing his computational methodologies. His research expertise spans advanced spatial modeling techniques, including Multilayer-Perceptron Markov-Chain models, fractal analysis, and geographically weighted regression, with proficiency in R, Python, MATLAB, ArcGIS, and various geospatial analysis tools. With substantial teaching experience across multiple French and Lebanese universities, Dr. Al-Shaar has delivered courses on transportation engineering, geomatics, urban planning, research methodology, and smart mobility. Prior to his academic career, he accumulated significant professional experience as a design and proposal engineer, working on major infrastructure projects across Lebanon, Saudi Arabia, and the Gulf region, including the Riyadh Metro and various urban water master plans.

* * *

Debika Chaudhuri obtained her B. Tech and M. Tech from West Bengal University of Technology, West Bengal. She is pursuing her Ph.D in VLSI Technology. She has more than 12 years of teaching experience. She has published 19 papers and 2 patents. She has also 1 UK design patent grant and 2 Indian design patent grant in her credit. Apart from her academic duties as Assistant Professor in the Electronics Department, School of Engineering & Technology, IFTM University she is also engaged in administrative work like coordinator, Departmental IQAC committee at IFTM University.

Makram Chehayeb is a geomatics engineer, computer scientist, and urban planning researcher specializing in the integration of artificial intelligence with spatial analysis and digital urbanism. He serves as Head of the Surveying Engineering

Department at the Islamic University of Lebanon, where he lectures, supervises final-year projects, and develops academic programs. With over a decade of teaching and training experience, he focuses on educational technology, data analytics, GIS, and smart city systems. His academic background includes a PhD in Urban Planning and Policies, a research master's degree in urbanism, a bachelor's degree in surveying engineering, and a second bachelor's degree in computer science. He has contributed to international conferences and peer-reviewed journals on topics such as remote sensing–based population estimation, environmental assessment, and spatial decision support. His work bridges research, technology, and policy to advance sustainable and data-driven urban development across diverse global contexts and interdisciplinary innovation initiatives and practice worldwide.

Seenivasan E. is currently working as an Assistant Professor, School of Commerce, Vel Tech Rangarajan Dr. Sagunthala R&D Institute of Science and Technology, Chennai.

Walid Fahs holds an M.Sc. in Networking and Telecommunications from Université de Technologie de Troyes, France, and a Ph.D. in Networking from Blaise Pascal University in Clermont-Ferrand, France, which he completed in 2008. His doctoral research focused on the evaluation and simulation of a novel access method for industrial 802.11 wireless networks, conducted in collaboration with the Network and Protocol team. He currently serves as a professor in the Computer Science Department within the Faculty of Sciences and Arts at the Islamic University of Lebanon. His research interests span advanced neural network software development, next-generation wireless communication systems, and heterogeneous network architectures. Professor Fahs has an experience for supervising Master's and Ph.D. students in both Lebanon and France. He has authored and co-authored up to 30 peer-reviewed papers published in international conferences and journals.

Vennila G. is a dedicated academic and researcher known for her work in advanced computing and modern technological frameworks. Her scholarly focus includes deep learning, artificial intelligence, data analytics, and the development of secure computing systems. She has actively supported student learning by supervising various academic projects and contributing research publications to recognized journals and conferences. Through her ongoing efforts in research, guidance, and academic development, Dr. Vennila plays a significant role in advancing innovation within the academic community.

Ibrahim Tarik Jabal, M.Si., is a Professor at the Faculty of Agriculture and Animal Husbandry, University of Muhammadiyah Malang. He is an expert in agricultural socioeconomics, actively campaigns for social engineering to improve agricultural commodities, and serves on the Advisory Board of the Indonesian Agribusiness Association (AAI) for the 2023-2027 Instagram period. Profile and Career: Position: Professor/Professor at the University of Muhammadiyah Malang (UMM). Expertise: Agricultural socioeconomics. Other Trusts: Appointed to the Advisory Board of the Indonesian Agribusiness Association (AAI) for the 2023-2027 Instagram period. Thoughts: Emphasizes the importance of social engineering to increase the productivity and welfare of farmers in Indonesia. Focus: Active in research and development of agribusiness and sustainable agriculture. Prof. Jabal Tarik Ibrahim frequently provides insights into agricultural policy, the importance of agricultural modernization, and social approaches in managing the national agricultural sector.

Sivasubramanian K. is working as Head of the Department of Economics at Kristu Jayanti Deemed to be University, Bangalore, India. Dr. Sivasubramanian is a recognized research supervisor for PhD programme. He has 16 years of experience in teaching, research, and industry. He earned a Ph.D. from the University of Madras. Both foreign and national evaluators highly commended his thesis. He has published 52 research papers in Scopus and 7 UGC CARE-listed journals and Book chapters. He has presented many research papers at national and international conferences. He has also received 3 best paper awards from various conferences at national and international levels. He has received the Research Award from Kristu Jayanti University for a significant contribution to research publications under the Platinum category in the 2023-2024 academic year. He is also the editorial and review board member of various international-level Q1 & Q2 Journals. He completed a minor research project funded by the Bothi Niketan Trust. Dr. Sivasubramanian is also the Board of Studies and Board of Examiners of various colleges in India.

Kutubuddin Kazi has in Top 2% Scientists list in the World for Year 2025. He completed his B.E., M.E., Ph.D. in E&TC Engineering and Post Doctorate in "IoT in Healthcare Applications", and is nowadays working as a Professor & Head of Department, E&TC Engineering Department and was Dean R&D. His area of Interest is IoT, IoRT, IoBT, AI, ML, and AIIoT. He had suggested KSK Approach, KSK1 Approach, KK Approach, KVS Approach and DL Approach in the arena of IoT and its security. He has published more than 270+ papers, including Book chapters with various renowned publishers like IGI Global, CRC, NOVA publishers. Also he had received various 56 Patents from Indian Patent House (Utility, Design and Copyright), UK Design Grant Patents, South African Design Grant Patents and Canadian Copyright Patents. He received funds for various projects on IoT applications. He published 87 books under different National and International publishers. He worked as Editor for 2 book. He worked as a Reviewer for Scopus Conferences and Journal and for IGI Global. Also, he works as Editorial Board Member for 6 Journals. He got 4 Best Researcher Awards, 1 Best Faculty Award, a appreciation letter form Nxtwave, 3 Best HoD Award and 2-Appreciation Letter from MoE's Innovation Cell, Government of India.

Koko Kusnanto is an independent researcher affiliated with the Department of Economics and Business, University of Prof. Dr Mustopo (Religious), Jakarta, Indonesia. His research focuses on management science and its practical applications.

Shilpa Mahajan is a distinguished Certified Ethical Hacker (CEHv11) and CISCO Certified Instructor with a notable career spanning over 16 years in research and education. Currently serving as an Associate Professor at the NorthCap University. Dr. Mahajan holds a Ph.D. in Wireless Sensor Networks from Guru Nanak Dev University, Amritsar, and graduated with distinction from Punjab Engineering College, Chandigarh. Her extensive contributions include authoring numerous papers published in prestigious international journals, books, conferences, and holding patents. In her current role, she guides doctoral scholars and successfully supervises M. Tech and B. Tech projects. Dr. Mahajan has designed courses focusing on Computer Networks, Network Security, and Cryptography. She actively participates in various academic activities, serving as a resource person for Faculty Development Programs (FDPs), workshops, guest lectures, invited talks, and panel discussions. Dr. Mahajan's expertise is underscored by her proactive involvement in chairing sessions at conferences, highlighting her standing within the academic community. Notably, she coordinated the ATAL FDP on Web Security in 2022.

Rimaira Sanaurairba Margono is an independent researcher and currently a student at the Department of Information Systems, University of National Development "Veteran" (UPN), Jakarta, Indonesia. Her research focuses on information science, with particular emphasis on applied research and practice.

Tri Margono is a senior researcher at the Research Centre for Social Welfare, Villages, and Connectivity, National Research and Innovation Agency (BRIN), Jakarta, Indonesia. Previously, he worked as a researcher at the Indonesian Institute of Sciences (LIPI). His research focuses on applied information, rural sociology, and development communication. ORCID ID: 0000-0002-8062-1653.

Duraimutharasan N. is presently associated with REVA University since August 2024 and his research interests are Artificial intelligence and data security.

Atanu Nag obtained his B. Sc., M.Sc. and Ph.D. degree from University of Kalyani, West Bengal and is presently working as Associate Professor and Head in the Department of Physics, School of Sciences, IFTM University, Moradabad, Uttar Pradesh. He is a recognized innovation ambassador by the Ministry of Education Innovation Cell and is currently working as the Convener of the Institution's Innovation Council at IFTM University with a mission to develop a robust start-up ecosystem there. He has been teaching since last 17 years. He has published more than 50 Research papers in high impact Journals including SCI, Scopus, WoS indexing and over 40 proceeding papers in conferences. One research scholar had completed her Ph.D. and five scholars are perusing their Ph.D. under his supervision. He is the author of more than 10 books written for engineering and science students and working as editorial and reviewer capacity for regular and special issue of reputed journals and conferences. He has 1 UK design patent grant and 2 Indian design patent grant in his credit. He had organized two national conferences and many workshops, faculty development programmes as convener or coordinator.

Rahayu Relawati is a senior lecturer and researcher at the Faculty of Agriculture and Animal Husbandry, University of Muhammadiyah Malang (UMM). She specializes in agribusiness and agricultural socioeconomics and actively teaches in the UMM Master of Agribusiness Postgraduate Program. Institution: University of Muhammadiyah Malang (UMM). Faculty: Agriculture and Animal Husbandry, Department of Agribusiness. Education/Degree: Ir., MM. Expertise: Agribusiness Management, Agricultural Socioeconomics. Academic Contributions: Rahayu Relawati focuses on research related to agricultural product marketing, consumer behavior, and local agribusiness development. She is also involved in collaborative research on agricultural socioeconomics in Indonesia, particularly within the research network of the University of Muhammadiyah Malang and other universities. She is a permanent lecturer in the UMM postgraduate program, mentoring students in the field of agribusiness.

Vighnesh S. is a MSc Economics Student at the Department of Economics, Kristu Jayanti University, Bangalore.

Heena Shaikh completed BE (E&TC), MTech (E&TC). She had 5 years of experience. Her 7 books on Sensor Technology are published. Also she received 3 Indian Patents. She has published more than 15 papers in different journals.

Supaprawat Siripipatthanakul is working as an Adjunct Professor at Manipal GlobalNxt University, Malaysia, and a Faculty Member at the Faculty of Education at Bangkok Thonburi University, Thailand. He is also a Visiting Professor at the Faculty of Economics and Business at Universitas Terbuka, Indonesia. He has published more than 90 papers and books on Google Scholar. He has a CEFR Level 5 (C1), an English Teacher certificate from AAARYA Business College in Singapore, TEFL & TESOL certificates. 9 certificates in research (Scopus & WOS), and 8 certificates of Ethics in Research. He lectures on research methodology (business/ education/ healthcare) for international universities. He has been invited to review the manuscripts of Q1 and Q2-Scopus, WOS journals/Scopus, and Springer books (240 verified reviews by WOS/ 393 verified reviews by ORCID) since 2022. He is an Editor of IGI Global Scopus Books and PLoS One, Scopus & WOS, SJR Q1 USA.

Triyani Sumiati is a senior lecturer at the School for Industrial Technology and Pharmacy (STTIF), Bogor, Indonesia. Her research focuses on Applied pharmaceutical sciences. ORCID ID: 0000-0002-7153-5374.

M. P. Sutawi is a distinguished academic and expert in Livestock Agribusiness at the University of Muhammadiyah Malang (UMM). He is widely recognized for his research on the intersection of agriculture, livestock management, and the socio-economic well-being of farmers.

Vinothina V. is currently working as an Associate Professor at the Department of Computer Science, Kristu Jayanti University, Bangalore, India.

Index

G

I

J

K

L

M

N

P

R

S

T

U

V

Y

www.ingramcontent.com/pod-product-compliance
Lightning Source LLC
LaVergne TN
LVHW060256011026
849161LV00011B/2723